DIGITAL SATELLITE COMMUNICATIONS

Information Technology: Transmission, Processing, and Storage

Series Editors: Robert Gallager
Massachusetts Institute of Technology
Cambridge, Massachusetts

Jack Keil Wolf
University of California at San Diego
La Jolla, California

Digital Satellite Communications
Giovanni Corazza, ed.

Immersive Audio Signal Processing
Sunil Bharitkar and Chris Kyriakakis

Digital Signal Processing for Measurement Systems: Theory and Applications
Gabriele D'Antona and Alessandro Ferrero

Coding for Wireless Channels
Ezio Biglieri

Wireless Networks: Multiuser Detection in Cross-Layer Design
Christina Comaniciu, Narayan B. Mandayam and H. Vincent Poor

The Multimedia Internet
Stephen Weinstein

MIMO Signals and Systems
Horst J. Bessai

Multi-Carrier Digital Communications:
Theory and Applications of OFDM, 2nd Ed
Ahmad R.S. Bahai, Burton R. Saltzberg and Mustafa Ergen

Performance Analysis and Modeling of Digital Transmission Systems
William Turin

Wireless Communications Systems and Networks
Mohsen Guizani

Interference Avoidance Methods for Wireless Systems
Dimitrie C. Popescu and Christopher Rose

Stochastic Image Processing
Chee Sun Won and Robert M. Gray

Coded Modulation Systems
John B. Anderson and Arne Svensson

Communication System Design Using DSP Algorithms:
With Laboratory Experiments for the TMS320C6701 and TMS320C6711
Steven A. Tretter

A First Course in Information Theory
Raymond W. Yeung

Nonuniform Sampling: Theory and Practice
Edited by Farokh Marvasti

Simulation of Communication Systems, Second Edition
Michael C. Jeruchim, Phillip Balaban and K. Sam Shanmugan

Principles of Digital Transmission
Sergio Benedetto, Ezio Biglieri

DIGITAL SATELLITE COMMUNICATIONS

Giovanni E. Corazza
University of Bologna, Italy

 Springer

Giovanni Corazza (Ed)
University of Bologna
Italy

Digital Satellite Communications

Library of Congress Control Number: 2006932582

ISBN 0-387-25634-2   e-ISBN 0-387-34649-X
ISBN 9780387256344   e-ISBN 9780387346496

Printed on acid-free paper.

9 8 7 6 5 4 3 2 1

springer.com

Cogito ergo sum (Descartes)
Ago ergo erigo (de Bono)

Contents

3 Satellite Channel Impairments 65

S. Scalise, M. Álvarez Díaz , J. Bito, M. Bousquet, L. Castanet, I. Frigyes, P. Horvath, A. Jahn, M. Krejcarek, J. Lemorton, M. Luglio, S. Morosi, M. Neri, and M.A. Vázquez-Castro

5 Modulation Techniques 175

P. T. Mathiopoulos, G. Albertazzi, P. Bithas, S. Cioni, G. E. Corazza, A. Duverdier, T. Javornik, S. Morosi, M. Neri, S. Papaharalabos, A. Ribes, N. C. Sagias

6 Parameter Estimation and Synchronization   219
C. Mosquera, M.-L. Boucheret, M. Bousquet, S. Cioni, W. Gappmair, R. Pedone, S. Scalise, P. Skoutaridis, M. Villanti

7 Distortion Countermeasures   263
A. A. Rontogiannis, M. Alvarez-Diaz, M. Casadei, V. Dalakas, A. Duverdier, F.-J. Gonzalez-Serrano, M. Iubatti, T. Javornik, L. Lapierre, M. Neri, P. Salmi

8 Diversity Techniques and Fade Mitigation 313

G. K. Karagiannidis, M. Bousquet, C. Caini, L. Castanet, M. A. Vázquez Castro, S. Cioni, I. Frigyes, P. Horvath, T. Javornik, G. Kandus, M. Luglio, P. Salmi, D. A. Zogas

11 Systems and Services

A. Duverdier, C. Bazile, M. Bousquet, B. G. Evans, I. Frigyes, M. Luglio, M. Mohorcic,
R. Pedone, A. Svigelj, M. Villanti, A. Widiawan

List of Figures

List of Tables

Contributing Authors

Gianni Albertazzi, University of Bologna (Italy)

Marcos Álvarez Díaz, University of Vigo (Spain)

Caroline Bazile, CNES (France)

Petros Bithas, National Observatory of Athens (Greece)

Janos Bito, Budapest University of Technology and Economics (Hungary)

Marie-Laure Boucheret, GET (France)

Michel Bousquet, Supaéro (France)

Carlo Caini, University of Bologna (Italy)

Antonio Cardilli, CNIT-University of Florence (Italy)

Marika Casadei, University of Bologna (Italy)

Laurent Castanet, ONERA (France)

Marco Chiani, University of Bologna (Italy)

Stefano Cioni, University of Bologna (Italy)

Giovanni E. Corazza, University of Bologna (Italy)

Vassilis Dalakas, National Observatory of Athens (Greece)

Alban Duverdier, CNES (France)

Johannes Ebert, Graz University of Technology (Austria)

Bertalan Eged, Budapest University of Technology and Economics (Hungary)

Harald Ernst, DLR (Germany)

Barry G. Evans, University of Surrey (UK)

Istvan Frigyes, Budapest University of Technology and Economics (Hungary)

Wilfried Gappmair, Graz University of Technology (Austria)

Francisco J. Gonzalez-Serrano, Universidad Carlos III de Madrid (Spain)

Balàzs Héder, Budapest University of Technology and Economics (Hungary)

Peter Horvath, Budapest University of Technology and Economics (Hungary)

Matteo Iubatti, University of Bologna (Italy)

Axel Jahn, TriaGnoSys (Germany)

Tomaž Javornik, Jožef Stefan Institute (Slovenia)

Gorazd Kandus, Jožef Stefan Institute (Slovenia)

George K. Karagiannidis, Aristotle University of Thessaloniki (Greece)

Wolfgang Kogler, Graz University of Technology (Austria)

Nikos V. Kokkalis, National Observatory of Athens (Greece)

Otto Koudelka, Graz University of Technology (Austria)

Mate Krejcarek, Budapest University of Technology and Economics (Hungary)

Luc Lapierre, CNES (France)

Joël Lemorton, ONERA (France)

Gianluigi Liva, University of Bologna (Italy)

Michele Luglio, University of Rome "Tor Vergata" (Italy)

P. Takis Mathiopolous, National Observatory of Athens (Greece)

Mihael Mohorčič, Jožef Stefan Institute (Slovenia)

Simone Morosi, CNIT-University of Florence (Italy)

Carlos Mosquera, University of Vigo (Spain)

Massimo Neri, University of Bologna (Italy)

Csaba Novak, Budapest University of Technology and Economics (Hungary)

Stylianos Papaharalabos, University of Surrey (UK)

Raffaella Pedone, University of Bologna (Italy)

Atta Quddus, University of Surrey (UK)

Andrè Ribes, CNES (France)

Luca S. Ronga, CNIT-University of Florence (Italy)

Athanasios A. Rontogiannis, National Observatory of Athens (Greece)

Nikos C. Sagias, National Observatory of Athens (Greece)

Paola Salmi, Alstom S.p.A., formerly with University of Bologna (Italy)

Sandro Scalise, DLR (Germany)

Gonzalo Seco-Granados, Universitat Autonoma de Barcelona (Spain)

Paris Skoutaridis, University of Surrey (UK)

Aleš Švigelj, Jožef Stefan Institute (Slovenia)

Alessandro Vanelli-Coralli, University of Bologna (Italy)

Maria Angeles Vázquez-Castro, Universitat Autonoma de Barcelona (Spain)

Marco Villanti, University of Bologna (Italy)

Anggoro Widiawan, University of Surrey (UK)

Manfred Wittig, European Space Agency

Dimitris A. Zogas, Aristotle University of Thessaloniki (Greece)

Preface

Dear Reader,

Thank you for investing your time into this book devoted to the fascinating world of *Digital Satellite Communications*. This is an Edited book which possesses some peculiarities that we believe make it different from other books of the kind. In fact, this book is the result of the joint effort of some sixty co-Authors who took part to a common European Research Project, named SatNEx, i.e. the European Network of Excellence (NoE) in Satellite Communications. This NoE is an instrument funded by the European Commission under the sixth Framework Programme (FP6), within the thematic priority *Mobile and Wireless Systems beyond 3G* of the Information Society and Technology (IST) area. In brief, the objectives of the SatNEx NoE are the integration of knowledge and facilities, the pursuit of jointly executed research, and the spreading of excellence beyond the boundary of the SatNEx community. Within SatNEx, several joint activities have been launched on different areas of interest; among these, the *Flexwave* joint activity has been devoted to the study of all aspects related to the physical layer of a satellite communication system. This book is the output of two years of collaboration among the *Flexwave* partners.

Interestingly, writing this book has revealed to be instrumental for achieving all the objectives that were initially set. In fact, the manuscript was the vehicle for the integration of the knowledge portfolio of the participating partners; it became the stage for setting up new research activities to be jointly carried out; and last, but not least, it is obviously an instrument for knowledge dissemination to an extended readership. Now, what we would like to underline is that the design of the book and its realization was really a joint effort, geared towards achieving as much as possible the homogeneity of style and the balance that would be typical for a single-Author book. These targets have been pursued by working side-by-side to extract the main messages that each Chapter should contain, and then to expand around these fundamental concepts to achieve a fluent and enjoyable text. Obviously, the experience that each co-Author brought into the project was the basic enabler to dwell in depth into the most advanced technical aspects of digital satellite communications. Indeed, editing this book has been a privilege and a unique experience; whether we were successful in reaching our ambitious objectives is left for you, the Reader, to judge.

The book provides state-of-the-art coverage on all aspects related to the physical layer of a satellite link, including channel propagation, coding, modulation, synchronization and parameter estimation, predistortion and equalization, diversity and combining, multiple access, and software radio. As such, the book is intended for post-graduate Students, Researchers, practicing Engineers, particularly in Communications and Signal Processing. We have chosen an approach that focuses on the latest advancements in the field, in hope to give lasting reference value to the manuscript. The necessary theoretical background is included in dedicated chapters, in order to have a self-standing book. The chapters can be read in sequence or by going directly to the topics of interest. The Reader will note that the sequence of chapters resembles somewhat the adventure undergone by a signal in passing through a transmission chain.

As book Editor and on behalf of all co-Authors, we would like to honor an ample list of entities and people that have helped and contributed in different forms in making this book a reality. First, we would like to thank the European Commission for its drive on European Research, which included granting us the opportunities and the instruments for working together effectively. Particular thanks go to the Project Officer for SatNEx, Pertti Jauhiainen, as well as to Joao da Silva, Augusto de Albuquerque, Bernard Barani, and Manuel Monteiro, for their continuous support throughout our project. Second, we want to acknowledge our Project Coordinator, Erich Lutz from DLR, and his management team members Markus Werner, Anton Donner, Filomena del Sorbo, and Dörthe Gottschalk. Only very good coordination allows the achievement of significant results in a large project such as a Network of Excellence, and DLR provided us with the ideal working environment. We also feel it is right to cite here all SatNEx participating entities, including those not directly involved in the development of the book, for their open and friendly collaboration: the Aristotle University of Thessaloniki (Greece), the Budapest University of Technology and Economics (Hungary), CNES (France), CNIT (Italy), DLR (Germany), the Graz University of Technology (Austria), the Fraunhofer Institute (Germany), GET-ENST (France), ISTI-CNR (Italy), the Jožef Stefan Institute (Slovenia), the National Observatory of Athens (Greece), the National Technical University of Athens (Greece), the Technical University of Aachen (Germany), TESA/SUPAERO/ONERA (France), the Universidad Carlos III de Madrid (Spain), the Universitat Autonoma de Barcelona (Spain), the University of Aberdeen (UK), the University of Bologna (Italy), the University of Bradford (UK), the University of Rome "Tor Vergata" (Italy), the University of Surrey (UK), and the University of Vigo (Spain). Definitely, we want to thank our Publisher that has given us the opportunity to come out with this manuscript in a highly competitive field of knowledge. Specific thanks go to

Alex Greene, Melissa Guasch, and Chris Simpson. Also, our thankful thought goes to Prof. Jack Wolf from UCSD, who gave us the first assessment of our project and opened for us the door of Springer.

On a more personal level, my direct collaborators within the University of Bologna deserve grateful recognition for all their supportive efforts in completing the manuscript: Gianni Albertazzi, Carlo Caini, Marika Casadei, Stefano Cioni, Rosario Firrincieli, Matteo Iubatti, Massimo Neri, Raffaella Pedone, Paola Salmi, Marco Villanti, and Alessandro Vanelli-Coralli. Finally, to my parents Giancarlo and Maddalena, to my wife Susy, my son Emanuele, and my daughter Martina: thank you for giving meaning to my life, and for bearing up with a busy son, husband and father.

Dear Reader, the book is yours to enjoy.

PROF. GIOVANNI E. CORAZZA
University of Bologna

Acronyms

3G Third Generation

AC Adaptive Coding

ACI Adjacent Channel Interference

ACM Adaptive Coding and Modulation

ACS Add-Compare-Select

ADC Analog-to-Digital Converter

ADSL Asynchronous Digital Subscriber Line

AFD Average Fade Duration

AGC Automatic Gain Control

ALC Automatic Level Control

ALRT Average Likelihood Ratio test

AM/AM Amplitude-to-Amplitude conversion

AM/PM Amplitude-to-Phase conversion

AM Adaptive Modulation

AoF Amount of Fading

APDI Average PDI

APP a-Posteriori Probability

APSK Amplitude Phase Shift Keying

ARQ Automatic Repeat reQuest

ASEP Average Symbol Error Probability

ASIC Application-Specific Integrated Circuit

ASK Amplitude Shift Keying

ATM Asynchronous Transfer Mode

AWGN Additive White Gaussian Noise

BC Broadcast Channel

BCH Bose-Chaudhuri-Hocquenghem

BCJR Bahl-Cocke-Jelinek-Raviv

BEC Binary Erasure Channel

BEP Bit Error Probability

BER Bit Error Rate

BFWA Broadband Fixed Wireless Access

BICM Bit Interleaved Coded Modulation

BITCM Bit Interleaved Turbo Coded Modulation

BSS Broadcast Satellite Services

BPSK Binary Phase Shift Keying

C/I Carrier to Interference Ratio

CA-MLSD Conventional Adaptive-MLSD

CCSDS Consultative Committee Space Data System

cdf cumulative distribution function

CDM Code Division Multiplex

CDMA Code Division Multiple Access

CF-DAMA Combined Free - Demand Assignment Multiple Access

CFAR Constant False Alarm Rate

chf characteristic function

CPFSK Continuous Phase Frequency Shift Keying

CPM Continuous Phase Modulation

CPU Central Processing Unit

CRB Cramér-Rao bound

CRC Cyclic Redundancy Check

CRMA Code Reuse Multiple Access

CSEP Conditional Symbol Error Probability

CSI Channel State Information

DA Data Aided

DAB Digital Audio Broadcasting

DAC Digital-to-Analog Conversion

DAMA Demand Assignment Multiple Access

DBFN Digital BeamForming Network

DBPSK Differential Binary Phase Shift Keying

DD Decision Directed

D&D Detection and Decision

DE Density Evolution

DEPSK Differential Encoding Phase Shift Keying

DFE Decision Feedback Equalizer

DLPC Down-Link Power Control

DMB Digital Multimedia Broadcast

DMC Discrete Memoryless Channel

DP Data Predistorter

DPC Dirty Paper Coding

DPDI Differential PDI

DRR Data Rate Reduction

DS-CDMA Direct Sequence CDMA

DSL Digital Subscriber Line

DSP Digital Signal Processor

DVB Digital Video Broadcast

DVB-H DVB for Handhelds

DVB-DSNG Digital Video Broadcasting - Digital Satellite News Gathering

DVB-RCS Digital Video Broadcast - Return Channel via Satellite

DVB-S Digital Video Broadcast - Satellite

DVB-S2 Digital Video Broadcasting - Satellite 2nd generation

EC-BAID Extended Complex Blind Adaptive Interference Detector

EEPC End-to-End Power Control

EGC Equal-Gain Combining

EIRP Effective Isotropic Radiated Power

ES Earth Station

ETSI European Telecommunications Standards Institute

EXIT EXtrinsic Information Transfer

FB Feed-Backward

FCM Fixed Coded Modulation

FDD Frequency Division Duplex

FDM Frequency Division Multiplex

FDMA Frequency Division Multiple Access

FEC Forward Error Correcting

FER Frame Error Rate

FF Feed-Forward

FFH Fast Frequency Hopping

FFT Fast Fourier Transform

FH Frequency Hopping

FH-CDMA Frequency Hopping CDMA

FIR Finite Impulse Response

FLOPS FLoating-Point Operations per Second

FLUTE FILE Delivery over Unidirectional Transport

FMT Fade Mitigation Technique

FPGA Field Programmable Gate Array

FSK Frequency Shift Keying

FSS Fixed Satellite Services

GCMAC Generalized Cerebellar Model Articulation Controller

GEO Geostationary Earth Orbit

GES Gateway Earth Station

GF Galois Field

GFBT Gallager First Bounding Technique

GLRT Generalized Likelihood Ratio Test

GMSK Gaussian Minimum Shift Keying

GPDI Generalized PDI

GPRS General Packet Radio Service

GPS Global Positioning System

GSBT Gallager Second Bounding Technique

GSC Generalized-Selection Combining

GSM Global System Mobile

HAP High Altitude Platform

HCC Hybrid Concatenated Convolutional Code

HEO Elliptical Orbit

HPA High Power Amplifier

i.i.d. independent identically distributed

IBO Input Back-Off

IC Interference Cancellation

IETF Internet Engineering Task Force

IF Intermediate Frequency

IM Intermodulation

IMR Intermediate Module Repeater

IMT2000 International Mobile Telephony 2000

IMUX Input MUltipleXer

IOWC Input-Output Weight Coefficient

IP Internet Protocol

IPL Inter-Platform Link

ISI Inter-Symbol Interference

ISL Inter Satellite Links

ISP Internet Service Provider

IUB Improved Upper Bound

LAN Local Area Network

LBF Local Basis Functions

LCR Level Crossing Rate

LDPC Low-Density Parity-Check

LE Linear Equalizer

LEO Low Earth Orbit

LLR Log-Likelihood Ratio

LLRT Logarithmic LRT

LMS Land Mobile Satellite

LMS Least Mean Square

LNB Low Noise Block-converter

LOS Line of Sight

L&R Luise and Reggiannini

LRT Likelihood Ratio Test

LUT Look-Up Table

MAC Multiple-Access Channel

MAP Maximum-A-Posteriori

MBMS Multimedia Broadcast Multicast Service

MC-CDMA Multi-Carrier CDMA

MCRB Modified Cramér-Rao bound

MDD Multiple Differential Detectors

MDS Maximum Distance Separable

MEO Medium Earth Orbit

MF-TDMA Multi Frequency TDMA

mgf moments generating function

MIMO Multiple-Input Multiple-Output

ML Maximum Likelihood

MLP Multi-Layer Perceptrons

MLSD Maximum Likelihood Sequence Detection

M&M Mengali and Morelli

MOE Minimum Output Energy

MPEG Moving Pictures Expert Group

MMSE Minimum Mean Square Error

MPEG-TS MPEG Transport Stream

MRC Maximal-Ratio Combining

MSK Minimum Shift Keying

MSE Mean Square Error

MSS Mobile Satellite Services

MUD Multi-User Detection

MV Minimum Variance

NACK Negative Acknowledgement

NASA National Aeronautics and Space Administration

NCC Network Control Center

NCPDI Non Coherent PDI

NCR Network Clock Reference

NDA Non-Data Aided

NN Neural Network

NP Neyman-Pearson

OBBS On-Board Beam Shaping

OBO Output Back-Off

OBP On-Board Processor

OFDM Orthogonal Frequency Division Multiplex

OFDMA Orthogonal Frequency Division Multiple Access

O&M Oerder and Meyr

OMUX Output MUltipleXer

OS Operative System

OQPSK Offset Quadrature Phase Shift Keying

PA Pilot Aided

PAM Pulse Amplitude Modulation

PAN Personal Area Network

PAPR Peak-to-Average Power Ratio

PC Power Control

PCCC Parallel Concatenated Convolutional Code

PCS Personal Communication System

pdf probability density function

PDI Post Detection Integration

PDP Power Delay Profile

PE Parameter Estimation

PEP Pairwise Error Probability

PER Packet Error Rate

PIC Parallel Interference Cancellation

PL Physical Layer

PLL Phase Locked Loop

PMP Point-to-Multi-Point

PN Pseudo Noise

PPAM Pulse Position and Amplitude Modulation

PPM Pulse Position Modulation

PRMA Packet Reservation Media Access

PSD Power Spectral Density

PSK Phase Shift Keying

PSP Per-Survivor Processing

QAM Quadrature Amplitude Modulation

QEF Quasi Error Free

QoS Quality of Service

QPSK Quadrature Phase Shift Keying

R&B Rife and Boorstyn

RBF Radial Basis Functions

RC Raised Cosine

REC Rectangular Pulse Shaping

RF Radio Frequency

RLN Rice-Lognormal

RLS Recursive Least Square

RMS Root Mean Square

ROC Receiver Operating Characteristics

RRM Radio Resource Management

RS Reed-Solomon

RSC Recursive Systematic Convolutional

RSSI Received Signal Strength Indicator

RTD Round Trip Delay

RTT Round Trip Time

rv random variable

SAMA Spread ALOHA Multiple Access

SB Sphere Bound

SC Selection Combining

SCCC Serial Concatenated Convolutional Code

S-DAB Satellite Digital Audio Broadcasting

SDMA Space Division Multiple Access

SDR Signal-to-Distortion Ratio

SDR Software-Defined Radio

SEP Symbol Error Probability

SER Symbol Error Ratio

SF-TDMA Single Frequency TDMA

SFH Slow Frequency Hopping

SIC Serial Interference Cancellation

SIMO Single Input-Multiple Output

SINR Signal to Interference plus Noise Ratio

SISO Soft-Input Soft-Output

SNIR Signal to Noise plus Interference Ratio

SNORE Signal to Noise Ratio Estimation

SNR Signal to Noise Ratio

SOS Second Order Statistics

SOVA Soft-Output Viterbi Algorithm

SP Signal Predistorter

SQNR Quantization Noise Ratio

SR Software Radio

SS Spread Spectrum

SSPA Solid State Power Amplifier

SS-TDMA self-stabilizing TDMA

STBC Space-Time Block Coding

ST Space-Time

STC Space Time Coding

STTC Space-Time Trellis Coding

S-DMB Satellite Digital Multimedia Broadcasting

S-UMTS Satellite UMTS

TB Tangential Bound

TC Threshold Crossing

TCM Trellis Coded Modulation

TCP Transfer Control Protocol

TD Total Degradation

TD Time Diversity

TDD Time Division Duplex

TDM Time Division Multiplexing

TDMA Time Division Multiple Access

TEE Turbo Embedded Estimation

TFB Transfer Function Bound

TH Time Hopping

TH-PPM Time-Hopped PPM

THP Tomlinson-Harashima Precoding

TNL Terrestrial Network Link

TS Transport Stream

TS Terminal Station

TSB Tangential Sphere Bound

TSC Time Share of Connections

TSF Time Share of Fades

TWTA Travelling Wave Tube Amplifier

UDL Up/Down Link

UDP User Datagram Protocol

UE User Equipment

UES User Earth Station

ULPC Up-Link Power Control

UMTS Universal Mobile Telecommunications System

US Uncorrelated Scattering

UW Unique Word

UWB Ultra Wide Band

VA Viterbi Algorithm

VCO Voltage Controlled Oscillator

VSAT Very Small Aperture Terminal

V&V Viterbi and Viterbi

W-CDMA Wideband CDMA

Wi-Fi Wireless Fidelity

WSS Wide-Sense Stationarity

ZF Zero-Forcing

Chapter 1

INTRODUCTION

G.E. Corazza

University of Bologna, Italy

The study and design of digital communication links and their application to satellite systems and networks is the focus of this book. The theoretical foundations of the selected universe of discourse are probability, statistics, electromagnetic field propagation, communications theory, and information theory. Considerations on fundamental limits, bounds, criteria, algorithms, complexity and implementation details are not dissimilar from those used for terrestrial digital communications. Therefore, the contents of this book are definitely also applicable to all those systems which do not include a space segment in their network architecture. There are, however, certain peculiarities of satellite links that make it essential to have a specifically devoted book. We will limit here the discussion to only two of them. The first fact is that deep-space communications are one of the few practical applications for which the perfectly symmetric additive white Gaussian noise (AWGN) model is an accurate model of the reality, and where an improved design translates directly into increased scientific value. It is precisely for this reason that the history of coding is tightly related to the history of deep-space communications; and it is for this reason that coding has central relevance in this book. The second fact is that for satellite links other than deep-space links, it is an unescapable requirement to drive the on-board power amplifiers into saturation, to maximize the exploitation of the scarce energy resources. This reality impacts dramatically on the design of the digital communication link, both in the transmitter and in the receiver. This book dwells into all of these aspects, to enable the reader to approach successfully the study and design of efficient and flexible physical layers for broadband mobile and fixed satellite links.

Efficiency should be intended both in terms of spectrum efficiency and in terms of power efficiency. In other words, for a given bandwidth the information rate must be maximized in conjunction to the minimization of the signal-to-noise ratio (SNR) required to achieve the specified quality of service. Essentially, this objective translates into the quest for achieving an operating point which is as close as possible to the limit that Claude Shannon has proved in his splendid formulation of information theory, both for power limited and for spectrum limited systems. To this end, very advanced coding and modulation techniques have been devised, possibly with the use of iterative detection, equalization, and/or estimation techniques. The pervasive use of iterative techniques transform the digital receiver into a network of soft-input-soft-output entities which inter-communicate, passing along messages which progressively refine the reliability of the decisions that will be finally made on the received data symbols. This new vision is pervasive into the book. One more word on the problem of synchronization and parameter estimation: by pushing the performance to the Shannon limit, very low SNR values are encountered, which renders the ancillary operations of synchronization and parameter estimation extremely difficult. Therefore, very efficient algorithms, reaching the Cramer-Rao bound performance, must be devised and adapted according to the selected modulation and coding pair, for the various channel conditions. These algorithms must work in harmony with equalization, demodulation, and decoding algorithms for enhanced performance.

On the other hand, flexibility is an unavoidable necessity for designing physical layers which not only can adapt dynamically to time varying channel conditions, but that can also be used in generic and somewhat unspecified systems. This may indeed sound rather strange, but it is truly a specific peculiarity of those satellite communication systems which used transparent transponder for relaying the signal to the ground. In this case, the most precious part of the system architecture is essentially agnostic to the adopted signal shape, which can therefore evolve through the years, thanks for the ever-growing progress in technology. If one is ready to adopt this point of view, then it is clear that the possible application scenarios and propagation channel conditions are virtually uncountable, and it is not conceivable to design anything less than a flexible system. Flexibility must be intended in a twofold way. First of all, the effective spectrum efficiency must be adaptable to the traffic and channel conditions in a finely granular way. As a consequence, the information rate carried by the physical layer correspondingly can span an extremely large set of possible values, which are to be selected in a dynamic fashion. Secondly, most of the signal processing functions must be performed in software, leading to a pure software radio design. Both aspects are tackled in the book.

There exists a clear demand for broadband communications. The achievable information data rates should be increased by one to two orders of magnitude above those presently achievable via satellite. This is true for broadcasting, mobile services, and fixed services, although the absolute values may change. In order to find available spectrum resources, several possible frequency allocations must be considered, along with their specific propagation characteristics, both for fixed and for mobile environments. In fact, the mobile and fixed scenarios differ considerably and not only for the propagation channel conditions: hardware and software constraints, traffic conditions, performance requirements, service characteristics are all relevant dimensions that require specific approaches. Yet, in view of the convergence that is becoming a reality in the global telecommunications network, it becomes more and more important to design physical layers having both scenarios in mind at the same time. And this is the state of mind with which the book is written.

For an edited book to have reference value, it is recognized that it must contain significant new material in terms of scientific research, along with a solid description of the state-of-the-art. Therefore, the intention is to position advanced scientific research issues within the general domain of future satellite networks, as a part of the interworking telecommunications architecture. An emphasis is given to the conception and development of novel and challenging ideas, in order to extend the reach and impact onto the long term vision.

Let's now complete this Introductory Chapter with an overview of the contents of the subsequent Chapters. The Reader can decide which topics are of interest and jump directly to the specific part of the book.

Chapter 2, **Theoretical Background**, gives the reader all of the fundamental conceptual tools which are needed to be able to really extract of the value from the material. It contains a thorough background on mathematical statistics, including the theory of probability and stochastic processes. Univariate, bivariate, and multivariate distributions are described, and a host of densities are included, such as Gaussian, Log-normal, Rayleigh, Exponential, Weibull, Gamma, Rice, Chi-square, and others. An introduction to information theory as applied to satellite communications is also addressed. This allows to rigorously define essential concepts such as system and channel capacity, both for single and multiuser scenarios, and the intrinsic limits on transmission performance. Finally, a review of detection and estimation theory provides a unified approach on performance analysis and algorithm design, principally based on the maximum likelihood criterion.

Chapter 3, **Satellite channel impairments**, aims at describing the characteristics of propagation impairments and the effects they have on the performance of the physical layer of satellite communications systems. This is an essential element to be taken into account in the design of air interfaces and in link budget analyzes. Three typologies of propagation effects are considered: mobile and indoor multipath effects, which must be taken into account in the design of mobile satellite systems operating above 1 GHz, also aiming at in-building penetration; atmospheric effects, to be considered for broadband satellite systems operating high frequencies; and finally clear sky effects that disturb optical satellite communications. Results from measurement campaigns for land-mobile, aeronautical, railway, maritime, and indoor environments are reported. The dynamic characteristics of the propagation impairments are presented, and the state-of-the-art of for propagation models (statistical, stochastic, physical models) is reviewed, including single-state and multi-state models. Finally, a characterization of the sources of interference in multiple access systems concludes the chapter.

The goal of Chapter 4, **Forward Error Correction**, is to firstly provide a unified review based on linear algebra of the most significant error correcting codes which were developed from the 1950's to the present, and that have found application in satellite communications. The description starts from Hamming codes, BCH, and Reed Solomon (RS) codes; then, it proceeds all the way to the description of the more challenging developments and improvements that took place in recent years, namely Turbo codes and Low-Density Parity-Check (LDPC) codes based on graph theory. Considerations related to the design, decoding and practical applications of these codes are provided. Two items at the end of this Chapter deserve particular attention: a performance comparison between Turbo and LDPC codes, and a treatment on the use of forward error correction techniques at higher layers, also known as packet-coding.

In Chapter 5, **Modulation techniques**, the possible modulation techniques are reviewed, including those employing constant and non-constant envelope signaling formats. APSK and continuous phase modulations are also described, which are now at the center of rapid developments for both narrowband and broadband satellite air interfaces. Then, coded modulation techniques are considered, dwelling on both trellis coded modulation and bit-interleaved coded modulation schemes. The third and final part of the Chapter considers the problem of detection with ideal channel state information, in both AWGN and fading channels. Bounds on error performance are also provided. The problem of estimating auxiliary parameters in a realistic digital receiver is left for the subsequent Chapter.

Chapter 6, **Parameter estimation and synchronization**, tackles all the major synchronization tasks faced by a receiver in practice. Symbol synchronization, phase, frequency, code epoch, frame epoch, gain and channel profile are all needed by the digital receiver to properly decode the incoming signal. All of these are considered in the book. Based on a Maximum Likelihood theoretical environment, algorithms are classified according to the role played by the coding/decoding process in the synchronization procedure, distinguishing amongst code aware and code unaware techniques. Classical algorithms are presented together with recent developments, and practical receiver architectures are discussed for the purpose of illustration.

Resolving the linear and non-linear distortions induced in a signal transmitted through a satellite link is the topic of Chapter 7, **Distortion countermeasures**. The main focus is on the non-linear effects caused by the satellite on board high power amplifier, modelled both as a memoryless machine and with memory. The distortion countermeasures techniques are then distinguished in two categories: transmitter techniques and receiver techniques. The first category comprises a number of linearization and predistortion methods which aim at compensating the HPA nonlinearity by properly inserting a non-linear processor, so that the overall system response is quasi-linear. The second category consists of a number of linear and non-linear equalization techniques which remove the signal distortion by processing the signal at the receiver. In both cases, several approaches are considered based on look-up tables, neural networks, Volterra methods, blind methods. In addition, three features are worth mentioning for this Chapter: an introduction to turbo equalization techniques, a description of precoding, and the yet open research topic of joint precoding and predistortion.

The focus of Chapter 8, **Diversity techniques and fade mitigation**, is on the powerful concept of diversity and other fade mitigation techniques. These techniques can be used to counteract blockage, multipath, and atmospheric effects encountered in satellite communication systems, as described in Chapter 4. More specifically, the following diversity domains are considered: satellite diversity; antenna diversity with Multiple Input Multiple Output channels and space time coding; site diversity; frequency diversity; power control; polarization diversity; adaptive coding and modulation. A specific Section is devoted to the problem of blockage and mulipath fading mitigation, which includes the use of higher layers and of terrestrial repeaters, also known as gap fillers. The Chapter is concluded by a review of possible centralized and distributed detection and decision algorithms for controlling the application of atmospheric effects mitigation techniques.

In Chapter 9, **Multiuser Satellite Communications**, we introduce a number of basic multiple access techniques and medium access control protocols that are specific to the satellite scenario, including connection-oriented and contention-oriented protocols. It should be noted that the satellite link poses major constraints on a access protocols performance, due to the large propagation delay, the remote control of on-board processing capabilities, and the power limitations that preclude the use of some protocols developed for a terrestrial scenario as they are. Among the connection oriented protocols, the reader will find FDMA, TDMA, CDMA, SDMA, OFDMA, MF-TDMA, and UWB. On the other hand, among the contention-oriented protocols, several random access and reservation algorithms are considered. A classification according to the most efficiently supported types of traffic is presented. The hybrid solution based on demand assignment is also discussed. We also present a brief introduction to some basic results of multiuser capacity discussing the effect of having channel side or state information. Finally, techniques able to mitigate or cancel the interference arising from the spectrum sharing are also introduced and their applications to current satellite networks are presented.

Chapter 10, **Software radio**, explores the opportunities related to the implementation of reconfigurable architectures for space communications, both on board and in the ground equipment. In fact, the trends toward the convergence of FPGA and DSP technologies, along with recent advances in space qualified signal processing devices, enable the implementation of software-based radio layers for satellite systems. Software Radio on-board will allow a full control over the satellite switching functionalities, enabling a new season of on-board processing payloads. As we have outlined above, flexibility in the air interface is a key requirement for present and future telecommunication systems, and software radio is the enabler to achieve it.

And finally, the objective of Chapter 11, **Systems and Services**, is to give an overview of the state-of-the-art in satellite systems and the relevant services, providing the context for the physical layer design which is addressed in this book. The Chapter shows how satellite communication systems are built around a mission, a system architecture and a specific satellite design. A general presentation of present and future fixed and mobile services is provided, introducing the fundamental concept of convergence of services. The Chapter is concluded with a discussion of future system architectures encompassing hybrid constellations, high altitude platforms, and terrestrial extensions.

In concluding this Introductory Chapter, we would like to express our firm belief that, notwithstanding all the advanced algorithms and architectures that we are able to design today, there are yet plenty of challenges to be faced, that will bring future innovations that are completely unseen today. And this is all the fascination of scientific research.

Chapter 2

THEORETICAL BACKGROUND

R. Pedone[1], G. Albertazzi[1], G. E. Corazza[1], P. T. Mathiopolous[2], C. Mosquera[3], N. C. Sagias[2], M. Villanti[1]

[1] *University of Bologna, Italy*

[2] *Institute of Space Applications and Remote Sensing, National Observatory of Athens, Greece*

[3] *University of Vigo, Spain*

2.1   Introduction

In all scientific fields, the formulation of basic theories is the universally recognized means for exploring and modelling the real world. The design of new concepts is thus usually guided by the application of these fundamental instruments.

This book focuses in particular on physical layer procedures and algorithms for efficient satellite communications. In this framework, since different design approaches are often guided by the same basic concepts, a unified vision on the theory underlying the communication design appears to be essential. This chapter aims at providing a uniform and rigorous treatment of the cardinal analytical concepts, which will be extensively utilized in the subsequent chapters, providing the reader with the required background for full comprehension of the rest of the book. Pursuing this objective, this chapter is valuable in itself as a concise, yet ample collection of tools to be used by the interested researcher for approaching new design issues. Surely, the ideas reported here go beyond the specific satellite applications considered in this book, and can be considered as a more general theoretical background.

In order to provide an up-to-date description of the theoretical tools, the content of this chapter comes both from classical references, such as [1], [2], [3], and state-of-the-art research publications. More specifically, three main theoretical areas are identified within the chapter. First of all, the main principles of probability and statistics are illustrated in section 2.2, defining the basic book symbology and introducing fundamental concepts in strict link with the following chapters, like for example the definition of the most commonly employed random variable distributions. Second, a collection of the most important ideas extracted from Information Theory and Network Information Theory is presented in section 2.3, providing the reader with essential concepts like the channel capacity both for single and multiuser scenarios, and inherent limits on transmission performance. Finally, an outlook on detection and estimation theory is considered in section 2.4 describing the most common detection and estimation criteria, and introducing theoretical intrinsic performance bounds, like the Cramer Rao bound.

2.2   Probability and statistics

This section presents a thorough review of probability and statistics theory. Basic definitions and tools are provided for univariate, bivariate, and multivariate statistics [1].

2.2.1   Types of probability

In statistics it is often necessary to make statements along the lines of "*the probability of observing a given event is p percent*". However, dealing with probabilistic statements is not often easy. One part of this difficulty stems from the innately abstract nature of probability, and a second from fundamental questions about the nature of probability. Nevertheless, the probabilistic point of view is a useful tool for inference, and hence, the term "probability" must first be defined. Three definitions of probability are provided and discussed in the following.

1 *Probability based on logic*. This view presupposes that we are dealing with a finite set of possibilities, which are randomly drawn. For example, in drawing from a deck of cards there is a finite set of possible hands. Such problems are easily solved if we can recognize all possible outcomes, and count the number of ways that a particular event may occur. Then, the probability of the event is the number of times it can occur divided by the total number of possibilities. To use a familiar example, we all recognize that there are 52 cards in a deck and 4 of these cards are Kings. Thus, the probability of drawing a King is equal to 4/52 = 0.0769.

2 *Probability based on experience*. This second view of probability assumes that if a process is repeated for a number of times n, and the event $\mathcal{A}$ occurs x out of these times, then the probability of the event $\mathcal{A}$ will converge on x/n, as n becomes large. If we flip a coin many times, we expect to see half of the flips turn up heads. Such estimates will become increasingly reliable as the number of replications n increases. For example, if a coin is flipped 10 times, there is no guarantee that exactly 5 heads will be observed. The proportion of heads may range from 0 to 1,

although in most cases we would expect it to be closer to 0.50 than to 0 or 1. However, if the coin is flipped 100 times, chances are better that the proportions of heads will be close to 0.50. With 1000 flips, the proportion of heads will be an even better reflection of the true probability.

3 *Subjective probability.* In this view, probability is treated as a quantifiable level of belief ranging from 0 (complete disbelief) to 1 (complete belief). For instance, an experienced physician may say "*this patient has a* 50% *chance of recovery*". Presumably, this is based on an understanding of the relative frequency of what might occur in similar cases. Although this view of probability is subjective, it permits a constructive way for dealing with uncertainty.

These different types of probability are not mutually exclusive: they all obey to the same mathematical laws, and their computation methods are similar. In fact, all types of probability can be expressed as the *relative frequency* of the occurrence of a specific event, i.e., as the number of times the considered event occurs divided by the total number of performed experiments. This observation allows to introduce *a general definition of probability*.

DEFINITION 2.1 *Given the event* $\mathcal{A}$, *its probability is defined as*

$$Pr\{\mathcal{A}\} = \frac{N_A}{N_{tot}} \tag{2.1}$$

where and N_{tot} *is the total number of possible outcomes, and* N_A *is the number of outcomes that are favorable to event* $\mathcal{A}$.

2.2.2 Properties of probability

Many texts on statistics [1] cover the laws and axioms of probability. Here, we present a less formal approach by introducing only selected and useful properties of probability.

1 *The range of possible probability values.* Probabilities cannot be lower than 0 and greater than 1.

2 *Complement of events.* The complement of an event is its "opposite," e.g. the event which is not happening when the event occurs. For example, if we consider the event of having a head from a coin flip, its complement is the event of having a tail. Given the event $\mathcal{A}$, its complement is indicated by the same symbol with a line overhead, $\bar{\mathcal{A}}$. The sum of the probabilities of an event and its complement is always equal to 1, i.e.,

$$Pr\{\mathcal{A}\} + Pr\{\bar{\mathcal{A}}\} = 1$$

Therefore, the probability of the complement of an event is equal to 1 minus the probability of the event, i.e.,

$$Pr\{\bar{\mathcal{A}}\} = 1 - Pr\{\mathcal{A}\}$$

3 *Probability distributions.* A basic concept of probability theory is represented by probability distributions, also denoted as probability functions, probability densities or masses. Probability distributions list and describe probability values for all possible occurrences of a certain experiment. There are in general two types of probability distributions, namely:

- *Discrete distributions*, which describe a finite set of possible occurrences, for discrete "count data". For example, the number of successful treatments out of 2 patients is discrete, because the number of successes can be only 0, 1, or 2. The probability of all possible occurrences, i.e., $Pr\{0 \text{ successes}\}$, $Pr\{1 \text{ success}\}$, $Pr\{2 \text{ successes}\}$, constitutes the probability distribution for this discrete problem.

- *Continuous distributions*, which describe a continuum of possible occurrences. For example, the probability of a given birth-weight can be reasonably anything from half a pound to more than 12 pounds. Thus, the problem at hand is continuous, with an infinite number of possible points between two boundary values (think in terms of the Xeno's Paradox).

In the following, definitions and properties are presented focusing mainly on continuous distributions.

2.2.3   Univariate statistics

Considering a reader already familiar with the basic quantities of probability and statistics, this section provides a short introduction of these concepts applied to univariate statistics.

2.2.3.1   Random variables.

DEFINITION 2.2 *A random variable (rv) X is a process of assigning a number $X(\zeta)$ to every outcome ζ of an experiment.*

For example, given the random process represented by a transmitted signal $s(t)$, the signal sampled at the instant t_1, $S = s(t_1)$ is a random variable. Indicating with x the possible values of X, the resulting function must satisfy the following two conditions:

1 The set $\{X \leq x\}$ is an event for every x.

2 The probabilities of the events $\{x = +\infty\}$ and $\{x = -\infty\}$ are equal to zero, i.e.,

$$Pr\{x = +\infty\} = 0 \ \ \text{and} \ \ Pr\{x = -\infty\} = 0$$

2.2.3.2 Cumulative distribution function. In addition to the probability of a given number of successes, it is often useful to quantify the probability of observing a number of events that is lower than or equal to a given threshold. This measure is identified as the cumulative probability of the event, and is expressed with the *cumulative distribution function (cdf)* of the rv X as [1]

$$F_X(x) = Pr\{X \le x\} \tag{2.2}$$

defined in the range of admissible values of x. The cdf has some interesting properties, as detailed in the following. Denoting as a and b two real values such that $a < b$,

1 $F_X(x = +\infty) = 1$ and $F_X(x = -\infty) = 0$

2 $F_X(\cdot)$ is a non decreasing function, i.e., $F_X(a) \le F_X(b)$

3 $Pr\{X > x\} = 1 - F_X(x)$

4 $Pr\{a \le X \le b\} = F_X(b) - F_X(a)$

5 For every $x \le \lambda$, if $F_X(\lambda) = 0$, then $F_X(x) = 0$

2.2.3.3 Probability density function. The first derivative of $F_X(x)$ with respect to x, i.e.,

$$f_X(x) = \frac{\partial F_X(x)}{\partial x} \tag{2.3}$$

is identified as the *probability density function (pdf)* of the rv X. Some properties related to the pdf are reported in the following. Being a and b two real values, such that $a < b$,

$$f_X(x) \ge 0, \ \forall x \in (-\infty, +\infty) \tag{2.4a}$$

$$\int_{-\infty}^{\infty} f_X(x)\,dx = 1 \tag{2.4b}$$

$$Pr\{a \le X \le b\} = F_X(b) - F_X(a) = \int_a^b f_X(x)\,dx \tag{2.4c}$$

2.2.3.4   **Moments.**   The average value of the rv X^k is identified as the *kth order moment* of X and is defined as

$$\mu_k(X) \overset{\triangle}{=} \mathrm{E}_X\left[X^k\right] = \int_0^\infty x^k\, f_X(x)\,\mathrm{d}x \tag{2.5}$$

where $\mathrm{E}_X[\cdot]$ denotes the expectation average with respect to X. For $k = 1$, the average value of X is obtained as

$$\mu_X = \mu_1(X) = \mathrm{E}_X\left[X\right] \tag{2.6}$$

while for $k = 2$, the power of the rv X is derived as

$$P_X = \mu_2(X) = \mathrm{E}_X\left[X^2\right] \tag{2.7}$$

For a non-central rv X, i.e., a rv with non null average value, an interesting figure is also provided by the *kth order central moments* of X, defined as

$$\nu_k(X) \overset{\triangle}{=} \mathrm{E}_X\left[(X - \mu_X)^k\right] = \int_0^\infty (x - \mu_X)^k\, f_X(x)\,\mathrm{d}x \tag{2.8}$$

For $k = 2$, the variance of the rv X is derived from (2.8) as

$$\sigma_X^2 = \nu_2(X) = \mathrm{E}_X\left[(X - \mu_X)^2\right] = P_X - \mu_X^2 \tag{2.9}$$

Using the binomial identity [4, eq. (1.111)], ν_k can be expressed in terms of μ_k as

$$\nu_k(X) = \sum_{l=0}^k \binom{k}{l} \mu_k(X)\, \mu_X^{k-l} \tag{2.10}$$

2.2.3.5   **Moment generating function.**   The *moments generating function (mgf)* of the rv X is defined as

$$\mathcal{M}_X(s) \overset{\triangle}{=} \mathrm{E}_X\left[\exp\{-s\,X\}\right] = \int_{-\infty}^\infty \exp\{-s\,x\}\, f_X(x)\,\mathrm{d}x \tag{2.11}$$

Equivalently, the mgf of X can be introduced with positive sign of s as $\mathcal{M}_X(s) = \mathrm{E}\left[\exp\{s\,X\}\right]$. The mgf is a very useful instrument for the computation of the kth order moment as

$$\mathrm{E}_X\left[X^k\right] = \left.\frac{\partial^k \mathcal{M}_X(x)}{\partial x^k}\right|_{x=0} \tag{2.12}$$

Moreover, the pdf of X can be derived by applying the inverse Laplace transform $\mathcal{L}^{-1}\{\cdot;\cdot\}$ of the mgf with according to

$$f_X(x) = \mathcal{L}^{-1}\{\mathcal{M}_X(s); x\} \tag{2.13}$$

while the corresponding cdf can be extracted as

$$F_X(x) = \mathcal{L}^{-1}\left\{\frac{\mathcal{M}_X(s)}{s}; x\right\} \tag{2.14}$$

2.2.3.6 **Characteristic function.** The *characteristic function (chf)* of the rv X is defined as

$$\Psi_X(\omega) \triangleq E_X\left[\exp\{-\jmath\omega X\}\right] = \int_{-\infty}^{\infty} \exp\{-\jmath\omega x\}\, f_X(x)\, dx \qquad (2.15)$$

Notably, the chf is structurally very similar to the mgf, except for the fact that the variable s in the Laplace domain is replaced by the variable $\jmath\omega$ in Fourier domain.

2.2.3.7 **Transformations of random variables.** Given the rv X with pdf $f_X(x)$, consider the rv Y equal to a real function $g(\cdot)$ of X, i.e., $Y = g(X)$. The pdf of Y is given by

$$f_Y(y) = \sum_{i=1}^{L} \frac{f_X(x_i)}{\left|\frac{dg}{dx}(x_i)\right|} \qquad (2.16)$$

where $x_i = g^{-1}(y)$, $i = 1,\dots,L$ are the L real solutions of the equation $y = g(x)$, being $g^{-1}(\cdot)$ the inverse function of $g(\cdot)$.

For example, if R is a rv representing a signal amplitude with pdf $f_R(r)$ defined for $r \geq 0$, and $\Gamma = g(R) = R^2$ is a new rv representing the signal power, the pdf of Γ can be obtained applying eq. (2.16) as

$$f_\Gamma(\gamma) = \frac{f_R(\sqrt{\gamma})}{2\sqrt{\gamma}}$$

2.2.4 Bivariate statistics

Starting from the basic definitions provided above for univariate statistics, this section extends their applicability to joint statistics of two rv's. This is often required to delineate a complete picture of the problem at hand. For example, the daily sale of ice cream is a rv with a certain pdf, and similarly is the daily temperature; however, these two variables are not independent, since ice cream sales are normally larger when the temperature is higher, and thus the two random effects require to be jointly evaluated.

2.2.4.1 **Bivariate joint cumulative distribution function.** Given two rv's X and Y with cdf's $F_X(\cdot)$ and $F_Y(\cdot)$ respectively, the *bivariate (joint) cdf* $F_{X,Y}(\cdot,\cdot)$ is the probability of the event $\{X \leq x, Y \leq y\}$, where x and y are two arbitrary real numbers, i.e.,

$$F_{X,Y}(x,y) = Pr\{X \leq x, Y \leq y\} \qquad (2.17)$$

Similarly to the univariate case, $F_{X,Y}(\cdot,\cdot)$ satisfies the following properties:

$$F_{X,Y}(x,-\infty) = F_{X,Y}(-\infty,y) = 0 \qquad (2.18a)$$

$$F_{X,Y}(+\infty, +\infty) = 1 \tag{2.18b}$$

$$F_X(x) = F_{X,Y}(x, +\infty) \quad \text{and} \quad F_Y(y) = F_{X,Y}(+\infty, y) \tag{2.18c}$$

2.2.4.2   Bivariate joint probability density function.   The *bivariate (joint) pdf* of the rv's X and Y is given by

$$f_{X,Y}(x, y) = \frac{\partial^2 F_{X,Y}(x, y)}{\partial x \, \partial y} \tag{2.19}$$

>From eq.(2.19), the corresponding joint cdf $F_{X,Y}(\cdot, \cdot)$ can be expressed as

$$F_{X,Y}(x, y) = \int_{-\infty}^{x} \int_{-\infty}^{y} f_{X,Y}(z, w) \, \mathrm{d}z \, \mathrm{d}w \tag{2.20}$$

The properties of $f_{X,Y}(\cdot, \cdot)$ are the following:

$$f_X(x) = \int_{-\infty}^{\infty} f_{X,Y}(x, y) \, \mathrm{d}y \quad \text{and} \quad f_Y(y) = \int_{-\infty}^{\infty} f_{X,Y}(x, y) \, \mathrm{d}x \tag{2.21a}$$

$$f_X(x) = \frac{\partial F_{X,Y}(x, +\infty)}{\partial x} \quad \text{and} \quad f_Y(y) = \frac{\partial F_{X,Y}(+\infty, y)}{\partial y} \tag{2.21b}$$

$$\int_{-\infty}^{\infty} \int_{-\infty}^{\infty} f_{X,Y}(x, y) \, \mathrm{d}x \, \mathrm{d}y = 1 \tag{2.21c}$$

2.2.4.3   Bivariate conditional probability density function and Bayes theorem.   The *conditional pdf* of the rv X with respect to the rv Y is defined as the function $f_{X|Y}(x|y)$ that verifies

$$f_{X,Y}(x, y) = f_{X|Y}(x|y) f_Y(y) \tag{2.22}$$

Symmetrically, the conditional pdf of Y with respect to X is the function $f_{Y|X}(y|x)$ that verifies

$$f_{X,Y}(x, y) = f_{Y|X}(y|x) f_X(x) \tag{2.23}$$

>From these definitions, the *Bayes theorem* can be derived as

$$f_{X|Y}(x|y) = \frac{f_X(x) f_{Y|X}(y|x)}{f_Y(y)} = \frac{f_{X,Y}(x, y)}{f_Y(y)} \tag{2.24}$$

2.2.4.4   Bivariate product moments.   The $(k + l)th$ *order average value* of the product of the rv's X and Y is defined as

$$\mu_{k+l}(X, Y) \triangleq \mathrm{E}_{X,Y}\left[X^k Y^l\right] = \int_{-\infty}^{\infty} \int_{-\infty}^{\infty} x^k y^l f_{X,Y}(x, y) \, \mathrm{d}x \, \mathrm{d}y \tag{2.25}$$

2.2.4.5 **Bivariate joint moment generating function.** The *bivariate* *(joint) mgf* of the rv's X and Y is defined as

$$
\begin{aligned}
\mathcal{M}_{X,Y}(s_1, s_2) &\triangleq \mathrm{E}_{X,Y}\left[\exp\{-s_1\,X - s_2\,Y\}\right] \\
&= \int_{-\infty}^{\infty}\int_{-\infty}^{\infty} \exp\{-s_1\,x - s_2\,y\}\,f_{X,Y}(x,y)\,\mathrm{d}x\,\mathrm{d}y
\end{aligned}
\tag{2.26}
$$

The joint mgf is very useful in finding the $(k+l)$th order moment as

$$
\mu_{k+l}(X,Y) = \left.\frac{\partial^{k+l}\mathcal{M}_{X,Y}(s_1, s_2)}{\partial s_1^k\,\partial s_2^l}\right|_{s_1=0,\,s_2=0}
\tag{2.27}
$$

Moreover, the joint pdf of X and Y can be derived by applying the inverse Laplace transform to the joint mgf, i.e.,

$$
f_{X,Y}(x,y) = \mathcal{L}^{-1}\left\{\mathcal{M}_{X,Y}(s_1, s_2)\,;x,y\right\}
\tag{2.28}
$$

while the corresponding joint cdf can be extracted as

$$
F_{X,Y}(x,y) = \mathcal{L}^{-1}\left\{\frac{\mathcal{M}_{X,Y}(s_1, s_2)}{s_1\,s_2}\,;x,y\right\}
\tag{2.29}
$$

2.2.4.6 **Bivariate joint characteristic function.** The *bivariate (joint)* *chf* of the rv's X and Y is defined as

$$
\begin{aligned}
\Psi_{X,Y}(\omega_1, \omega_2) &\triangleq \mathrm{E}_{X,Y}\left[\exp\{-\jmath\omega_1\,X - \jmath\omega_2\,Y\}\right] \\
&= \int_{-\infty}^{\infty}\int_{-\infty}^{\infty} \exp\{-\jmath\omega_1\,x - \jmath\omega_2\,y\}\,f_{X,Y}(x,y)\,\mathrm{d}x\,\mathrm{d}y
\end{aligned}
\tag{2.30}
$$

2.2.4.7 **Correlation coefficient and covariance.** The *correlation co-efficient* ρ between the rv's X and Y is defined as [1]

$$
\rho = \frac{C_{X,Y}}{\sigma_X\,\sigma_Y}
\tag{2.31}
$$

where σ_X^2 and σ_Y^2 are the variance of the rv X and Y respectively, and $C_{X,Y}$ is the *covariance* of the two rv's X and Y defined as

$$
C_{X,Y} = \mathrm{Cov}\,[X,Y] = \mathrm{E}_{X,Y}\left[(X - \mu_X)\,(Y - \mu_Y)\right]
\tag{2.32}
$$

being μ_X and μ_Y the mean values of X and Y, respectively. Notably it holds that $|\rho| \leq 1$. The two rv's are said to be *uncorrelated* if $\rho = 0$, i.e., if their covariance $C_{X,Y} = 0$.

2.2.4.8   Statistical independence between two random variables.
This section introduces the basic concept of *statistical independence* between
two rv's. If two rv's X and Y are statistical independent a number of interesting
properties holds, as detailed in the following.

1 The joint pdf of X and Y can be written as the product of the single pdf's,
    i.e.,
    $$f_{X,Y}(x,y) = f_X(x)\, f_Y(y) \tag{2.33}$$

2 The joint cdf of X and Y can be written as the product of the single cdf's,
    i.e.,
    $$F_{X,Y}(x,y) = F_X(x)\, F_Y(y) \tag{2.34}$$

3 The joint mgf of X and Y can be written as the product of the single
    mgf's, i.e.,
    $$\mathcal{M}_{X,Y}(s,d) = \mathcal{M}_X(s)\, \mathcal{M}_Y(d) \tag{2.35}$$
    and the joint chf of X and Y can be written as the product of the single
    chf's as
    $$\Psi_{X,Y}(s,d) = \Psi_X(s)\, \Psi_Y(d) \tag{2.36}$$

4 The product moments of X and Y can be written as
    $$\mathrm{E}_{X,Y}\left[X^k\, Y^l\right] = \mathrm{E}_X\left[X^k\right] \mathrm{E}_Y\left[Y^l\right] \tag{2.37}$$

5 The correlation coefficient associated to two independent rv's is equal to
    zero, i.e., $\rho = 0$, which means that *independence implies uncorrelation*.
    The opposite implication (uncorrelation $\Rightarrow$ independence) is not true in
    general, except for Gaussian variates.

2.2.4.9   Bivariate transformations.   Consider the two rv's X and Y
with joint pdf $f_{X,Y}(\cdot,\cdot)$, and a set of two rv's Z and W so that $Z = g_1(X,Y)$
and $W = g_2(X,Y)$, where $g_1(\cdot,\cdot)$ and $g_2(\cdot,\cdot)$ are two arbitrary real functions.
If the set of equations $z = g_1(x,y)$, $w = g_2(x,y)$ admits no then solution
$f_{Z,W}(z,w) = 0$, otherwise if K real roots (x_i,y_i) there exist, then

$$f_{Z,W}(z,w) = \sum_{i=1}^{K} \frac{f_{X,Y}(x_i,y_i)}{|J(x_i,y_i)|} \tag{2.38}$$

where

$$J(x,y) = \begin{pmatrix} \frac{\partial g_1(x,y)}{\partial x} & \frac{\partial g_1(x,y)}{\partial y} \\ \frac{\partial g_2(x,y)}{\partial x} & \frac{\partial g_2(x,y)}{\partial y} \end{pmatrix} \tag{2.39}$$

is the jacobian matrix of the set of equations, which is computed in correspondence of the i-th root (x_i,y_i) in eq.(2.38).

2.2.5   Multivariate statistics

This section generalizes the concepts we have presented for univariate and bivariate statistics to the case with more than two rv's, presenting the framework of multivariate statistics.

2.2.5.1   Multivariate joint cumulative distribution function.   Let X_1, $X_2, \ldots, X_L$ be L rv's with cdf's $F_{X_\ell}(\cdot)$, $\ell = 1, 2, \ldots, L$. The *multivariate joint cdf* $F_{X_1, X_2, \ldots, X_L}(x_1, x_2, \ldots, x_L)$ is defined as the probability of the event $\{X_1 \le x_1, X_2 \le x_2, \ldots, X_L \le x_L\}$ where x_ℓ, $\ell = 1, 2, \ldots, L$ are arbitrary real numbers, i.e.,

$$F_{X_1, X_2, \ldots, X_L}(x_1, x_2, \ldots, x_L) = Pr\{X_1 \le x_1, X_2 \le x_2, \ldots, X_L \le x_L\} \tag{2.40}$$

Similarly to the bivariate case, the properties of $F_{X_1, X_2, \ldots, X_L}(\cdot, \cdot, \ldots, \cdot)$ are the following:

$$F_{X_1, X_2, \ldots, X_L}(x_1, \ldots, x_{k-1}, -\infty, x_{k+1}, \ldots, x_L) = 0 \tag{2.41a}$$

$$F_{X_1, X_2, \ldots, X_L}(+\infty, +\infty, \ldots, +\infty) = 1 \tag{2.41b}$$

$$F_{X_1, \ldots X_{k-1}, X_{k+1}, \ldots, X_L}(x_1, \ldots x_{k-1}, x_{k+1}, \ldots, x_L) = \\ F_{X_1, X_2, \ldots, X_L}(x_1, \ldots, x_{k-1}, +\infty, x_{k+1}, \ldots, x_L) \tag{2.41c}$$

2.2.5.2   Multivariate joint probability density function.   The *multivariate joint pdf* of the rv's X_1, $X_2, \ldots, X_L$ is given by

$$f_{X_1, X_2, \ldots, X_L}(x_1, x_2, \ldots, x_L) = \frac{\partial^L F_{X_1, X_2, \ldots, X_L}(x_1, x_2, \ldots, x_L)}{\partial x_1 \, \partial x_2 \cdots \partial x_L} \tag{2.42}$$

With the aid of eq. (2.42), $F_{X_1, X_2, \ldots, X_L}(\cdot, \cdot, \ldots, \cdot)$ can be expressed as an L-fold integral according to

$$F_{X_1, X_2, \ldots, X_L}(x_1, x_2, \ldots, x_L) = \\ \int_{-\infty}^{x_1} \int_{-\infty}^{x_2} \cdots \int_{-\infty}^{x_L} f_{X_1, X_2, \ldots, X_L}(y_1, y_2, \ldots, y_L) \, dy_1 \, dy_2 \cdots dy_L \tag{2.43}$$

Again, similarly to the bivariate case, the properties of $f_{X_1, X_2, \ldots, X_L}(\cdot, \cdot, \ldots, \cdot)$ are the following. For $1 \le k \le L$, it holds that

$$f_{X_k}(x_k) = \\ \int_{-\infty}^{\infty} \cdots \int_{-\infty}^{\infty} f_{X_1, X_2, \ldots, X_L}(x_1, x_2, \ldots, x_L) dx_1 \cdots dx_{k-1} \, dx_{k+1} \cdots dx_L \tag{2.44a}$$

$$f_{X_1,\ldots X_{k-1},X_{k+1},\ldots,X_L}(x_1,\ldots x_{k-1},x_{k+1},\ldots,x_L) =$$
$$\frac{\partial^{L-1} F_{X_1,X_2,\ldots,X_L}(x_1,\ldots x_{k-1},+\infty,x_{k+1},\ldots,x_L)}{\partial x_1 \cdots \partial x_{k-1}\, \partial x_{k+1} \cdots \partial x_L} \qquad (2.44b)$$

$$\int_{-\infty}^{\infty} \cdots \int_{-\infty}^{\infty} f_{X_1,X_2,\ldots,X_L}(x_1,x_2,\ldots,x_L)\, \mathrm{d}x_1\, \mathrm{d}x_2 \cdots \mathrm{d}x_L = 1 \quad (2.44c)$$

2.2.5.3   Multivariate conditional probability density function.   The *multivariate conditional pdf* of the rv X_n with respect to the rv's $X_1,\ldots,X_{n-1}$ is defined as the function $f_{X_n|X_1,\ldots,X_{n-1}}(x_n|x_1,\ldots,x_{n-1})$ that verifies:

$$f_{X_1,\ldots,X_n}(x_1,\ldots,x_n) =$$
$$f_{X_n|X_1,\ldots,X_{n-1}}(x_n|x_1,\ldots,x_{n-1})\, f_{X_1,\ldots,X_{n-1}}(x_1,\ldots,x_{n-1}) \qquad (2.45)$$

By iterative application of the Bayes criterion on the residual cumulative pdf, the following property can be identified

$$f_{X_1,\ldots,X_n}(x_1,\ldots,x_n) = f_{X_n|X_1,\ldots,X_{n-1}}(x_n|x_1,\ldots,x_{n-1})\cdot$$
$$f_{X_{n-1}|X_1,\ldots,X_{n-2}}(x_{n-1}|x_1,\ldots,x_{n-2}) \ldots f_{X_2|X_1}(x_2|x_1)\, f_{X_1}(x_1)$$
$$(2.46)$$

2.2.5.4   Multivariate product moments.   The $\left(\sum_{i=1}^{L} k_i\right)$ *th order moment* of the product of the rv's $X_1, X_2, \ldots, X_L$ is defined as

$$\mu_{k_1+k_2+\cdots+k_L}(X_1,X_2,\ldots,X_L) \triangleq \mathrm{E}_{X_1,X_2,\ldots,X_L}\left[\prod_{i=1}^{L} X_i^{k_i}\right] =$$
$$= \int_{-\infty}^{\infty} \cdots \int_{-\infty}^{\infty} \prod_{i=1}^{L} x_i^{k_i}\, f_{X_1,X_2,\ldots,X_L}(x_1,x_2,\ldots,x_L)\, \mathrm{d}x_1\, \mathrm{d}x_2 \cdots \mathrm{d}x_L$$
$$(2.47)$$

2.2.5.5   Multivariate joint moment generating function.   By definition, the *multivariate joint mgf* of the rv $X_1,X_2,\ldots,X_L$ is given by

$$\mathcal{M}_{X_1,X_2,\ldots,X_L}(s_1,s_2,\ldots,s_L) \triangleq \mathrm{E}_{X_1,X_2,\ldots,X_L}\left[\exp\left\{-\sum_{i=1}^{L} s_i\, X_i\right\}\right]$$
$$= \int_{-\infty}^{\infty} \cdots \int_{-\infty}^{\infty} \exp\left\{-\sum_{i=1}^{L} s_i\, x_i\right\} f_{X_1,X_2,\ldots,X_L}(x_1,x_2,\ldots,x_L)\, \mathrm{d}x_1\, \mathrm{d}x_2 \cdots \mathrm{d}x_L$$
$$(2.48)$$

The mgf ia a useful means for computing the $\left(\sum_{i=1}^{L} k_i\right)$th order moment as

$$\mu_{\sum_{i=1}^{L} k_i}(X_1, X_2, \ldots, X_L) =$$

$$= \left. \frac{\partial^{\sum_{i=1}^{L} k_i} \mathcal{M}_{X_1, X_2, \ldots, X_L}(s_1, s_2, \ldots, s_L)}{\partial s_1^{k_1} \partial s_2^{k_2} \cdots \partial s_L^{k_L}} \right|_{s_1 = s_2 = \cdots s_L = 0}$$

$$(2.49)$$

Moreover, the multivariate joint pdf can be derived by applying the inverse Laplace transform, i.e.,

$$f_{X_1, X_2, \ldots, X_L}(x_1, x_2, \ldots, x_L) =$$
$$\mathcal{L}^{-1}\left\{ \mathcal{M}_{X_1, X_2, \ldots, X_L}(s_1, s_2, \ldots, s_L); x_1, x_2, \ldots, x_L \right\}$$

$$(2.50)$$

while the corresponding cdf can be extracted as

$$F_{X_1, X_2, \ldots, X_L}(x_1, x_2, \ldots, x_L) =$$
$$\mathcal{L}^{-1}\left\{ \frac{\mathcal{M}_{X_1, X_2, \ldots, X_L}(s_1, s_2, \ldots, s_L)}{s_1\, s_2 \cdots s_L}; x_1, x_2, \ldots, x_L \right\}$$

$$(2.51)$$

2.2.5.6   Multivariate joint characteristic function.   The *multivariate joint chf* of the rv's $X_1, X_2, \ldots, X_L$ is defined as

$$\Psi_{X_1, X_2, \ldots, X_L}(\omega_1, \omega_2, \ldots \omega_L) \overset{\triangle}{=} E_{X_1, X_2, \ldots, X_L}\left[\exp\left\{ -\jmath \sum_{i=1}^{L} \omega_i X_i \right\} \right]$$

$$= \int_{-\infty}^{\infty} \cdots \int_{-\infty}^{\infty} \exp\left\{ -\jmath \sum_{i=1}^{L} \omega_i x_i \right\} f_{X_1, X_2, \ldots, X_L}(x_1, x_2, \ldots, x_L)\, dx_1\, dx_2 \cdots dx_L$$

$$(2.52)$$

2.2.5.7   Covariance matrix.   Given L rv's $X_1, X_2, \ldots, X_L$, the *covariance matrix* is defined as

$$\Upsilon_{X_1, \ldots, X_L} \overset{\triangle}{=} \begin{pmatrix} C_{1,1} & C_{1,2} & \cdots & C_{1,L} \\ C_{2,1} & C_{2,2} & \cdots & C_{2,L} \\ \vdots & \vdots & \ddots & \vdots \\ C_{L,1} & C_{L,2} & \cdots & C_{L,L} \end{pmatrix}$$

$$(2.53)$$

where the generic element $C_{i,j}$ of the matrix is the covariance between the rv's X_i and X_j as defined in section 2.2.4.7, i.e.,

$$C_{i,j} \overset{\triangle}{=} E_{X_i, X_j}\left[(X_i - \mu_{X_I})(X_j - \mu_{X_J}) \right]$$

$$(2.54)$$

being μ_{X_i} and μ_{X_j} the mean values of the rv's X_i and X_j respectively. As defined in section 2.2.4.7, $C_{i,j}$ is in correspondence with the correlation coefficient between X_i and X_j according to

$$\rho_{i,j} = \frac{C_{i,j}}{\sigma_i \, \sigma_j} \tag{2.55}$$

where $\sigma_i^2 = \mathrm{E}_{X_i}\left[X_i^2\right] - \mu_{X_I}^2$ and $\sigma_j^2 = \mathrm{E}_{X_j}\left[X_j^2\right] - \mu_{X_J}^2$ are the variances of X_i and X_j, respectively.

2.2.5.8   Statistical independence between L random variables.   This section generalizes the basic concept of *statistical independence* to the case of $L > 2$ rv's. If the rv's $\{X_\ell\}_{\ell=1}^L$ are statistical independent, a number of interesting properties holds, as detailed in the following. In particular, the joint pdf, cdf, mgf, and product moments of $X_1, X_2, \ldots, X_L$ can be written as the product of the univariate quantities, i.e.,

$$f_{X_1,X_2,\ldots,X_L}(x_1, x_2, \ldots, x_L) = \prod_{i=1}^L f_{X_i}(x_i) \tag{2.56a}$$

$$F_{X_1,X_2,\ldots,X_L}(x_1, x_2, \ldots, x_L) = \prod_{i=1}^L F_{X_i}(x_i) \tag{2.56b}$$

$$\mathcal{M}_{X_1,X_2,\ldots,X_L}(s_1, s_2, \ldots, s_L) = \prod_{i=1}^L \mathcal{M}_{X_i}(s_i) \tag{2.56c}$$

$$\Psi_{X_1,X_2,\ldots,X_L}(\omega_1, \omega_2, \ldots, \omega_L) = \prod_{i=1}^L \Psi_{X_i}(\omega_i) \tag{2.56d}$$

$$\mathrm{E}_{X_1,X_2,\cdots,X_L}\left[X_1^{k_1} X_2^{k_2} \cdots X_L^{k_L}\right] = \prod_{i=1}^L \mathrm{E}_{X_i}\left[X_i^{k_i}\right] \tag{2.56e}$$

Similarly to the bivariate case, if the rv's $\{X_\ell\}_{\ell=1}^L$ are independent, then $\rho_{i,j} = 0, \forall\, i, j = 1, 2, \ldots, L$, i.e., *independence implies uncorrelation*. The opposite implication is again not true in general, except for Gaussian variates.

2.2.5.9   Multivariate transformations.   Given the set of L rv's $X_1, X_2, \ldots, X_L$ with multivariate joint pdf $f_{X_1,X_2,\ldots,X_L}(\cdot, \cdot, \ldots, \cdot)$, consider another set of rv's $Y_1, Y_2, \ldots, Y_L$ given by the L equations $Y_\ell = g_\ell(X_1, X_2, \ldots, X_L)$, where $g_\ell(\cdot)$, $\ell = 1, \ldots, L$ are L real functions. If the set of L equations $y_\ell = g_\ell(x_1, x_2, \ldots, x_L)$ admits no solution, then the multivariate joint pdf

$f_{Y_1, Y_2,...,Y_L}(y_1, y_2, \ldots, y_L) = 0$; otherwise, if the set of equations has a single solution $(\bar{x}_1, \bar{x}_2, \ldots, \bar{x}_L)$, then

$$f_{Y_1, Y_2,...,Y_L}(y_1, y_2, \ldots, y_L) = \frac{f_{X_1, X_2,...,X_L}(\bar{x}_1, \bar{x}_2, \ldots, \bar{x}_L)}{|J(\bar{x}_1, \bar{x}_2, \ldots, \bar{x}_L)|} \qquad (2.57)$$

where

$$J(x_1, x_2, \ldots, x_L) = \begin{pmatrix} \frac{\partial g_1}{\partial x_1} & \frac{\partial g_1}{\partial x_2} & \cdots & \frac{\partial g_1}{\partial x_L} \\ \frac{\partial g_2}{\partial x_1} & \frac{\partial g_2}{\partial x_2} & \cdots & \frac{\partial g_2}{\partial x_L} \\ \vdots & \vdots & \ddots & \vdots \\ \frac{\partial g_L}{\partial x_1} & \frac{\partial g_L}{\partial x_2} & \cdots & \frac{\partial g_L}{\partial x_L} \end{pmatrix} \qquad (2.58)$$

is the jacobian matrix of the set of equations, which is computed in correspondence of the solution $(\bar{x}_1, \bar{x}_2, \ldots, \bar{x}_L)$ in eq.(2.57).

2.2.6 Gaussian distribution

The *Gaussian (or Normal) distribution* has been favorably used in several fields of communication theory. For example, it is used for modelling Additive White Gaussian Noise (AWGN) present in digital receivers. Moreover, when modelling fading effects, the designer is often interested in distributions, which are generated from Gaussian random processes. If the rv X is Gaussian distributed with mean $\mu_X = \mathrm{E}_X[X]$ and standard deviation $\sigma_X = \sqrt{\mathrm{E}_X[(X - \mu_X)^2]}$, its pdf is given by

$$f_X(x) = \frac{1}{\sqrt{2\pi}\sigma_X} \exp\left\{ -\frac{(x - \mu_X)^2}{2\sigma_X^2} \right\} \qquad -\infty < x < +\infty \qquad (2.59)$$

The corresponding cdf is

$$F_X(x) = Q\left(\frac{x - \mu_X}{\sigma_X} \right) \qquad (2.60)$$

where $Q(\cdot)$ is the standard one-dimensional Gaussian Q-function defined by $Q(x) = \int_x^\infty \exp\{-u^2/2\} \, du/\sqrt{2\pi}$, which can also be expressed in terms of the complementary error function as $Q(x) = \mathrm{erfc}(x/\sqrt{2})/2$.

In the following, we will use the notation $\mathcal{N}(\mu_X; \sigma_X)$ to indicate that a rv X is Normal with pdf as in eq. (2.59).

2.2.7 Log-normal distribution

Consider a rv Z described by a Normal distribution in dB. Converting decibels into the linear scale, the *Log-normal distribution* is obtained, and the pdf of Z is given by

$$f_Z(z) = \frac{\xi}{\sqrt{2\pi}\,\sigma_{Z,\text{dB}}\,z}\,\exp\left\{-\frac{[10\log_{10}z - \mu_{Z,\text{dB}}]^2}{2\,\sigma_{Z,\text{dB}}^2}\right\} \qquad (2.61)$$

where $\xi = 10/\ln(10)$, and $\mu_{Z,\text{dB}}$ and $\sigma_{Z,\text{dB}}$ are the mean and the standard deviation of the associated normal variate $10\log_{10}Z$, respectively. An example of rv described by the Log-normal distribution is provided by the large-scale shadowing.

Accordingly, the cdf of Z is

$$F_Z(z) = Q\left[\frac{10\log_{10}z - \mu_{Z,\text{dB}}}{\sigma_{Z,\text{dB}}}\right] \qquad (2.62)$$

and, by substituting eq. (2.61) in eq. (2.5), the kth order moment of Z can be derived as

$$\mathrm{E}_Z\left[Z^k\right] = \exp\left\{\frac{k}{\xi}\mu_{Z,\text{dB}} + \frac{1}{2}\left(\frac{k}{\xi}\right)^2\sigma_{Z,\text{dB}}^2\right\} \qquad (2.63)$$

2.2.8   Rayleigh distribution

The *Rayleigh distribution* is one of the most well-known distributions used in channel modelling of short-term fading, where a large number of multiple reflective paths are present and there is no Line of Sight (LOS) signal component. Let C be a complex rv defined as

$$C = C_I + \jmath\,C_Q \qquad (2.64)$$

where C_I and C_Q are in-phase (I) and quadrature-phase (Q) zero mean Gaussian rv's with variance σ^2, i.e., $\mathcal{N}(0;\sigma)$. By transforming the rv C into polar coordinates, it can also be written as

$$C = R\,\exp\{\jmath\Theta\} \qquad (2.65)$$

where its magnitude R and argument Θ are given respectively by

$$R = \sqrt{C_I^2 + C_Q^2} \qquad (2.66)$$

$$\Theta = \arctan\left(\frac{C_Q}{C_I}\right) \qquad (2.67)$$

In particular, the magnitude R results to be Rayleigh distributed according to

$$f_R(r) = \frac{r}{\sigma^2}\,\exp\left\{-\frac{r^2}{2\sigma^2}\right\}, \qquad r \geq 0 \qquad (2.68)$$

and its argument Θ is uniformly distributed as

$$f_\Theta(\theta) = \frac{1}{2\pi}, \qquad \theta \in [0, 2\pi) \tag{2.69}$$

The cdf and the kth moment of the Rayleigh distributed rv R are

$$F_R(r) = 1 - \exp\left\{-\frac{r^2}{\omega^2}\right\} \tag{2.70}$$

$$\mathrm{E}_R\left[R^k\right] = \omega^k\,\Gamma\left(1 + \frac{k}{2}\right) \tag{2.71}$$

where $\omega^2 = \mathrm{E}_R\left[R^2\right] = 2\sigma^2$, and $\Gamma(\cdot)$ is the Gamma function.

2.2.9 Exponential distribution

Consider a Rayleigh distributed rv R. Then, the rv represented by its power, i.e.,

$$\Gamma = R^2 \tag{2.72}$$

is distributed according to a *exponential distribution* . An example of exponential distributed rv is provided by the power of short-term fading. By substituting eq. (2.72) in (2.70), the cdf of Γ can be derived as

$$F_\Gamma(\gamma) = 1 - \exp\left\{-\frac{\gamma}{\mu_\Gamma}\right\} \tag{2.73}$$

where

$$\mu_\Gamma = \mathrm{E}_\Gamma\left[\Gamma\right] = \mathrm{E}_R\left[R^2\right] = \omega^2 \tag{2.74}$$

After taking the first derivative of the above equation with respect to γ, the corresponding pdf of the exponential distribution can be obtained as

$$f_\Gamma(\gamma) = \frac{1}{\mu_\Gamma}\exp\left\{-\frac{\gamma}{\mu_\Gamma}\right\} \tag{2.75}$$

and substituting eq. (2.72) and (2.74) in eq. (2.71), the kth order moments of Γ can be also obtained as

$$\mathrm{E}_\Gamma\left[\Gamma^k\right] = \mu_\Gamma^k\,\Gamma(1 + k) = k!\,\mu_\Gamma^k \tag{2.76}$$

where $\Gamma(\cdot)$ indicates the Gamma function. Using eq. (2.11) and (2.75), the mgf can be also obtained as

$$\mathcal{M}_\Gamma(s) = \frac{1}{1 + s\,\mu_\Gamma} \tag{2.77}$$

2.2.10   Weibull distribution

The Weibull distribution was first introduced by Walodi Weibull at the Royal Institute of Technology of Sweden back in 1949 for estimating machinery lifetime [5]. Nowadays, the Weibull distribution is by far the world most popular statistical model in the field of reliability engineering and failure data analysis [6] [7]. It is also used in many other applications, such as weather forecasting and fitting data of all kinds, while it is widely applied in radar systems to model the dispersion of the received signals level produced by some types of clutter [8]. Concerning wireless communication theory, the Weibull model exhibits an excellent fit to experimental fading channel measurements, for both indoor [9] [10] and outdoor [11] [12] environments, with the required physical justification to be given in [13].

Consider a rv $R > R_0$ distributed according to a *Weibull distribution*, with R_0 an arbitrary real constant value. The three-parameter Weibull pdf is given by

$$f_R(r) = \frac{\beta}{\omega} \left(\frac{r - R_0}{\omega}\right)^{\beta-1} \exp\left\{-\left(\frac{r - R_0}{\omega}\right)^{\beta}\right\} \qquad (2.78)$$

where $\omega > 0$ is the scaling parameter, $\beta > 0$ is the shaping parameter (or slope), and R_0 is the location parameter. Moreover, the cdf of R is given by

$$F_R(r) = 1 - \exp\left\{-\left(\frac{r - R_0}{\omega}\right)^{\beta}\right\} \qquad (2.79)$$

A special case for the Weibull distribution is for $R_0 = 0$, in correspondence of which R can be obtained by a power transformation of a Rayleigh distributed rv X. In fact, considering the following transformation

$$R = g(X) = X^{2/\beta} \qquad (2.80)$$

using eq.(2.16), and recalling the Rayleigh pdf of eq.(2.68), the pdf of R results to be exactly the expression (2.78) with $R_0 = 0$, where

$$\omega = \sqrt{\frac{E_R[R^2]}{\Gamma\left(1 + \frac{2}{\beta}\right)}} \qquad (2.81)$$

and similarly for the cdf. For the special cases of $(R_0 = 0, \beta = 2)$ and $(R_0 = 0, \beta = 1)$, the Weibull distribution further simplifies to the Rayleigh and exponential distributions, respectively.

In addition, for the particular case of $R_0 = 0$, the nth moment of the Weibull distribution is given by

$$E_R[R^n] = \omega^n \Gamma\left(1 + \frac{n}{\beta}\right) \qquad (2.82)$$

An interesting property of the Weibull distribution is that the instantaneous square power of R, $\gamma = R^2$, is again Weibull distributed. In fact the mean value of γ is

$$\mu_\gamma = \mathrm{E}_R\left[R^2\right] = \omega^2 \Gamma\left(1 + \frac{2}{\beta}\right) \qquad (2.83)$$

and by defining $a = 1/\Gamma(1 + 2/\beta)$, the cdf of γ can be computed as

$$F_\gamma(\gamma) = 1 - \exp\left\{-\left(\frac{\gamma}{a\,\mu_\gamma}\right)^{\beta/2}\right\} \qquad (2.84)$$

which confirms that the rv $\gamma = R^2$ is also Weibull distributed. This is a general property of Weibull distributions: the kth power of a Weibull distributed rv with parameters (β, ω) results in another Weibull distributed rv with parameters $(\beta/k, \omega)$.

The first derivative of eq. (2.84) with respect to γ, provides the corresponding pdf as

$$f_\gamma(\gamma) = \frac{\beta}{2a\,\mu_\gamma}\left(\frac{\gamma}{a\,\mu_\gamma}\right)^{\beta/2-1} \exp\left\{-\left(\frac{\gamma}{a\,\mu_\gamma}\right)^{\beta/2}\right\} \qquad (2.85)$$

and the kth order moments can be derived from eq. (2.82) yielding

$$\mathrm{E}_\gamma\left[\gamma^k\right] = (a\,\mu_\gamma)^k\,\Gamma\left(1 + \frac{2k}{\beta}\right) \qquad (2.86)$$

The mgf of γ can be evaluated applying the definition (2.11), yielding to an integral in the form $\int_0^\infty x^{\beta/2-1}\exp\{-s\,x\}\exp\{-\xi\,x^{\beta/2}\}\,dx$, where ξ is a positive value. Closed-form solution of this integral [14], [15] provides

$$\mathcal{M}_\gamma(s) = \frac{\beta}{2}\frac{k^{1/2}\,l^{(\beta-1)/2}}{(2\pi)^{\frac{k+l}{2}-1}\,(a\,\mu_\gamma s)^{\beta/2}} G_{l,k}^{k,l}\left[\frac{l^l\,k^{-k}}{(a\,\mu_\gamma s)^l}\;\middle|\;\begin{array}{l}\frac{1-\beta/2}{l},\frac{2-\beta/2}{l},\ldots,\frac{1-\beta/2}{l}\\ \frac{0}{k},\frac{1}{k},\ldots,\frac{k-1}{k}\end{array}\right] \qquad (2.87)$$

where $G_{l,k}^{k,l}(\cdot)$ is the Meijer's G-function [4, eq. (9.301)], and (k, l) is a pair of positive integers chosen such that $l/k = \beta/2$. For example, for $\beta = 3.5$, $k = 4$ and $l = 7$, while for integer value of β, $k = 2$ and $l = \beta$. It is worth mentioning, that for given values of k and l the Meijer's G-function can be written in terms of Hypergeometric functions [4, sections 9.1 and 9.2] using properties which can be found in [16].

2.2.11 Nakagami-m distribution

Another well-known distribution, which is very useful for the statistical characterization of short-term fading envelopes is the *Nakagami-m distribution* .

Given a Nakagami-m distributed rv R, its pdf is given by

$$f_R(r) = \frac{2m^m}{\Omega^m \, \Gamma(m)} \, r^{2m-1} \, \exp\left\{ -\frac{m}{\Omega} \, r^2 \right\} \tag{2.88}$$

where $\Gamma(\cdot)$ is the Gamma function, $\Omega = \mathrm{E}_R\left[R^2\right]$, and $m \geq 1/2$ is a positive number also identified as the shape factor. Notably, the Nakagami-m distribution includes the cases of Rayleigh ($m = 1$) and one-sided Gaussian ($m = 1/2$) distributions as special cases.

2.2.12   Gamma distribution

Consider a Nakagami-m distributed rv R, then the rv $\Gamma = R^2$ is *Gamma distributed* with pdf given by

$$f_\Gamma(\gamma) = \frac{m^m}{\mu_\Gamma^m \, \Gamma(m)} \, \gamma^{m-1} \, \exp\left\{ -\frac{m}{\mu_\Gamma} \, \gamma \right\} \tag{2.89}$$

where $\mu_\Gamma = \mathrm{E}_R\left[R^2\right]$, and $\Gamma(\cdot)$ is the Gamma function. For $m = 1$, eq. (2.89) reduces to the exponential distribution. The corresponding cdf and the kth order moment are respectively given by

$$F_\Gamma(\gamma) = \frac{\gamma\left(m, \frac{m}{\mu_\Gamma}\gamma\right)}{\Gamma(m)} \tag{2.90}$$

$$\mathrm{E}_\Gamma\left[\Gamma^k\right] = \frac{\Gamma(k+m)}{\Gamma(m)\, m^k} \, \mu_\Gamma^k \tag{2.91}$$

where $\gamma\left(\cdot,\cdot\right)$ is the lower incomplete Gamma function [4, eq. (8.350/1)]. Moreover, the mgf of Γ is given by

$$\mathcal{M}_\Gamma(s) = \left(1 + \frac{\mu_\Gamma}{m}\, s\right)^{-m} \tag{2.92}$$

For m integer, the Gamma distribution is also known as the *Erlang distribution*.

2.2.13   Rice distribution

In rural and sub-urban environments, where a direct component exists (LOS), small-scale fading can be modeled according to the *Rice (or Nakagami-n) distribution* , which assumes a constant mean power for both direct and diffuse components. This distribution Z is mathematically constructed by adding the constant magnitude A of the direct component to the zero mean complex Gaussian distributed rv C representing the diffuse component, with average power $2\sigma^2$, where σ^2 is the average power of the I and Q Gaussian components (see Section 2.2.8). The resulting magnitude $R = |Z| = |A + C|$ is a Rice distributed rv, and its argument $\Phi = \arg[Z]$ is not uniformly distributed as it is for the phase of C. The joint pdf of the magnitude and phase of the resulting distribution can be compactly expressed as

$$f_{R,\Phi}(r, \phi) = \frac{r}{2\pi \sigma^2} \exp\left\{ -\frac{A^2 + r^2 - 2Ar \cos \phi}{2\sigma^2} \right\} \quad (2.93)$$

where the magnitude $R \geq 0$ and the phase $0 \leq \Phi < 2\pi$ are no longer statistically independent rv's. By averaging eq. (2.93) with respect to the phase, i.e., by solving the integral $f_R(r) = \int_0^{2\pi} f_{R,\Phi}(r, \phi)\, d\phi$, the Rice pdf is determined as

$$f_R(r) = \frac{r}{\sigma^2} \exp\left\{ -\frac{r^2 + A^2}{2\sigma^2} \right\} I_0\left(\frac{rA}{\sigma^2} \right) \quad (2.94)$$

where $I_0(\cdot)$ is 0th order modified Bessel function of the first kind. It results that $\Omega = \mathrm{E}_R[R^2] = A^2 + 2\sigma^2$. A frequently used parameter is the power ratio between the direct and the diffuse components, identified as the *Rice factor*, K equal to

$$K = n^2 = \frac{A^2}{2\sigma^2} \quad (2.95)$$

By introducing K, the Rice pdf can be rewritten as

$$f_R(r) = \frac{2r(K+1)e^{-K}}{\Omega} \exp\left\{ -\frac{K+1}{\Omega} r^2 \right\} I_0\left(2r\sqrt{\frac{K(K+1)}{\Omega}} \right) \quad (2.96)$$

which by normalizing $\Omega = 1$ becomes simply

$$f_R(r) = 2r(K+1)\exp\left\{ -r^2(K+1) - K \right\} I_0\left(2r\sqrt{K(K+1)} \right) \quad (2.97)$$

It can be easily recognized that for $K = 0$ (i.e., $A = 0$), the Rice distribution reduces to the Rayleigh pdf.

Similarly to the magnitude, the distribution of the phase Φ can be obtained by applying $f_\Phi(\phi) = \int_0^\infty f_{R,\Phi}(r, \phi)\, dr$, yielding

$$f_\Phi(\phi) = \frac{e^{-K}}{2\pi} \left\{ 1 + \sqrt{K\pi} \cos \phi\, e^{K \cos^2 \phi} \left[1 + \mathrm{erf}\left(\sqrt{K} \cos \phi \right) \right] \right\} \quad (2.98)$$

Again, it can be easily recognized that for $K = 0$, eq. (2.98) reduces to the uniform distribution shown in section 2.2.8.

2.2.14   Chi-square distribution

The *Chi-square distribution* is the distribution of the sum of square Gaussian random variables. According to the centrality of the composing Gaussian variates, it is possible to distinguish between central and non-central Chi-square distributions, as detailed in the following.

2.2.14.1   Central Chi-square distribution.   A rv X described by the *central Chi-square distribution* $\chi^2(0)$ has pdf given by

$$f_X(x) = \frac{x^{(\nu-2)/2}\exp\{-x/2\}}{2^{\nu/2}\Gamma(\nu/2)} \qquad x \geq 0 \qquad (2.99)$$

where ν is called the number of degrees of freedom of the $\chi^2(0)$ distribution, and $\Gamma(\cdot)$ is the Gamma function. Notably, the Chi-square distribution with ν degrees of freedom is equal to the Gamma variate with $m = \nu/2$ and $\mu_\Gamma = \nu$. The mean of the $\chi^2(0)$ with ν degrees of freedom is equal to ν, and its variance is 2ν. The associated mgf is given by

$$\mathcal{M}_\gamma(s) = (1 + 2s)^{-\nu/2} \qquad (2.100)$$

Interestingly, the $\chi^2(0)$ pdf with ν degrees of freedom of eq. (2.99) is equal to the pdf of the sum of the squares of ν independent unit normal variates $\mathcal{N}(0; 1)$. If the sum of the squares of ν independent normal variates with variance σ^2 is considered, i.e., $\mathcal{N}(0; \sigma)$, the central Chi-square pdf can be generalized as

$$f_X(x) = \frac{x^{(\nu-2)/2}\exp\{-x/2\sigma^2\}}{\sigma^\nu 2^{\nu/2}\Gamma(\nu/2)} \qquad x \geq 0 \qquad (2.101)$$

and its mean value is equal to $\nu\sigma^2$, and its variance is $2\nu\sigma^4$. The associate cdf is

$$F_X(x) = 1 - \exp\left\{-\frac{x}{2\sigma^2}\right\} \sum_{k=0}^{\nu/2-1} \frac{1}{k!}\left(\frac{x}{2\sigma^2}\right)^k \qquad x \geq 0 \qquad (2.102)$$

2.2.14.2   Non central Chi-square distribution.   A rv X described by the *non central Chi-square distribution* $\chi^2(\delta)$ has pdf given by

$$f_X(x) = \frac{1}{2}\left(\frac{x}{\delta}\right)^{\frac{\nu-2}{4}}\exp\left\{-\frac{\delta+x}{2}\right\} I_{\frac{\nu}{2}-1}\left(\sqrt{x\delta}\right) \qquad x \geq 0 \qquad (2.103)$$

where ν is called the number of degrees of freedom of the $\chi^2(0)$ distribution, δ is the non centrality parameter, and $I_n(\cdot)$ is the modified Bessel function of nth order. The mean value is in this case equal to $\nu + \delta$, and the variance is $2(\nu + 2\delta)$.

Interestingly, the $\chi^2(\delta)$ pdf with ν degrees of freedom of eq. (2.103) is equal to the pdf of the sum of the squares of ν independent Gaussian variates with unit variance and mean value δ_i, $i = 1, \ldots, \nu$, i.e., $\mathcal{N}(\delta_i; 1)$, and the non centrality parameter δ is related to the Gaussian mean values according to

$$\delta = \sum_{i=1}^{\nu} \delta_i^2 \qquad (2.104)$$

If the sum of the squares of ν independent normal variates with variance σ^2 is considered, i.e., $\mathcal{N}(\delta_i; \sigma)$, the non central Chi-square pdf can be generalized as

$$f_X(x) = \frac{1}{2\sigma^2} \left(\frac{x}{\delta}\right)^{\frac{\nu-2}{4}} \exp\left\{-\frac{\delta+x}{2\sigma^2}\right\} I_{\frac{\nu}{2}-1}\left(\frac{\sqrt{x\delta}}{\sigma^2}\right) \qquad x \geq 0 \quad (2.105)$$

In this case, the mean value is equal to $\nu\sigma^2 + \delta$, and the variance is $2(\nu\sigma^4 + 2\delta\sigma^2)$. The computation of the associate cdf provides

$$F_X(x) = 1 - Q_{\nu/2}\left(\frac{\sqrt{\delta}}{\sigma}, \frac{\sqrt{x}}{\sigma}\right) \qquad (2.106)$$

where $Q_M(a,b)$ is the generalized Marcum-Q function of order M, defined as [17]

$$Q_M(a,b) = \int_b^\infty x \left(\frac{x}{a}\right)^{M-1} \exp\left\{-\frac{x^2+a^2}{2}\right\} I_{M-1}(ax)\mathrm{d}x \qquad (2.107)$$

2.2.14.3   Bounds for the Marcum Q function.   The computation of the generalized Marcum-Q function of order M, $Q_M(a,b)$, defined in eq. (2.107), and in particular of the popular case of $M = 1$, referred to as Marcum-Q function, $Q(a,b)$, is frequent in many problems of signal detection.

To allow evaluations with no too costly numerical computations, it is often useful to have simple bounds for $Q_M(a,b)$ [18]. Since $Q_M(a,b)$ can be expressed as

$$Q_M(a,b) = Q(a,b) + \exp\left\{-\frac{a^2+b^2}{2}\right\} \sum_{k=1}^{M-1} \left(\frac{b}{a}\right)^k I_k(ab) \qquad (2.108)$$

it is possible to bound only the Marcum-Q function $Q(a,b)$ in terms of the 0th order Bessel function, and the generalized $Q_M(a,b)$ is automatically bounded.

Two cases are considered according to the relative values of a and b. The first case is for $b \geq a$, and the second is for $b < a$.

PROPOSITION 2.2.1 *Upper bound for $b \geq a$:*

$$Q(a,b) \leq \frac{I_0(ab)}{\exp\{ab\}} \left\{\exp\left\{-\frac{(b-a)^2}{2}\right\} + a\sqrt{\frac{\pi}{2}} \operatorname{erfc}\left(\frac{b-a}{\sqrt{2}}\right)\right\} \qquad (2.109)$$

PROPOSITION 2.2.2 *Lower bound for $b \geq a$:*

$$Q(a,b) \geq \sqrt{\frac{\pi}{2}} \frac{I_0(ab)b}{e^{ab}} \operatorname{erfc}\left(\frac{b-a}{\sqrt{2}}\right) \qquad (2.110)$$

PROPOSITION 2.2.3 *Upper bound for* $b < a$:

$$
Q(a,b) \leq 1 - \frac{I_0(ab)}{e^{ab}} \left\{ \exp\left\{ -\frac{a^2}{2} \right\} - \exp\left\{ -\frac{(b-a)^2}{2} \right\} + \right.
$$
$$
\left. + a\sqrt{\frac{\pi}{2}} \left[\mathrm{erfc}\left(-\frac{a}{\sqrt{2}} \right) - \mathrm{erfc}\left(\frac{b-a}{\sqrt{2}} \right) \right] \right\}
$$

(2.111)

PROPOSITION 2.2.4 *Lower bound for* $b < a$:

$$
Q(a,b) \geq 1 - \exp\left\{ -\frac{a^2 - \zeta^2}{2} \right\} \left\{ \exp\left\{ -\frac{\zeta^2}{2} \right\} - \exp\left\{ -\frac{(b-\zeta)^2}{2} \right\} + \right.
$$
$$
\left. + \zeta\sqrt{\frac{\pi}{2}} \left[\mathrm{erfc}\left(-\frac{\zeta}{\sqrt{2}} \right) - \mathrm{erfc}\left(\frac{b-\zeta}{\sqrt{2}} \right) \right] \right\}
$$

(2.112)

where $\zeta = (\log I_0(ab))/b$.

These bounds are valid also for large values of a and b without suffering numerical problems. Considering a value of the Marcum-Q function around 10^{-5}, for increasing a, the bounds improve. Moreover, in the case $b \geq a$, the tightness improves quickly for increasing b.

2.2.15 Hoyt distribution

The *Hoyt (or Nakagami-q) distribution* is useful for modelling strong ionospheric scintillations observed on satellite links. The pdf of a Hoyt distributed rv R is

$$
f_R(r) = \frac{1+q^2}{q\,\Omega} r \exp\left\{ -\frac{(1+q^2)^2}{4\,q^2\,\Omega} r^2 \right\} I_0\left(\frac{1-q^4}{4\,q^2\,\Omega} r^2 \right)
$$

(2.113)

where $\Omega = E_R\left[R^2\right]$ and $0 < q \leq 1$ is the shaping parameter. It can be easily recognized that for $q \to 0$ the one-sided Gaussian distribution is obtained, and for $q = 1$, eq. (2.113) reduces to the Rayleigh pdf of eq. (2.68). Following the standard method for the rv transformations, the pdf of $\Gamma = R^2$ is given by

$$
f_\Gamma(\gamma) = \frac{1+q^2}{2\,q\,\mu_\Gamma} \exp\left\{ -\frac{(1+q^2)^2}{4\,q^2\,\mu_\Gamma} \gamma \right\} I_0\left(\frac{1-q^4}{4\,q^2\,\mu_\Gamma} \gamma \right)
$$

(2.114)

with $\mu_\Gamma = E_R\left[R^2\right]$. The corresponding kth order moment and the mgf are respectively given by

$$
E_\Gamma\left[\Gamma^k\right] = \Gamma(k+1)\,{}_2F_1\left[-\frac{k-1}{2}, -\frac{k}{2}; 1; \left(\frac{1-q^2}{1+q^2} \right)^2 \right] \mu_\Gamma^k
$$

(2.115)

$$\mathcal{M}_\Gamma(s) = \left[1 + 2\,s\,\mu_\Gamma + \left(\frac{2\,q\,s\,\mu_\Gamma}{1+q^2}\right)^2\right]^{-1/2} \tag{2.116}$$

where $\Gamma(\cdot)$ is the Gamma function, and $_2F_1\,(\cdot\,;\cdot\,;\cdot)$ is the Gauss hypergeometric function [4, eq. (9.100)].

2.2.16   Bernoulli distribution

A *Bernoulli trial* is a probabilistic experiment that can have two possible outcomes labeled as $n = 0$ ("success") and $n = 1$ ("failure") with probabilities $1 - p$ and p, respectively. The resulting variate is a discrete rv, and thus it is completely characterized by the probability value associated to each possible outcome, i.e., by its probability distribution. The associated discrete rv N is *Bernoulli distributed* and its probability distribution is given by

$$p_N(n) = \begin{cases} 1 - p & n = 0 \\ p & n = 1 \end{cases} \tag{2.117}$$

The mgf and the the nth moment of N are respectively given by

$$\mathcal{M}_N(s) = 1 - p + p\,\exp\{s\} \tag{2.118}$$

$$\mathrm{E}_N\left[N^n\right] = p \tag{2.119}$$

The distribution of heads and tails in coin tossing is an example of a Bernoulli distribution with $p = 1/2$.

The Bernoulli distribution is the simplest discrete distribution, and it the building block for other more complicated discrete distributions. In particular, the binomial, geometric, and negative binomial variates are based on sequences of independent Bernoulli trials, which are curtailed in various ways, for example after N trials or x successes. Table 2.1 summarizes these distributions.

Table 2.1.   Distributions obtained by the Bernoulli distribution as special cases

Distribution	Curtailment parameter
Binomial distribution	Number of successes in N trials
Geometric distribution	Number of failures before the first success
Negative binomial distribution	Number of failures before the xth success

2.3   Information theory

The understanding of the limits of any practical communication system requires the use of a basic set of quantitative measures and tools. We will start this section with the presentation of the main quantities used in Information Theory, which will be used subsequently to show the maximum performance of a transmission for different scenarios. For a more in-depth study, we refer the interested reader to the excellent book by Cover and Thomas [2].

2.3.1 Basic definitions

Let X denote a discrete rv, with probability of occurrence given by $p(x) = Pr\{X = x\}$.

DEFINITION 2.3 *The self-information of the event* $X = x_i$ *is given by*

$$\mathcal{I}(x_i) = \log_a \frac{1}{p(x_i)} = -\log_a p(x_i) \tag{2.120}$$

The units of $\mathcal{I}(x)$ are determined by the base of the logarithm. If the base a of the logarithm is 2, which is the usual case, the information is measured in *bits*. In the following, we will use the symbol $\log(\cdot)$ to refer to $\log_2(\cdot)$ in order to simplify the notation. Note that an event $X = x_i$ with probability equal to one conveys no information, since in this case $\mathcal{I}(x_i) = 0$. In fact, low probability events are associated with higher values of information. If we weight the self-information by the probability of occurrence of the event and sum over all possible events, we obtain the average self-information known as *entropy*.

DEFINITION 2.4 *The entropy* $H(X)$ *of the discrete rv* X, *which takes values in the alphabet* $\mathcal{X}$ *is given by*

$$H(X) = \mathrm{E}_X\left[\mathcal{I}(x)\right] = \sum_{x \in \mathcal{X}} p(x)\mathcal{I}(x) \tag{2.121}$$

which can be also written as

$$H(X) = \mathrm{E}_X\left[\log \frac{1}{p(x)}\right] = -\sum_{x \in \mathcal{X}} p(x) \log p(x) \tag{2.122}$$

It can be proved that for an alphabet $\mathcal{X}$ with cardinality n, the entropy $H(X)$ is bounded as $0 \le H(X) \le \log n$. The maximum value $H(X) = \log n$ is achieved by the uniform rv, for which $p(x) = 1/n$ for all $x \in \mathcal{X}$. Thus, the entropy of a discrete source taking values in a set with finite cardinality is maximum when all its outcomes are equally probable. The entropy can be considered also as the average number of questions necessary to find out the outcome of a rv: the more concentrated the probability over a reduced subset of values, the lower the entropy.

EXAMPLE 2.5 *Consider the discrete rv* X, *which can take on five different values with the following probabilities:* $Pr\{X = a\} = 1/2$, $Pr\{X = b\} = 1/4$, $Pr\{X = c\} = 1/8$, $Pr\{X = d\} = 1/16$, $Pr\{X = e\} = 1/16$. *Thus,* we have that $H(X) = 1.875$: *half of the times one question would solve the value of X, one quarter of the times two questions would be enough, and so on. Overall, 1.875 is the average number of necessary questions to obtain the value of X.*

There exists a quantitative way of measuring the difference between the probability distributions describing a given rv: the *relative entropy or Kullback-Leibler distance*.

DEFINITION 2.6 *Let $p(x), q(x)$ denote two different probability distributions of a discrete rv X taking values in the alphabet $\mathcal{X}$. The relative entropy, or Kullback-Leibler distance, between $p(x)$ and $q(x)$ is given by*

$$D(p\|q) = \sum_{x \in \mathcal{X}} p(x) \log \frac{p(x)}{q(x)} = \mathrm{E}_X \left[\log \frac{p(x)}{q(x)} \right] \tag{2.123}$$

The Kullback-Leibler distance measures the inefficiency of assuming a given probability distribution for describing a given phenomenon with respect to the actual one. For a rv X with probability distribution $p(x)$, a code can be constructed with average length $H(X)$ to describe X. However, if the code is designed assuming a different distribution for X, i.e., $q(x)$, the number of bits needed on average to describe X results equal to $H(X) + D(p\|q)$.

Strictly speaking, $D(p\|q)$ does not correspond to the classical definition of a distance measurement because $D(p\|q) \neq D(q\|p)$. However, the Kullback-Leibler distance happens to be highly useful and provides much insight in many information theory problems.

A very useful extension of the entropy is the *conditional entropy*, which can be interpreted as the information or uncertainty contained in a rv after a related rv is observed.

DEFINITION 2.7 *Let X, Y denote two discrete rv's with probabilities respectively given by $p(x)$ and $p(y)$, and with joint probability $p(x, y)$. The conditional entropy $H(Y|X)$ is the average conditional self-information defined as*

$$\begin{aligned} H(Y|X) &= \mathrm{E}_{X,Y}\left[\mathcal{I}(y|x)\right] = \sum_{x \in \mathcal{X}} \sum_{y \in \mathcal{Y}} p(x)p(y|x) \log \frac{1}{p(y|x)} \\ &= \sum_{x \in \mathcal{X}} \sum_{y \in \mathcal{Y}} p(x, y) \log \frac{1}{p(y|x)} \end{aligned} \tag{2.124}$$

The entropy can also be defined for a set of rv's, even unlimited, by introducing the *joint entropy*, as the following definitions illustrate.

DEFINITION 2.8 *The joint entropy of the rv's* $(X_1, X_2, \ldots, X_n) \in \mathcal{X}^n$ *is given by*

$$H(X_1, X_2, \ldots, X_n) = - \sum_{x_1, x_2, \ldots, x_n \in \mathcal{X}^n} p(x_1, x_2, \ldots, x_n) \log p(x_1, x_2, \ldots, x_n)$$

(2.125)

where $p(x_1, x_2, \ldots, x_n)$ *is the multivariate joint probability.*

DEFINITION 2.9 *Considering* n *rv's extracted from the random process* $\mathbf{X} = \{X_i\}$, *it is possible to define the entropy of the random process or entropy rate, which is given by the following limit, provided it exists*

$$H(\mathbf{X}) = \lim_{n \to \infty} \frac{1}{n} H(X_1, X_2, \ldots, X_n)$$

(2.126)

To fully understand the information transmission flow, it is required to determine the amount of information that the outcome of a received rv Y provides about the outcome of a transmitted rv X. This measure is known as *mutual information*, as defined in the following.

DEFINITION 2.10 *Let* X, Y *denote two discrete rv's with probabilities respectively given by* $p(x)$ *and* $p(y)$. *The information content provided by the outcome* y *about the occurrence of the outcome* x *is defined as the mutual information between* x *and* y, *and is defined as*

$$\mathcal{I}(x; y) = \log \frac{p(x|y)}{p(x)}$$

(2.127)

where $p(x|y)$ *is the conditional probability of* X *subject to the event* $Y = y$.

In particular, if the rv's X and Y are statistically independent, the mutual information between any two of their outcomes is null. In any other case, the knowledge of the outcome of Y will provide information about X, by changing the *a priori* probabilities of X.

It can be readily seen that the information provided by the occurrence of the event $Y = y$ about the probability of occurrence of the event $X = x$ is equal to the information provided by the occurrence of the event $X = x$ about the probability of occurrence of the event $Y = y$, i.e, $\mathcal{I}(x; y) = \mathcal{I}(y; x)$.

By weighting the mutual-information $\mathcal{I}(x, y)$ through the probability of occurrence of the joint event $(X = x, Y = y)$, and summing over all possible joint events, we obtain the average mutual-information, identified as the *mutual information*.

DEFINITION 2.11 *The mutual information* $\mathcal{I}(X; Y)$ *between the rv's* X *and* Y *is given by*

$$\mathcal{I}(X; Y) = \sum_{x \in \mathcal{X}} \sum_{y \in \mathcal{Y}} p(x, y) \log \frac{p(x|y)}{p(x)} = \sum_{x \in \mathcal{X}} \sum_{y \in \mathcal{Y}} p(x, y) \log \frac{p(x, y)}{p(x)p(y)}$$

(2.128)

It results that $\mathcal{I}(X;Y) = \mathcal{I}(Y;X) \geq 0$, with $\mathcal{I}(X;Y) = 0$ for independent rv's X and Y. Equivalently, it can be shown that $\mathcal{I}(X;Y) = D(p(x,y)\|p(x)p(y))$, i.e, the mutual information is equal to the relative entropy of the joint distribution of X,Y with respect to the product of the single distributions.

Interestingly, it holds that

$$\mathcal{I}(X;Y) = H(X) - H(X|Y) = H(Y) - H(Y|X) \tag{2.129}$$

which allows to provide a very useful meaning to the mutual information. In fact, given that $H(X|Y)$ is the average amount of uncertainty in X after having observed Y, and $H(X)$ is the average amount of uncertainty in X prior to any observations, it follows that $\mathcal{I}(X;Y)$ is the average reduction level of uncertainty about X provided by the observation of Y.

All the quantities defined above for discrete rv's can be extended to continuous rv's with some cautions, as we show in the following. First, we replace the probability distribution $p(x)$ with the continuous rv pdf $f_X(x)$. Then, the following definitions can be introduced.

DEFINITION 2.12 *The entropy $H(X)$ of the continuous rv X with pdf $f_X(x)$ is defined as*

$$H(X) = -\mathrm{E}_X\left[\log f_X(x)\right] = -\int_{\mathcal{X}} f_X(x) \log f_X(x)\mathrm{d}x \tag{2.130}$$

where $\mathcal{X}$ denotes the domain where $f_X(x) > 0$.

This is an extension of the entropy for a discrete rv as defined in eq. (2.122), also known as *differential entropy*. Similarly to the discrete case, other quantities such as joint entropy, conditional entropy, and entropy of a random process are defined in similar terms. In particular, the *conditional entropy* is given by

$$H(X|Y) = -\int_{\mathcal{X}}\int_{\mathcal{Y}} f_{X,Y}(x,y) \log f_{X|Y}(X|Y)\mathrm{d}x\mathrm{d}y \tag{2.131}$$

where $\mathcal{X}$ and $\mathcal{Y}$ are the domains of x and y for which the logarithm is well defined.

However, the entropy lacks the same physical meaning as its discrete counterpart; in particular, for continuous rv's $H(X)$ is not guaranteed to be positive[1]. Moreover, differently from the discrete case, where the maximum entropy was in correspondence of the uniform rv, the continuous rv that maximizes the differential entropy is the Gaussian rv. The entropy of the Gaussian rv with variance σ^2 is equal to $h(\mathcal{N}) = 1/2 \log_2(2\pi e \sigma^2)$. The mutual information $\mathcal{I}(X;Y)$

---

[1]Consider for example a uniform rv X such that $f_X(x) = 2$ for $0 < x < 0.5$ and $f_X(x) = 0$ otherwise. In this case, $H(X) = -1$.

for continuous rv's has instead the same meaning as in the discrete case, which makes it so useful in communication theory: namely, it can be considered as a measure of the reduction of uncertainty.

DEFINITION 2.13 *Let X, Y denote two continuous rv's with pdf's respectively given by $f_X(x)$ and $f_Y(y)$, and joint pdf equal to $f_{X,Y}(x,y)$. The information content provided by the observation of the outcome of Y about the outcome of X is defined as the mutual information between X and Y, and is given by*

$$\mathcal{I}(X;Y) = \int_{\mathcal{X}} \int_{\mathcal{Y}} f_{X,Y}(x,y) \log \frac{f_{X,Y}(x,y)}{f_X(x)f_Y(y)} \mathrm{d}x\mathrm{d}y \qquad (2.132)$$

where $\mathcal{X}$ and $\mathcal{Y}$ are the domains of x and y for which the logarithm is well defined.

Following the corresponding definition for discrete rv's, also in the continuous case the mutual information can be expressed in terms of entropy according to eq. (2.129).

2.3.2   Channel capacity fundamentals

As outlined from the previous definitions, the mutual information $\mathcal{I}(X;Y)$ quantifies the average reduction of the uncertainty on X provided by the observation of the rv Y. Thus, if X represents the channel input and Y is the corresponding output, a reliable communication can take place if the receiver is able to precisely determine the outcome of X from the observation of the outcome of Y. Equivalently, if $\mathcal{I}(X;Y) = H(X)$, the uncertainty present in X and quantified by $H(X)$ is null once Y is observed. The concept of *channel capacity* states this idea in a precise way, as detailed in the following.

First, we consider the simplest and more general discrete channel case, i.e., the Discrete Memoryless Channel (DMC). Let X_i denote a discrete rv with probability distribution $p(x)$, drawn independently for the instants $i = 1, \ldots, n$, and Y_i a related discrete rv with conditional probability distribution $p(y|x)$. The DMC channel is characterized by the fact that each output value y_i is dependent only on the input value x_i at the same instant. Thus, if we focus on a specific instant i in the DMC, we can omit the subscript to compact the notation.

DEFINITION 2.14 *The channel capacity of a DMC channel is defined as the maximum mutual information between the input rv X and the output rv Y at a generic instant, with respect to all possible input distributions, as*

$$C = \max_{p(x)} \mathcal{I}(X;Y) \qquad (2.133)$$

The units of the capacity C are *bits per input symbol* or *bits per channel use*. This can be translated into bits/s provided that we normalize by the rate at which symbols enter the channel.

A first extension of the DMC case is the channel where outputs are not quantized. Due to its practical relevance, an interesting example in this field is provided by the discrete-time AWGN channel depicted in fig. 2.1. In this case, the rv's under consideration are continuous in values. At the instant i, it holds

$$Y_i = X_i + N_i \tag{2.134}$$

where N_i denotes a zero mean Gaussian rv with variance σ_N^2. This channel is also memoryless, and for a given X_i, the output Y_i is Gaussian distributed. Similarly to the DMC case, we cam omit the subscript i in the following, for simplicity. In this case the channel capacity is in the same form as in eq.

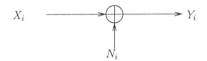

Figure 2.1. Discrete-time AWGN channel

(2.133), but since the input variate is continuous the capacity expression has to be slightly modified into

$$C = \max_{f_X(x):\text{constraint}} \mathcal{I}(X;Y) \tag{2.135}$$

where the constraint can take on different forms.

EXAMPLE 2.15 *For the discrete-time AWGN channel, the mutual information $\mathcal{I}(X;Y)$ is maximized when X_i is Gaussian distributed. In this case, it is common to refer to a power constraint in the form*

$$\mathrm{E}_X\left[x_i^2\right] \le P \tag{2.136}$$

and the associated channel capacity is given by

$$C = \max_{f_X(x):\mathrm{E}_X[x_i^2]\le P} \mathcal{I}(X;Y) = \frac{1}{2}\log\left(1 + \frac{P}{\sigma_N^2}\right) \qquad bit/channel\ use \tag{2.137}$$

where we have used the result on the differential entropy of the Gaussian rv.

This result can be extended to band-limited continuous-time channels. In this case, the capacity is given per unit time.

THEOREM 2.16 *By considering a continuous-time AWGN channel in the form $y(t) = x(t) + n(t)$, and assuming channel bandwidth B, signal power P,* *and one-sided noise power spectral density equal to N_0 Watt/Hz, the channel capacity is given by*

$$C = B \log \left(1 + \frac{P}{BN_0} \right) \qquad bits/s \qquad (2.138)$$

Again the capacity is obtained as the mutual information for a Gaussian distributed input signal $x(t)$.

There are different ways to derive this result due to Claude Shannon. The simplest one, not the most rigorous though, stems from the capacity for the discrete-time AWGN channel, $C = 1/2 \log(1 + P/\sigma_N^2)$, and uses the fact that $2B$ samples per second can be ideally recovered at the receiver. The noise power BN_0 follows immediately.

In 1948, Shannon presented his *Noisy Channel Coding Theorem*. In essence, this theorem states that *there exist channel codes that allow to achieve reliable communications with an error probability as small as desired, provided that the transmission rate is lower than the channel capacity*. This is possible by introducing some form of channel coding, i.e, by grouping the transmitted symbols to surmount the channel effects. The generic channel encoder is characterized by the code rate r defined as $r = k/n$, being k the number of bits in the input message, and n the number of coded symbol in the output sequence.

THEOREM 2.17 *Discrete-time Shannon Noisy Channel Coding Theorem.*

Considering a channel with capacity C bit/channel use, all rates r below C ($r < C$) are achievable, i.e., there exists a coding scheme such that the output of the source can be transmitted over the channel with an arbitrarily small error probability (Sufficient condition).

Conversely, any coding scheme such that the output of the source can be transmitted over the channel with an arbitrarily small error probability must have $r \leq C$ (Necessary condition).

A similar theorem can be introduced in the continuous time domain, by comparing the information rate R in input to the encoder to the continuous time capacity value in bit/s.

THEOREM 2.18 *Continuous-time Shannon Noisy Channel Coding Theorem.*

Considering a channel with capacity C bit/s, all rates R below C ($R < C$) are achievable, i.e., there exists a coding scheme such that the output of the source can be transmitted over the channel with an arbitrarily small error probability (Sufficient condition).

Conversely, any coding scheme such that the output of the source can be transmitted over the channel with an arbitrarily small error probability must have $R \leq C$ (Necessary condition).

2.3.3 Capacity of Fading Channels

The single user AWGN channel was the first case analyzed by Shannon with practical significance for the practitioner communication engineers. Despite its simplicity, it covers a wide range of applications, although the passing of years has opened the interest for the understanding of more complex scenarios, such as selective channels both in time and/or frequency, multiuser channels, and MIMO systems. Notably, a satellite communication link may fall under different categories for example depending on its bandwidth, user mobility, and type of application. Thus, we do consider of paramount importance the understanding of the rate limitations and coding strategies of a number of channels types, which are frequently encountered in the practice. The optimal input distributions, provided they are known, suggest practical signaling and coding strategies.

In the following, we first examine single user channels, where the transmitter and receiver have a single antenna. Capacity is in general a complex expression in terms of the channel variations in time and/or frequency, and also depends upon the knowledge that the transmitter and/or receiver have of the channel state (Channel State Information (CSI)). We will handle discrete-time systems, given that most continuous-time systems can be converted to the discrete-time case with capacity scaled accordingly. Thus, in the following capacity expressions are provided keeping in mind the underlying continuous-time system.

DEFINITION 2.19 *Flat-fading channel. Let the rv G denote the channel gain following a given distribution $f_G(g)$. A flat-fading channel is described by the following expression*

$$Y_k = G_k X_k + N_k \qquad (2.139)$$

where N_k denotes the additive noise, commonly assumed to be AWGN with one-sided power spectral density N_0 Watt/Hz, X_k and Y_k represent the transmitted and received rv's respectively, and G_k is the channel multiplicative factor at the instant k. All quantities are assumed to be complex.

The flat-fading model, also known as *frequency non-selective model*, describes the cases for which all frequency components undergo the same attenuation and phase shift in the channel, as explained in Chapter 4. The channel gain G_k can change at each time instant k, or remain constant over a block length, with either correlated or independent variations. The capacity of this channel depends on the knowledge of G_k at the transmitter and receiver. In all cases, we assume that the statistical distribution of G is known to the transmitter and receiver, or equivalently the distribution of the received Signal to Noise Ratio (SNR) is known.

EXAMPLE 2.20 *Memoryless Rayleigh fading channel [19]. Consider the channel described by eq. (2.139), where G_k and N_k are independent identically* *distributed (i.i.d.) circular complex Gaussian rv's, and* $\{X_k\}$ *is average-power limited. This channel is identified as the memoryless Rayleigh fading channel, because the channel amplitude rv* G_k *follows a Rayleigh distribution (see Section 2.2.8). In the absence of any CSI at both transmitter and receiver, the distribution able to achieve the capacity is discrete with a finite number of mass points [19], [20].*

The rapidly varying channel considered in this example is the idealized model of a very pessimistic scenario, for which the channel entropy rate is comparable to the source entropy, and this makes unfeasible the estimation of channel parameters for decoding purposes. The capacity and its corresponding associated coding strategy must be computed numerically. When the SNR goes to zero, the capacity of the memoryless Rayleigh fading channel approaches the AWGN case with the same SNR value. In this case, capacity is achieved with only two signaling values, one of them being zero. As a consequence, in order to handle rapidly fading channels in very low SNR scenarios, it is required to use on-off keying. At high SNR, the loss in performance due to fading effects grows rapidly [20].

If *perfect CSI is available at the receiver*, then it is possible to perform coherent reception. The Shannon capacity (also known as ergodic capacity or average capacity) is given in the following for this case [21].

DEFINITION 2.21 *Average Capacity. The Shannon (ergodic) capacity of the fading channel of eq.(2.139), for which the channel gain power* $\{\gamma_k\} = \{|G_k|^2\}$ *is a stationary ergodic process perfectly known at the receiver, is given by*

$$C = B \int_0^{\infty} \log\left(1 + \frac{P\gamma}{BN_0}\right) f_{\gamma}(\gamma) \mathrm{d}\gamma \qquad bits/s \qquad (2.140)$$

where B is the bandwidth, P is the input power constraint ($\mathrm{E}_X\left[|x_k|^2\right] < P$), N_0 *is the AWGN one-sided power spectral density, and* $f_{\gamma}(\gamma)$ *is the pdf of each* γ_k *rv.*

Notably, the codewords able to achieve the capacity must be long enough to suffer all possible fading states. Although it is tempting to think of the previous expression as a justification for the use of variable-rate codes, this is not the case, because the channel state is not known at the transmitter [21].

EXAMPLE 2.22 *In Nakagami-m fading, the pdf of the received signal is given by eq. (2.89), where* m *is the fading severity parameter ranging from* $1/2$ *to* ∞. *The average capacity has been solved for* m *integer in [22]. A more general solution for arbitrary* m *of the average channel capacity is*

$$C = \frac{B}{\ln 2} \frac{1}{\Gamma(m)} \left(\frac{m}{\zeta}\right)^m G_{2,3}^{3,1}\left[\frac{m}{\zeta} \,\middle|\, \begin{matrix} -m, 1-m \\ 0, -m, -m \end{matrix}\right] \qquad (2.141)$$

where ζ is the average signal-to-noise ratio, defined as $\zeta = PE_\gamma\left[\gamma\right]/BN_0$, and $G^{k,l}_{l,k}(\cdot)$ is the Meijer's G-function [4, eq. (9.301)]. Note that for $m = 1$, using [16, /06.35.26.0001.01], eq. (2.141) reduces to the average capacity of the well-known Rayleigh model [23, eq. (5)].

EXAMPLE 2.23 *In Rice fading, the pdf of the received signal has been defined in section 2.2.13, where $K = n^2$ is the Rice factor. The Rice average channel capacity can be obtained as*

$$C = \frac{B\,e^{-K}}{\ln 2}\sum_{n=0}^{\infty}\frac{1}{(n!)^2}\frac{K^n(1+K)^{n+1}}{\zeta^{n+1}}G^{3,1}_{2,3}\left[\frac{K+1}{\zeta}\,\middle|\,\begin{matrix}-1-n\,,\,-n\\0\,,\,-1-n\,,\,-1-n\end{matrix}\right]$$

(2.142)

where ζ is the average signal-to-noise ratio.

EXAMPLE 2.24 *In a Weibull fading environment, the pdf of the received signal is given by eq.(2.85), where $\beta \geq 1$ is the Weibull fading severity parameter [15]. Then, the average channel capacity is [24], [25]*

$$C = \frac{B}{\ln 2}\frac{\beta}{2\,(a\,\zeta)^{\frac{\beta}{2}}}\frac{\sqrt{k}\,l^{-1}}{(2\pi)^{\frac{k+2l-3}{2}}}$$

$$\times\,G^{k+2l,l}_{2l,k+2l}\left[\frac{(a\,\zeta)^{-\frac{\beta k}{2}}}{k^k}\,\middle|\,\begin{matrix}\Upsilon(l,-\frac{\beta}{2})\,,\,\Upsilon(l,1-\frac{\beta}{2})\\\Upsilon(k,0)\,,\,\Upsilon(l,-\frac{\beta}{2})\,,\,\Upsilon(l,-\frac{\beta}{2})\end{matrix}\right]$$

(2.143)

where ζ is the average signal-to-noise ratio, $\Upsilon(n,\xi) = \xi/n,\,(\xi+1)/n,\ldots,\,(\xi+n-1)/n$, with ξ an arbitrary real value, and $n,\,k,\,l$ are positive integers with

$$\frac{l}{k} = \frac{\beta}{2}$$

(2.144)

Depending upon the value of β, a set with minimum values of k and l can be properly chosen (e.g. for $\beta = 2.5$, $l = 5$ and $k = 4$).

If no significant channel variability occurs during the transmission, then the use of the term *slow-fading channel* is customary, as in the case of geostationary satellite channels and slow moving terminals. A simple and practical way of describing slow fading evolution is to work under the hypothesis of a piecewise constant SNR at the receiver. Accordingly, the received power γP remains constant for a number of channel uses, and then changes to a new i.i.d. value (block fading). This approach produces a capacity which is a rv itself: for a fixed transmitted rate, given that the transmitter does not know the CSI, there is a non-negligible probability that the channel cannot support the necessary capacity at a given instant. This is commonly referred to as *outage capacity or capacity with outage.*

DEFINITION 2.25 *The capacity with outage denotes the achievable rate in a block fading channel with a given probability. Outage occurs for those fading states for which it is not possible to keep the specified rate.*

Thus, capacity decreases for small outage probability, given the necessity of decoding correctly the bits under severe fading conditions [21]. For delay constrained applications, the zero outage capacity is a meaningful value. It indicates the maximum instantaneous data rate in all fading conditions[2]. For those channel states for which the channel is in outage for the specified capacity, retransmissions will be necessary.

EXAMPLE 2.26 *Consider the flat Rayleigh block fading channel given by eq. (2.139), where G_k is known at the receiver and is constant for a period of time long enough to cover the length of the codewords. The channel power gain $\gamma = |G|^2$ is exponentially distributed provided that the channel fading is Rayleigh (Section 2.2.9). In such a case, the outage probability P_{out} for a given capacity C, with a received average signal to noise ratio ζ and bandwidth B is equal to*

$$P_{out} = 1 - \exp\left\{ -\frac{2^{C/B} - 1}{\zeta} \right\} \tag{2.145}$$

If *perfect CSI is present at both transmitter and receiver*, then the transmitter can accordingly adapt the power and rate of the transmitted signal. *Transmitter side information* can be achieved through the use of a return channel signaling the SNR values estimated at the receiver. For capacity analysis purposes, we assume the SNR estimation procedure to be ideal; however, more on this topic can be found in Chapter 7. The unavoidable delay associated to the low rate feedback channel is considered negligible with respect to the channel fading rate. The channel under consideration can be again described through eq. (2.139), where now G_k is instantaneously known with no error at both transmitter and receiver. The objective is the maximization of the capacity given by [26]

$$C = B \int_0^\infty \log\left(1 + \frac{P(\gamma)\gamma}{BN_0}\right) f_\gamma(\gamma)\,\mathrm{d}\gamma \tag{2.146}$$

with $\gamma = |G|^2$ subjected to the average power constraint

$$\int_0^\infty P(\gamma)f_\gamma(\gamma)\,\mathrm{d}\gamma \leq P \tag{2.147}$$

---

[2]Zero-outage capacity is sometimes referred to as delay-limited capacity.

The optimum strategy adapts the power following a waterfilling approach with time [26], i.e.,

$$\frac{P(\gamma)}{N_0 B} = \begin{cases} \frac{1}{\gamma_0} - \frac{1}{\gamma} & \gamma > \gamma_0 \\ 0 & \gamma < \gamma_0 \end{cases} \qquad (2.148)$$

where the cutoff value γ_0 must be obtained numerically from the following relation

$$\int_{\gamma_0}^{\infty} \left(\frac{1}{\gamma_0} - \frac{1}{\gamma} \right) d\gamma = \frac{P}{N_0 B} \qquad (2.149)$$

As a consequence, for good channel conditions a larger amount of power is allocated and higher data rates are transmitted. Time diversity is thus used by the different encoders, each one linked to a different fading state[3]. If the instantaneous SNR is too small, transmission is halted till channel conditions improve. As expected, the exploitation of the CSI at the transmitter improves the capacity, although only marginally; improvements are noticeable only for very low SNR. Then, one could wonder what is the point of adapting transmitter parameters. If the transmitter power is not adapted, then the capacity corresponds to the case with CSI at the receiver only, given by eq. (2.140). However, rate adaptation at the transmitter makes the decoding phase easier; a non-adaptive strategy requires the use of channel correlation statistics in the code design, and the decoder complexity can be very large. In other words, the power adaptation yields only a slight capacity gain, and from the implementation point of view it appears more attractive to adapt only the rate at the transmitter, as a function of the SNR experienced by the receiver and signalled through a feedback channel[4]. Note that long-term strategies may not be useful for delay sensitive applications, since users with poor channel conditions would not receive data until their channels improve. For these cases, coding strategies that keep constant rates are more beneficial.

DEFINITION 2.27 *Frequency-selective fading channels. Fading channels that are characterized by a gain in the frequency domain that is not constant over the entire signal bandwidth are identified as frequency-selective fading channels. Frequency selectivity is usually caused by multipath and can evolve with time due to relative motions of the transmitter, receiver and reflective environment (see Chapter 4).*

---

[3]Each encoder uses a power and rate, which are a consequence of the waterfilling solution. This is the case of any type of flat-fading, irrespective of its rate of change.

[4]Capacity can also be achieved with fixed rate codes and a power allocation of the waterfilling type[27]. In this case standard Gaussian codes can be employed, provided that they are long enough to match the ergodicity of the channel. Suboptimal strategies based on the inversion of the channel are also possible: in the simplest one, the emitted power is given by $P(\gamma) = c/\gamma$, where c is adjusted for a given average power constraint.

For a *time-invariant frequency-selective channel* with input X_k, output Y_k, and AWGN component N_k with one-sided power spectral density N_0, indicating with H_k the channel impulse response, we can describe the input-output relation as

$$Y_k = H_k * X_k + N_k \qquad (2.150)$$

The optimal input distribution of $\{X_k\}$ required to achieve the capacity is related to the channel frequency response $H(f)$. If ideal CSI is available at both transmitter and receiver side, the optimal power distribution follows a water-filling profile over frequency, similar to the water-filling over time power allocation for time-variant flat fading channels [17], i.e.,

$$P(f) = \begin{cases} \frac{1}{\gamma_0} - \frac{1}{\gamma(f)} & \gamma(f) > \gamma_0 \\ 0 & \gamma(f) < \gamma_0 \end{cases} \qquad (2.151)$$

where $\gamma(f) = |H(f)|^2/N_0$ and γ_0 a constant guaranteeing that

$$\int_B P(f)df = P \qquad (2.152)$$

This expression, originally derived by Shannon, makes it clear than a larger amount of power has to be allocated to those frequencies with higher SNR, and as a consequence higher rates are available. The resulting capacity can be written as

$$C = \int_B \log\left(1 + \frac{|H(f)|^2 P(f)}{N_0}\right) df \qquad (2.153)$$

The capacity of *time-varying frequency-selective fading channels* is usually unknown. If perfect CSI is known at both transmitter and receiver, then capacity can be achieved through a two-dimensional joint water-filling in time and frequency [21].

2.3.4 Multiuser Channels Capacity

Multiuser channels are characterized by several users that have to share the medium. In this necessarily short presentation, we focus on two main types of channels that are defined in the following: the broadcast channel and the multiple access channel, respectively.

DEFINITION 2.28 *Broadcast Channel (BC). The BC channel is characterized by one transmitter sending towards two or more receivers, as depicted in Figure 2.2. It is representative of a downlink or forward link.*

DEFINITION 2.29 *Multiple-Access Channel (MAC). The MAC channel is characterized by one receiver that receives signals from two or more transmitters, as depicted in Figure 2.3. It is representative of an uplink or reverse link.*

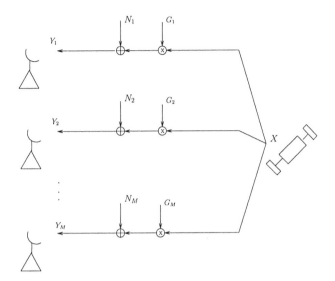

Figure 2.2. The broadcast channel

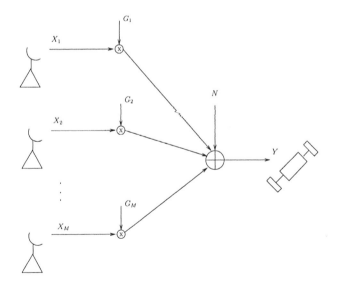

Figure 2.3. The multiple access channel

In the following, we present the most relevant results on capacity for multiuser channels. This particular part of Information theory is also identified as *Network Information Theory*. We no longer have a single communication channel, since we have many-to-one or one-to-many links, and as a consequence, we are interested in the achievable rates for all established communications, i.e., in determining a *capacity region*.

DEFINITION 2.30 *The capacity region of a multiuser channel is the set of all possible achievable rates.*

A single parameter that can be usefully introduced is the *sum-rate capacity*, which expresses the maximum global achievable rate. This parameter must be handled with care. In fact, the allocation of all resources (bandwidth and power) to the user with the best channel conditions effectively allow to achieve the maximum overall rate, but only one user actually exploits the benefits of the communication.

Consider the broadcast channel depicted in Figure 2.2. Without loss of generality, we assume that the channel power gains $\gamma_m = |G_m|^2$ are ordered in such a way that $0 \leq \gamma_1 \leq \gamma_2 \leq, \ldots, \leq \gamma_M$, and the AWGN noise has the same power $N_0 B$ at every receiver[5]. For a transmitted power P and bandwidth B, the possible achievable rates are given by the following theorem.

THEOREM 2.31 *Broadcast channel capacity [2]. The capacity region of the Gaussian BC is given by the rate vectors R_m, $m = 1, \ldots, M$ satisfying*

$$R_m \leq B \log \left(1 + \frac{\gamma_m P_m}{N_0 B + \gamma_m \sum_{n=m+1}^{M} P_n} \right) \qquad (2.154)$$

for any power combination such that $\sum_{m=1}^{M} P_m = P$.

If all users experience the same channel quality, i.e., have the same values of γ_m, the capacity region can be achieved with *time, frequency or code orthogonal division* (see Chapter 10). Otherwise, for unbalanced users, the maximization of the transmitted rates is achieved through the use of *superposition coding with successive interference cancellation* (see Chapter 10) [29] or through *Dirty Paper Coding (DPC)* [30][6]. In both cases, a smart coding is adopted, which takes into account the different SNR values.

In *fading broadcast channels* the capacity region depends on the channel knowledge that transmitter and receiver may have. If both base station and

---

[5]Different noise levels can be handled normalizing the channel gain accordingly. For example, co-channel interference coming from other beams must be included together with noise [28].
[6]The transmitter performs a modulo precoding, pre-substracting other users codewords in order to increase channel gains.

receivers know perfectly the channel, then optimal coding strategies as those mentioned above can be used, provided that fading is slow enough. In this case, the codes can be designed as if the channel were static. If all channels towards the receivers fade independently, the corresponding diversity can be used to maximize the sum-rate capacity, i.e., the maximum overall achievable rate. As a consequence, sum-rate capacity is achieved by devoting all resources at any given fading state to the user experiencing the best channel, assuming that there are no priorities assigned to the different users [31]. For delay-constrained applications, capacity with outage is a more reasonable measure, similarly to single user systems. *The capacity with outage region is the set of fixed rates that can be achieved in all non-outage fading states.* As opposed to the ergodic capacity approach, a larger amount of power is transmitted in bad channels to guarantee a fixed rate [32].

If different users at non coincident locations communicate with the same base station, we have a MAC channel as shown in Figure 2.3. There are some important differences with respect to BC. For example, in the MAC case each transmitter has an *individual power constraint*, and the desired signal and the interference component come from different transmitters, being in essence multiplied by different channel gains. However, there are some elegant connections between the Gaussian MAC and the Gaussian BC that strictly link the capacity results for both scenarios [33].

Considering a MAC channel with constant channel power gains γ_m, $m = 1, \ldots, M$, assuming that each user m has the average power constraint given by P_m, and the one-sided Gaussian noise power spectral density at the central receiver is given by N_0, for a bandwidth B the possible achievable rates are given by the following theorem.

THEOREM 2.32 *Multiple access channel capacity. The capacity region of the Gaussian MAC channel is given by the rate vectors R_m, $m = 1, \ldots, M$ satisfying simultaneously the following equations [34]:*

$$\sum_{m \in S} R_m \leq B \log \left(1 + \frac{\sum_{m \in S} \gamma_m P_m}{N_0 B} \right) \qquad (2.155)$$

for all possible subsets $S \subset \{1, 2, \ldots, M\}$.

Thus, in the MAC case, the sum capacity of any subset of users must be below the capacity of an equivalent single user with the whole received power at its disposal. Figure 2.4 shows the capacity region for the two-users case. Interestingly, one user can transmit at the same rate as if no other users were present, and there is still room for non-zero rates from other users. In fact, by successive interference cancellation at the receiver it can be proved that it is

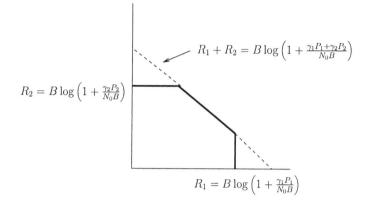

Figure 2.4. Capacity region for a two-user multiple access channel

possible to achieve the following set of rates

$$R_m = B \log \left(1 + \frac{\gamma_m P_m}{N_0 B + \sum_{n=m+1}^{M} \gamma_n P_n} \right), \quad m = 1, \ldots, M \quad (2.156)$$

In this set of rates the signal coming from user M, which has the largest gain, is decoded first. For any other decoding order the overall rate is the same, and given by the sum rate capacity of eq. (2.155) with $S = \{1, 2, \ldots, M\}$, i.e.,

$$C = B \log \left(1 + \frac{\sum_{m=1}^{M} \gamma_m P_m}{N_0 B} \right) \quad (2.157)$$

although the rate contribution from each link is different.

We now consider the *AWGN MAC with varying channel gains*, assuming perfect CSI at both transmitters and receiver. The maximization of the sum-rate capacity calls for choosing only one user at any given fading state. In a symmetric scenario with similar statistics and with the same power for all users, the user experiencing the best channel can transmit, while the others remain silent. Each transmitter must distribute the power in time following a time water-filling approach. Differently, if the users have different channel statistics or different power constraints, the optimal criterion is still the allocation of the whole channel to a single transmitter at any given state, although the channel gains must be weighted by the power constraint of each user [35]. However, for strongly asymmetric channels, overall rate maximization is not the best strategy. Consider, for example, a user located very far from the base station with respect to the other users. The sum-rate maximization criterion would translate into a low rate achievable for this user. Regarding delay-sensitive applications, capacity with outage has to be considered, and it results relatively complex to obtain the capacity region for a given outage probability. In this sense, the application of the *duality of the Gaussian BC and MAC* is very useful [33]. Capacity results and coding strategies can be applied both to BC and MAC with a clear symmetry . In fact, the BC capacity region can be derived as the union of MAC capacity regions, while dually the MAC capacity region is the intersection of BC capacity regions. In other words, the same principles analyzed before for the BC case still apply: for the ergodic capacity approach, a larger amount of power is allocated for better channels, whereas for outage driven applications, the opposite trend is selected.

The previous considerations apply also to the case of *multiuser frequency-selective channels*, including both BC and MAC cases. If the signal bandwidth is divided into sub-bands, for which the channel gains can be considered constant, then the strategies exposed above still apply: at each fading state and for each sub-band, the best channel is the only used. Power must be allocated following the water-filling criterion over time and frequency [34].

2.3.5   MIMO systems

The use of several antennas at transmitter and/or receiver enhances the achievable rates due to the additional degrees of freedom, which provide spatial diversity. Space plays now a similar role as time and frequency. These systems, usually referred to as *Multiple-Input Multiple-Output (MIMO) system*, are usually described through a vectorial notation. Thus, in the *frequency non-selective time-invariant case*, we have

$$y = Hx + n \qquad (2.158)$$

where H is the MIMO channel transfer matrix. If no CSI is available at the transmitter side, the capacity of the MIMO link with informed receiver is given by the following theorem [36].

THEOREM 2.33 *The capacity of a time-invariant MIMO channel with input covariance matrix $P = \mathrm{E}\left[xx^H\right]$ is equal to*

$$C = B \log \left| I + \frac{HPH^H}{N_0 B} \right| \qquad (2.159)$$

where $|\cdot|$ denotes the matrix determinant operator. This capacity is achieved for Gaussian x.

If the transmitter knows the channel H, then the transmit covariance matrix can be optimized subject to the constraint on the power $\mathrm{tr}\{P\} < P$ [34], being $\mathrm{tr}\{\cdot\}$ the matrix trace operator. The capacity is achieved by using waterfilling power allocation in the spatial domain: the transmit power is distributed according to the eigenvectors of the MIMO channel by assigning a larger amount of power to the dominant eigenvectors. Eigen-modes with a SNR below a given threshold are not excited.

If fading is present, the channel matrix H varies randomly in time and, as a consequence, the channel capacity is also a rv with average value known as the *ergodic or average capacity*, defined as

$$C = \mathrm{E}\left[B \log \left| I + \frac{HPH^{\mathrm{H}}}{N_0 B} \right| \right] \qquad (2.160)$$

EXAMPLE 2.34 *Consider a Rayleigh fading channel, for which H is a random Gaussian matrix with i.i.d. components. The number of transmit and receive antennas is n_t and n_r respectively. Thus, the average (ergodic) capacity is obtained for*

$$P = \frac{P}{n_t} I \qquad (2.161)$$

and applying eq. (2.160) is given by

$$C = \mathrm{E}\left[B \log \left| I + \frac{P}{N_0 B n_t} HH^{\mathrm{H}} \right| \right] \qquad (2.162)$$

Thus, the maximization of the transmission rates with an uninformed transmitter requires to transmit uncorrelated streams with the equal power.

It can be shown that capacity increases linearly with $\min(n_t, n_r)$ for high SNR, being n_t and n_r the number on transmitter and receiver antennas respectively. Moreover capacity increases linearly with n_r for low SNR, and linearly with n for all SNR if $n_t = n_r = n$.

If perfect CSI is available at the transmitter, then transmitter diversity provides a capacity increase equal to n_t/n_r for low SNR [34]. The outage performance of MIMO systems offers useful insights in some cases. Thus, it is known that for i.i.d. slow Rayleigh fading channels the gain is equal to the product of n_t and n_r.

The extra degrees of freedom provided by the use of more than one antenna can play also a relevant role in multiuser channels. Consider for example the MAC channel. If ideal CSI is available at the terminals and base station, each user has to vary its transmit power as a function of the channel states of all users, similarly to the single antenna MAC case. However, although no closed form solution is known for the maximization of the sum rate capacity, it can be seen than due to the use of several antennas, more than one user must transmit simultaneously [34]. With uninformed transmitters, the spatial diversity can provide important gains, especially for high SNR. In particular, orthogonal signalling strategies are far from optimal even for the case of power balanced users.

The codes able to achieve the capacity require the use of successive cancellations in the uplink base station and in the receiver of a point-to-point MIMO channel. This successive cancellation technique is no longer possible in the receivers of a MIMO downlink channel[7], which must decode their respective received signals independently. Under intuitive terms, it seems that the interference coming from other users should be cancelled before being transmitted, since the base station has access to all signals. The precoding scheme which achieves capacity for the AWGN MIMO channel is called *Dirty Paper Coding (DPC)* from the paper by H. Costa [37]. Costa studied transmission with side information or encoding for the interfering channel, and quantified the capacity of a channel with interference known at the transmitter side. The channel input-output relation is in this case given by

$$Y_k = X_k + S_k + N_k \qquad (2.163)$$

where the same structure of the AWGN channel of eq. (2.134) is maintained, except for the addition of the term S_k which represents the Gaussian i.i.d. interference component, assumed to be perfectly known at the transmitter. Costa proved that if the transmitter non-causally knows S_k, the capacity is still given by eq. (2.137), which is here rewritten for the reader convenience:

$$C = \frac{1}{2} \log \left(1 + \frac{P}{\sigma_N^2} \right) \qquad \text{bit/channel use} \qquad (2.164)$$

In this case, capacity is achieved through the use of random codewords, for which there is not a one-to-one correspondence with the transmitted message. The selected codeword depends on the interference state, as if "the writer knew the dirt in the paper before writing the message", to cite the analogy that gives the name to the procedure [37]. DPC turns out to be an essential method to compute the capacity region of the MIMO BC channel, which can be written as

$$y_m = H_m x + n_m, \qquad m = 1, 2, \ldots, M \qquad (2.165)$$

THEOREM 2.35 Capacity of the MIMO BC *[38]*. *The capacity region of the MIMO BC channel is given by the convex hull of the union of all rate vectors in the form*

$$R_m = B \log \frac{|N_0 B I + H_m \sum_{n=m}^{M} P_n H_m^H|}{|N_0 B I + H_m \sum_{n=m+1}^{M} P_n H_m^H|} \qquad m = 1, \ldots, M \qquad (2.166)$$

---

[7]As already noticed, the single-input-single-output downlink capacity can be still achieved through superposition coding and successive interference cancellation, since stronger users can decode signals of weaker users.

over all possible order permutations and positive semi-definite covariance matrices $P_1, P_2, \ldots, P_M$, such that $tr\{P_1+P_2+\ldots+P_M\} = tr\{\mathrm{E}\left[xx^{\mathrm{H}}\right]\} \leq P$.

Practical signalling techniques have still to be developed for the MIMO multi-user scenario in order to exploit all degrees of freedom provided by the use of more than one antenna [21].

2.4 Detection and estimation theory

This section describes the basic concepts of detection and estimation theory that are instrumental for the theoretical characterization of communication systems [3].

DEFINITION 2.36 *Detection theory encompasses the measurement of a discrete parameter belonging to an M-ary alphabet and typically assesses the detection quality through the evaluation of an error probability.*

DEFINITION 2.37 *Estimation theory entails the measurement of a continuous parameter which can assume values within a continuous interval, and quantifies the merit figure in terms of estimation error, ε.*

Detection and estimation theory can be in general applied both to a single parameter or to a parameter multiplicity, often requiring in the latter case a joint application of the two. Detection/estimation can then refer to a single value, to a series of discrete values (e.g. data flow demodulation with memory), or to a continuum (e.g. parameter tracking in time). The receiver design based on detection and estimation theory is an example of *non structured approach*, which does not assume any a priori pragmatic structure, but formulates optimality criteria for specific objective functions.

2.4.1 Detection theory

Consider a source S that produces a discrete rv, each value of which is identified as the *hypothesis H_i*, with $i = 0, \ldots, M - 1$ being M the alphabet cardinality. Each hypothesis has a priori probability π_i, $\sum_{i=0}^{M-1} \pi_i = 1$. A probabilistic transition mechanism (e.g. the channel) maps each H_i into a vector of received samples, which contains the information associated to H_i plus noise. Into a finite N-dimensional observation space the *observed vector* is $r = [r_1, \ldots, r_N]^{\mathrm{T}}$. The conditional pdf $f_{r|H_i}(r|H_i)$ is the *likelihood or transition function* of the received vector, and is the basic instrument to perform detection. A decision rule subdivides the observation space into *detection regions*, $Z_j, j = 0, \ldots, M'$, filling-in the entire observation space with no intersections, and provides the selected hypothesis $\hat{H}_j$, when r is in the region Z_j. In general $M' \geq M$, where equality holds for *hard detection* (i.e., decision for one of the initial hypotheses), while the greater holds for *soft detection* (also reliability information is provided to the cascade blocks).

In the following, we assume $M' = M$, and in particular we first consider a simple binary detection problem ($M = 2$), leaving to a next step the generalization to general M-ary detection problems.

2.4.1.1   Binary detection.   A binary detection problem has to discriminate between the two hypotheses H_0 and H_1. Four cases can thus occur:

1 $\hat{H}_0|H_0$, i.e., choose H_0 with H_0 true, which corresponds to the *correct rejection* event, that occurs with probability $P_s=P_{00}=\int_{Z_0} f_{r|H_0}(r|H_0)\mathrm{d}r$

2 $\hat{H}_1|H_0$, i.e., choose H_1 with H_0 true, which corresponds to the *false alarm* event, with probability $P_{fa} = P_{10} = \int_{Z_1} f_{r|H_0}(r|H_0)\mathrm{d}r = 1 - P_s$

3 $\hat{H}_1|H_1$, i.e., choose H_1 with H_1 true, which corresponds to the *correct detection* event, with probability $P_d = P_{11} = \int_{Z_1} f_{r|H_1}(r|H_1)\mathrm{d}r$

4 $\hat{H}_0|H_1$, i.e., choose H_0 with H_1 true, which corresponds to the *missed detection* event, with probability $P_{md} = P_{01} = \int_{Z_0} f_{r|H_1}(r|H_1)\mathrm{d}r = 1 - P_d$

The specific processing method for the observed vector r depends on the adopted *detection criterion*. In the following, the most common criteria are presented.

2.4.1.1.1   Bayes criterion.   The Bayes criterion is based on the assumptions that the a-priori probabilities, π_i, $i = 0, 1$ characterize the source outputs, and a *cost* is assigned to each of the possible detection event, namely C_{00}, C_{10}, C_{11}, and C_{01} respectively for $\hat{H}_0|H_0$, $\hat{H}_1|H_0$, $\hat{H}_1|H_1$, and $\hat{H}_0|H_1$.

DEFINITION 2.38 *The risk is the average cost defined as*

$$\mathcal{R} = C_{00}P_{00}\pi_0 + C_{10}P_{10}\pi_0 + C_{11}P_{11}\pi_1 + C_{01}P_{01}\pi_1 \qquad (2.167)$$

DEFINITION 2.39 *The Bayes detection criterion is optimum in the sense of the minimum risk $\mathcal{R}$. The minimum risk is called the Bayes risk.*

THEOREM 2.40 *Assuming that the cost of a wrong decision is higher than the cost of a correct decision, i.e., $C_{10} > C_{00}$ and $C_{01} > C_{11}$, the detection regions Z_0 and Z_1 that minimize $\mathcal{R}$ are defined by*

$$\pi_1(C_{01} - C_{11})f_{r|H_1}(r|H_1) \underset{\hat{H}_0}{\overset{\hat{H}_1}{\gtrless}} \pi_0(C_{10} - C_{00})f_{r|H_0}(r|H_0) \qquad (2.168)$$

where $C_{01} - C_{11}$ and $C_{10} - C_{00}$ are also identified as relative costs of the wrong decisions with respect to the correct ones.

DEFINITION 2.41   *The likelihood ratio associated to the observed vector is*

$$\ell(\boldsymbol{r}) = \frac{f_{\boldsymbol{r}|H_1}(\boldsymbol{r}|H_1)}{f_{\boldsymbol{r}|H_0}(\boldsymbol{r}|H_0)} \tag{2.169}$$

DEFINITION 2.42   *The Bayes criterion threshold is $\xi = \frac{\pi_0(C_{10}-C_{00})}{\pi_1(C_{01}-C_{11})}$.*

THEOREM 2.43   *The Bayes risk minimization implies to solve the Likelihood Ratio Test (LRT)*

$$\ell(\boldsymbol{r}) \underset{\hat{H}_0}{\overset{\hat{H}_1}{\gtrless}} \xi \tag{2.170}$$

DEFINITION 2.44   *The Logarithmic LRT (LLRT) is*

$$\Lambda(\boldsymbol{r}) = \ln \ell(\boldsymbol{r}) \underset{\hat{H}_0}{\overset{\hat{H}_1}{\gtrless}} \ln \xi \tag{2.171}$$

THEOREM 2.45   *Introducing the likelihood ratio conditional pdf's $f_{\ell|H_1}(\ell|H_1)$ and $f_{\ell|H_0}(\ell|H_0)$, the correct detection and false alarm probabilities can also be expressed as*

$$P_d = \int_\xi^\infty f_{\ell|H_1}(\ell|H_1)\mathrm{d}\ell \tag{2.172}$$

$$P_{fa} = \int_\xi^\infty f_{\ell|H_0}(\ell|H_0)\mathrm{d}\ell \tag{2.173}$$

2.4.1.1.2   MAP criterion.

DEFINITION 2.46   *The Maximum-A-Posteriori (MAP) detection criterion is obtained from the Bayes criterion by imposing that the correct decisions have no cost, while the wrong decisions have cost equal to 1, i.e., $C_{00} = C_{11} = 0$ and $C_{10} = C_{01} = 1$ (decisions symmetry). In this case, the Bayes risk coincides with the average error probability.*

DEFINITION 2.47   *The MAP criterion is optimum in the sense of the minimum average error probability.*

THEOREM 2.48   *The MAP criterion corresponds to the test*

$$Pr\{H_1|\boldsymbol{r}\} \underset{\hat{H}_0}{\overset{\hat{H}_1}{\gtrless}} Pr\{H_0|\boldsymbol{r}\} \tag{2.174}$$

where $Pr\{H_i|\boldsymbol{r}\}$, $i = 0, 1$, is the a posteriori conditional pdf of the hypothesis H_i.

2.4.1.1.3   ML criterion.

DEFINITION 2.49 *The Maximum Likelihood (ML) detection criterion is obtained by the Bayes criterion by imposing $C_{00} = C_{11} = 0$ and $C_{10} = C_{01} = 1$ (decisions symmetry) and assuming uniform distribution for H_0 and H_1 ($\pi_0 = \pi_1 = 1/2$). Thus, when the hypotheses distribution is actually uniform, the ML criterion is a particular case of the MAP criterion. But in general its applicability can be considered wider.*

THEOREM 2.50 *The ML criterion corresponds to the LRT*

$$\ell(r) \underset{\hat{H}_0}{\overset{\hat{H}_1}{\gtrless}} 1 \qquad (2.175)$$

The quality of the ML criterion depends on how the actual hypotheses distribution is similar to the uniform distribution.

2.4.1.1.4   Minimax criterion.   The minimax criterion is a more appropriate criterion than ML to be employed when the hypotheses are not uniformly distributed. Differently from Bayes and MAP, it does not require to know the a-priori probabilities π_i, but it is sub-optimal.

DEFINITION 2.51 *The minimax criterion selects π_1 equal to the a-priori value that maximizes the Bayes risk. As a consequence, minimax is suboptimum with respect to Bayes.*

THEOREM 2.52 *The minimax criterion is thus represented by the minimax equation*

$$C_{11} - C_{00} + P_{md}(C_{01} - C_{11}) - P_{fa}(C_{10} - C_{00}) = 0 \qquad (2.176)$$

THEOREM 2.53 *Imposing null costs for the correct decisions ($C_{11} = C_{00} = 0$), the minimax equation is simply*

$$\frac{P_{md}}{P_{fa}} = \frac{C_{10}}{C_{01}} \qquad (2.177)$$

2.4.1.1.5   Neyman-Pearson criterion.   The Neyman-Pearson (NP) criterion interestingly derives a solution similar to the Bayes criterion requiring the minimal a-priori information.

DEFINITION 2.54 *The NP criterion maximizes the correct detection probability, under the constraint of a maximum value, α, for the false alarm probability, i.e., $P_d \to \max | P_{fa} \le \alpha$.*

This lead to a LRT formally identical to eq. (2.170) except for the threshold which is a-posteriori computed imposing the constraint over P_{fa}.

2.4.1.1.6   Receiver operating characteristics.

DEFINITION 2.55 *Performance of a binary detector is typically in terms of Receiver Operating Characteristics (ROC), i.e., correct detection vs. false alarm probability for equal values of the detection threshold ξ. Each point of a ROC corresponds to a specific value of the threshold ξ.*

PROPERTY 2.4.1 *The threshold value $\xi=0$ corresponds to the point $(P_{fa}, P_d)=(1,1)$; the threshold value $\xi=\infty$ corresponds to $(P_{fa}, P_d)=(0,0)$.*

PROPERTY 2.4.2 *All ROC concavity is downward.*

PROPERTY 2.4.3 *All ROC are always above the diagonal line $P_{fa} = P_d$.*

PROPERTY 2.4.4 *The slope of the ROC in each point is equal to the threshold value necessary to achieve (P_{fa}, P_d) of that point.*

2.4.1.1.7   Sufficient statistic.
In general, the observed vector can be expressed as a couple $r = (x, y)$, with $x \in \mathbb{R}$, and $y \in \mathbb{R}^{N-1}$.

THEOREM 2.56 *The scalar x is the sufficient statistic of $r \Leftrightarrow f_{y|x,H_1}(y|x,H_1) = f_{y|x,H_0}(y|x,H_0)$*

THEOREM 2.57 *If x is the sufficient statistic of r, the following identity holds for the likelihood ratio*

$$\ell(r) = \ell(x) \tag{2.178}$$

2.4.1.2   *M*-ary detection.
The same assumptions as the binary case holds, with $i = 0, \ldots, M - 1$. The criteria already defined for the binary detection problem can be extended to the M-ary case as in the following.

DEFINITION 2.58 *The Bayes risk for an M-ary detection problem is defined as*

$$\mathcal{R} = \sum_{i=0}^{M-1} \sum_{j=0}^{M-1} C_{ij}\pi_j f_{r|H_j}(r|H_j) \tag{2.179}$$

DEFINITION 2.59 *The marginal cost of the decision $\hat{H}_i$ is the quantity*

$$\beta_i = \sum_{j=0}^{M-1} C_{ij}\pi_j f_{r|H_j}(r|H_j) \tag{2.180}$$

THEOREM 2.60 *Extension of the Bayes criterion. The minimization of the Bayesian risk for the M-ary detection is obtained by selecting $\hat{H}_k$ with*

$$\beta_k = \min_i \beta_i \tag{2.181}$$

THEOREM 2.61 *Extension of the MAP criterion. The minimization of the error probability for the M-ary detection is obtained by selecting $\hat{H}_k$ with*

$$Pr\{H_k|\boldsymbol{r}\} = \max_i Pr\{H_i|\boldsymbol{r}\} \tag{2.182}$$

THEOREM 2.62 *Extension of the ML criterion. The ML criterion for the M-ary detection selects $\hat{H}_k$ if*

$$f_{\boldsymbol{r}|H_k}(\boldsymbol{r}|H_k) = \max_i f_{\boldsymbol{r}|H_i}(\boldsymbol{r}|H_i) \tag{2.183}$$

2.4.1.3 Performance bounds.

The most famous and commonly adopted upper bound for the error probability of a digital communication system is the union bound. The union bound belongs to the class of Bonferroni-type inequalities typical of the probability theory [39] [40], which are universally true regardless of the underlying probability space and for all choices of the basic events.

THEOREM 2.63 *General bounding approach based on Bonferroni type inequalities. Consider a finite set of events $A_1, A_2, , \ldots, A_M$, in a probability space, where M is the set cardinality. Then, the probability of the union of the events A_j, $j = 1, \ldots, M$ can be exactly expressed as*

$$P_U = Pr\left\{\bigcup_{j=1}^{M} A_j\right\} = \sum_{j=1}^{M}(-1)^{j+1}S_j = S_1 - S_2 + \ldots + (-1)^{M+1}S_M \tag{2.184}$$

where

$$S_j = \sum Pr\{A_{i_1} \cap A_{i_2} \cap \ldots \cap A_{i_j}\} \tag{2.185}$$

with the summation taken over all $1 \leq i_1 < i_2 < \ldots < i_j \leq M$. If the sum in eq. (2.184) is truncated at a particular S_j, a lower or upper bound is obtained depending on the sign of the last term.

The sum in eq. (2.184) is said to be an alternating sum that satisfies the alternating inequalities. Due to the complexity of S_j computation for $j > 2$, it is very common in actual applications to use the first- or the second-order terms, i.e., S_1 and S_2.

THEOREM 2.64 *Union bound. Given the events $A_1, A_2, \ldots, A_M$, the simplest form of the Bonferroni-type inequality is the ubiquitous union bound, which computes the probability of these events considering only the first-order probabilities $Pr\{A_i\}$, $i = 1, 2, \ldots, M$, i.e.,*

$$Pr\left\{\bigcup_{i=1}^{M} A_i\right\} \leq Pr\{A_1\} + Pr\{A_2\} + \ldots + Pr\{A_M\} \tag{2.186}$$

DEFINITION 2.65 *Considering a N-dimensional Euclidean space signal set with cardinality* M, $\mathcal{S} = \{s_1, s_2, \ldots, s_M\}$, *the error probability is defined as*

$$P_e = \sum_{i=1}^{M} Pr\{E|s_i\} Pr\{s_i\} \tag{2.187}$$

where the probability of the error event E conditioned on the transmission of s_i, $Pr\{E|s_i\}$, is equal to the probability of selecting s_j, for all $j \neq i$.

THEOREM 2.66 *Gallager First Bounding Technique (GFBT). The error probability* [41] *is bisected into the joint probability of error and noise residing in the volume region $\mathcal{R}$ around the transmitted symbol, identified as the Gallager region, plus the joint probability of error and noise residing in the complement of $\mathcal{R}$, i.e.,*

$$\begin{aligned} P_e &= P(E, r \in \mathcal{R}) + P(E, r \notin \mathcal{R}) \\ &= P(E, r \in \mathcal{R}) + P(E|r \notin \mathcal{R}) P(r \notin \mathcal{R}) \\ &\leq P(E, r \in \mathcal{R}) + P(r \notin \mathcal{R}) \end{aligned} \tag{2.188}$$

where the Gallager region $\mathcal{R}$ is defined as

$$\mathcal{R} = \{r | D(s_i, r) \leq \xi\} \tag{2.189}$$

with ξ a real-valued detection threshold dependent on the signal space that has to be optimized to fix the bound tightness, and

$$D(s_i, r) = \ln \left(\frac{g(r)}{Pr\{r|s_i\}} \right) \tag{2.190}$$

is defined as the discrepancy between the received sequence r and the transmitted one s_i, which involves a function $g(\cdot)$ that must be optimized.

For any observation vector r, $g(r)$ has the same sign as the channel conditional transition probability $Pr\{r|s_i\}$. Besides, the discrepancy function does not need to be calculated at the receiver side, and represents an analytical tool to generalize the bounding technique. As a matter of fact, for most cases $D(\cdot, \cdot)$ is not computable.

THEOREM 2.67 *Gallager Second Bounding Technique (GSBT). For an ML-decoded sequence, the word-error probability is given by* [42]

$$P_e \leq \sum_r Pr\{r|s_i\} \cdot \left\{ \sum_{j \neq i} \left[\frac{Pr\{r|s_j\}}{Pr\{r|s_i\}} \right]^\lambda \right\}^\rho \tag{2.191}$$

where λ and ρ are nonnegative quantities which have to be optimized.

Starting by these general definitions for the error probability, a more specific expression can be derived taking into account the specific geometry of the signal constellation. The P_e evaluation requires in general the computation of the terms $Pr\{E|s_i\}$ over a complex decision region identified as the *Voronoi* region. In general, each constellation point in the signal space identifies a decision region $\mathcal{D}_i$ constituted by all points in $\mathbb{R}^N$ that are closer to that constellation point than to any other. In order to quantify these distances from the constellation points, the Euclidean metric is commonly employed, thus

$$\mathcal{D}_i = \{r \in \mathbb{R}^N : \delta(r, s_i) \leq \delta(r, s), \forall s \in \mathcal{S}\} \qquad (2.192)$$

where $\delta(r, s) = \|r - s\|$. It can be seen that $\mathcal{D}_i$ is a volume delimited by N inequalities, each one specifying a half-space which corresponds to a one-sided hyper-plane in $\mathbb{R}^N$. The resulting decision region is a convex polytope, and the error probability can be computed integrating a N-dimensional Gaussian pdf over the polytope. If the angles among the polytope faces are not equal to $\pi/2$, this computation is very costly so the search for tight bounding techniques is very challenging in order to compute P_e.

2.4.2 Estimation theory

The source S produces in this case a continuous parameter a, that has to be estimated after the probabilistic transition mechanism through the observation of the received vector r and an appropriate processing rule dependent on the nature of the parameter. In particular, it is necessary to distinguish between random and deterministic parameters. The estimate is in general indicated as $\hat{a}(r)$, and is obviously a function of the observed vector. Estimation is based on the knowledge of the transition or likelihood function $f_{r|a}(r|a)$.

2.4.2.1 Estimation of random parameters.
The estimation theory for random parameter is guided by the Bayes criterion. Similarly to the Bayes detection that assumes to know the a-priori probabilities π_i, the Bayes estimation presumes the knowledge of the a-priori pdf of the random parameter, $f_a(a)$.

DEFINITION 2.68 *The cost function with two arguments, $C(a, \hat{a})$, is in general a function from $\mathbb{R}^2$ to $\mathbb{R}$, which depends on the actual parameter, a, and on its estimate, $\hat{a}$.*

The complexity associated to the cost function with two arguments can be often large, so typically the degrees of freedom is limited through the introduction of the estimation error.

DEFINITION 2.69 *The estimation error is defined as $\varepsilon = \hat{a} - a$.*

DEFINITION 2.70 *The cost function with one argument, $C(\varepsilon)$, is a function from $\mathbb{R}$ to $\mathbb{R}$, which depends on the estimation error.*

DEFINITION 2.71 *The Bayes risk for estimation is the quantity*

$$\mathcal{R} = \mathrm{E}_{a,\hat{a}}\left[C(a,\hat{a})\right] \tag{2.193}$$

DEFINITION 2.72 *The Bayes estimation criterion provides the estimate of the parameter a that minimizes the risk, i.e.,*

$$\hat{a}_B = \arg\left\{\min_{\hat{a}}\ \mathrm{E}_{a,\hat{a}}\left[C(a,\hat{a})\right]\right\} \tag{2.194}$$

and the corresponding optimum risk in the Bayes sense is

$$\mathcal{R}_B = \min_{\hat{a}} \mathrm{E}_{a,\hat{a}}\left[C(a,\hat{a})\right] \tag{2.195}$$

Differently from the detection theory where the Bayes criterion provides a general solution independent of the actual costs, with estimation theory the Bayes criterion necessitates to specify the cost function in order to obtain more specialized solutions. In the following, we consider some typical cases for cost functions.

THEOREM 2.73 *Considering the square error cost function, i.e., $C(\varepsilon) = \varepsilon^2$, the Bayes risk has the meaning of mean square error, and the Bayes criterion corresponds to the Minimum Mean Square Error (MMSE) criterion, which provides as estimate the mean value of the a-posteriori distribution, i.e.,*

$$\hat{a}_{MMSE} = \mathrm{E}_{a|\boldsymbol{r}}\left[a|\boldsymbol{r}\right] \tag{2.196}$$

The corresponding MMSE risk is

$$\mathcal{R}_{MMSE} = \mathrm{E}_{\boldsymbol{r}}\left[\mathrm{Var}\left[a|\boldsymbol{r}\right]\right] \tag{2.197}$$

THEOREM 2.74 *Considering the absolute error cost function, i.e., $C(\varepsilon) = |\varepsilon|$, the Bayes criterion provides as estimate the median of the a-posteriori distribution, $\hat{a}_{ABS}$, that satisfies*

$$\int_{-\infty}^{\hat{a}_{ABS}} f_{a|\boldsymbol{r}}(a|\boldsymbol{r})\mathrm{d}a = \int_{\hat{a}_{ABS}}^{\infty} f_{a|\boldsymbol{r}}(a|\boldsymbol{r})\mathrm{d}a \tag{2.198}$$

THEOREM 2.75 *Considering the uniform cost function, i.e., $C(\varepsilon) = 1 - \delta(\varepsilon)$, the Bayes criterion corresponds to the MAP criterion, which provides as estimate*

$$\hat{a}_{MAP} = \arg\left\{\max_{a} f_{a|\boldsymbol{r}}(a|\boldsymbol{r})\right\} \tag{2.199}$$

THEOREM 2.76 *Necessary condition for the MAP estimate is the satisfaction of the MAP equation*

$$\left.\frac{\partial \ln f_{r|a}(r|a)}{\partial a}\right|_{a=\hat{a}} + \left.\frac{\mathrm{d}}{\mathrm{d}a}\ln f_a(a)\right|_{a=\hat{a}} = 0 \qquad (2.200)$$

THEOREM 2.77 *If the a-posteriori rv, $a|r$, is Gaussian distributed, then it holds:* $\hat{a}_{MMSE} = \hat{a}_{ABS} = \hat{a}_{MAP}$.

PROPERTY 2.4.5 *If the cost function is even, i.e., $C(\varepsilon) = C(-\varepsilon)$, and convex with upward concavity, and the a posteriori pdf $f_{a|r}(a|r)$ is symmetric with respect to its mean value, then the MMSE estimate, $\hat{a}_{MMSE}$, is optimum.*

PROPERTY 2.4.6 *If the cost function is even, i.e., $C(\varepsilon) = C(-\varepsilon)$, and non-decreasing, and the a posteriori pdf $f_{a|r}(a|r)$ is symmetric with respect to its mean value and unimodal satisfying the condition $\lim_{x\to\infty} C(x)f_{a|r}(x|r)$, then the MMSE estimate, $\hat{a}_{MMSE}$, is optimum.*

2.4.2.2 Estimation of deterministic parameters.

Modelling the unknown parameter as rv can be often unrealistic. For these cases, estimation is typically performed through the maximum likelihood estimation criterion, as more detailed in the following. To cope with the absence of a-priori information, the measure of the estimate quality necessitates some specific figures that are related to the resulting estimation error, ε.

DEFINITION 2.78 *The first measure of quality is the expectation of the error $\mathrm{E}\,[\varepsilon]$. The actual value of this figure classifies the corresponding estimate into three classes, as follows.*

- *If $\mathrm{E}\,[\varepsilon] = 0$, the estimate is said to be unbiased.*

- *If $\mathrm{E}\,[\varepsilon] = b$, with b constant with respect to the unknown parameter a, the estimate is said to be biased with a known bias. We can always obtain an unbiased estimate by subtracting b.*

- *If $\mathrm{E}\,[\varepsilon] = b(a)$, with b function of the unknown parameter a, the estimate is said to be biased.*

The estimate has good quality if it is unbiased (or has constant bias). However, this is not sufficient for a good estimate.

DEFINITION 2.79 *The second measure of quality is the variance of the error $\mathrm{Var}\,[\varepsilon]$ that quantifies the spread of the estimation error.*

In general, the optimum estimate is unbiased with small variance of the estimation error. However, this variance can not be reduced indefinitely, as there exists a theoretical lower bound to its value, namely the Cramer Rao bound, as described in the following.

2.4.2.2.1   Maximum Likelihood estimation.

DEFINITION 2.80 *The ML estimation criterion selects the estimate that maximizes the likelihood or transition function, $f_{r|a}(r|a)$, namely*

$$\hat{a}_{ML} = \arg \left\{ \max_a \ln f_{r|a}(r|a) \right\} \tag{2.201}$$

or equivalently

$$\hat{a}_{ML} = \arg \left\{ \max_a f_{r|a}(r|a) \right\} \tag{2.202}$$

THEOREM 2.81 *Necessary condition for the ML estimate is satisfaction of the ML equation*

$$\frac{\partial \ln f_{r|a}(r|a)}{\partial a} = 0 \tag{2.203}$$

THEOREM 2.82 *The ML equation is the limiting case of the MAP equation with $f_a(a) \to uniform$.*

PROPERTY 2.4.7 *The ML estimate converges in probability to the correct value of a for the number of observations, N, approaching infinite.*

PROPERTY 2.4.8 *The ML estimate is asymptotically Gaussian for $N \to \infty$, i.e., $\mathcal{N}(a, \text{Var}[\varepsilon])$.*

2.4.2.2.2   Cramer Rao bound.   The Cramér-Rao bound (CRB) provides the minimum value for the variance of the estimation error of an unbiased estimate.

THEOREM 2.83 *If $\hat{a}$ is unbiased, i.e., $\mathrm{E}[\varepsilon] = 0$, and $\partial f_{r|a}(r|a)/\partial a$ and $\partial^2 f_{r|a}(r|a)/\partial a^2$ exist and are absolutely integrable, then*

$$\text{Var}[\varepsilon] \geq \mathrm{E}_{r|a} \left[\left(\frac{\partial \ln f_{r|a}(r|a)}{\partial a} \right)^2 \right]^{-1} \tag{2.204}$$

$$\text{Var}[\varepsilon] \geq -\mathrm{E}_{r|a} \left[\frac{\partial^2 \ln f_{r|a}(r|a)}{\partial a^2} \right]^{-1} \tag{2.205}$$

DEFINITION 2.84 *An estimate that satisfies the* CRB *with the equality is said to be an efficient estimate.*

THEOREM 2.85 *If an efficient estimate exists, it coincides with the ML estimate.*

PROPERTY 2.4.9 *The ML estimate is asymptotically efficient for $N \to \infty$.*

2.4.2.2.3   Modified Cramer Rao bound.   The Modified Cramér-Rao bound (MCRB) is the dual of the CRB in the presence of *multiple parameter estimation*. The parameters collection is composed by an *unknown deterministic parameter,* a, and by a set of *random parameters* represented by the vector u, disturbing the estimation of a. The a-priori distribution $f_u(u)$ is known. The CRB is applicable also in this case, but in general results to be very computationally complex. To avoid this drawback, the MCRB is introduced.

THEOREM 2.86 *If $\hat{a}$ is unbiased, i.e.,* $\mathrm{E}\,[\varepsilon_a] = 0$, *and* $\partial \ln f_{r|a}(r|a)/\partial a$, $\partial^2 \ln f_{r|a}(r|a)/\partial a^2$ *and* $\partial \ln f_{r|a,u}(r|a,u)/\partial a$ *exist and are absolutely integrable, then*

$$\mathrm{Var}\,[\varepsilon_a] \geq \mathrm{E}_{r,u|a} \left[\left(\frac{\partial \ln f_{r|a,u}(r|a,u)}{\partial a} \right)^2 \right]^{-1} \qquad (2.206)$$

The MCRB is a less tight lower bound with respect to the classic CRB.

2.4.3   Detection with parameter uncertainty

This section extends binary detection theory to the case of composite hypotheses [3], i.e., where the discrete unknown hypothesis, H_i, to be detected is coupled with a set, Θ, of unknown continuous parameters. This problem is often identified also as signal detection in the presence of a parameter uncertainty. The theoretical approach to handle the composite hypotheses detection depends on the nature of the continuous parameters, and in particular is different for a random or deterministic Θ.

DEFINITION 2.87 *Average likelihood ratio testing. If Θ is a set of random parameters, the optimal detection in the sense of the Bayes criterion is provided by the Average Likelihood Ratio test (ALRT), defined as*

$$\ell_\Lambda(r) = \frac{\int_{\mathcal{D}_\Theta} f_{r|H_1,\Theta}(r|H_1,\Theta)f_\Theta(\Theta)\mathrm{d}\Theta}{\int_{\mathcal{D}_\Theta} f_{r|H_0,\Theta}(r|H_0,\Theta)f_\Theta(\Theta)\mathrm{d}\Theta} = \frac{f_{r|H_1}(r|H_1)}{f_{r|H_0}(r|H_0)} \underset{H_0}{\overset{H_1}{\gtrless}} \xi \qquad (2.207)$$

where $\mathcal{D}_\Theta$ is the Θ domain, and $f_\Theta(\Theta)$ is the distribution of Θ that is assumed to be known.

The average over the variable Θ allows to achieve a simple binary hypothesis-testing problem. However, the ALRT does not contain any estimate of Θ, thus it corresponds to the receiver insensitive to the parameters.

DEFINITION 2.88 *Generalized likelihood ratio testing. If Θ is a set of unknown deterministic parameters, the optimal detection is provided by the Gen-*

eralized Likelihood Ratio Test (GLRT), defined as

$$\ell_G(r) = \frac{\max_\Theta f_{r|H_1,\Theta}(r|H_1,\Theta)}{\max_\Theta f_{r|H_0,\Theta}(r|H_0,\Theta)} \underset{H_0}{\overset{\hat{H}_1}{\gtrless}} \xi \qquad (2.208)$$

The maximization is performed separately in the fraction, so that the numerator provides the ML estimate of Θ under H_1, while the denominator produces the ML estimate under H_0. Thus, the GLRT provides a number of estimates equal to the number of discrete hypotheses.

A more general likelihood ratio testing technique can be finally derived as a combination of the two previous methods, in the presence of both deterministic and random parameters. The resulting tests are identified as AGLRT or GALRT according to the fact that the average or the generalized likelihood ratio test is the first to be performed.

2.5 Conclusions

In this chapter, a collection of the basic theoretical concepts necessary for the comprehension of the rest of the book has been presented, responding to the twofold objective of introducing the symbology that will be utilized in subsequent chapters, and providing an ample background to approach satellite communications design. In particular, the focus has been devoted to probability and statistics fundamentals, introducing the basic instruments that will be frequently referenced in the following; information theory definitions and results, with application to the most interesting channels for digital communications via satellite, highlighting the inherent limits on achievable performance; detection and estimation theory fundamentals, which are essential concepts in the receiver design and performance evaluation.

Chapter 3

SATELLITE CHANNEL IMPAIRMENTS

S. Scalise[1], M. Álvarez Díaz [5], J. Bito[6], M. Bousquet[2], L. Castanet[2], I. Frigyes[6],
P. Horvath[6], A. Jahn[1], M. Krejcarek[6], J. Lemorton[2], M. Luglio[3], S. Morosi[7],
M. Neri[4], and M.A. Vázquez-Castro[8]

[1] *DLR (German Aerospace Center), Germany*

[2] *ONERA/TeSA, France*

[3] *University of Rome "Tor Vergata", Italy*

[4] *University of Bologna, Italy*

[5] *University of Vigo, Spain*

[6] *Budapest Technical University, Hungary*

[7] *CNIT, Italy*

[8] *Universitat Autònoma de Barcelona, Spain*

3.1   Introduction

A good understanding of the impairments introduced by the propagation channel is of paramount importance in the design of a satellite-based communications system.

The relevant sources of channel impairments depend upon several factors, among them the considered scenario (e.g. mobile vs. fixed terminals), the carrier frequency and the employed access scheme. For instance, effects like shadowing and blockage are typical of mobile reception, to which section 3.2.1 is mainly dedicated. On the other hand, atmospheric attenuation phenomena, presented in section 3.3 can be safely disregarded at L- (1-2 GHz) and S-Band (2-4 GHz) frequencies, whereas they starts to become relevant at higher frequency bands such as Ku (10-12 GHz), Ka (20-30 GHz) and above. Finally, the contribution of interference to achievable Signal to Noise plus Interference Ratio (SNIR) shall not be forgotten: section 3.4 presents typical interference estimations for different access schemes. Each of the aforementioned sections has been however organized so that it can be read independently from the rest of the chapter.

In general, the main focus of this chapter is on specific results and considerations derived by models and measurements, rather than on reviewing the well-known theory of channel characterization and modelling. Nevertheless, the most important definitions are recalled for the sake of clarity and adequate references are provided for the interested readers.

3.2 Mobile Satellite Propagation Channel

This section will first review the basics of linear time-varying communication channels, then present some models typically adopted in satellite communications and finally report the results of some measurements and experiments.

3.2.1 Channel Characterization

The baseband equivalent impulse response of a linear time-varying communication channel will be indicated throughout this chapter as $h(t, \tau)$, being t the observation time and τ the delay, i.e. the arrival time of the diffused signal components normalized with respect to that of the direct ray. $h(t, \tau)$ is related to the transmitted signal $s(t)$ and to the received signal $r(t)$ by the well-known relationship

$$r(t) = \int\limits_{-\infty}^{+\infty} s(t - \tau) h(t, \tau) d\tau. \qquad (3.1)$$

3.2.1.1   Channel Characteristic functions.   From the *Time-Varying Impulse Response* $h(t, \tau)$, characterizing the channel behavior in the time-delay domain, three additional *characteristic functions*, also known as Bello's functions [43], may be derived by means of the Fourier transform, namely: the *Time-Varying Transfer Function* $H(t, \nu) \triangleq \mathcal{F}_\tau \{h(t, \tau)\}$, characterizing the channel in the time-frequency domain, the *Spreading Function* $s(f, \tau) \triangleq \mathcal{F}_t \{h(t, \tau)\}$ for the Doppler-delay domain and the *Doppler Resolved Transfer Function* $T(f, \nu) \triangleq \mathcal{F}_t \{H(t, \nu)\} \triangleq \mathcal{F}_\tau \{s(f, \tau)\}$ for the Doppler-frequency domain.

In most of the cases of practical interest, however, the variations of the channel response with the time t are random, meaning that $h(t, \tau)$ and all the above defined characteristic functions are random processes. A set of four correlation functions is hence employed to characterize random channels: *Time-Delay Correlation Function* $C_h(t_1, t_2, \tau_1, \tau_2) \triangleq \mathrm{E}_{t, \tau} [h^*(t_1, \tau_1), h(t_2, \tau_2)]$, and the *Time-Frequency Correlation Function* $C_H(t_1, t_2, \nu_1, \nu_2)$, *Doppler-Delay Correlation Function* $C_s(f_1, f_2, \tau_1, \tau_2)$ and *Doppler-Frequency Correlation Function* $C_T(f_1, f_2, \nu_1, \nu_2)$ who are defined likewise.

One of the most popular assumptions for a physically reasonable simplification of random channels characterization is that of Wide-Sense Stationarity (WSS) [43] in the time domain. As a consequence of this assumption, the time-delay and the time-frequency correlation functions depends only on the time difference $\Delta t = t_2 - t_1$ and not on the absolute time t, i.e.

$$C_h(t_1, t_2, \tau_1, \tau_2) = C_h(\Delta t, \tau_1, \tau_2)$$
$$C_H(t_1, t_2, \nu_1, \nu_2) = C_H(\Delta t, \nu_1, \nu_2). \tag{3.2}$$

The WSS assumption is normally accompanied by the dual assumption of Uncorrelated Scattering (US) [43], stating that echoes with different delays are statistically uncorrelated (in other words, that the process is WSS also in the frequency domain). Formally, this implies that

$$C_h(\Delta t, \tau_1, \tau_2) = R_h(\Delta t, \tau_1)\delta(\tau_1 - \tau_2). \tag{3.3}$$

Starting from the *Spaced-Time Delay Correlation Function* $R_h(\Delta t, \tau)$, a simplified set of functions, mutually related by the cyclic Fourier transforms as shown in fig. 3.1, is then derived under WSS-US assumption:

■ the *Spaced-Time Spaced-Frequency Correlation Function*

$$R_H(\Delta t, \Delta \nu) \triangleq \mathcal{F}_\tau \{R_h(\Delta t, \tau)\}$$

■ the *Doppler-Delay Power Spectrum* or *Scattering Function*

$$R_s(f, \tau) \triangleq \mathcal{F}_{\Delta t} \{R_h(\Delta t, \tau)\}$$

■ the *Doppler Spaced-Frequency Power Spectrum*

$$R_T(f, \Delta \nu) \triangleq \mathcal{F}_{\Delta t} \{R_H(\Delta t, \Delta \nu)\} \triangleq \mathcal{F}_\tau \{R_s(f, \tau)\}.$$

Throughout the rest of the chapter, WSS-US channels will be considered if not otherwise specified.

To conclude this section, the definitions of one-dimensional (hence simpler but coarser) functions often employed in the practice are recalled:

■ the *Power Delay Profile*

$$\phi(\tau) \triangleq R_h(\Delta t = 0, \tau)$$

■ the *Spaced-Frequency Correlation Function*

$$c(\Delta \nu) \triangleq R_H(\Delta t = 0, \Delta \nu) = \mathcal{F}_\tau \{\phi(\tau)\}$$

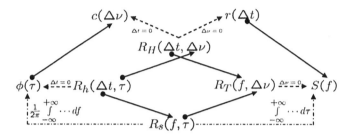

Figure 3.1.   Relationship among characteristic functions under WSS-US assumption: continuous arrows indicate Fourier transforms

- the *Spaced-Time Correlation Function*

$$r(\Delta t) \overset{\Delta}{=} R_H(\Delta t, \Delta \nu = 0)$$

- the *Doppler Spectrum*

$$S(f) \overset{\Delta}{=} R_T(f, \Delta \nu = 0) \overset{\Delta}{=} \mathcal{F}_{\Delta t} \{r(\Delta t)\}.$$

As depicted in fig. 3.1, the solely knowledge of scattering function $R_s(f, \tau)$ is sufficient to derive the complete set of functions defined above.

3.2.1.2   Channel Parameters and Classification.   From this last set of functions, the following channel parameters can be extracted:

- the *Coherence Time* $T_c \overset{\Delta}{=} \mathcal{S}\{r(\Delta t)\}$

- the *Coherence Bandwidth* $B_c \overset{\Delta}{=} \mathcal{S}\{c(\Delta \nu)\}$

- the *Delay Spread* $\sigma_\tau \overset{\Delta}{=} \mathcal{S}\{\phi(\tau)\}$

- the *Doppler Shift* $f_d \overset{\Delta}{=} \mathcal{A}\{S(f)\}$

- the *Doppler Spread* $\sigma_f \overset{\Delta}{=} \mathcal{S}\{S(f)\}$

where

$$\mathcal{A}\{f(x)\} = \frac{\int\limits_{-\infty}^{+\infty} x f(x) dx}{\int\limits_{-\infty}^{+\infty} f(x) dx}, \quad \mathcal{S}\{f(x)\} = \sqrt{\frac{\int\limits_{-\infty}^{+\infty} (x - \mathcal{A}\{f(x)\})^2 f(x) dx}{\int\limits_{-\infty}^{+\infty} f(x) dx}} \qquad (3.4)$$

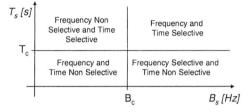

Figure 3.2.   Time and Frequency Selectivity

are the integral mean and Root Mean Square (RMS) of an arbitrary function $f(x)$.

The coherence time T_c and the coherence bandwidth B_c respectively indicate the time interval and the frequency shift over which the channel is correlated. Depending on the relationships between the coherence time T_c and the time duration of the transmitted pulse T_s and between the coherence bandwidth B_c and the transmitted signal bandwidth B_s, the following definitions, graphically exemplified in fig. 3.2, apply:

DEFINITION 3.1 *The channel is said selective in frequency if B_c is smaller or comparable to B_s. This means that frequency components of the received signal separated by more than B_c Hz can suffer significantly different levels of attenuation and delay.*

DEFINITION 3.2 *Conversely, if B_c is much larger than B_s, all the frequency components of the received signal undergo almost the same attenuation and delay and the channel is said non-selective in frequency.*

DEFINITION 3.3 *The channel is said selective in time if T_c is smaller or comparable to T_s. This means that the dynamics of the channel is typically faster than the duration of one pulse.*

DEFINITION 3.4 *On the other hand, if T_c is much larger than T_s, the channel can be assumed stationary within one or more transmitted pulse(s), hence non selective in time.*

The Doppler shift f_d is the frequency offset introduced by the channel, whereas the Doppler spread σ_f, related to the coherence time by the approximate relation $\sigma_f \approx 1/T_c$, is an RMS measure of the one-sided bandwidth of the Doppler spectrum. Alternatively the *3 dB Fading Bandwidth*, defined as the frequency interval in which the spectrum presents a decay less or equal than 3 dB from its maximum, is sometime used. For time non selective channels, σ_f is hence much smaller than B_s, meaning that the frequency dispersion caused by the Doppler effect is negligible.

The delay spread σ_τ, related to the coherence bandwidth by the approximate relation $\sigma_\tau \approx 1/B_c$, represents the delay range where significant signal echoes may be found. For frequency non selective channels, σ_τ is hence much smaller than T_s, meaning that possible echoes arriving with delay $\tau \neq 0$ cannot be resolved. For this reason, this type of channels are also referred to as non time dispersive. In this case, the time-varying impulse response can be written as $h(t,\tau) = a(t)\delta(\tau)$, where $a(t)$ is often referred to as *Channel Coefficient*.

DEFINITION 3.5 $R(t) \triangleq |a(t)|$, *representing the variation of the channel amplitude with the time, is called fading envelope. The explicit dependence on the time t will be often omitted throughout the rest of this chapter in order to simplify the notation.*

To summarize, for the well known duality properties introduced by the Fourier transform, selectivity in the time/frequency domain corresponds to a dispersive behavior in the frequency/time domain and viceversa.

3.2.1.3   First and Second Order Channel Statistics.   Assuming a frequency non selective channel, the fading envelope R can be characterized by first order statistics such as the probability density function (pdf) $f_R(r)$ and the complementary cumulative distribution function (cdf) $1 - F_R(r)$ (see 2.2.3.2). However, the first order statistics are unable to characterize events occurring at specific pairs of time samples, frequency samples, or spatial samples. Therefore, second order characterization becomes mandatory when memory effects are present in various domains. For frequency non-selective channels, time and space domain correlation are the only relevant quantities. Time-related second order statistics (induced by the relative motion between the terminal and the satellite) are important in the design of e.g. Forward Error Correcting (FEC) schemes, interleaving strategies, Automatic Repeat reQuest (ARQ) mechanisms and other techniques aiming to improve the link robustness, as shown in Chapter 4. In the following, the most commonly used second order statistics are defined.

DEFINITION 3.6 *The Level Crossing Rate (LCR), N_X is the expected rate at which the envelope R crosses a specific level X with positive slope [44]:*

$$N_X = \int_0^\infty \dot{r} f_{R,\dot{R}}(r = X, \dot{r}) d\dot{r} \qquad (3.5)$$

where $\dot{R}$ is the time derivative of the signal envelope, and $f_{R,\dot{R}}(r,\dot{r})$ is the joint pdf of the envelope R and its time derivative $\dot{R}$.

DEFINITION 3.7 *The Average Fade Duration (AFD), L_X, is defined as the average length of the time intervals corresponding to the envelope falling below a certain level X. The AFD is related to the LCR by the relation*

$$L_X = F_R(r = X)/N_X. \qquad (3.6)$$

Other less frequently used second order statistics are the *Time Share of Connection* and the *Time Share of Fades*, whose definitions can be found in [45].

3.2.2 Channel Models

As for many other types of communication channels, the modelling of satellite channels can follow either a deterministic or a statistical approach.
Deterministic Channel Models provide the knowledge of the channel impulse response or transfer function with a very high accuracy, at the expense of a large computational complexity. The most time-consuming part of this approach is the determination of all the relevant paths from transmitter to receiver, considering either the Universal Theory of Diffraction (UTD) or simplified/empirical diffraction models. Although in recent times, algorithms have been proposed that reduce the model complexity, their application to satellite systems is still practically unfeasible, due to the vast area covered by a single satellite beam.
On the other hand, *Statistical Channel Models*, to which this section is mainly devoted, attempt to model the relevant propagation phenomena such as distance-dependent attenuation, diffraction, absorption, and scattering by means of suitable statistical distributions.
In general, the received signal contains a *direct component*, and a number of multiple paths, which are usually identified as the *diffuse component*. Fading effects, i.e. variations of the signal amplitude and phase with the time t are typically divided into *long-term* and *short-term* fading.

DEFINITION 3.8 *The long-term fading, commonly referred to as* shadowing, *models the attenuation caused by the orography and large obstacles, such as hills, buildings, trees, etcetera, through absorption and diffraction mechanisms.*

In order to measure a significant power level variation due to long-term fading, the mobile terminal must typically travel several hundreds of wavelengths. No phase rotation is usually associated to shadowing.

DEFINITION 3.9 *The short-term fading models variations in the signal amplitude due to constructive and destructive interference in the sum of multiple rays, mainly caused by reflections over surrounding surfaces.*

In this case, power fluctuations are measurable over distances comparable to the signal wavelength.

3.2.2.1   Single-State First-Order Characterization for Frequency Non-Selective Channels.   Under the assumption that no multipath component is prevalent, and applying the central limit theorem, the channel coefficient can be written as $a = X + jY + s$, where X and Y are zero-mean independent Gaussian variates with variance σ^2 characterizing the diffused components and s is the amplitude of direct component. The fading envelope R follows a *Rice* distribution (2.2.13) with parameters s and σ, whereas arg(a) is distributed according to (2.98).

DEFINITION 3.10 *The Rice Factor $K \stackrel{\Delta}{=} s^2/2\sigma^2$, represents the ratio between the power of the direct component and the average power of the multipath components.*

If $s = 0$, i.e. only diffused components are present, the Rice distribution reduces to a *Rayleigh* distribution (2.2.8) of parameter σ. An alternative to the Rayleigh distribution is the *Nakagami* distribution, also known as m-distribution (2.2.11). Although, in contrast to terrestrial wireless channel, the case of a standalone Rayleigh distribution is almost never used for satellite channels, this distribution is still very useful to model certain components of the received signal, as it will be shown next. On the other hand, the Rice distribution, despite its apparent simplicity, is surprisingly good in fitting measurement campaigns with satisfactory accuracy.

As far as long-term shadowing is concerned, this can be accurately described through a *lognormal* distribution (2.2.7). In order to jointly describe large and short-term fading, composite distributions can be employed.

- The *Suzuki* distribution [46] is widely accepted for urban terrestrial mobile channels and combines Rayleigh and lognormal distributions assuming that the short term fading is Rayleigh distributed around a mean value varying in the long term according to a lognormal distribution. Formally,

$$f_{R,\text{Suzuki}}(r) = \int_0^\infty f_{R,\text{Rayleigh}}(r|2\sigma^2 = s_0) f_{S_0,\text{lognormal}}(s_0) ds_0.$$
(3.7)

- Alternatively, Loo [47] proposed a distribution suitable for rural environments, specifically accounting for shadowing due to roadside trees. The *Loo* distribution assumes that the lognormal shadowing affects only the direct component, while the diffuse components have constant average power. The resulting channel coefficient can be written as $a = S_0 + X + jY$, where X and Y are still zero-mean independent Gaussian variates with constant variance σ^2 whereas S_0 obeys a lognormal distribution. The resulting pdf of the fading envelope R, conditioned to a certain value of S_0 is Rice distributed, hence the envelope pdf can be written as

$$f_{R,\text{Loo}}(r) = \int_0^\infty f_{R,\text{Rice}}(r|s^2 = s_0) f_{S_0,\text{lognormal}}(s_0) ds_0. \qquad (3.8)$$

It can be shown that the Loo distribution tends to a lognormal/Rayleigh distribution for very large/small values of the average Rice Factor $\mathrm{E}[K] = \mathrm{E}[S_0]/2\sigma^2$.

- The *Rice-Lognormal* (RLN) distribution [48] is a composition of Rice and lognormal, with shadowing affecting now both direct and diffuse components. As a matter of fact, the diffuse component power is no longer constant, since it suffers the same variations as the direct component. This is based on the observation that long-term fading is caused by major obstacles which are likely to affect both direct and multipath components. The channel coefficients can hence be written as $a = S_0(X + jY + s)/\sqrt{s^2 + 2\sigma^2}$, i.e. as the product of two independent random variables, respectively lognormal and a Rice distributed, being the latter normalized so that it has unitary power. The resulting distribution for the fading envelope is

$$f_{R,\text{RLN}}(r) = \int_0^\infty f_{R,\text{Rice}}(r|s^2 + 2\sigma^2 = s_0) f_{S_0,\text{lognormal}}(s_0) ds_0. \quad (3.9)$$

If the Rice Factor $K = 0$, the RLN reduces to a Suzuki distribution, as it can be seen by letting $s = 0$ in (3.9). If $K \to \infty$, $f_{R,\text{Rice}}(r)$ tends to a Dirac delta located at $r = 1$ and hence R follows a lognormal distribution. Finally, if the parameter δ_{dB} of the lognormal distribution tends to 0, the channel is purely Rician (Rayleigh if additionally $K = 0$).

- An even more general distribution is the *Generalized Rice-Lognormal* (GRLN) which contains as particular cases the RLN and the Loo distributions. The ensemble of the diffuse components is subdivided in two parts, respectively shadowed and unshadowed, resulting in a channel coefficient equal to $a = S_0(X + jY + s)/\sqrt{s^2 + 2\sigma^2} + X_1 + jY_1$, where X_1 and Y_1 are zero-mean independent Gaussian variates with variance σ_1^2 representing the unshadowed part of the diffuse components. Introducing the mean power ratio between the shadowed and unshadowed diffuse components, $\xi = \sigma^2/\sigma_1^2$, it can be proved formally that the GRLN tends to a Loo distribution if $\xi \to 0$ and to a RLN if $\xi \to \infty$.

The comparison between the distributions presented so far, including also others found in the literature, is summarized in tab. 3.1, where the last column describes the statistical correlation between direct and diffuse components. As it has been already pointed out, there is no unique choice in favor of a specific distribution,

Table 3.1. Comparison between first order single-state statistical models.

Model	Year	Direct component	Diffuse component	Correlation direct/diffuse
Rice	1945	const.	Rayleigh	zero
Loo	1985	LN	Rayleigh	zero
RLN	1994	LN	LN-Rayleigh	unity
GRLN	1995	LN	part1: Rayleigh part2: Rayleigh-LN	variable
Xie et al.	2000	LN	LN-generalized Rayleigh	unity
Pätzold et al.	1998	LN	LN-Rayleigh	unity
Hwang et al.	1997	LN	LN-Rayleigh	zero
Tjhung et al.	1998	LN	LN-Nakagami	unity
Abdi et al.	2003	Nakagami	Rayleigh	zero

since the achievable accuracy is strictly dependent on the characteristic of the propagation environment (e.g. rural, suburban, etc...). Furthermore, models based on single-state distributions apply to WSS fading conditions, and are thus unable to properly model non-stationary transitions, e.g. due to the terminal moving from urban to sub-urban environment, or to a change in elevation angle for a non-geostationary link. In order to be able to describe these variations in the statistical nature of the channel, multi-state statistical models have been introduced.

3.2.2.2 Multi-State Markov Chain Models for Frequency Non-Selective Channels.

A customary classification sub-divides the multi-state models according to the discreteness or continuousness of time and state. Whilst time and state continuous models reflect more closely the real propagation effects, time and state discrete models are more suitable for analytic and simulation purposes.

Some examples are proposed in the following, assuming in general that an underlying Markov process determines the channel state transitions. Each state in the Markov chain is described by one of the single-state statistical models presented in the previous section. A Markov chain model is characterized by the *State Transition Probability Matrix* $\mathbf{P}$ ($M \times M$), where M is the overall number of states. Each element in the matrix $\mathbf{P}$, P_{ij}, represents the transition probability from state i to state j. From $\mathbf{P}$, the *State Probability Vector* $\boldsymbol{\pi}$ ($1 \times M$), whose elements π_i represent the absolute probability of being in state i, can be derived by π solving the following linear system of equations:

$$\begin{bmatrix} \mathbf{P} - \mathbf{I_M} \\ 1 \dots 1 \end{bmatrix} \boldsymbol{\pi} = \begin{bmatrix} 0 & \dots & 0 & 1 \end{bmatrix}'. \qquad (3.10)$$

Letting $f_{R,i}(r)$ be the pdf describing the i-th state, the application of the total probability theorem leads to a total pdf of the form

$$f_{R,\text{tot}}(r) = \sum_{i=1}^{M} \pi_i f_{R,i}(r). \qquad (3.11)$$

The various multi-state models differ in the number of states M and in the pdfs describing each state.

EXAMPLE 3.11 *Lutz Model.* Lutz et al. [45] introduced a two-state statistical model, based on an extensive measurement campaign over Europe at elevation angles in the range $13° - 43°$. Channel states are identified either as *good* or *bad*. The good channel state, obeying a Rice distribution with Rice Factor K, corresponds to areas with unobstructed direct component from the satellite, whereas the bad channel state, following a Suzuki distribution with parameters μ_{dB} and δ_{dB}, corresponds to shadowed areas:

$$f_{R,\text{Lutz}}(r) = (1 - A) \cdot f_{R,\text{Rice}}(r) + A \cdot f_{R,\text{Suzuki}}(r). \qquad (3.12)$$

The most significant parameter of this model is the shadowing time-share A, that determines the probability to be in either bad or good state. For different satellite elevations, environments and antennas, the parameters A, K, μ_{dB} and δ_{dB} have been evaluated from the statistics of the recordings through a least square curve-fitting procedure [45]. Transitions between states are described by the first-order Markov chain, as shown in fig. 3.3 (a). Transition probabilities from state i to state j are indicated as P_{ij}, with $i, j = g, b$ according to good or bad states. It results

$$P_{gb} = vR/D_g, \quad P_{bg} = vR/D_b, \qquad (3.13)$$

where D_g, D_b are the average distances in meters over which the good and the bad states tend to persist, v denotes the terminal speed and R is the transmission data rate. Furthermore, $P_{gg} = 1 - P_{gb}$ and $P_{bb} = 1 - P_{bg}$. The time share of shadowing can be consequently determined as

$$A = \frac{D_b}{D_b + D_g} \qquad (3.14)$$

As it can be noticed, A is independent of the speed and data rate of the user. In [49], Saunders et al. estimate the time share of shadowing A through an hybrid geometrical-statistical approach considering parameters such as street width and building height distributions. This allows predictions to be made for systems operating in areas where no direct measurements are available, and permits existing measurements to be scaled for new parameter ranges.

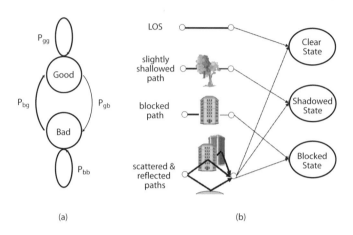

Figure 3.3. Lutz (a) and Karasawa (b) mu ...., .., ... ution channel models

A generalization of the Lutz model to include satellite diversity was proposed in [50], assuming statistical interdependence among the different satellite links at different elevation and azimuth angles. The model has four states for a twofold satellite diversity, representing all possible combinations of good and bad states.

EXAMPLE 3.12 *Karasawa Model.* Karasawa et al. [51] proposed a three-state statistical model, namely clear, shadowed, and blocked state. The Karasawa model is represented pictorially in fig. 3.3 (b). The clear state, with time share C, is described by the Rice distribution, corresponding to an unblocked direct component; the shadowed state, with time share S, is described by the Loo pdf; the blocked state, with time share B, is described by a Rayleigh distributed fading. Thus, the resulting pdf is a weighted linear combination of Rice, Loo and Rayleigh distributions:

$$f_{R,\text{Karasawa}}(r) = C f_{R,\text{Rice}}(r) + S f_{R,\text{Loo}}(r) + B f_{R,\text{Rayleigh}}(r), \quad (3.15)$$

where the distribution parameters and the associated weights are determined by fitting measurement campaigns. Note that when $B = 0$ the model reduces to the Barts et al. model.

Tab. 3.2 summarizes the main features of other multi-states statistical channel models developed in the literature.

3.2.2.3   Doppler Spectrum Models.   The Doppler spectrum arises due to the different angles of arrival of the various multipath components to/from a mobile terminal. In order to characterize the Doppler Spectrum, statistical

Table 3.2. Comparison among multi-state statistical models.

Model	Year	Number of states	Single state distributions
Lutz et al.	1991	2 (4 with diversity)	Rice, Suzuki
Barts et al.	1992	2	Rice, Loo
Vucetic et al.	1992	M	Rayleigh, Rice, Loo
Rice et al.	1997	3	Rice (2 states), Suzuki
Karasawa et al.	1997	3	Rice, Loo, Rayleigh
Fontan et al.	1997	3	Loo (3 states)
Wakana	1997	5 (2 main)	Rayleigh (2 states), Rice (3 states)

models for the angle of arrival, the intensity of the multipath components and for the terminal motion are needed.

In [52], Clarke assumes two-dimensional (2D) horizontal propagation with uniform angles of arrival over the azimuth, equal multipath amplitude, and constant terminal speed and direction. Under these assumptions, the Spaced-Time Correlation Function $r(\Delta t)$ is given by

$$r(\Delta t) = 2\sigma^2 J_0(2\pi f_{d,m} \Delta t) \tag{3.16}$$

where $J_0(\cdot)$ is the zero order Bessel function of the first kind, $2\sigma^2$ is the mean power of the Rayleigh distributed diffuse component, and $f_{d,m}$ is the maximum Doppler shift associated to the terminal motion, which is a function of the terminal speed v according to $f_{d,m} = v f_0/c$, where c is the speed of light and f_0 the carrier frequency. By taking the Fourier transform of the above correlation function, the *2D isotropic Doppler spectrum*, also known as Jakes Doppler spectrum [53], is obtained

$$S_{2D}(f) = \begin{cases} \dfrac{2\sigma^2}{\pi f_{d,m}\sqrt{1-\left(\frac{f}{f_{d,m}}\right)^2}} & \text{for } |f| \leq f_{d,m} \\ 0 & \text{elsewhere} \end{cases} \tag{3.17}$$

This spectrum is continuous and symmetrical, tending to infinity for $|f| = f_{d,m}$. Two modifications on the 2D isotropic spectrum are necessary to extend this model, typically employed for land-mobile terrestrial wireless channel, to the case of the Land Mobile Satellite (LMS) channel. First, if the satellite is non-geostationary with apparent velocity v_{sat}, then an additional Doppler shift is present over the entire spectrum and equals $f_{d,sat} = f_0 v_{sat}/c$. Second, if a direct component with amplitude s is present, the resulting Doppler spectrum consists in the superposition of a continuous and a discrete component. The continuous component is again the shifted 2D isotropic spectrum, whereas the discrete contribution is a delta centered in $f_{d,sat} + f_{d,d}$, where $f_{d,d}$ is the additional Doppler shift depending on the terminal motion and on the angle of arrival of the direct component.

The extension from 2D to a 3D model was tackled by various works in the literature. Assuming an isotropic distribution of the angles of arrival distribution over an hemisphere, it can be shown rigorously that the corresponding Doppler Spectrum is simply rectangular [54]:

$$S_{3D}(f) = \frac{\sigma^2}{f_{d,m}} \text{rect} \left\{ \frac{f - f_{d,sat}}{2f_{d,m}} \right\} + s^2 \delta(f - f_{d,sat} - f_{d,d}) \qquad (3.18)$$

where rect$\{x\} = 1$ for $x \in [-1/2, 1/2]$ and zero otherwise. Evidently, this Doppler spectrum shape is much more amenable to numerical simulation, and it is additionally more meaningful for LMS channels, where 3D propagation is certainly the case.

3.2.2.4   Modelling of Frequency Selective Channels.   When the signal bandwidth B_s is larger than the coherence bandwidth B_c of the channel, diffuse components may have a significant delay with respect to the direct component. In this case time dispersion associated to the presence of diffuse components shall be taken into account. By discretizing the diffuse components, the channel impulse response can be written as

$$h(t, \tau) = \sum_{i=0}^{N-1} A_i(t) \delta[t - \tau_i(t)] e^{j[2\pi f_{d,i} t + \theta_i(t)]}, \qquad (3.19)$$

where $A_i(t)$, $\tau_i(t)$, $f_{d,i}$, and $\theta_i(t)$ are respectively the amplitude, delay, Doppler shift, and phase of the ith discrete components and $\delta(t)$ is the Dirac delta function. This type of model is also referred to as the tapped-delay line model, in which the term of index $i = 0$ corresponds to the direct component, while the terms for $i \neq 0$ correspond to multipath echoes, each of them suitable of an independent statistical characterization thanks to the US assumption.

A remarkable example of tapped-delay line model is the so called *DLR Model* [55], shown in fig. 3.4. In order to adapt the channel model to the attitudes of the channel, $h(t, \tau)$ is divided into three regions:

1   The **direct path** A_0. The amplitude distribution depends on the channel state and can be modelled as in case of a frequency non-selective channel.

2   The region of **near echoes**. The number $N^{(n)}$ of near echoes in the vicinity of the receiver obeys a Poisson distribution with parameter λ, i.e. $N^{(n)} \sim \text{Poisson}(\lambda) = (\lambda^n/n!) \, e^{-\lambda}$. The delays lie typically in an interval $0 < \tau_i^{(n)} \leq \tau_e$, $\tau_e \approx 400 \dots 600$ ns and their distribution obeys

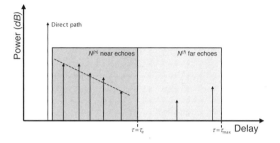

Figure 3.4. DLR Wideband Channel Model

an exponential pdf of parameter b, i.e. $\tau_i^{(n)} \sim \mathrm{Exp}\,(b) = e^{-\frac{\tau}{b}}/b$. In case of the land mobile scenario, most of the echoes typically appear in the near region. The power delay profile $\phi(\tau)$, i.e. the envelope of the mean power $\mathrm{E}\left[A_i^2\right]$ decreases in this region according to an exponential law, $\phi(\tau) = \phi_0 \cdot e^{-\delta\tau}$. The instantaneous echoes powers $A_i(t)^2$ vary around their mean values according to a Rayleigh-distributed pdf of parameter $\sigma^2 = \phi(\tau_i)$. The reader should note that the statistics of the near echoes depend on the near environment of the mobile receiver.

3 The region of **far echoes**. The number of far echoes $N^{(f)} = N - N^{(n)} - 1$ follows again a Poisson distribution. The echoes appear uniformly distributed in an interval $\tau_e < \tau_i^{(f)} \leq \tau_{\max}$, though only few echoes can be normally observed. A Rayleigh distribution is again used to characterize the instantaneous power $A_i(t)^2$ of the far echoes. It should be noted that the statistics of the far echoes depend on the background environment of the mobile receiver.

An alternative method proposed by Saunders et al. [56] considers the LMS channel as the combination of two cascaded processes: the satellite process, associated with the satellite to Earth path, and the terrestrial process, associated with the effects of the mobile motion relative to terrestrial scatterers. The satellite process is characterized by a certain path loss (including free-space loss, antenna radiation and pattern and atmospheric absorption effects), a time delay associated to the satellite-to-ground path length and a Doppler shift due to satellite motion relative to ground. The terrestrial process comprises a direct part and a multipath part. The latter is modelled according to a tapped-delay line where the taps gain are Rice distributed, in accordance with the Doppler shift induced by terminal motion.

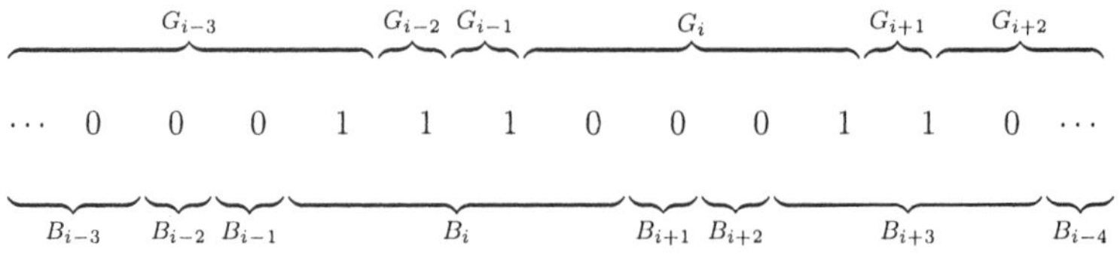

Figure 3.5.   Typical segment of an error process in a channel with memory

3.2.2.5   From Analog to Digital Channel Models.   All the models
presented so far are analog channel models, since they attempt to reproduce the
fluctuations of the received signal level. In the following sections, we briefly
introduce *digital* Markov-chain based channel models (see [57] for a detailed
analysis).

The first step is to introduce a *digital error process* by defining a threshold for
the Signal to Noise Ratio (SNR) depending on the actual modulation (provided
that hard demodulation is performed in the receiver). Assuming a constant noise
floor, the series of the received signal strength can be hence converted into an
error process as shown in fig. 3.5. The channel model is based on a Markov
chain: in each state it generates either a "1", representing a faulty symbol, or a
"0" corresponding to a correctly received symbol. If soft demodulation is used,
then also the adopted FEC scheme shall be considered and the error process
refereed to the decoded bits. In order to characterize the channel models, error
statistics shall be defined.

DEFINITION 3.13 *The error gap G_i is the distance between two faulty symbols:*

$$G_i = \{e_{j+1}, e_{j+2}, \cdots, e_{j+\ell-1}, e_{j+\ell}\}, \tag{3.20}$$

where $e_{j+1}, e_{j+2}, \cdots, e_{j+\ell-1}$ equal to 0 are, $e_j = 1$ and $e_{j+\ell} = 1$.

According to this, the pdf of an error gap can be defined as follows:

$$f_G(\ell) = Pr\left(\bigcap_{m=j+1}^{j+\ell-1} e_m = 0 \bigcap e_{j+\ell} = 1 \mid e_j = 1\right). \tag{3.21}$$

DEFINITION 3.14 *The error burst B_i is the distance between two undisturbed
symbols:*

$$B_i = \{e_{j+1}, e_{j+2}, \cdots, e_{j+\ell-1}, e_{j+\ell}\}, \tag{3.22}$$

where $e_{j+1}, e_{j+2}, \cdots, e_{j+\ell-1}$ equal to 1 are, $e_j = 0$ and $e_{j+\ell} = 0$.

By analogy to the above, the pdf of an error burst can be written as:

$$f_B(\ell) = Pr \left(\bigcap_{m=j+1}^{j+\ell-1} e_m = 1 \bigcap e_{j+\ell} = 0 \mid e_j = 0 \right). \tag{3.23}$$

DEFINITION 3.15 *The block error rate $P(m,n)$ denotes the probability of having n faulty symbols in the series of m symbols.*

EXAMPLE 3.16 (GILBERT-ELLIOTT MODEL) . It is a Markov-chain with two states; in the good state G the symbol error probability is e_G, in the bad state B, the error probability is e_B[58]. The block error rate can be written for a Gilbert-Elliott model as follows:

$$P(m,n) = H_G(m,n) + H_B(m,n), \tag{3.24}$$

where the functions $H_G(m,n)$ and $H_B(m,n)$ represent the block error rate under the initial assumption of starting from state "G" or state "B", weighted with the stationary state probability. They can be defined recursively as:

$$
\begin{aligned}
H_G(m,n) &= (1 - e_G) \cdot [H_G(m,n-1) \cdot p_{GG} + H_B(m,n-1) \cdot p_{BG}] \\
&\quad + e_G \cdot [H_G(m-1,n-1) \cdot p_{GG} + H_B(m-1,n-1) \cdot p_{BG}], \\
H_B(m,n) &= (1 - e_B) \cdot [H_B(m,n-1) \cdot p_{BB} + H_G(m,n-1) \cdot p_{GB}] \\
&\quad + e_B \cdot [H_B(m-1,n-1) \cdot p_{BB} + H_G(m-1,n-1) \cdot p_{GB}], \tag{3.25}
\end{aligned}
$$

with the initial values

$$
\begin{aligned}
H_G(0,1) &= \pi_G \cdot (1 - e_G) & H_G(1,1) &= \pi_G \cdot e_G \\
H_B(0,1) &= \pi_B \cdot (1 - e_B) & H_B(1,1) &= \pi_B \cdot e_B. \tag{3.26}
\end{aligned}
$$

In case of channels suffering from very deep fades, where signal synchronization losses may be often expected, the latencies due to the reacquisition times shall be taken into account in order to increase the accuracy of the digital model. For this reason, adaptation to the channel statistical variation realized by *adaptive digital models* can be applied as described in [57].

3.2.3 Measurements and Models Parameters Estimation

A very large amount of measurements and experiments have been carried out in the last decades to characterize the behavior of the land-mobile and indoor satellite channel. For the sake of brevity, this section include a very limited subset of the most relevant results, although adequate references are provided.

3.2.3.1 Satellite to Indoor. The results of a rather detailed, experimental and theoretical investigation on the characteristics of satellite-to-indoor

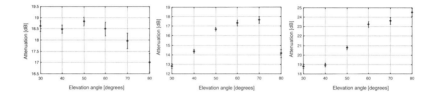

Figure 3.6. Penetration loss vs. elevation angle at 7th, 4th and 1st floor. Frequency 1620 MHz [59] ©1999 Springer

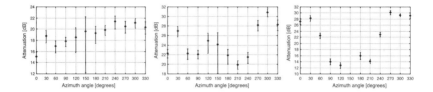

Figure 3.7. Penetration loss vs. azimuth angle at 7th floor, 60° elevation; 1st floor, 60° elevation; 1st floor, 30° elevation. Frequency 1620 MHz [59] ©1999 Springer

propagation are published in [59]. Low Earth Orbit (LEO) satellites were mostly dealt with, as propagation characteristics of the LEO-to-indoor links are the most complex and the most hostile ones. However, some of the results can be extrapolated to other types of satellite orbits.

Satellite was simulated by a helicopter: orbital data did correspond to that of a satellite with an altitude of 1000 km. Operating frequencies were 2420 MHz and 1620 MHz (the former only for narrowband measurements).

The average indoor penetration loss together with its range of variation as a function of the elevation angle in floors 1, 4 and 7 is shown in fig. 3.6. As it can be seen, no unique relationship between the penetration loss and the elevation angle can be appreciated. The same consideration applies to the penetration loss for different azimuth angles shown in fig. 3.7. For lower floors, the significant fluctuations show a strong dependence on the internal building structure. The cdfs of the short-term fading are shown in fig. 3.8: the leftmost plot refers to the entire database and the rightmost one to an azimuthal sector of 15°. As it can be seen, the cdf of the entire database follows a Suzuki-like distribution, that of a sector is rather close to a Rayleigh distribution, and the cdf of a single azimuth position (not shown here) follows nearly exactly a Rayleigh distribution. The impact of antennas with different gains and po-

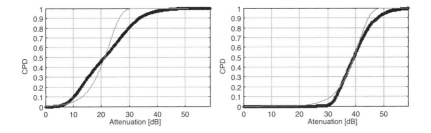

Figure 3.8. cdf of the short-term fading for the entire database (a) and for an azimuthal sector (b). Thick curves: empirical cdf, thin curves: theoretical cdf [59] ©1999 Springer

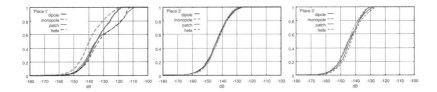

Figure 3.9. cdf at different locations: window (a), hall center (b), close to the wall (c) [59] ©1999 Springer

larization characteristics was also investigated. The cdfs determined at three different positions at the 6th floor are shown in fig. 3.9. As it can be seen, close to the window the difference in received fields corresponds more or less to the difference in the antenna gains and polarizations, proving that the field is close to a plane wave. In the hall center, the cdfs do not depend on the antenna characteristics, proving that in this position the field is diffuse and with random polarization. Close to the wall of the hall, an intermediate situation (partially diffuse field) can be appreciated. As far as 2nd order statistics are concerned, fig. 3.10 shows the measured power delay profiles and the correspondent RMS delay spreads. The results were taken at the 6th floor and averaged over 5 sec, while the helicopter speed was 60 km/h. As it can seen, the delay spread is in both cases rather low, and somewhat (though not significantly) smaller at the window-side, where Line of Sight (LOS) conditions can be accounted for. Several other registered curves had very similar characteristics. Finally, the Doppler spectrum was determined using an unmodulated (narrowband) signal. Some examples are shown in fig. 3.11. Although the nominal radial speed of

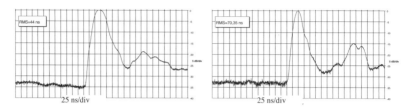

Figure 3.10. Exemplary power delay profiles at window side (a) and non-window side (b). Elevation 60° [59] ©1999 Springer

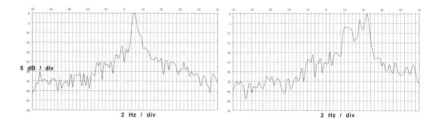

Figure 3.11. Sample Doppler spectra [59] ©1999 Springer

the helicopter was zero (circular orbits around the building were intended), in practice this was not realized, mainly due to the heavy wind present during the measurements, as confirmed by Global Positioning System (GPS) recordings. This caused nonzero *average* Doppler shifts, measured between 5 and 35 Hz, corresponding to radial speeds between 3 and 22 km/h. The measured Doppler *spread* is between about 3 and 14 Hz. Furthermore, some spectra are unimodal and some are multimodal. It is likely that the multimodal spectra, more frequent with long averaging time, are resulting from the superposition of several unimodal spectra, each containing a different Doppler *shift*. The true Doppler spread was estimated to be about 2 Hz. Notice that since the angular speed of the helicopter corresponded to that of a real LEO satellite, the estimated Doppler spread is also realistic.

3.2.3.2   Land Mobile and Railway.   This subsection reports the results of some direct measurements of the LMS channel, to show by means of some concrete examples how the key parameters characterizing some of models previously described can be estimated. For a more exhaustive overview

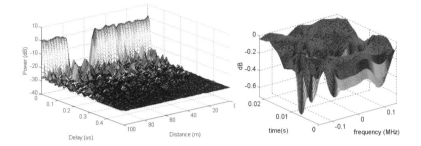

Figure 3.12. Exemplary Channel Impulse Response at L-Band, urban environment (a) and Time-Varying Transfer Function at S-Band, sub-urban environment (b) [60]

Table 3.3. 3 dB-Coherence Bandwidth and Delay Spread for several environments at 1.82 GHz

Environment	3 dB-Coherence Bandwidth $B_c^{(3dB)}$ in MHz	Delay Spread σ_τ in ns
highway	11.6	22
rural	8.7	29
urban, LOS	11.3	22
urban, non LOS	7.4	36
indoor, LOS	0.23	1065
indoor, non LOS	4.7	53

of available measurement results, the reader is however invited to refer to the bibliographical references provided throughout the text.

3.2.3.2.1   Frequency Selective Model for the LMS Channel at L and S Bands.

Frequency selectivity is indeed the most relevant phenomenon characterizing the LMS Channel at lower frequency bands, as it can be appreciated from fig. 3.12. Notice that the dependence on the time t of the channel impulse response $h(t, \tau)$ has been replaced in fig. 3.12 (a) by the dependence on the covered space $s(t)$. The measured coherence bandwidth for different environments is presented in tab. 3.3 [55]. In contrast to the RMS-coherence bandwidth defined in section 3.2.1, here the 3 dB-coherence bandwidth, defined as the frequency span $\Delta\nu$ such that $c(\Delta\nu) = 1/2 \cdot c(0)$, is used. In this case, the approximate relationship with the delay spread is $B_c^{(3dB)} = 1/4 \cdot \sigma_\tau$. As it can be seen, the channel has to be considered frequency selective for signal bandwidths of practical interest, expect in the special subcase of indoor reception in LOS conditions. As an example, the parameters for the DLR wideband model presented in section 3.2.2 are presented in tab. 3.4 for handheld reception

Table 3.4. Parameters of the near-range echoes of the DLR Model in different environments at different elevation angles [55]

environment	parameter / elevation	λ / -	τ_e / ns	b / μs	ϕ_0 / dB	$d = 10\delta \cdot \log_{10}(e)$ / dB/μs
open	15°	1.6	400	0.033	-28.5	3.0
	25°	1.2	400	0.03	-28.6	1.0
	35°	1.2	400	0.027	-25.7	9.5
	45°	0.5	400	0.027	-29.0	1.1
	55°	-	400	-	-	-
rural road	15°	1.8	400	0.061	-25.9	10.7
	25°	1.5	400	0.055	-24.9	19.2
	35°	1.6	400	0.043	-25.3	14.1
	45°	1.8	400	0.051	-24.5	13.4
	55°	1.6	400	0.047	-21.7	36.8
suburban	15°	1.2	400	0.037	-22.6	-21.9
	25°	1.4	400	0.038	-23.8	23.7
	35°	1.2	400	0.039	-24.9	19.4
	45°	1.5	400	0.027	-24.4	23.0
	55°	1.6	400	0.033	-24.7	18.7
urban	15°	1.2	600	0.118	-16.5	11.0
	25°	4.0	600	0.063	-17.0	26.2
	35°	3.5	600	0.069	-23.6	6.5
	45°	3.6	600	0.081	-23.5	8.5
	55°	3.8	600	0.079	-26.1	6.3
highway	15°	1.2	600	0.072	-27.0	6.4
	25°	2.2	600	0.077	-25.8	7.3
	35°	2.8	600	0.091	-26.8	30.6
	45°	1.8	600	0.043	-27.1	29.5
	55°	-	-	-	-	-

at 1.82 GHz: echoes with long delays are quite rarely since multipath signals hit the ground more likely for higher elevation angles. However, the number of near echoes is higher. Many echoes appear in the highway environment because of the good metallic reflectors of the car bodies. As far as non-frequency selective models are concerned, a complete collection for the LMS Channel at L and S bands can be found in [61].

3.2.3.2.2   Frequency Non-Selective Model for the LMS Channel at Ku, Ka and EHF Bands.   Due to the higher frequency and to the usage of directive antennas, filtering out most of the multipath components, frequency non-selectivity is normally assumed above 10 GHz for all data-rates of practical interest. Fig. 3.13 shows a comparison between the first order statistics of the

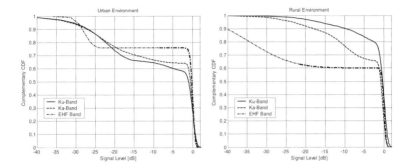

Figure 3.13. Complementary cdf for the LMS Channel at Ku, Ka and EHF Bands in the Urban and Rural Environments [63] © 2005 IEEE

received signal level in different environments for Geostationary Earth Orbit (GEO) satellite links using different frequency bands. The results have been obtained using models derived from direct measurements. For the sake of briefness, we the complete sets of parameters for the different models is not reproduced here, since they can be found in the references provided hereafter. For Ku-band (10-12 GHz), a three-state model assuming a Ricean pdf within each state, whose parameters can be found in [62], has been considered. Also for Ka-band (20-30 GHz), a three state model extracted from [61] has been used. Since for the rural case, three different sub-environments (leaf tree, pine tree and tree alley) and six different road orientations with respect to the satellite positions are considered in [61], we assumed that the three sub-environments and the six road orientations are equally likely in order to derive a generic model valid for the rural environment. Finally, for EHF-band (40 GHz), the two-state Lutz model assuming a Rice pdf in LOS state and a Suzuki pdf in non-LOS state taken from [55] has been adopted. As it can be noticed, the behavior of the propagation channel tends to become closer to an on-off channel, especially in the urban environment, as the carrier frequency increases. At EHF band, this is also a consequence of the fact that a two state only model has been considered. Nevertheless, the same effect can be noticed by comparing the results for Ku and Ka bands, where both models have three states. By inspecting the complementary cdfs in the rural case, it can be noticed how the effect of vegetative shadowing tends to become more remarkable as the carrier frequency increases: at Ku-band, 90% link availability could be obtained with 10 dB margin, whereas between 15 and 20 dB are required at Ka-band and more than 40 dB at EHF-band. As already pointed out, in the urban case, where most of the fades are due to buildings, no such an effect can be appreciated and the channel tends to assume an on-off behavior: around 30 dB margin are required in all cases to achieve 90% link availability.

3.2.3.2.3   The Railroad Satellite Channel.   This can be considered as a very special subcase of the LMS channel where, in addition to the presence of isolated obstacles such as bridges, trees and small building that may appear along the railroad, although with relatively low occurrence with respect to normal roads, one has to consider along electrified lines the presence of approximately equally spaced metallic structures (such as electrical trellises and post with or without brackets) to supply power to the train. Even if their layout and exact geometry can change depending on the considered railway path, it turned out that the attenuation introduced by these kind of obstacles can be accurately modelled using knife-edge diffraction theory: in presence of an obstacle having one infinite dimension (e.g. mountains or high buildings), the knife-edge attenuation can be computed as the ratio between the received field in presence of the obstacle and the received field in free space conditions. In the case addressed here, as shown in fig. 3.14, the obstacle has two finite dimensions, and the received field is hence the sum of the contributions coming from both sides of the obstacle. Therefore the resulting attenuation can be written as follows:

$$A_s(h) = \frac{1}{\sqrt{2}G_{\max}} \left(G\left(\alpha_1\left(h\right)\right) \left| \int_{Kh}^{\infty} e^{-j\frac{\pi}{2}v^2}\, dv \right| + G\left(\alpha_2\left(h\right)\right) \left| \int_{-\infty}^{K(h-d)} e^{-j\frac{\pi}{2}v^2}\, dv \right| \right)$$

$$K = \sqrt{\frac{2}{\lambda}\frac{a+b}{a \cdot b}},$$

$$\text{(3.27)}$$

where λ is the wavelength, a is the distance between the receiving antenna and the obstacle, b is the distance between the obstacle and the satellite, h is the height of the obstacle above the LOS and d is the width of the obstacle. Finally, the usage of a directive antenna with radiation pattern $G(\alpha)$ has to be considered. This implies an additional attenuation due to the fact that whenever the two diffracted rays reach the receiving antenna with angles α_1 and α_2 as in fig. 3.14, the antenna shows a gain less than the maximum achievable ($G_{\max}$) and depending on the variable h. The corresponding space-varying attenuation has been computed with values for a, b and d that have been extracted from a typical layout of the Italian railway and assuming a circular aperture antenna. Notice that the value of a depends on both the elevation angle and on the orientation of the railroad. The curve reported in fig. 3.14 refers to the worst case scenario. As it can be seen, the attenuation due to the electrical trellises is relevant for values of h in a range of about 0.5 m. The fluctuations around the free space loss condition (0 dB attenuation) are due to the fact that, as the train passes under the electrical trellis, this intersects Fresnel zones of different order. These may be either "in phase", causing constructive interference or "out of phase", causing destructive interference. The results obtained by ap-

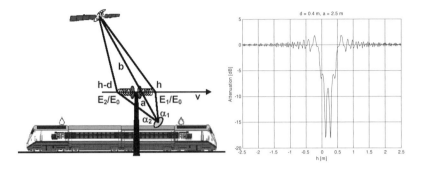

Figure 3.14. Geometry of the knife-Edge diffraction applied to electrical trellises and predicted attenuation at Ku-Band according to (3.27) [64] © 2006 IEEE

plying (3.27) have been validated against measured data showing an excellent agreement, e.g. calculations for a frequency of 1.5 GHz has been performed, resulting in attenuations between 3 and 5 dB, matching very well with the experimental result reported in [65].

As a concluding remark, the effect of the catenaries, according to dedicated measurements performed at Ku-band, turned our to be almost negligible, i.e. not exceeding 2 dB extra attenuation.

3.2.3.3   Aeronautical and Maritime.   Measurements in the aeronautical and maritime environments are less common than those in the land-mobile scenario. Nevertheless, given the inherent difficulties in covering those kind of areas by means of terrestrial signals, the results presented below are extremely important in the field of satellite communications.

3.2.3.3.1   Aircrafts. [1] Results for the Aeronautical Satellite Channel at L-Band frequency (1.6 GHz), obtained by software simulations using the models described in ITU-R P.682.1 for the analysis of non-frequency selective effects and in [66] for the analysis of frequency selective effects are presented and commented in this section. In both cases, an omnidirectional terminal antenna has been considered and it has been assumed that the aircraft is overflying the sea, since only in this case significant reflections from the earth can be expected. An RMS sea slope $\beta_0 = 0.07$ and a significant wave height $H = 1.5$ m denote a calm sea state.

Fig. 3.15 shows the first and second order statistics of the fading envelope under

---

[1]The pictures and the results shown in this section are a courtesy of Inmarsat Ltd.

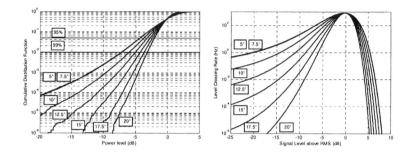

Figure 3.15. cumulative distribution function and Level Crossing Rate for the Aeronautical
Satellite Channel (L-Band) ©2003 Inmarsat

non-frequency selectivity assumption. As it can be noticed, the considered elevation angles, starts to plays a very important role if link availability greater of
equal than 99% is required, as a consequence of the fact that the LCR increases
of around two order of magnitude when the elevation reduces from 20° down
to 5°. Accordingly, the estimated Rice Factor K (see 3.2.2.1) reduces from
14.3 dB down to 8.5 dB. On the other hand, the fading bandwidth (see 3.2.1.2),
calculated considering the worst case of an aircraft flying directly towards the
LOS direction, decreases from 5.3 to 15.5 Hz. This result may appear inconsis-
tent with the aforementioned increasing amount of multipath, but the peculiar
geometry of the aeronautical scenario plays here a key role. In fact, the only
source of multipath is the sea surface, whose path difference with respect to
the direct ray is very small at low elevation angles. As far as frequency selective effects are concerned, the delay difference between the arrival of the
LOS component and the arrival of the first multipath component is a pure geometrical problem. The first multipath components to arrive will always be the
ones coming from the zone around the specular reflection point, since this is
geometrically the shortest reflected path from the satellite to the antenna on
the aircraft. In this case, since the aircraft altitude has been fixed to 28000 ft,
equal to ca. 8.5 km, the only parameter influencing the differential delay is the
elevation angle, as shown in fig. 3.16.

3.2.3.3.2   Stratospheric Platforms.   High Altitude Platforms (HAPs)
represent an interesting possibility to complement satellite connectivity over
limited and dense populated areas, such as large cities. According to the outcomes of [67], where the HAP-to-ground channel has been investigated, the
LOS component is assumed to be much stronger than the multipath ones, and
time-varying phenomena are nearly negligible. The value of delay spread is

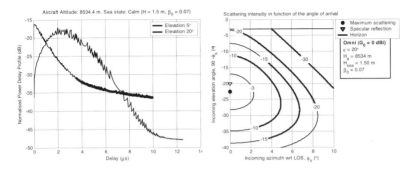

Figure 3.16. Power Delay Profile and Scattering Intensity for the Aeronautical Satellite Channel (L-Band) ©2003 Inmarsat

rather small and the temporal dispersion is negligible: therefore, the channel does not result to be frequency selective. On the other hand, the frequency dispersion results to be more severe because of the combination of the movements of the platform with those of the mobile receiver. The channel coefficients can be obtained as the superposition of several branches, between which the LOS component is characterized by a higher amplitude. All components are modelled in [67] with a Gaussian distribution.

3.2.3.3.3   Maritime.   Results for the Maritime Satellite Channel at L-Band frequency (1.6 GHz), obtained using the same models for the aeronautical case, are qualitatively very similar to those shown in figs. 3.15 and 3.16, since only different speed and altitude have to be taken into account.

3.3   Atmospheric Effects

Satellite telecommunication links operating at frequencies above 10 GHz are greatly disturbed by tropospheric phenomena which can degrade link availability and service quality [68]. The aim of this section is then to present a state-of-the-art review of the corresponding propagation impairments for satellite telecommunication systems operating at Ku, Ka and Q/V bands.

3.3.1   Propagation Effects

Satellite signals at frequencies above 20 GHz are perturbed when passing through the troposphere due to various phenomena. Four kinds of effects have to be considered: attenuation, scintillation, depolarization and increase of the antenna temperature in the receiving Earth Stations (ESs). Ionospheric effects can be considered as negligible for these frequency bands.

3.3.1.1   Attenuation (Oxygen, Water, Cloud, Rain, Melting Layer).

Attenuation is the strongest propagation effect that impacts satellite links at high frequency bands. Tropospheric attenuation is caused by atmospheric gases, clouds and precipitation. At Ku-band, rain attenuation is the main effect to be considered, at Ka-band and above the other contributions are not negligible anymore. In addition to the free space loss, only depending on the link path length and on the carrier frequency, the, the overall attenuation experienced by the link depends also on the presence of the different components in the atmosphere. At frequencies lower than 100 GHz, dry air attenuation is essentially caused by the oxygen present in the atmosphere. The molecular oxygen spectrum is characterized by a series of close absorption lines between 50 and 70 GHz, which do not allow operational satellite links around 60 GHz to be established. These lines form a continuum of absorption at low altitudes (corresponding to high pressures). Another absorbtion line is located at 118.7 GHz and other lines are centered at frequencies higher than 300 GHz. Several phenomena such as the pressure and the Doppler line broadening effects affect the molecular natural line widths. Therefore the line shapes of these molecules result to be complex functions of the meteorological parameters such as pressure and temperature. Oxygen attenuation increases with decreasing temperature, whose typical average values for different geographic areas can be obtained for instance from ITU-R P.1510. At frequencies higher than 20 GHz, oxygen attenuation, although much lower than rain attenuation, must be accurately calculated because in contrast to rain attenuation, this contribution is always present. Although oxygen attenuation becomes noticeable only for low elevation links ($E_{sat} \leq 10°$) between 20 and 40 GHz, this effect degrades considerably clear sky link budgets between 40 and 50 GHz (EHF and V bands uplinks). Oxygen concentration, is almost constant during the day and during the year and slightly varies over the globe, so as for oxygen attenuation.

Water vapor is a polar molecule with resonance peaks located at 22.235, 67.8, 120 and 183 GHz, and at several other frequency values in the millimetric wave and infrared spectral regions. As for oxygen attenuation, several phenomena such as the pressure and the Doppler line broadening effects affect the molecular natural line widths. Therefore the line shapes of these molecules result to be complex functions of the meteorological parameters such as pressure, temperature and water vapor density. Water vapor attenuation depends on humidity which can be expressed in terms of absolute humidity or in terms of Integrated (or total) Water Vapor Content of the atmosphere (IWVC, measured in kg/m^2), defined as the quantity of water vapor inside a column of 1 m^2. Values of this parameter can be obtained everywhere in the world

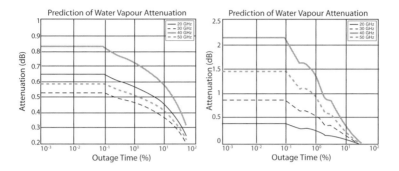

Figure 3.17. Water vapor (a) and Cloud (b) attenuations for an Earth station located in Milan ($E_{\text{sat}} = 38°$)

from maps given in ITU-R P.836. The attenuation spectrum of water vapor exhibits an absorption line at 22.235 GHz, which affects systems operating at frequencies higher than 20 GHz. In particular, at Ka-band, downlinks will be more affected than uplinks by water vapor attenuation, whereas both links will be more or less equally affected at Q/V and EHF bands (see fig. 3.17). This effect can be noticeable with respect to the total attenuation for high percentages of time, either in tropical or equatorial areas (characterized by a strong humidity) or for low elevation links, i.e. below 10°. As oxygen attenuation, water vapor attenuation is present at anytime in the atmosphere, although the amount of water vapor in the atmosphere is as a function of temperature and of atmospheric conditions. Hence, the attenuation due to water vapor is submitted to some variability due to the total water vapor content which fluctuates with the period of the day (daylight, night), with the seasons and exhibits significant geographical variations.

Clouds are constituted of suspended drops which are clearly smaller in size than wavelengths typically employed by satellite communication systems. Hence, the Rayleigh approximation of the Mie scattering theory can be used to calculate cloud attenuation that, below 300 GHz, depends only on the liquid water content and on the droplets temperature. Cloud attenuation depends directly on the amount of liquid water, which is a function of the type of cloud present on the propagation path. Cloud attenuation increases with the Integrated (or total) Liquid Water Content (ILWC, measured in kg/m^2) of the cloud and becomes noticeable starting from Ka-band. Values of the ILWC can be obtained everywhere in the world from maps given in ITU-R P.840. Cloud attenuation, exhibiting a stronger variability than water vapor attenuation, is typically present only for time percentages lower than 50% of an average year.

While cloud attenuation can be neglected at Ku-band, it has to be considered at Ka-band, especially in tropical or equatorial climates or for low elevation links (see fig. 3.17). At V-band, cloud attenuation can reach large values especially on the uplink, strongly degrading the link quality. Attenuation due to clouds presents normally a temporal variability in the order of 1 min and a relatively high spatial correlation (several tens to hundreds of kilometers). This attenuation fluctuates with the period of the day and with the seasons.

Rain attenuation is the strongest impairment which affects satellite communications systems at Ku-band and above. Rain specific attenuation (in dB/km) increases with the quantity of water and can be determined from the rainfall rate R (in mm/h). At frequency bands below Ku, this parameter is sufficient to calculate rain attenuation because all droplets are smaller than the wavelength (so the Rayleigh approximation is valid). On the contrary, for frequencies higher than 20 GHz, the drop size distribution can have an impact on attenuation: two different distributions are able to produce two different rain attenuations for the same rain rate. However, statistical information on drop size distribution is not available. Therefore, statistical models are used to characterize rain attenuation, with the rain rate as input parameter [69]. As rain presents a strong variability in space and time, the rain rate is a function of the percentage of time and depends on the climatic area. Statistical studies show that the probability for the rain intensity to be greater than a specified level follows a lognormal distribution. Nevertheless, for heavy rain, a Gamma distribution can be also used. Global information on the cumulative distribution of rainfall rate is available in ITU-R P.837. Spatial characteristics of precipitation influence the total attenuation on a link. Two major types of rain may be identified for temperate regions: stratiform (or widespread) and convective (or showery). Two more specific types have to be considered for tropical and equatorial regions and refer to tropical storms (hurricanes, typhoons) and monsoon precipitation. The dimensions of the rainy volume and the rain intensity depend on the type of rain. The rain height is another important parameter for calculating its influence, and is assimilated to the height of the -2°C isotherm in ITU-R P.839. Stratiform precipitation is composed of large areas with low rainfall rates, which may include some small intense showers. This type of precipitation spreads over several hundreds of km^2 in winter and in spring, presents vertical extents between 3 and 4 km and is very significant in Europe. Convective precipitation consists of small areas with intense rainfall (up to 50 mm/h for 0.01% of an average year in temperate areas). Its vertical extent can reach more than 10 km in equatorial regions with, a horizontal extent lower than 10 km^2. Attenuation due to rain presents a high temporal variability (in the order of 1 sec), which makes this phenomenon appear as random. Furthermore, it affects the satellite links for time-intervals long enough to degrade the most popular services, such as TV and radio programs as well as internet sessions. At Ku-band, rain atten-

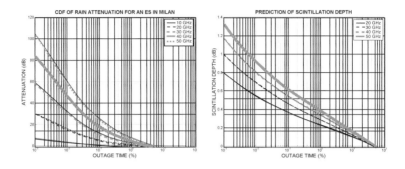

Figure 3.18. Rain (a) and Scintillation (b) attenuations for an Earth station located in Milan ($E_{\text{sat}} = 38°$)

uation can be overcome through the use of static margins, always associated to a given service availability in % of an average year. At higher frequency bands, this is not the case anymore (see fig. 3.18), especially if the service requires a sufficiently high availability or a certain level of priority, and Fade Mitigation Techniques (FMTs) have to be implemented in order to meet the service requirements.

3.3.1.1.1   Modelling of Rain Attenuation.   A lot of propagation models for prediction of rain attenuation have been proposed in the literature for more than 30 years, 16 of them, considered as the best performing, are reviewed in [70]. Two categories of models can be isolated: on the one hand empirical models and on the other hand physical-oriented models. Empirical models relies on the calculation of a specific attenuation (expressed in dB/km) and on the estimation of a reduction factor in order to take into account the fact that rain cells have a limited spatial extension. The latter calculation is the empirical part. Physical-oriented models rely instead on a description of the spatial structure of rain cells and require the introduction of the whole cdf of rain rate (that can be obtained from ITU-R P.837). Comparisons between these rain attenuation models have been carried out in the framework of [70] and the obtained accuracies are around 35%. These comparisons showed that the differences between the RMS error of the various models are often negligible compared to their absolute values and to the year-to-year variability of precipitation which is around 30%.

3.3.1.2   Scintillation.   Scintillation corresponds to rapid fluctuations of the received signal amplitude, phase and angle of arrival. Depending on the frequency band, ionospheric and/or tropospheric scintillation may affect the the satellite link: above 10 GHz, tropospheric scintillation is the unique phenomenon responsible for rapid signal fluctuations, since ionospheric scintillation decreases with the frequency whereas tropospheric scintillation increases. Tropospheric scintillation is caused by small-scale refractive index inhomogeneities induced by atmospheric turbulence along the propagation path. In satellite links, the significant scintillation effects are mainly attributed to strong turbulence in clouds and the highest intensity of scintillation are strongest in Summer around noon. Scintillation effects are higher for very cloudy sky conditions, in particular in the presence of cumulus clouds. However, the scintillation of the transmitted signal may also appear in clear sky and in rainy conditions. From the meteorological point of view, tropospheric scintillation increases essentially with temperature and humidity (quantifiable using the wet term of the refractive index that can be obtained from ITU-R P.453, as well as the ILWC of convective clouds [70]. Consequently, a higher intensity of scintillation is observed in hot seasons, during wet days and in the hottest hours of the day. In a similar way, scintillation is higher in humid areas (tropical and equatorial areas) than in dry regions. However, local characteristics such as localized convection phenomena with cumulus clouds and orographic effects can also produce strong scintillation.

From a radiofrequency point of view, the intensity of tropospheric scintillation has been proved to increase with the frequency and decrease with the elevation angle and the antenna size (see fig. 3.18). Scintillation fades could have a major impact on the performance of low-margin communications systems above 10 GHz, for which the long-term availability is sometimes predominantly governed by scintillation effects rather than by rain (for time percentages higher than 1%). In particular, at Ka or V-bands where FMTs have to be designed, the dynamics of the scintillating fades may interfere with the adopted FMTs. On the other hand, for high availability systems, scintillation fades are significantly weaker than rain attenuation at conventional elevation angles (higher than $15°$), so that scintillation can be neglected in the link budgets. However, this is not the case for low elevation links, where the impairments produced by tropospheric scintillation are particularly severe: peak-to-peak scintillation amplitudes exceeding 15 dB have been for instance observed at 10 GHz for elevation angles below $5°$.

3.3.1.3   Depolarization.   Dual polarization transmission is a way to double the system capacity in a given frequency band. However, some atmospheric effects limit system performances in terms of isolation between polarizations. These phenomena have to be carefully studied and accurately modelled in order to be able to design dual polarization satellite communication systems, since they are different in the ionosphere and in the troposphere. For frequencies above 10 GHz, ionospheric effects (such as Faraday rotation which affects mainly system behavior in linear polarization) can be neglected. Tropospheric effects are caused by the presence of non-spherical hydrometeors (rain drops, ice crystals, snow flakes) with a non-vertical falling direction. It leads to different attenuations and phase shifts on each of the main orthogonal polarization directions, induces a rotation of the polarization plane and then an increase of the coupling between both orthogonal polarizations. Generally, at Ku-band, it is possible to assume that depolarization is mainly produced by events like showers or storms, producing also simultaneous strong attenuation. However, at Ka-band and especially at V-band, ice depolarization can be very strong (ice crystals in high altitude clouds which simultaneously produce weak attenuation or ice/snow/water in the melting layer of stratiform rain which simultaneously produce moderate attenuation) [71].

3.3.2 Influence of Propagation Impairments on Link Budgets

Link budgets will be affected differently if propagation impairments occur either on the uplink or on the downlink [72]. Throughout this section, subindexes ES and SAT denotes the Earth Station and the Satellite, subindex FS additionally denotes the free-space case, whereas A_P is the overall attenuation caused by all the atmospheric effects presented so far, also referred to as fading, due to its dependence on the time t.

3.3.2.1 Uplink Analysis. The influence of uplink fading on the carrier-to-noise ratio $(\frac{C}{N_0})$ is quite simple and it corresponds actually to a linear variation with respect to uplink fades. In fact:

$$
\begin{aligned}
\left(\frac{C}{N_0}\right)_{\text{up}} [\text{dB}] &= \text{EIRP}_{\text{ES}} - A_{\text{FS,up}} - A_{P,\text{up}} + \left(\frac{G}{T}\right)_{\text{SAT}} - k_B = \\
&= \left(\frac{C}{N_0}\right)_{\text{FS,up}} - A_{P,\text{up}}.
\end{aligned}
\tag{3.28}
$$

3.3.2.2   Downlink Analysis.   In the case of fading occurring on the downlink, the downlink carrier-to-noise ratio is affected in the following way:

$$
\begin{aligned}
\left(\frac{C}{N_0}\right)_{\text{down}} [\text{dB}] &= (1 - \Delta_1)\,\text{EIRP}_{\text{SAT}} - A_{\text{FS,down}} - A_{P,down} \\
&\quad + (1 - \Delta_2)\left(\frac{G}{T}\right)_{\text{ES}} - k_B = \\
&= \left(\frac{C}{N_0}\right)_{\text{FS,down}} - A_{P,\text{down}} - \Delta_1 \cdot \text{EIRP}_{\text{SAT}} - \Delta_2 \cdot \left(\frac{G}{T}\right)_{\text{ES}}.
\end{aligned}
\tag{3.29}
$$

In the latter formula, atmospheric propagation impairments act directly through the the term $A_{P,\text{down}}$, but also via a possible decrease Δ_1 of the satellite Effective Isotropic Radiated Power (EIRP) due to an uplink fade (only for transparent repeaters) and through a degradation Δ_2 of the ES figure-of-merit, due to an increase of the antenna noise temperature in presence of an atmospheric perturbation on the downlink. In fact, as the attenuation increases, so does the emission noise. For ESs with low-noise front-ends, the increase of the antenna noise temperature T_A may have a great impact on the resulting signal-to-noise ratio. Recommendation ITU-R P.618 gives a method to estimate the increase of the antenna noise temperature due to an atmospheric perturbation. Accordingly, the antenna noise temperature is given by:

$$
T_A\,[\text{K}] = \frac{T_{\text{sky}}}{A_{p,\text{down}}} + T_M\left(1 - \frac{1}{A_{p,\text{down}}}\right) + T_{\text{ground}},
\tag{3.30}
$$

where T_M is the effective temperature of the medium, T_{sky} is the sky brightness temperature as seen by the antenna (see fig. 3.19) and T_{ground} is the noise temperature from the ground in the vicinity of the ES and captured by the sides lobes of the antenna radiation pattern and partly by the main lobe when the elevation angle is small. T_M depends on the contribution of scattering phenomena to the attenuation, on the physical extent of clouds and rain cells, on the vertical variation of the physical temperature of the scatterers and, to a lesser extent, on the antenna beamwidth. By comparing radiometric observations and simultaneous beacon attenuation measurements, the effective temperature of the medium has been determined to lie in the range 260-280 K in presence of rain and clouds along the path in a frequency range between 10 and 30 GHz. The empirical formula $T_M\,[\text{K}] = 1.12 \cdot T_{\text{amb}} - 50$ can be used to estimate T_M.

Fig. 3.20 (a) presents an example of the degradation of the figure-of-merit of an ES located in Milan with respect to time percentage for the downlinks of Ka-band and Q/V-band.

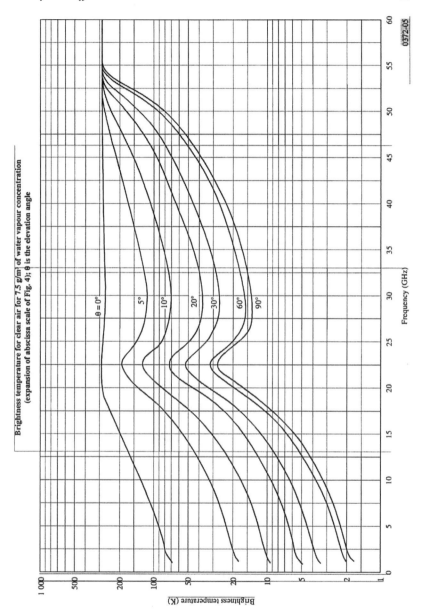

Figure 3.19. Sky brightness temperature from Recommendation ITU-R P.372 (Reproduced with the kind permission of ITU)

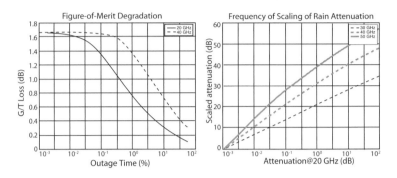

Figure 3.20. Figure-of-merit loss for an ES located in Milan ($E_{sat} = 38°$) (a) and example of long-term frequency scaling of rain attenuation (b)

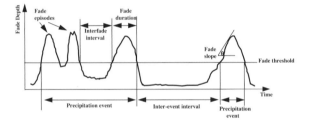

Figure 3.21. Features characterising the dynamics of fade events from Recommendation ITU-R P.1623 (Reproduced with the kind permission of ITU)

3.3.3   Dynamic Behavior of the Propagation Channel

Most of the models mentioned earlier describe the propagation channel on a first-order statistical basis. However, the dynamical behavior and the second order statistics of the propagation channel are also useful for the system design and in particular for the design of FMTs [73]. The temporal dynamics of the propagation channel is an important information because it directly influences some aspects of satellite system design and FMTs optimization. Although we will focus next on the dynamics of rain events, which represent an especially meaningful example, the definitions given below are valid for any type of fading.

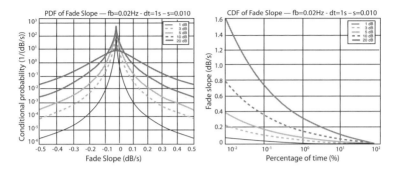

Figure 3.22. Distributions of fade slope for different attenuation thresholds

3.3.3.1 Fade Slope. The fade slope is defined as the rate of change in dB/s of rain attenuation, as shown in fig. 3.21. Being able of quantifying the fade slope is important for satellite communications systems operating at Ka-band and above, in which FMTs have to be implemented. The knowledge of the fade slope is in fact useful either to design a control loop that can follow the signal variations, or to allow a better short term prediction of the propagation conditions. In both cases, the relevant information is the slope of the slow varying component of the signal. It has already been demonstrated that fast fluctuations of the signal, due in particular to tropospheric scintillation, are unpredictable, since the spaced-time correlation function decreases very rapidly so that the scintillation component is completely uncorrelated after only two seconds [71]. Several parameters on which the fade slope depends have been clearly identified, namely the attenuation level, the elevation angle and the bandwidth of the low pass filter used to remove the fast fluctuations due to scintillation. Moreover, other meteorological parameters such as e.g. the climatic zone are supposed to influence the statistics of the fade slope. The cdf of the fade slope can be predicted using ITU-R P.1623, as exemplified in fig. 3.22.

3.3.3.2 Fade Duration. The fade duration is defined as the period of time between two consecutive crossing of a given attenuation threshold (fig. 3.21). The fade duration is an important parameter to be taken into account in the system design because of the following reasons:

- the fade duration gives information on the outage period or system unavailability due to propagation over a given link,

- when using FMTs, the fade duration is useful to define statistical duration for the system to remain in a certain mitigation configuration before it comes back to its nominal mode,

- it is important for the operator point of view to have an insight into the statistical duration of an event resulting in a link outage, in order to efficiently share the available resources between the different users,

- the fade duration is a key element in the process of choosing FEC codes and best modulation schemes: for satellite communication systems, the propagation channel does produce bursts of errors rather than isolated errors. The fade duration impacts directly the choice of the coding scheme, e.g. the size of the codeword in block codes or the interleaving depth in concatenated codes.

The fade duration can be described by two different types of cdfs [74]:

- $P(d > D, a > A)$, defined as the probability of occurrence of fades of duration d longer than D s, given that the attenuation a is greater than A dB. This probability can be estimated from the ratio of the number of fades of duration $d > D$ to the total number of fades observed, given that the threshold A is exceeded.

- $F(d > D, a > A)$, the cumulative exceedance probability, or, equivalently, the total fraction (between 0 and 1) of fade time due to fades of duration d longer than D s, given that the attenuation a is greater than A dB. This probability can be estimated from the ratio of the total fading time due to fades of duration longer than D given that the threshold A is exceeded, to the total exceedance time of the threshold.

Fade slope cdfs can be predicted using ITU-R P.1623, as shown in fig. 3.23.

3.3.3.3   Frequency Scaling.   The objective of frequency scaling methods is to estimate the magnitude of a propagation effect (attenuation, scintillation, depolarization) at a given frequency from the knowledge of its magnitude at another frequency, generally a lower one. Two general interests of using such kind of methods can be identified:

- study the influence of propagation effects on the design of systems operating at high frequency bands (for instance Ka or V-band) from the knowledge of the performances of current systems operating at conventional frequency bands (Ku-band),

- design FMTs in an open-loop configuration such as uplink power control where the attenuation on the link (for instance the uplink) is estimated

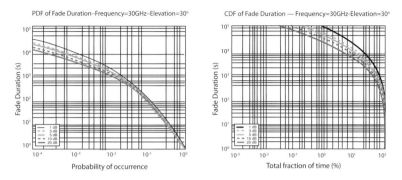

Figure 3.23. Distributions of fade duration for several attenuation thresholds

from measurements at another frequency (for instance, beacon tracking at a lower frequency band).

Predictions of long-term frequency scaling can be performed with ITU-R P.618. Fig. 3.20 presents scaled attenuation obtained at 30, 40 and 50 GHz from the introduction of 20 GHz rain attenuation. Two kinds of frequency scaling methods can actually be used. The first one is related to long-term variations and uses cdfs of propagation effects for the same percentage of exceedance. The second one aims at studying the instantaneous ratio of attenuation, scintillation log-amplitude or discrimination in polarization at two distinct frequencies. As seen in previous sections, attenuation is the combination of several contributions such as rain, oxygen, water vapor and clouds, depending on the frequency in different ways. This is the main cause of inaccuracy for long-term frequency scaling predictions. As far as instantaneous frequency scaling is concerned, it exhibits strong fluctuations coming from the natural variability of the phenomena (hysteresis effect) [71][75].

3.4   Interference Characterization

Within this section, all interference sources are first described. Then, a comprehensive model for the particularly meaningful case of co-channel interference in multi-beam systems is presented.

3.4.1   Interference Sources

Four types of interference can be identified: adjacent channel interference, co-channel interference, cross-channel interference and adjacent system interference. In the uplink (see fig. 3.24), these interference are produced at the input of the satellite receiver, by signals transmitted by ESs either belonging

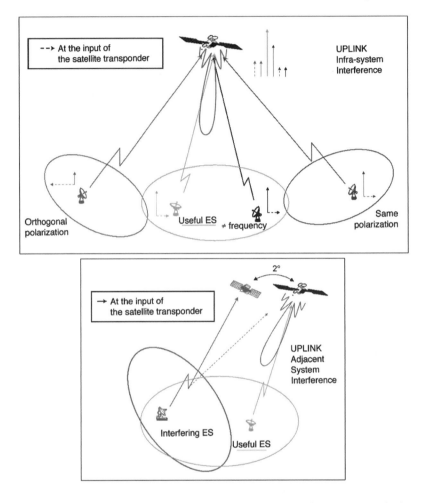

Figure 3.24. Uplink intra-system (a) and adjacent system (b) interference (no polarization re-use)

to the same system or to other systems. In the downlink (see fig. 3.25), these interference are produced at the input of the ES receiver, by signals transmitted by either the satellite(s) belonging to the considered system or by satellites of other systems. In an ideal system, all interfering signals should be strongly attenuated, thanks to the appropriate selection of access scheme, frequency and polarization plan, antenna pattern and filter mask.

Adjacent channel interference is produced in the uplink by signals belonging to ESs of the same system as the useful ES and in the downlink by signals

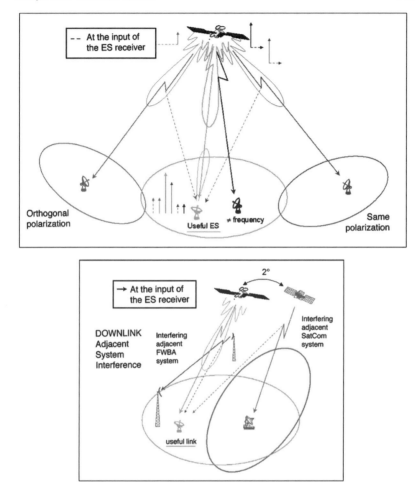

Figure 3.25. Downlink intra-system (a) and adjacent system (b) interference (no polarisation re-use)

transmitted by the satellite to other ESs of the same system. In both cases, all considered ESs are located in the same spot-beam and interfering signals are transmitted at different frequencies but with the same polarization. In Frequency Division Multiple Access (FDMA) and Time Division Multiple Access (TDMA) access schemes, interference is produced because of the non-ideal performance of the transmit and receive filters.

In contrast to the previous case, co-channel interference is produced by signals belonging to ESs (uplink) and satellites (downlink) of the same system, using this time the same carrier frequency and the same polarization as the useful signal. These interfering signals are transmitted from/to different spot beams with respect to the useful ES (inter-beam interference), and in case of Code Division Multiple Access (CDMA) systems also from/to the same spot-beam (intra-beam interference). In the uplink, the inter-beam co-channel interference is conditioned by the satellite antenna pattern in the direction of the interferers, whereas the intra-beam interference in CDMA is limited by code orthogonality properties. In the downlink, the inter-beam interference is conditioned by the satellite antenna roll-off of the adjacent spot beams in the direction of the useful ES whereas the intra-beam interference in CDMA is limited by code correlation properties.

Cross-channel interference is produced by signals belonging to ESs (uplink) and satellite (downlink) of the same system, using also in this case the same carrier frequency as the useful signal, but an orthogonal polarization. Interfering signals are transmitted to/from ESs located in different spot-beams with respect to the useful ES if single polarization is used, and located also in the same spot-beam as the useful ES for dual polarization systems. In the uplink of single polarization systems, this interference is limited by the satellite antenna roll-off in the direction of the interferers and by the cross-polarization isolation of the satellite antenna, whereas in the case of polarization re-use, the interference is limited by the cross-polarization isolation of both the satellite and the ES antennas. Similarly, in the downlink of single polarization systems, the cross-channel interference is limited by the satellite antenna roll-off of the adjacent spot-beam in the direction of the useful ES and by the cross-polarization isolation of the satellite antenna. In the case of polarization re-use, cross-channel interference is limited by the cross-polarization isolation of both the satellite and the ES antennas.

Adjacent system interference is produced by signals belonging to a different communication system, transmitted at the same frequency and polarization as the useful signal. In the uplink, this interference is dependent on the angular separation of the two satellites from the considered interfering ES (see fig. 3.24) and on its antenna pattern. Interference from terrestrial systems such as broadband fixed wireless access or radar systems can be considered as negligible apart from specific cases at very low elevation links. In the downlink, this interference is also limited by the angular separation of the interfering satellite(s) with respect to the useful one(s), as seen from the useful ES position (see fig. 3.25). Interference from the aforementioned terrestrial systems have to be evaluated, especially for low elevation links.

Finally, even if it can not be rigourously defined as interference, multicarrier effects (intermodulation noise, carrier suppression) can be considered as a fifth source in the downlink interference budget. Multicarrier effects are produced by satellite power amplifiers operated at or close to saturation. In this configuration, due to non-linearity effects, intermodulation products are generated and the satellite output power is shared between the useful carriers and the intermodulation noise. On the other hand, carriers having a different power are amplified in a different way, leading to the suppression of weak carriers by stronger ones.

3.4.2   Co-channel Interference in Multi-Beam Multi-Satellite Systems

In a multi-beam system, the co-channel interference caused by adjacent beams is the main component of interference. For this reason, a comprehensive model presented in [76] will be described in this section.

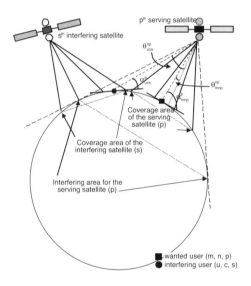

Figure 3.26. Geometrical structure for uplink [76] ©2002 John Wiley and Sons Limited. Reproduced with permission.

In a satellite system not all the ESs (hereafter also referred to as users) or all the satellites are actual interferers. Due to the spherical shape of the Earth and to Earth morphology, in fact, the interfering set is composed in the uplink only by those users visible to the satellite, as shown in fig. 3.26. Similarly, in the downlink the interfering set is composed by only those users served by any of the satellites in visibility conditions from the useful user location, as depicted in fig. 3.27. Fig. 3.26 and fig. 3.27 show also the main angular parameters involved in the model presented in the next section.

Contributions to interference can be generated whenever signals are not perfectly orthogonal [77]. In the uplink, both the useful and interfering signals, generated by ESs and received at the satellite serving the useful user, have a power level depending on each ESs positions and on the respective propagation conditions. In the downlink, interference is experienced at the ES receiver input. Concerning the serving satellite, both the useful and the interfering signals are radiated from the satellite antenna through the same path. As a consequence, the fluctuations experienced by all signals, mainly due to downlink propagation channel, are highly correlated. The main lobe radiates both useful and interfering signals (if any, in CDMA case) of the serving cell, while interfering signals from other cells are radiated by the side lobes of corresponding radiation pattern overlapping the main lobe serving the useful user. When more than one satellite contributes to interference, the signals from non-serving satellites coming from main or side lobes are pure interference. Signals coming from different satellites go through different paths thus experiencing different propagation conditions. Without power control, both the useful and interfering signal levels as received by the useful ES depend on its position with respect to all involved satellites, but not on the positions of the interfering terminals. Finally, if satellite diversity strategies are adopted [78], more than one serving satellite shall be considered.

3.4.2.1   The model.   The analysis of the interference scenario can be performed either on a statistical basis due to the intrinsic random behavior of several parameters (e.g. propagation channel, user distribution [79]) or on a deterministic basis considering both ESs and satellites in fixed positions [80]. To characterize interference in a multisatellite multibeam scenario the Carrier to Interference Ratio (C/I) is commonly used: toward its estimation, different system choices (multiple access, ESs antennas, payload architecture, etc.) and several parameters (orthogonality among signals, activity factor, polarization isolation, channel impairments, actual angular distribution of all the users, etc.) must be taken into account.

Starting from several models presented in [80][81][82][83], and from the more comprehensive model proposed in [76], which identifies and details the subset of interfering users for both CDMA and FDMA uplink and downlink, a new model is presented, including also the effects of the troposphere. CDMA and FDMA cases are dealt with separately because the selection of actual interferers is performed with different criteria. In case of TDMA access no *resource reuse* concept is implemented (as in FDMA). Thus no interference can arise from simultaneous use of the same time slot by more than one users. Theoretically, interference could arise from lack of synchronization but this event is taken into account and severely prevented by means of opportune time guard

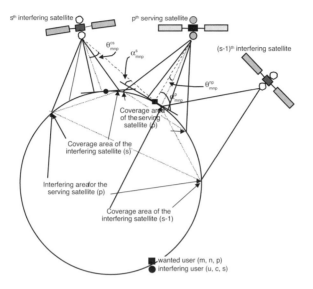

Figure 3.27. Geometrical structure for downlink [76] ©2002 John Wiley and Sons Ltd. Reproduce with permission

bands. Therefore, no specific model has been developed. When, as very often, TDMA is used in conjunction with FDMA, thus offering hybrid multiple access, namely Multi Frequency TDMA (MF-TDMA), the interference arises in the frequency domain.

The common assumption is that, in general, the set of interferers can be divided into three subsets:

1 users belonging to the same cell of the useful user

2 users belonging to the other cells of the same satellite of the useful user

3 users employing other satellites.

The respective interference levels will be indicated with I_1, I_2 and I_3. The resulting Carrier to Interference Ratio can be written as:

$$\left(\frac{C}{I}\right)_{\text{uplink}} = \frac{C_{\text{up}}}{\sum_{i=1}^{3} I_{i-\text{up}}}, \qquad (3.31)$$

where C_{up} is the carrier power and the denominator is the sum of the three contributions to interference. However, quantifying each of the three contributions separately makes it possible to evaluate the interference spatial distribution

Table 3.5. Relevant symbols and parameters for interference calculation

Symbol	Parameter
α_{ucs}^{s}	angle between the tangent to the earth in the location of the user (u, c, s) and the line between the satellite s and the user (u, c, s)
A_{mnp}	supplementary (e.g. due to atmospheric effects) attenuation experienced by the mnp user
CD	cluster dimension
d_{ucs}^{s}	slant range for the link from the user (u, c, s) to the satellite s (the one serving the useful user)
e	type of environment (rural, suburban, urban)
$f_{ucs}\left(e, \alpha_{ucs}^{s}\right)$	mobile channel fading experienced by the user (u, c, s) depending on the environment e and on the angle α_{ucs}^{s}
$g_{[ucs]}\left(\alpha_{ucs}^{s}\right)$	antenna gain of the mobile terminal (u, c, s) in the direction α_{ucs}^{s}
γ_{ucs}^{mnp}	orthogonality factor between the user (u, c, s) and the user (m, n, p)
$G_{[cs]}\left(\theta_{ucs}^{cs}\right)$	satellite antenna gain in the cell c of the satellite s in the direction of θ_{ucs}^{cs}
k	number of full cluster in the antenna pattern
l_{mnp}	latitude of the mnp user
L_{mnp}	longitude of the mnp user
$\lambda_{[ucs]}$	wave length of the user (u, c, s)
mnp	useful user index (m-th user in the n-th cell of the p-th satellite)
μ_{ucs}	activity factor of the user (u, c, s)
n	cell number in one satellite coverage ($n = \text{CDq}_{\text{rn}} + \text{t}$ in FDMA)
NC	number of cells in a satellite
NI	number of interfering cells ($1 \leq \text{NI} \leq \text{N}_{cell}$), where N_{cell} is the number of satellite antenna beams
NS	number of satellites
NU_{cs}	number of interfering users in the cell c of satellite s
q	cluster index ($0 \leq \text{q} \leq k - 1$)
ρ_{cs}^{np}	polarization isolation factor between the cell (c, s) and the cell (n, p)
t	cell number in the cluster ($1 \leq t \leq \text{CD}$)
θ_{ucs}^{cs}	angle between the boresight of the cell c of the satellite s and the line between the satellite s and the user (u, c, s)
ucs	user number u ($1 \leq u \leq \text{NU}_{cs}$) of the cell number c ($1 \leq c \leq \text{NS}$) of the satellite number s ($1 \leq s \leq \text{NS}$)
$w_{[ucs]}$	power transmitted by the mobile terminal (u, c, s)
$W_{[cs]}$	power transmitted to cell c by the satellite s

around the useful user (for example as a function of the elevation angle) and to allow the implementation of local interference reduction strategies if uneven distribution should result. The adopted notation is detailed in tab. 3.5. The square brackets mean that the parameter may not depend on the subscripts in between. The useful user is identified through subscript mnp which refers to the user m located in cell n served by satellite p, whereas the indexes u, c and s are used for generic users, cells and satellites. The supplementary attenuation, mainly due to the effects of the propagation in the troposphere, depends on the frequency, on the user geographical position (latitude and longitude) and on the elevation angle, as discussed in section 3.3.

3.4.2.1.1 CDMA case. In this case, all the three subsets provide contributions to total interference.

The factor γ is introduced [81] to take into account the correlation properties among the users, i.e. among the spreading sequences assigned to each user. γ ranges from zero to one depending on the code assignment strategy and it equals zero if two signals are perfectly orthogonal and synchronized.

CDMA Uplink. The carrier power for the uplink can be written as follows:

$$C_{\text{up}} = \frac{w_{[mnp]}g_{[mnp]}\left(\alpha_{mnp}^{p}\right)G_{[np]}\left(\theta_{mnp}^{np}\right)}{\left(\frac{4\pi d_{mnp}^{p}}{\lambda}\right)^{2}f_{mnp}\left(e,\alpha_{mnp}^{p}\right)A_{mnp}\left(\lambda,l_{mnp},L_{mnp},\alpha_{mnp}^{p}\right)}. \tag{3.32}$$

If power control is implemented the user power w assumes different values among users.

The fading channel attenuation $f(\cdot)$ can be evaluated according to any of the channel models presented in section 3.2.

In the case of ESs equipped with an omnidirectional antenna, the gain $g(\cdot)$ is independent on the angle α, otherwise the value depends on the antenna radiation pattern . Anyway the above expression allows to consider also more sophisticated adaptive array antenna that can be employed to achieve space selectivity.

The intra-cell interference contribution is:

$$I_{1,\text{up}} = \sum_{\substack{u=1 \\ u \neq m}}^{\text{NU}_{np}} \frac{\gamma_{unp}^{mnp}\mu_{unp}w_{[unp]}g_{[unp]}\left(\alpha_{unp}^{p}\right)G_{[np]}\left(\theta_{unp}^{np}\right)}{\left(\frac{4\pi d_{unp}^{p}}{\lambda}\right)^{2}f_{unp}\left(e,\alpha_{unp}^{p}\right)A_{unp}\left(\lambda,l_{unp},L_{unp},\alpha_{unp}^{p}\right)}. \tag{3.33}$$

All the interfering users are located in the main lobe of the antenna beam which serves the useful user too. The number of users per cell NU_{cs} is different for each cell and is time variant.

The activity factor μ is usually assumed constant if all the users utilize the same service. In case of voice service, it typically assumes values ranging from 0.4 up to 0.5. In case of multimedia services, it assumes different values for different pair of users depending on the service. In particular if the ratio between data traffic and voice traffic increases, C/I is expected to degrade due to the increased activity factor [80]. For asymmetric services such as Internet, it is unbalanced in the two communication directions.

The second term represents the inter-cell interference:

$$I_{2,\text{up}} = \sum_{\substack{c=1 \\ c \neq n}}^{NC} \sum_{u=1}^{NU_{cp}} \frac{\gamma_{ucp}^{mnp} \mu_{ucp} w_{[ucp]} g_{[ucp]} \left(\alpha_{ucp}^p \right) G_{[np]} \left(\theta_{ucp}^{np} \right) \rho_{cp}^{np}}{\left(\frac{4\pi d_{ucp}^p}{\lambda} \right)^2 f_{ucp} \left(e, \alpha_{ucp}^p \right) A_{ucp} \left(\lambda, l_{ucp}, L_{ucp}, \alpha_{ucp}^p \right)} \quad (3.34)$$

and it is due to the satellite antenna side lobes and to the beam overlapping areas. Once the coverage antenna pattern of each satellite is defined, NC is constant unless beam turn-off technique is adopted.

To improve spectral efficiency the use of orthogonal polarization in different beams is implemented to increase isolation among beams using the same frequency. The reduction of the interference level due to the polarization isolation is taken into account introducing the factor ρ.

In the third contribution, the users served by other satellites are considered:

$$I_{3,\text{up}} = \sum_{\substack{s=1 \\ s \neq p}}^{NS} \sum_{c=1}^{NC} \sum_{u=1}^{NU_{cs}} \frac{\gamma_{ucs}^{mnp} \mu_{ucs} w_{[ucs]} g_{[ucs]} \left(\alpha_{ucs}^p \right) G_{[np]} \left(\theta_{ucs}^{np} \right) \rho_{cs}^{np}}{\left(\frac{4\pi d_{ucs}^p}{\lambda} \right)^2 f_{ucs} \left(e, \alpha_{ucs}^p \right) A_{ucs} \left(\lambda, l_{ucs}, L_{ucs}, \alpha_{ucs}^p \right)} .$$
$$(3.35)$$

NS is time variant depending on the instantaneous geometry of the constellation.

The interfering set is composed of the user with $\alpha_{ucs}^p > 0$. In this case interfering users can be in the main lobe or in the side lobes of the satellite receiving antenna radiation pattern.

The value assumed by the factor γ depends on the correlation properties of the assigned code families and on the degree of synchronization that is achieved. Without any particular reference, the same value for each pair of users can be assumed, although the value used for $I_{1,\text{up}}$ may differ to those for $I_{2,\text{up}}$ and/or $I_{3,\text{up}}$. If no particular code assignment strategy is adopted and the users are not synchronized a unique (average) value for all the factors γ can be assumed in all the equations above. If the users in the same cell are assumed to be quasi-synchronous, the factors γ can be ≈ 0 inside the cell and it will assume different values from cell to cell in the different terms I_1, I_2 and I_3.

CDMA Downlink.   The received power at the useful ES is:

$$C_{\text{down}} = \frac{W_{np} g_{[nnp]} \left(\alpha_{mnp}^p \right) G_{[np]} \left(\theta_{mnp}^{np} \right)}{\left(\frac{4\pi d_{mnp}^p}{\lambda} \right)^2 f_{mnp} \left(e, \alpha_{mnp}^p \right) A_{mnp} \left(\lambda, l_{mnp}, L_{mnp}, \alpha_{mnp}^p \right)}, \quad (3.36)$$

where W is the power transmitted by the satellite. Power control can be implemented to counteract the near-far-effect.

The intra-cell interference is:

$$I_{1,\text{down}} = \sum_{\substack{u=1 \\ u \neq m}}^{NU_{np}} \frac{\gamma_{unp}^{mnp} \mu_{unp} W_{[np]} g_{[mnp]} \left(\alpha_{mnp}^{p}\right) G_{[np]} \left(\theta_{mnp}^{np}\right)}{\left(\frac{4\pi d_{mnp}^{p}}{\lambda}\right)^{2} f_{mnp} \left(e, \alpha_{mnp}^{p}\right) A_{mnp} \left(\lambda, l_{mnp}, L_{mnp}, \alpha_{mnp}^{p}\right)}.$$

(3.37)

If synchronous transmission in the same cells is achieved, the factor γ is equal to the inverse of the processing gain.

The inter-cell interference is due to the presence of side lobes and overlapping areas of the satellite antenna radiation pattern and it can be written as:

$$I_{2,\text{down}} = \sum_{\substack{c=1 \\ c \neq n}}^{NC} \sum_{u=1}^{NU_{cp}} \frac{\gamma_{ucp}^{mnp} \mu_{ucp} W_{[cp]} g_{[mnp]} \left(\alpha_{mnp}^{p}\right) G_{[cp]} \left(\theta_{mnp}^{cp}\right) \rho_{np}^{cp}}{\left(\frac{4\pi d_{mnp}^{p}}{\lambda}\right)^{2} f_{mnp} \left(e, \alpha_{mnp}^{p}\right) A_{mnp} \left(\lambda, l_{mnp}, L_{mnp}, \alpha_{mnp}^{p}\right)}.$$

(3.38)

The channel attenuation due to fading effect $f(\cdot)$ is the same experienced by all the interfering signals contributing to I_1 and I_2 and by the useful signal. Finally, the inter-satellite interference is:

$$I_{3,\text{down}} = \sum_{\substack{s=1 \\ s \neq p}}^{NS} \sum_{c=1}^{NC} \sum_{u=1}^{NU_{cs}} \frac{\gamma_{ucs}^{mnp} \mu_{ucs} W_{cs} g_{[mnp]} \left(\alpha_{mnp}^{s}\right) G_{[cs]} \left(\theta_{mnp}^{cs}\right) \rho_{np}^{cs}}{\left(\frac{4\pi d_{mnp}^{s}}{\lambda}\right)^{2} f_{mnp} \left(e, \alpha_{mnp}^{s}\right) A_{mnp} \left(\lambda, l_{mnp}, L_{mnp}, \alpha_{mnp}^{s}\right)}.$$

(3.39)

A satellite provides interference to user mnp if $\alpha_{mnp}^{s} > 0$. The use of sophisticated directive array antenna can reduce the interference from other satellites.

3.4.2.1.2 FDMA case. Only two contributions are envisaged in this case for the total interference: I_2 and I_3. $I_1 \equiv 0$ because no more than one user per cell is assumed to use the same frequency and ideal rectangular filters are assumed. To characterize this scenario with a certain number of cells NC and a certain cluster dimension CD a generalized model is proposed.

Assuming frequency reuse implementation compliant with the cellular concept only a subset of users actually contributes to interference (excluding contributions from adjacent channels). The selection must take into account the frequency reuse pattern. We assume regular cellular structure and that the number of cells may not be a multiple of the cluster size.

The set of interfering users corresponds to those utilizing the same frequency of the useful user. Assuming that the whole band is assigned to one cluster and that in the cluster the band is equally divided between cells, the number of interfering cells is:

$$\text{NI} = \begin{cases} k \\ k\text{-}1 \end{cases} \quad \begin{cases} NC = k\text{CD} \\ \forall t \leq \nu \\ \forall t > \nu \end{cases}$$

(3.40)

where $k = \left\lceil \frac{NC}{CD} \right\rceil$ is the number of full clusters in the pattern, $\nu = NC - k \cdot CD$ addresses the possibility that NC is not a multiple of CD, and t is the cell number in the cluster ($1 \leq t \leq CD$).

FDMA Uplink.   The uplink carrier power is:

$$C_{\text{up}} = \frac{w_{[mnp]} g_{[mnp]} \left(\alpha^p_{mnp}\right) G_{[np]} \left(\theta^{np}_{mnp}\right)}{\left(\frac{4\pi d^p_{mnp}}{\lambda_{mnp}}\right)^2 f_{mnp} \left(e, \alpha^p_{mnp}\right) A_{mnp} \left(\lambda_{mnp}, l_{mnp}, L_{mnp}, \alpha^p_{mnp}\right)}. \quad (3.41)$$

The inter-cell interference is

$$I_{2,\text{up}} = \sum_{\substack{q=0 \\ q \neq q_m}}^{\text{NI}} \frac{\mu_{m(CDq+t)p} w_{[m(CDq+t)p]} g_{[m(CDq+t)p]} \left(\alpha^p_{m(CDq+t)p}\right)}{\left(\frac{4\pi d^p_{m(CDq+t)p}}{\lambda_{mnp}}\right)^2 f_{m(CDq+t)p} \left(e, \alpha^p_{m(CDq+t)p}\right)} \cdot$$

$$\cdot \frac{G_{[np]} \left(\theta^{np}_{m(CDq+t)p}\right) \rho^{np}_{(CDq+t)p}}{A_{m(CDq+t)p} \left(\lambda_{mnp}, l_{m(CDq+t)p}, L_{m(CDq+t)p}, \alpha^p_{m(CDq+t)p}\right)},$$

$$(3.42)$$

where the serving cell n is written as CDq+t to take into account the adopted frequency reuse scheme. The wavelength λ is the same of the useful user because, by definition, all the interferers use the same frequency.
The inter satellite interference is:

$$I_{3,\text{up}} = \sum_{\substack{s=1 \\ s \neq p}}^{\text{NS}} \sum_{q=0}^{\text{NI}} \frac{\mu_{m(CDq+t)s} w_{[m(CDq+t)s]} g_{[m(CDq+t)s]} \left(\alpha^p_{m(CDq+t)s}\right)}{\left(\frac{4\pi d^p_{m(CDq+t)s}}{\lambda_{mnp}}\right)^2 f_{m(CDq+t)s} \left(e, \alpha^p_{m(CDq+t)s}\right)} \cdot$$

$$\cdot \frac{G_{[np]} \left(\theta^{np}_{m(CDq+t)s}\right) \rho^{np}_{(CDq+t)s}}{A_{m(CDq+t)s} \left(\lambda_{mnp}, l_{m(CDq+t)s}, L_{m(CDq+t)s}, \alpha^p_{m(CDq+t)s}\right)}.$$

$$(3.43)$$

FDMA Downlink.   The downlink carrier power is:

$$C_{\text{down}} = \frac{W_{[np]} g_{[mnp]} \left(\alpha^p_{mnp}\right) G_{[np]} \left(\theta^{np}_{mnp}\right)}{\left(\frac{4\pi d^p_{mnp}}{\lambda_{mnp}}\right)^2 f_{mnp} \left(e, \alpha^p_{mnp}\right) A_{mnp} \left(\lambda_{mnp}, l_{mnp}, L_{mnp}, \alpha^p_{mnp}\right)}.$$

$$(3.44)$$

The first interference term accounts for, as in the previous case, the inter-cell interference:

$$I_{2,\text{down}} = \sum_{\substack{q=0 \\ q \neq q_m}}^{\text{NI}} \frac{\mu_{m(\text{CDq+t})p} W_{[(\text{CDq+t})p]} g_{[mnp]}\left(\alpha_{mnp}^p\right)}{\left(\frac{4\pi d_{m(\text{CDq+t})p}^p}{\lambda_{mnp}}\right)^2 f_{mnp}\left(e, \alpha_{mnp}^p\right)} \cdot$$
$$\cdot \frac{G_{[(\text{CDq+t})p]}\left(\theta_{mnp}^{(\text{CDq+t})p}\right) \rho_{np}^{(\text{CDq+t})p}}{A_{mnp}\left(\lambda_{mnp}, l_{mnp}, L_{mnp}, \alpha_{mnp}^p\right)}, \tag{3.45}$$

whereas the second term represents the inter-satellite interference:

$$I_{3,\text{down}} = \sum_{\substack{s=1 \\ s \neq p}}^{\text{NS}} \sum_{q=0}^{\text{NI}} \frac{\mu_{m(\text{CDq+t})s} W_{[(\text{CDq+t})s]} g_{[mnp]}\left(\alpha_{mnp}^s\right)}{\left(\frac{4\pi d_{mnp}^s}{\lambda_{mnp}}\right)^2 f_{mnp}\left(e, \alpha_{mnp}^s\right)} \cdot$$
$$\cdot \frac{G_{[(\text{CDq+t})s]}\left(\theta_{mnp}^{(\text{CDq+t})s}\right) \rho_{np}^{(\text{CDq+t})s}}{A_{mnp}\left(\lambda_{mnp}, l_{mnp}, L_{mnp}, \alpha_{mnp}^s\right)}. \tag{3.46}$$

3.5   Conclusions

The different elements contributing to determine the final SNIR of a satellite link and its temporal variations have been widely discussed throughout this chapter, namely fading due to shadowing and blockage in case of mobile and satellite-to-indoor links, atmospheric impairments for high carrier frequencies and interference sources.

The main focus has been constantly devoted on highlighting which are the relevant effects to be taken into account for different satellite systems and scenarios in order to have a realistic estimation of the channel impairments to be faced.

Whenever possible, theoretical results have been exemplified by means of outcomes from a variety of direct measurements and experiments, testifying the great effort devoted in the last decades to achieve a complete understanding of the propagation and interference effects characterizing satellite links.

This achievement is indeed of great support in facing novel ambitious challenges, deriving from the always increasing bandwidth demand, like turning into a concrete reality the possibility to exploit higher frequency bands, e.g. above Ka for fixed broadband services and in Ku-Ka bands for mobile services.

Chapter 4

FORWARD ERROR CORRECTION

G. Albertazzi[1], M.Chiani[1], G.E. Corazza[1], A. Duverdier[2], H. Ernst[3], W. Gappmair[4], G. Liva[1], S. Papaharalabos[5]

[1] *University of Bologna, Italy*

[4] *Graz University of Technology, Austria*

[5] *University of Surrey, U.K.*

[3] *Germany Aerospace Center (DLR), Germany*

[2] *Centre National d'Etudes Spatiales (CNES), France*

4.1   Introduction to Coding

Coding theory is the study of error-correcting codes, which are used to transmit digital information in the presence of noise. Their application ranges from deep-space communications to the recovery from packet loss on the Internet. Part of the mathematical background, e. g., number theory or algebra over finite fields, has also an impact on complexity theory and cryptography.

The purpose of Forward Error Correcting (FEC) is to improve the capacity of a transmission channel by adding to the source data some carefully designed redundant information - a process known as channel coding with convolutional and block (algebraic) codes as the two major forms. For both of them, there exists a lot of useful encoding as well as decoding algorithms. In fact, any designer of a FEC scheme should take into account three main issues: construction, encoding, and decoding. The first issue (construction) refers to the creation of a smart code that satisfies a limited ensemble of code parameters. Then encoding is related to the conveyance of the source message to a particular codeword. Finally, by taking into account the receiver samples, decoding refers to the choice of one of all possible codewords through an appropriately selected algorithm. All these issues have a well-defined goal: to achieve the Shannon bound (see section 2.3.2). The metrics to assess the progress towards this objective are namely the Signal to Noise Ratio (SNR), the error performance and the computational complexity.

However, the history of error-correcting techniques is fairly young, beginning with Claude E. Shannon and his landmark paper on "A Mathematical Theory of Communication" published in 1948 [84], [85]. He showed that, even in a noisy channel, there exist ways to encode messages such that they have an arbitrarily good chance to be transmitted safely, provided that one does not exceed the *capacity* of the channel by trying to transmit as much information in a fast way. Shortly thereafter, Richard Hamming developed the first error correcting code tolerating the loss or distortion of one bit. Codes similar to this are called Hamming codes. Soon after, Marcel Golay generalized Hamming's approach via a code based on non-binary algebra. He proposed two triple-error correcting schemes, which carry his name. Nevertheless, the first practical application of channel coding was in deep-space communications. In fact, in the mid-1960's, National Aeronautics and Space Administration (NASA) engineers understood that deep-space communications and coding is a "marriage made in heaven" [86]. There are many reasons showing that this statement is true: first of all, the deep-space communication channel is accurately described by the Additive White Gaussian Noise (AWGN) model introduced by Shannon in 1948. Then, bandwidth is not a problem for this kind of application, so binary transmission can be used efficiently. The only complex part of the FEC algorithm is the decoder, but in the downlink it is located in the earth station, where computational complexity is not a problem. In addition, every dB gained in the downlink of the deep-space communication channel is really important. Even such a small benefit is a really strong incentive to develop and implement efficient coding schemes. In this way, thanks to deep-space communications, both convolutional and concatenated codes found their first practical usage in Voyager and Mariner missions in the later sixties as well as Turbo codes in the Cassini-Huygens mission to Saturn in 1997.

In this Chapter, we first provide a quick review to the most significant error correcting codes based on linear algebra, which were developed from the 1950's to the 1970's (i.e. starting from Hamming codes to BCH and RS codes). This is followed by the more challenging developments and improvements that took place in the recent years, namely Turbo codes and Low-Density Parity-Check (LDPC) codes, based on graph theory. Issues related to the design, decoding and practical applications of these codes are provided, followed by a comparison on their performance and complexity at the end of this Chapter.

4.2   Algebraic coding

Error-correcting codes constitute one of the key elements to achieve a higher degree of reliability as it is required for modern data transmission and storage systems [87], [88], [89]. This section introduces the reader to the theory of algebraic coding [90], i. e., finite (Galois) fields and their relationship to linear block as well as convolutional codes. Prominent examples are provided in the sequel, with particular emphasis on Bose-Chaudhuri-Hocquenghem (BCH) and Reed-Solomon (RS) codes. Powerful solutions for soft-decision and frequency-domain decoding of BCH and RS codes conclude the discussion.

4.2.1 Basic definitions

Coding theory, sometimes also called algebraic coding theory, was basically born to design codes for reliable transmission of information across noisy channels as already predicted by Shannon's theorem (see section 2.3.2). Since the early 1950's, when the first error-correcting codes were introduced, coding theory has used classical and modern algebraic techniques involving finite fields, group theory, and polynomial algebra, exploiting some areas of discrete mathematics, number theory, as well as the theory of experimental designs. However, almost all codes are based on the notion of the *field* which is to be detailed in the sequel.

DEFINITION 4.1 *A field is any set of elements which satisfy the axioms for both addition and multiplication (commutative division algebra). A finite field, i. e., a field with a finite number of elements, is also called a* Galois field.

DEFINITION 4.2 *Let $\mathcal{F}_q$ be a* Galois field *of cardinality q. Then there is a prime integer p such that $\mathcal{F}_q$ can be viewed as a finite-dimensional vector space over $\mathbb{Z}/p\mathbb{Z} = \mathcal{F}_p$. The prime p denotes the* field *characteristic. The m-dimensional extension of $\mathcal{F}_p$ is given by $\mathcal{F}_q$, where $q = p^m$. With respect to multiplicative operations "$\times$", the non-zero elements of $\mathcal{F}_q$ form a cyclic group $\mathcal{F}_q^*$ of order $q - 1$. A generator of $(\mathcal{F}_q^*, \times)$ is said* primitive. *If $\alpha \in \mathcal{F}_q^*$ is primitive, then $\mathcal{F}_q = \{\alpha^k,\ 0 \leq k \leq q - 1\}$.*

Let $\mathcal{F}$ be a finite set called the alphabet which is, throughout this chapter, supposed to be a finite field.

DEFINITION 4.3 *With I as some index set, a* code over $\mathcal{F}$ *is any subset of $\mathcal{F}^I$.*

DEFINITION 4.4 *The* code construction *refers to the design of a smart code which satisfies a limited ensemble of code parameters.*

DEFINITION 4.5 Encoding *is related to the conveyance of a message to a particular codeword. With* systematic codes, *the message to be conveyed appears as such in the codeword which is not the case with* non-systematic codes.

DEFINITION 4.6 Decoding *refers to the choice of one of the possible codewords through an appropriately selected algorithm.*

4.2.1.1   Block codes.   A *block code* C of length n over $\mathcal{F}$ is any subset of $\mathcal{F}^n$. The code C is *linear*, if it is a k-dimensional subspace of $\mathcal{F}^n$. We say that C is an (n, k) code and the ratio $r = k/n$ is called the *code rate*.

DEFINITION 4.7 *Any (k, n) matrix $\boldsymbol{G}$ whose rowspace equals C is called a* generator matrix *for C.*

A very simple encoding rule which maps messages $\boldsymbol{u} = (u_0, u_1, \ldots, u_{k-1})$ onto codewords $\boldsymbol{c} = (c_0, c_1, \ldots, c_{n-1})$ is provided by $\boldsymbol{c} = \boldsymbol{u} \cdot \boldsymbol{G}$. If $\boldsymbol{G} = (\boldsymbol{I}_k, \boldsymbol{P})$, where $\boldsymbol{I}_k$ specifies the k-dimensional identity matrix and $\boldsymbol{P}$ is some $(k, n - k)$ matrix, the encoding rule describes a systematic code. The $n - k$ symbols added by the encoder are called *redundancy*.

DEFINITION 4.8 *Let $\boldsymbol{x} = (x_0, x_1, \ldots, x_{n-1})$ and $\boldsymbol{y} = (y_0, y_1, \ldots, y_{n-1})$ be arbitrary elements in $\mathcal{F}^n$, then $(\boldsymbol{x}, \boldsymbol{y}) = \sum_{i=0}^{n-1} x_i \, y_i$ denotes the inner product of $\boldsymbol{x}$ and $\boldsymbol{y}$ in $\mathcal{F}^n$. Taking into account these specifications, we can define the* dual code *of C as*

$$C^\perp = \{ \boldsymbol{x} \in \mathcal{F}^n : (\boldsymbol{x}, \boldsymbol{c}) = \boldsymbol{0}, \ \forall \boldsymbol{c} \in C \} \tag{4.1}$$

Note that $C^\perp$ is an $(n, n - k)$ linear code. The generator matrix $\boldsymbol{H}$ of $C^\perp$ is called the *parity check matrix* of C, i .e.,

$$\boldsymbol{c} \in C \quad \Leftrightarrow \quad \boldsymbol{H} \cdot \boldsymbol{c}^T = \boldsymbol{0}$$

DEFINITION 4.9 *If $1 \leq \tau < k$, the code C_τ whose generator matrix is $\boldsymbol{G}_\tau = (\boldsymbol{I}_{k-\tau}, \boldsymbol{P}_\tau)$, where $\boldsymbol{P}_\tau$ is the $(k - \tau, n - k)$ matrix obtained by deleting the first τ rows of $\boldsymbol{P}$, is called a* shortened *version of C, i. e., C_τ is the set of all the codewords of C whose first τ coordinates are equal to zero.*

DEFINITION 4.10 *Let $1 \leq \tau < k$. If we delete in each codeword of C the same τ coordinates, we get a code C_τ^* of length $n - \tau$ and dimension greater than $k - \tau$. The code C_τ^* is called a* punctured *version of C.*

DEFINITION 4.11 *An (n', k) linear code C' is called an* extended *version of C, when C is the set of all the n-tuples obtained by deleting the last $n' - n$ coordinates of the codewords of C'. Extension is most often obtained by appending an overall* parity bit *to each codeword of C.*

4.2.1.2   Cyclic codes.   Let C be a linear code of length n over a finite field $\mathcal{F}$. It is said *cyclic*, when every cyclic shift $s^k(\boldsymbol{c}) = (c_{k+1}, \ldots, c_{n-1}, c_0, \ldots, c_k)$ of a codeword $\boldsymbol{c} = (c_0, c_1, \ldots, c_{n-1})$ is also a codeword. To describe cyclic codes, it is common practice to consider the n-tuple $\boldsymbol{a} = (a_0, a_1, \ldots, a_{n-1}) \in \mathcal{F}^n$ as a polynomial $a(X)$ of degree $n-1$, i. e., $a(X) = \sum_{i=0}^{n-1} a_i X^i$. The cyclic shift $s^k(\boldsymbol{c})$ reduces to a multiplication by X^k in the ring $(\mathcal{F}[X]/(X^n - 1), +, \times)$.

It turns out that C is cyclic iff C is an ideal of this ring. Since this ring is a principal domain, we have the following characterization:

THEOREM 4.12 *C is cyclic if and only if there is a polynomial $g(X)$ in $\mathcal{F}[X]$ such that $\boldsymbol{c} \in C$ iff $g(X)$ divides any $\boldsymbol{c}(X)$ without remainder.*

The polynomial $g(X)$ is called the *generator polynomial* of the cyclic code C and the dimension of C is $k = n - \deg(g(X))$. As a consequence, cyclic codes are easily encoded in a systematic way via a simple polynomial division. To this end, let $\boldsymbol{u}(X) = \sum_{i=0}^{k-1} u_i X^i$ denote the sequence to be encoded. The encoder generates $n - k$ parity symbols $\boldsymbol{p}(X) = \sum_{i=0}^{n-k-1} p_i X^i$ as the remainder of the Euclidean division of $X^{n-k}\boldsymbol{u}(X)$ by $g(X)$. The transmitted codeword is

$$\boldsymbol{c}(X) = X^{n-k}\boldsymbol{u}(X) + \boldsymbol{p}(X) \tag{4.2}$$

In order to obtain an $(n - \tau, k - \tau)$ shortened version of C, we can run the systematic encoder with $u_0 = \ldots = u_{\tau-1} = 0$ and send the coordinates $c_\tau, c_{\tau+1}, \ldots, c_{n-1}$ only.

4.2.1.3   Convolutional codes.   Convolutional codes might be defined from different points of view, e. g., employing the polynomial matrix approach related to the above definition of cyclic codes. An (n, k) convolutional code C over the field $\mathcal{F}$ can be characterized by a (k, n) generator matrix $\boldsymbol{G} = (g_{ij}(X))$ over $\mathcal{F}[X]$. The $g_{ij}(X)$ are the generator polynomials. Associated to $\boldsymbol{G}$, there are three important figures of merit:

1. *Memory $M = \sum_{i=0}^{k-1} M_i$, $M_i = \max\{\deg(g_{ij}(X)), 0 \le j < n\}$*

2. *Constraint length $K = M + 1$*

3. *Coding rate $r = k/n$*

In order to use $\boldsymbol{G}$ for encoding purposes, the information is mapped onto the coefficients of a k-tuple of the formal series $\boldsymbol{u}(X) = (u_0(X), \ldots, u_{k-1}(X))$ and the corresponding codeword is provided by

$$\boldsymbol{c}(X) = \boldsymbol{u}(X) \cdot \boldsymbol{G}(X)$$

This general encoder scheme is realized by k shift registers, the i-th of length M_i, with the output formed as a linear combination of the appropriate shift-register contents evaluated over $\mathcal{F}_q$. This approach which encoder makes use of solely *feedforward* FF shift registers produces *non-recursive* codes. In contrast to them, encoders based on *feedback* FB structured shift registers, referred as Recursive Systematic Convolutional (RSC) encoders, produce *recursive* codes.

In [91], it is shown that every FF code is equivalent to a FB code and vice versa, i. e., the same set of codewords is generated. Given an FF code, unambiguous conversion to FB is possible; the conversion from FB to FF, however, involves multiple solutions. For higher coding rates and low SNR values, it has been shown that a convolutional code with recursive-systematic encoding exhibits a better BER performance than a code with non-recursive non-systematic encoding [92].

Another important feature of these codes is the possibility to vary their code-rate through the periodic deletion of the encoder output. Given a mother code with rate r, this operation, referred as *puncturing*, is used to achieve coding rates larger than r. Originally, puncturing has been suggested in [93] to simplify both encoder and decoder designs for $r > 1/2$. In [94], the concept is extended to rate compatibility, i. e., puncturing is organized such that all code bits of higher-rated codes are used by the lower-rated codes as well.

4.2.2   Hamming geometry and decoding problem

As mentioned previously, the future role of FEC schemes was already predicted in 1948 by Shannon in his channel coding theorem. However, he did not offer any algorithms or recipes to meet that objective. As a consequence, the coding theory emerged in the course of time as a separate area of investigation in the digital communications domain. It was Richard W. Hamming who introduced in the early 1950's such useful definitions like minimum distance or detection/correction capability, which are even today necessary for a basic understanding of algebraic coding.

4.2.2.1   Hamming distance.   Let $\mathcal{F}$ be a finite field with q elements. Furthermore, let $\boldsymbol{x} = (x_0, x_1, \ldots, x_{n-1})$ and $\boldsymbol{y} = (y_0, y_1, \ldots, y_{n-1})$ be two vectors in $\mathcal{F}^n$. Then the vector space $\mathcal{F}^n$ can be turned into a metric space, if we introduce the *Hamming distance* as

$$d_H(\boldsymbol{x}, \boldsymbol{y}) = |\{i \in \{0, \ldots, n-1\}, \ x_i \neq y_i\}| \tag{4.3}$$

and the *Hamming weight* of $\boldsymbol{x}$ as $w_H(\boldsymbol{x}) = d_H(\boldsymbol{x}, \boldsymbol{0})$.

DEFINITION 4.13 *With $\mathcal{C}$ as a subset of $\mathcal{F}^n$, the* minimum distance *of $\mathcal{C}$ is defined as $d_{min}(\mathcal{C}) = \min\{d_H(\boldsymbol{x}, \boldsymbol{y}) : \boldsymbol{x}, \boldsymbol{y} \in \mathcal{C}, \boldsymbol{x} \neq \boldsymbol{y}\}$. Any two Hamming spheres of radius $(d_{min} - 1)/2$ around $\boldsymbol{x}, \boldsymbol{y} \in \mathcal{C}$, $\boldsymbol{x} \neq \boldsymbol{y}$ are non-intersecting. The subset $\mathcal{C}$ is* perfect, *when the spheres of radius $(d_{min} - 1)/2$ for all $\boldsymbol{x} \in \mathcal{C}$ cover $\mathcal{F}_q^n$.*

Let $\mathcal{C}_\tau$ be a shortened version of a linear code $\mathcal{C}$, then $d_{min}(\mathcal{C}_\tau) \geq d_{min}(\mathcal{C})$. Similarly, if $\mathcal{C}'$ is an extended version of $\mathcal{C}$, then $d_{min}(\mathcal{C}') \geq d_{min}(\mathcal{C})$.

DEFINITION 4.14 *The* detection capability *of a code $\mathcal{C}$, characterized by the minimum distance d_{min}, is equal to $d_{min} - 1$. On the other hand, its error correction capability is determined by $t = \lfloor (d_{min} - 1)/2 \rfloor$.*

The minimum Hamming distance is certainly an appropriate measure for the quality of a code. It is to be noticed, however, that most block codes are able to correct not only all patterns of fewer than t errors but many patterns of more than t errors as well.

DEFINITION 4.15 *A linear block code* C *is said to be a t-error-correcting perfect code if the set of t-spheres, centered at the codewords of the code, fill the whole finite field space (referred also as* Hamming space) *without overlapping.*

Hamming geometry concepts can be extended to infinite sequences as well. To this end, let $a = \sum_{i\in\mathbb{Z}} a_i X^i$ and $b = \sum_{i\in\mathbb{Z}} b_i X^i$ be two formal series over $\mathcal{F}^n$. Then, the Hamming distance between a and b is given by $d_H(a,b) = \sum_{i\in\mathbb{Z}} d_H(a_i, b_i)$. The minimum distance of a set C of formal series over $\mathcal{F}^n$, like a convolutional code, is then defined as $d_{min}(C) = \min\{d_H(c,c') : c, c' \in C, c \neq c'\}$.

4.2.2.2   Bounds.   Since performance analysis rarely admits exact expressions, tight analytical bounds are welcome as a useful engineering tool for gaining the necessary insight to the effect of various code parameters. The most applied bounds are evolutions of the classical union upper bound (see section 2.4.1.3). They establish a relationship between the minimum Hamming distance d_{min} or, equivalently, the error correction capability t on the one hand, and the redundancy $n - k$ on the other, where k and n denote the number of information and block symbols, respectively. The following theorems on upper and lower bounds are given without proofs which can be found in the open literature [90][87][89].

THEOREM 4.16 Singleton bound. *The minimum distance of* (n, k) *codes is upper-bounded by*

$$d_{min} \leq n - k + 1 \tag{4.4}$$

Maximum Distance Separable (MDS) codes satisfy eq. (4.4) with equality. Apart from trivial schemes like parity-check and repetition codes, Reed-Solomon codes are the most prominent example in this respect. Note also that MDS codes have a weight structure given in closed form.

THEOREM 4.17 Hamming bound. *The error correction capability of* (n, k) *codes is upper-bounded by*

$$n - k \geq \log_q \left[\sum_{i=0}^{t} \binom{n}{i}(q-1)^i \right] \tag{4.5}$$

For perfect codes, the Hamming (sphere-packing) bound is satisfied with equality. Binary repetition codes of odd block length, Hamming codes, binary and ternary Golay codes are the only examples known with this property.

THEOREM 4.18 Plotkin bound. *The minimum distance of (n, k) codes is upper-bounded by*

$$d_{min} \leq \frac{q-1}{q^k - 1} q^{k-1} n \qquad (4.6)$$

The *Elias bound* is an improvement of the Plotkin bound for codes with $n \gg 1$ and a relatively small minimum distance.

THEOREM 4.19 Gilbert bound. *The minimum distance of (n, k) codes is lower-bounded by*

$$n - k > \log_q \left[\sum_{i=0}^{d_{min}-1} \binom{n}{i} (q-1)^i \right] \qquad (4.7)$$

Varsharmov provided a slightly stronger version of the Gilbert bound. Therefore, although not totally correct, relationship (4.7) is often referred to as Varsharmov-Gilbert bound.

4.2.2.3   The decoding problem.   The symbols c_i as part of the codeword c are transmitted across a memoryless channel. We assume that all the codewords are used with equal probability. Channel input and output are random variables. The optimum value of the soft information for any symbol $y \in \mathcal{F}$, given that r_i has been received, is $\pi(i, y) = \Pr(y|r_i)$. The $n \times q$ reliability matrix Π can be derived from the $\pi(i, y)$ values. The hard decision vector $h = (h_0, h_1, \dots, h_{n-1}) \in \mathcal{F}^n$ is such that $h_i = \arg\max\{\pi(i, y), y \in \mathcal{F}\}$.

DEFINITION 4.20 *The hard decision decoder uses h as the input and solves the following problem: Find $\hat{c} \in \mathcal{C}$ such that $d_H(\hat{c}, h) = \min\{d_H(c, h), c \in \mathcal{C}\}$.*

DEFINITION 4.21 *The ϵ-finite distance hard decoder solves the following problem: Find all codewords $c \in \mathcal{C}$ such that $d_H(c, h) \leq \epsilon$.*

If $\epsilon > (d_{min} - 1)/2$, the ϵ-finite distance hard decoder outperforms the hard decoder providing a list of candidates, with the selected candidate maximizing the a posteriori likelihood. To this end, let $y = (y_0, y_1, \dots, y_{n-1}) \in \mathcal{F}^n$ such that the logarithmic score of y may be determined as

$$S_\Pi(y) = \sum_{i=0}^{n-1} \log(\pi(i, y_i)) \qquad (4.8)$$

DEFINITION 4.22 *The maximum a posteriori likelihood codeword decoder, often referred to as the* ML *decoder, solves the following problem: Find $\hat{c} \in \mathcal{C}$, where $S_\Pi(\hat{c}) = \max\{S_\Pi(c), c \in \mathcal{C}\}$.*

DEFINITION 4.23 *The maximum a posteriori likelihood* symbol *decoder, often referred to as the* MAP decoder, *solves the following problem: Find* $\hat{c}_j = \arg\max\{Pr(c_j = a|r), a \in \mathcal{F}\}$, *where* $j = 0, 1 \ldots, n-1$.

Let $a \in \mathcal{F}$ and $C_j^{(a)} = \{c \in C, c_j = a\}$, i.e., $C_j^{(a)}$ is the coset of all codewords whose j-th coordinate is a. Using Bayes rule, we can write

$$Pr(c_j = a|r) \propto \sum_{c \in C_j^{(a)}} \exp\{S_{\mathrm{II}}(c)\} \qquad (4.9)$$

DEFINITION 4.24 *A Soft-Input Soft-Output (SISO) decoder is an appropriate decoder that accepts soft input values and produces soft output values.*

The *soft output* of a binary SISO decoder, at a given time instant k, is based on the estimation of the probability that the transmitted information bit is one to the probability that the transmitted information bit is zero, given the observation of the received sequence of bits. When the logarithm of this ratio is obtained, the soft output is usually referred to as *Log-Likelihood Ratio (LLR)*:

$$LLR = L(\hat{u}_k) = \log \frac{P(u_k = 1|\mathbf{r})}{P(u_k = 0|\mathbf{r})} \qquad (4.10)$$

This is the estimation of the *a posteriori probability* (APP) of the transmitted bit, given the observation of the received sequence of bits. The sign of the LLR value corresponds to the *hard decision* of the transmitted bit. If it is positive, then bit "1" is assumed to be transmitted, otherwise if it is negative, then bit "0" is assumed to be transmitted. The magnitude of the LLR value corresponds to the *reliability* of this decision, which is a measure of the certainty of the transmitted bit i.e., *soft decision*.

Typical SISO decoders are based on either the Viterbi algorithm (i.e. Soft-Output Viterbi Algorithm (SOVA)) or the Maximum-A-Posteriori (MAP) probability or Bahl-Cocke-Jelinek-Raviv (BCJR) algorithm and its approximations e.g. Log-MAP and Max-Log-MAP algorithms. These algorithms, used to decode trellis codes, are described in more detail in sections 4.3.2.2, 4.3.2.3 and 4.3.2.4.

4.2.3   Golay code

The (23,12) binary Golay code is a cyclic code which represents a unique example of perfect codes with error correction capability $t > 1$. It is defined by the generator polynomial

$$g(X) = X^{11} + X^9 + X^7 + X^6 + X^5 + X + 1$$

The 2^{12} codewords are the centers of non-overlapping Hamming spheres of radius 3, each containing 2048 vectors in $\mathcal{F}_2^{23}$. Among all codes of length 23 and minimum distance 7, the (23,12) Golay code offers the highest possible rate. The (24, 12) Golay code is obtained from the (23, 12) Golay code by adding to each codeword an overall parity check. The extended code is self-dual, i. e., equal to its dual. Due to its highly symmetric structure, this code has attracted much attention and many constructions have been proposed. Several ML soft decoders have been proposed [95]. Based on Pless' construction of the (24,12) Golay code, an efficient symbol-by-symbol SISO decoding algorithm has been published in [96] .

Note also that the Golay code, in order to protect the power control field, is recommended in the GEO-mobile Radio Interface Specification [97].

4.2.4   Binary Hamming and Reed-Müller codes

DEFINITION 4.25 *Let H be an $m \times (2^m - 1)$ binary matrix such that the columns of H are the $2^m - 1$ non-zero vectors of $\mathcal{F}_2^m$ in some order. The related linear code over $\mathcal{F}_2$ is the binary Hamming code of length $n = 2^m - 1$, henceforth denoted by H_m.*

The dimension of H_m is $k = 2^m - 1 - m$, its minimum distance $d_{min} = 3$. Moreover, it is a perfect code with error correction capability $t = 1$. The extended Hamming code of length 2^m is obtained from the mother code H_m by appending an overall parity bit.

DEFINITION 4.26 *Let $(v_0, v_1, \ldots, v_{n-1})$, $n = 2^m$, denote a list of all vectors in $\mathcal{F}_2^m$ following some order. Furthermore, let $\mathcal{F}_{2,d}$ denote the vector space over $\mathcal{F}_2$ of all polynomials of degree lower or equal than d in m variables over $\mathcal{F}_2$. Then the d-th order Reed-Müller code of length 2^m over $\mathcal{F}_2$, denoted by $RM(m, d)$, is the range of the evaluation map*

$$\Phi_{m,d} \begin{cases} \mathcal{F}_{2,d}[X_1, \ldots, X_m] & \to & \mathcal{F}_2^n \\ P & \mapsto & (P(v_0), \ldots, P(v_{n-1})) \end{cases} \tag{4.11}$$

$RM(m, d)$ is a linear code of length n, dimension $k = \sum_{i=0}^{d} \binom{m}{i}$, and minimum distance $d_{min} = 2^{m-d}$. Note also that $RM(m, d)^{\perp} = RM(m, m-d-1)$ and $RM(m, m - 2) = RM(m, 1)^{\perp}$ is the extended binary Hamming code of length 2^m. Based on fast Hadamard transforms, a symbol-by-symbol SISO decoder for first-order RM codes and their duals can be easily designed with complexity $n \log_2 n$ [98]. Finally, the DVB-S2 standard uses $RM(2^5, 1)$ with minimum distance 2^4 to convey modulation and coding schemes to the receiver end.

4.2.5 BCH and Reed-Solomon codes

Reed-Solomon (RS) and Bose-Chaudhuri-Hocquenghem (BCH) schemes are a subset of the linear-cyclic block codes. Normally, they are used in a systematic manner, i. e., the redundant symbols – generated via an appropriate encoding algorithm to be explained in the sequel – are appended to the source data. The error correction capability is given by the generator polynomial. RS and BCH codes are strictly related; in fact, the former may be regarded as non-binary BCH codes, where symbols are elements in $\mathcal{F}_2^m$, instead of $\mathcal{F}_2$, and hence specified by m bits. Note also that RS codes are part of both the DVB-S and CCSDS standard [99], [100], while BCH algorithms are recommended in the DVB-S2 standard [101].

4.2.5.1 Code construction and Berlekamp algorithm. Let $\mathcal{F}$ be a finite field of cardinality q, where q is a power of 2. Also, let α denote the primitive element of $\mathcal{F}$ and $\mathcal{F}_s[X]$ be the set of all polynomials of degree at most s. Finally, $n = q - 1$ and t, e are two integers such that

$$0 < t \leq \lfloor n/2 \rfloor, \; 0 \leq e < n \tag{4.12}$$

DEFINITION 4.27 *The Reed-Solomon code $RS(n, t, e)$ is defined as the set of all n-tuples $\mathbf{c} = (c_0, c_1, \ldots, c_{n-1})$ in $\mathcal{F}^n$ such that the corresponding polynomial $C(X) = \sum_{i=0}^{n-1} c_i X^i$ has $\alpha^e, \alpha^{e+1}, \ldots, \alpha^{e+2t-1}$ as roots. In other words, $RS(n, t, e)$ is cyclic with $G(X) = \prod_{i=e}^{e+2t-1} (X - \alpha^i)$ as the generator polynomial.*

THEOREM 4.28 *$RS(n, t, e)$ is a k-dimensional subspace of $\mathcal{F}^n$, where $k = n - 2t$ with minimum distance $d_{min} = 2t + 1$.*

DEFINITION 4.29 *$RS(n, t, e) \cap \mathcal{F}^n$ is a BCH code of length n and constructed distance $d = 2t + 1$. The actual minimum distance of the BCH code is lower-bounded by d.*

Since RS and BCH codes are linear and cyclic, decoding is rather simple. The Berlekamp algorithm [90] provides a very efficient hard decoder with complexity $O\left(n \log^2 n\right)$ comprising three major steps:

1. Compute the syndrome polynomial $S(X)$, where $\deg(S(X)) \leq 2t - 1$ and $t = (n - k)/2$.

2. Determine iteratively the polynomials $\Lambda(X)$ and $\Omega(X)$ satisfying the modular congruence $\Lambda(X)S(X) = \Omega(X) \mod X^r$ for $r = 1, 2, \ldots, 2t$.

3. Find the roots of the error locator $\Lambda(X)$ identifying the error locations (Chien search). For non-binary schemes, calculate the error values with Forney's formula and subtract them from the hard decision vector.

4.2.5.2   List decoding.   Because of the related trellis complexity, MAP decoding of RS codes is beyond reach; even suboptimal solutions like Chase's algorithms are limited to fairly small code lengths. Lately, interpolation-based list decoders with reasonable complexity have been devised, which show a significant performance improvement compared to Berlekamp's algorithm. In order to outline the related background, we generalize the definition of RS codes using an evaluation map suited for interpolation. If $f(X)$ is a polynomial in $\mathcal{F}_{k-1}[X]$, it is easily verified that

$$c_f = (\alpha^{0(1-e)}f(\alpha^0), \alpha^{1(1-e)}f(\alpha^1), \ldots, \alpha^{(n-1)(1-e)}f(\alpha^{n-1})) \qquad (4.13)$$

is in $RS(n, t, e)$. Besides, comparing the dimensions, we have

$$RS(n, t, e) = \{c_f, \; f(X) \in \mathcal{F}_{k-1}[X]\}. \qquad (4.14)$$

DEFINITION 4.30 *Following [102], generalized Reed-Solomon codes of length n and dimension k, associated to the subset $\mathcal{D} = \{x_0, x_1, \ldots, x_{n-1}\}$ of $\mathcal{F}$ and to the weighting $\boldsymbol{v} = (v_0, v_1, \ldots, v_{n-1}) \in \mathcal{F}^n$, are defined by the set*

$$GRS(\mathcal{D}, \boldsymbol{v}, k) = \{(v_0 f(x_0), \ldots, v_{n-1} f(x_{n-1}))\} \qquad (4.15)$$

with $f(X) \in \mathcal{F}_{k-1}[X]\}$. Thus, we have $RS(n, t, e) = GRS(\mathcal{D}, \boldsymbol{v}, k)$, where $\mathcal{D} = \mathcal{F} \setminus \{0\}$, $x_i = \alpha^i$, and $v_i = \alpha^{i(1-e)}$.

Since the weights v_i can be compensated at the receiver, we assume that $v_i = 1$ and refer to this as $GRS(\mathcal{D}, k)$ codes giving up the weighting with $\boldsymbol{v}$. This method of generating an RS code is called the evaluation map method. It provides the necessary insight for interpolation-based decoding algorithms.

4.2.5.3   Bivariate polynomials.   The Guruswami-Sudan algorithm [102] and its soft counterpart [103] make extensive use of bivariate polynomials, which require some more notations in order to proceed. Throughout this section, $P(X, Y)$ is a bivariate polynomial over $\mathcal{F}$, i. e.,

$$P(X, Y) = \sum_{i,j} a_{ij} X^i Y^j \qquad (4.16)$$

DEFINITION 4.31 *The $(1, m)$-weighted degree of $X^i Y^j$ is defined as $i + mj$ such that the $(1, m)$-weighted degree of $P(X, Y)$ is given by*

$$\deg_{(1,m)}(P(X, Y)) = \max\{i + mj, \; (i, j) \in \mathbb{N}^2, \; a_{ij} \neq 0\} \qquad (4.17)$$

Let δ be some integer and $N(\delta)$ the number of monomials with $(1, m)$-weighted degrees smaller than δ, i. e.,

$$N(\delta) = \left|\{X^i Y^j, \; (i, j) \in \mathbb{N}^2, \; i + mj \leq \delta\}\right| \qquad (4.18)$$

which is the dimension over $\mathcal{F}$ of the vector space of bivariate polynomials of with $(1, m)$-weighted degrees less than δ.

DEFINITION 4.32 *Furthermore, we define the weighted lexicographic order over* $\mathbb{N}^2$ *as*

$$(u, v) \leq \text{ord}_{(1,m)}(i, j) \text{ when } \begin{cases} u + mv < i + mj \\ \text{or} \\ u + mv = i + mj \text{ and } v \leq j \end{cases} \tag{4.19}$$

This is a total. We denote by $\text{ord}_{(1,m)}(i, j)$ the position of (i, j) in the list of elements in $\mathbb{N}^2$ enumerated in ascending order. If $P(X, Y) \neq 0$, its order is defined as

$$\text{ord}_{(1,m)} P(X, Y) = \max\{\text{ord}_{(1,m)}(i, j), \ a_{ij} \neq 0\} \tag{4.20}$$

If $P(X, Y) = 0$, the order is set to $-\infty$.

DEFINITION 4.33 *The* (α, β) *Hasse derivative of* $P(X, Y)$ *is defined as*

$$D_H^{(\alpha,\beta)} P(X, Y) = \sum_{i \geq \alpha, j \geq \beta} C_i^\beta C_j^\beta a_{ij} X^{i-\alpha} Y^{j-\beta} \tag{4.21}$$

The close relationship to the Taylor expansion about $(x, y) \in \mathcal{F}^2$ shows the role of the Hasse derivative, i. e.,

$$P(X, Y) = \sum_{\alpha, \beta} D_H^{(\alpha,\beta)} P(x, y)(X - x)^\alpha (Y - y)^\beta \tag{4.22}$$

DEFINITION 4.34 *A point* $(x, y) \in \mathcal{F}^2$ *is said to be a root of multiplicity* m *of the bivariate polynomial* $P(X, Y)$ *when*

$$\alpha + \beta < m \quad \Rightarrow \quad D_H^{(\alpha,\beta)} P(x, y) = 0 \tag{4.23}$$

Requiring that (x, y) be root of multiplicity m of a polynomial $P(X, Y)$ imposes $m(m + 1)/2$ linear constraints on the vector space of bivariate polynomials.

4.2.5.4   The Guruswami-Sudan decoding algorithm.   Guruswami and Sudan have proposed an ϵ-finite distance hard decoder using an interpolative solution [102]. If $\epsilon > (n - k)/2$, the decoder outperforms the Berlekamp algorithm. Providing a list of candidates, the appropriately selected candidate maximizes the a posteriori likelihood. The decoder developed by Guruswami and Sudan can correct up to $n - \sqrt{nk}$ symbol errors. The hard decision vector

$h = (h_0, h_1, \ldots, h_{n-1})$, the decoder is fed with, is a list of elements in $\mathcal{F}^2$, i. e.,

$$L = \{(x_0, h_0), (x_1, h_1), \ldots, (x_{n-1}, h_{n-1})\} \tag{4.24}$$

The algorithm is now as follows:

1. *Interpolation*: given the list L and a positive integer m, compute the non- zero bivariate polynomial $P(X, Y) \in F[X, Y]$ with minimal $(1, k)$-weighted degree (or minimal order) that has all the points of L as zeros with multiplicity m.

2. *Factorization*: given the interpolation polynomial $P(X, Y) \in F[X, Y]$, factor out all the polynomials of the form $Y{-}g(X)$ such that $\deg(g(X)) < k$.

3. *Selection*: select the codeword $(g(x_0), g(x_1), \ldots, g(x_{n-1}))$ with the highest likelihood.

If c_f has been sent, where $f(X) \in F_{k-1}[X]$, Guruswami and Sudan have shown that under certain conditions $Y - f(X)$ can be factored out of the interpolation polynomial.

4.2.5.5   The Koetter-Vardy decoding algorithm.   Koetter and Vardy [103] have extended the Guruswami and Sudan method to an algebraic soft-decision decoding algorithm which significantly outperforms the list decoding algorithm. They proposed a method to translate soft information into multiplicities via a list of points in $\mathcal{F}^2$, where unequal multiplicities are assigned to those points according to their relative reliability.

DEFINITION 4.35 *A multiplicity matrix M is an $n \times q$ matrix with positive integer entries. The interpolation problem associated to M is the computation of a non-zero bivariate polynomial $P_M(X, Y) \in F[X, Y]$ with minimal $(1, k - 1)$-weighted degree (or minimal order) such that $P_M(X, Y)$ vanishes at all the points $(x_i, y) \in \mathcal{D} \times \mathcal{F}$ with multiplicity $M(i, y)$.*

The interpolation problem associated to M is tantamount to the solution of a linear system with

$$C_M = \frac{1}{2} \sum_{(x_i, y) \in \mathcal{D} \times \mathcal{F}} M(i, y)(1 + M(i, y)) \tag{4.25}$$

equations and the degree of $P_M(X, Y)$ upper-bounded by

$$\Delta_M = \min\{\delta \in \mathbb{N}, \ N(\delta) > C_M\} \tag{4.26}$$

The following theorem provides the main idea of the Koetter-Vardy decoding procedure.

THEOREM 4.36 *Let $f(X) \in F_{k-1}[X]$. If $\sum_{i=0}^{n-1} M(i, f(x_i)) > \Delta_M$, then $Y - f(X)$ can be factored out of $P_M(X, Y)$.*

DEFINITION 4.37 *For $f(X) \in F_{k-1}[X]$, we define the score of $f(X)$ relative to M as*

$$S_M(f) = \sum_{i=0}^{n-1} M(i, f(x_i)) \qquad (4.27)$$

The relationship stems from the theorem that, should there be factors of the form $Y - f(X)$ in $P_M(X, Y)$, they correspond to polynomials $f(X)$ in $F_{k-1}[X]$ (i. e. codewords) with the highest scores. Koetter and Vardy have translated the maximization of the score relative to Π into a maximization of the score relative to M by setting

$$M(i, y) = [m \times \Pi(i, y)] \qquad (4.28)$$

where m is the maximum permitted multiplicity. The most likely codewords c_f correspond to the factors $Y - f(X)$ in $P_M(X, Y)$ if there are any. Related to C_M, the interpolation costs are an increasing function of m. Therefore, the Koetter-Vardy algorithm can be summarized as [103]:

1. Initialization : choose the maximal multiplicity m.

2. Upon reception of r, compute $\Pi(i, y)$ for $y \in \mathcal{F}$, $0 \leq i \leq n - 1$. Let $M(i, y) = [m \times \Pi(i, y)]$.

3. Interpolation step : find $P_M(X, Y)$.

4. Partial factorization : find the factors of $P_M(X, Y)$ of the form $Y - f(X)$ where $\deg(f(X)) < k$.

5. Selection of the best codeword: assuming that s factors have been generated, re-encode the different $f_0, f_1, \ldots, f_{s-1}$, and select the factor $\hat{f} \in \{f_0, f_1, \ldots, f_{s-1}\}$ such that

$$\sum_{i=0}^{n-1} \Pi(i, \hat{f}(x_i)) = \max\left\{ \sum_{i=0}^{n-1} \Pi(i, f_j(x_i)), \ 0 \leq j \leq s - 1 \right\}. \qquad (4.29)$$

Letting $y \in \{0, 1\}$ when the multiplicity matrix is filled, the Koetter-Vardy algorithm can be used to decode BCH codes.

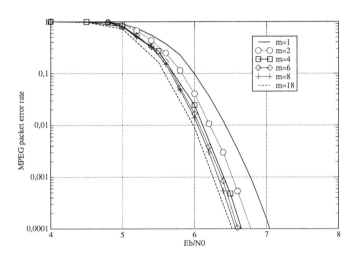

Figure 4.1. RS code for DVB-S : RS($n = 204$, $t = e = 8$, $K = 188$)

4.2.5.6 Performance evaluation.

For evaluation purposes, the implementation of the Koetter-Vardy algorithm can be avoided since, upon the reception of r, Π, M and Δ_M is derived; next, the score $S_M(c)$ of the emitted codeword with respect to the multiplicity matrix M is computed and, if $S_M(c) > \Delta_M$, then c belongs to the decoding list according to the fundamental theorem. Figure 4.1 shows for various values of m the achievable performance when the Koetter-Vardy algorithm is applied to the RS code specified by the Digital Video Broadcast - Satellite (DVB-S) standard. Note also that the case $m = 1$ corresponds to the bounded decoding of up to $\delta = \left\lfloor \frac{n-k-1}{2} \right\rfloor$ errors.

4.2.5.7 Frequency-domain decoding of BCH and RS codes.

Although the prime-factor Fourier transform (PFT) is less known than the radix-2 variant, most times simply addressed as fast Fourier transform (FFT), both techniques are well-established in digital signal processing [104]. With respect to PFT, the transform length N is decomposed into prime-factor powers, i. e., $N = \prod_{i=1}^{K} N_i^{l_i}$, $N_i \in \mathbb{P}$, where $\mathbb{P}$ denotes the set of prime numbers. The one-dimensional problem of length N is then mapped onto a K-dimensional space with the sample values in each dimension processed by a discrete Fourier transform (DFT). From this point of view, the most extreme case is the radix-2 FFT with $K = 1$, $N_i = 2$, and $l_i = \log_2 N$.

Nevertheless, Fourier transforms are not exclusively restricted to complex numbers, they apply also to finite fields $\mathcal{F}_q$. In this context, the DFT of a sequence of N samples $v_i \in \mathcal{F}_q$ is defined as $V_k = \sum_{i=0}^{N-1} v_i \alpha^{ik}$, which may elegantly be rewritten as

$$V_k = \mathbf{v}(\alpha^k); \quad k = 0, 1, 2, \ldots, N - 1 \qquad (4.30)$$

if $(v_0, v_1, v_2, \ldots, v_{N-1})$ is mapped to a polynomial $\mathbf{v}(X)$. According to the properties of finite fields [105, 106], the transform length N is determined by the order of the kernel $\alpha \in \mathcal{F}_q$, i. e., $\alpha^N \equiv 1 \mod p$, $\alpha \notin \{0, 1\}$, where N must be an integer divisor of $q - 1$; if $N = q - 1$, then α is the primitive root of $\mathcal{F}_q$. Hence, with radix-2 FFT's, it is required that $q = 2^m + 1 \in \mathbb{P}$ which holds only true for the so-called Fermat primes $p_\mu = 2^{2^\mu} + 1$, $\mu \in \{0, 1, 2, 3, 4\}$. Fermat transforms, however, suffer from two drawbacks which makes them less attractive in practice: as first, the number of symbols in $\mathcal{F}_{p_\mu}$ is not a power of two; then, integer operations have to be carried out, in contrast to bit operations used for binary-extended Galois fields $\mathcal{F}_{2^m}$.

As seen previously, efficient decoding algorithms have been devised for BCH and RS codes consisting, basically, of the following steps [87][89][90]:

1. Computation of syndromes
2. Solution of the key equation
3. Evaluation of error locations/values, error correction

Some of the steps may be omitted throughout the decoding process; binary codes, for instance, require no evaluation of the error values. In [107], it is shown that the computational complexity of step 1 (Horner's rule) and the identification of the error locations in step 3 (Chien's search) is of the order $\mathcal{O}(tn)$, while step 2 (Berlekamp-Massey or Euclidian algorithm) and error evaluation in step 3 (Forney's formula) need $\mathcal{O}(t^2)$ operations. The computation of syndromes and error locations is certainly the most time-consuming task if $t \ll n$, which is normally the case with BCH and RS codes used in practice.

Applying prime-factor or radix-2 algorithms, eq. (4.30) can be calculated most efficiently [108]. It is obvious that this is an issue for BCH and RS decoding, e. g., with respect to the evaluation of the *syndromes* $S_k = \mathbf{r}(\alpha^k)$, $k = e, e + 1, \ldots, 2t + e - 1$, where $\mathbf{r}(X)$ denotes the receiver polynomial of degree $n - 1$. But also Chien's search to identify the *error locations* calls for advanced DFT techniques, i. e., $\lambda_i = n^{-1} \Lambda(\alpha^{-i})$, $i = 0, 1, \ldots, n - 1$, where $\Lambda(X)$ symbolizes the error locator polynomial and n^{-1} is implicitly defined as $n \cdot n^{-1} \mod p \equiv 1$.

Assuming a finite field $\mathcal{F}_{2^m}$, a *regular* and *uniform* DFT algorithm for prime transform lengths $N \geq 3$ has been developed in [109] saving at least a quarter of the multiplications compared to the brute-force implementation (4.30). Figure 4.2 depicts the number of multiplications as a function of the error correction capability t, if this procedure is used for PFT evaluation of syndromes and error locations of an RS code with $n = 255$. For $n = 256$, the radix-2 FFT is shown as well, although the comparison is somewhat hypothetical since it requires integer operations over the Fermat prime field $\mathcal{F}_{257}$. Note that the PFT

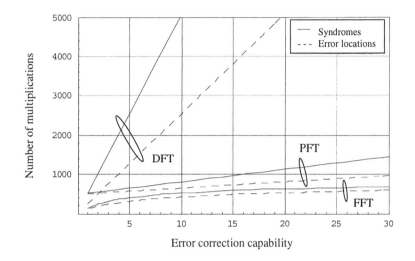

Figure 4.2. Computational complexity of RS decoders

complexity in Figure 4.2 develops rather flatly such that it does not diverge very much from the FFT variant, even if $t \gg 1$. The big slope of a DFT solution demonstrates the benefit of the PFT approach. Although not shown explicitly, the number of additions is reduced to the same extent.

4.3   Trellis and Turbo codes

Convolutional and Turbo codes are a class of high-performance error correcting codes finding wide-spread of use in deep-space satellite communications and other applications, where designers would like to achieve maximal information transfer over a bandwidth-efficient link in the presence of strong channel impairments. In this section, the basic concepts behind these code classes are presented, with particular emphasis on problems related to their decoding. Furthermore, the state-of-the-art codes used for satellite communications are described in more detail.

4.3.1   Trellis overview

Let $\mathcal{C}$ be a (n, k) binary block code with parity check matrix H of type $(n - k, n)$. In the sequel, we denote the j-th column of H by h_j.

$\mathcal{C}$ can be defined as a set of paths in a trellis of length $n + 1$ with 2^{n-k} states, each represented by an $(n - k)$ tuple of bits. The state s_k at position $0 \leq j < n$ is connected to two states at position $j + 1$: the state s_k with an edge labeled as

"0" and the state $s_k + h_j$ with an edge labeled as "1". A codeword is a sequence of labels attached to a path, which starts from state "0" and ends at state "0". Each section of the trellis may be different. Its description is tantamount to providing H.

A similar description can be given for binary convolutional codes. Since they are translation invariant, all the sections of the trellis are identical and the code is fully described by one section only.

Figure 4.3 depicts the trellis diagram of a 4-states non-recursive Feed-Forward (FF) convolutional encoder with rate 1/2 and generator polynomials $G(D) = (g_1(D), g_2(D)) = (1 + D + D^2, 1 + D^2)$. The *input* bits correspond to either "0" (shown in solid lines) or "1" (shown in dashed lines). The *output* bits are also shown in every state transition.

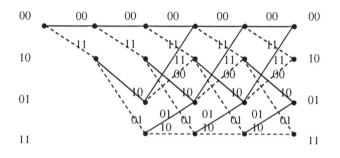

Figure 4.3. Trellis diagram of a convolutional encoder

A trellis diagram represents the state diagram of an encoder in time. It follows that each codeword is represented by a unique path through the trellis.

4.3.2 Non-iterative trellis-based decoding

Most of what follows may be applied to block codes but is widely used for convolutional codes.

4.3.2.1 Viterbi algorithm (VA). Assume a (n, k) convolutional encoder with memory order K and 2^K states. The encoder is assumed to be terminated, i.e., it starts from and returns to the zero state.

For an information sequence of length $K^* = kh$, with h being the length of the information bit sequence, there are 2^k branches leaving and entering each state and 2^{K^*} distinct paths through the trellis, corresponding to 2^{K^*} codewords. The codeword length is $N = n(h + m)$, where m is the number of terminating bits.

The decoding problem is based on finding an estimation $\hat{\mathbf{c}}$ of the codeword $\mathbf{c}$ by observing the received sequence $\mathbf{r}$. In addition, the ML decoder maximizes the log-likelihood function $\log P(\mathbf{r}|\mathbf{c})$. We define the *path metric* $M(\mathbf{r}|\mathbf{c})$, associated with the path (codeword) $\mathbf{c}$, as

$$M(\mathbf{r}|\mathbf{c}) = \log P(\mathbf{r}|\mathbf{c}) = \log \prod_{l=0}^{h+m-1} P(\mathbf{r}_l|\mathbf{c}_l) = \log \prod_{l=0}^{N-1} P(r_l|u_l) \quad (4.31)$$

or

$$M(\mathbf{r}|\mathbf{c}) = \sum_{l=0}^{h+m-1} \log P(\mathbf{r}_l|\mathbf{c}_l) = \sum_{l=0}^{N-1} \log P(r_l|u_l) \quad (4.32)$$

where $P(r_l|u_l)$ is the channel transition probability. The term $\log P(\mathbf{r}_l|\mathbf{c}_l)$ in (4.32) is called *branch metric*, whereas $\log P(r_l|u_l)$ in (4.32) is called *bit metric*. When the path metric is processed over the first t branches, then the *partial* path metric is obtained.

The objective of the Viterbi Algorithm (VA) [88] is to find the path through the trellis with the largest metric, by observing the received sequence $\mathbf{r}$. This is the *maximum-likelihood* path (codeword). The process of the VA is based on a recursive manner. The basic computation is the add-compare-select (ACS) operation. At each time unit, the VA *adds* 2^k branch metrics to each previously stored path metric, it *compares* the metrics of all 2^k paths entering each state and it *selects* the path with the largest metric (*survivor*). The survivor of each state is then stored along with its metric.

The process of the VA may be summarized in three steps:

1. At time unit $t = m$, compute the partial metric for the single path entering each state. Store the surviving path and its metric for each state.

2. Increase t by 1. Compute the partial metric for all 2^k paths entering a state by adding the branch metric entering that state to the metric of the connecting survivor at the previous time unit. For each state, compare the metrics of all 2^k paths entering that state, select the path with the largest metric (survivor), store it along with its metric and eliminate all other paths.

3. If $t < h + m$, repeat step 2; otherwise stop.

From time unit m to time unit h, there exist 2^K survivors, one for each state. After time unit h, there are fewer survivors and finally, at time unit $h + m$, one survivor remains.

THEOREM 4.38 *The final survivor $\hat{\mathbf{c}}$ in the VA is the maximum-likelihood path, characterized by $M(\mathbf{r}|\hat{\mathbf{c}}) \geq M(\mathbf{r}|\mathbf{c})$ for all $\mathbf{c} \neq \hat{\mathbf{c}}$.*

From the above theorem it is concluded that the VA is an optimum decoding algorithm. In practical implementations of the VA there are several factors that affect its performance. These are:

- Decoder memory: the decoder stores 2^K survivor paths and their corresponding metrics. That makes the storage requirements to increase exponentially with the constraint length of the encoder. Usually, the constraint length is limited to a small number, although there are some rare applications with a constraint length equal to 14. That yields the maximum coding gain with soft-decision decoding to be around 7 dB.

- Path memory: for each of the 2^K survivors, kh bits must be stored. This introduces difficulties, especially when the information sequence is long. The usual approach is to *truncate* the path memory of the decoder by storing only the most recent τ information bit blocks for each survivor, where $\tau \ll h$. If $\tau = 5K$, then the decoding approximates the ML performance.

- Complexity: the computational load is proportional to the 2^{k+K} branches per trellis section. High-speed decoding can be achieved by parallel processing. In this case, the throughput can be increased by a factor of 2^K, although the hardware increases to this extent too.

4.3.2.2 Soft-output Viterbi algorithm (SOVA). The SOVA is an extension to the standard VA that generates also reliability values on bits by observing the estimated codeword sequence. After the VA process is finished, SOVA stores on its memory two paths, the *survivor* and the *concurrent* path. The later one is the path which had diverged at a past time $j = k - \delta m$ and merged to the same state as the survivor path, i.e., at time $j = k$. The path metric difference $\Delta = M_1 - M_2$ between these two paths is stored and SOVA starts the process from the end of the trellis by tracing back.

The reliability value (LLR) of a bit is produced by an *updating rule* based on the estimated bits of the survivor path $\hat{u}_s$ and the concurrent path $\hat{u}_c$. All the LLR values of the survivor sequence are first initialized to $L_j(\hat{u}_k) = +\infty$ and then computed as

$$L_j(\hat{u}_k) = \min\left\{L_j(\hat{u}_k), \Delta\right\}, \ \hat{u}_{s,j} \neq \hat{u}_{c,j} \qquad (4.33)$$

only when the estimated bits of survivor and concurrent path are different from each other.

This algorithm was described in [110]. A *modified version* of it, which had been proposed earlier by Battail [111], updates the reliability values in case of $\hat{u}_{s,j} = \hat{u}_{c,j}$ as well by

$$L_j(\hat{u}_k) = \min\{L_j(\hat{u}_k), \Delta + L_c\}, \; \hat{u}_{s,j} = \hat{u}_{c,j} \qquad (4.34)$$

where L_c represents the reliability of the concurrent path. The extended updating rule makes this modified version of SOVA superior to the proposed algorithm from [110].

4.3.2.3 Maximum a posteriori (MAP) probability decoding.

The MAP algorithm, referred also as BCJR, is a well-known algorithm that estimates the transmitted sequence of bits. This is the main difference to the VA, which estimates the transmitted codeword sequence. For a single convolutional code, these two algorithms have more or less identical performance, although the former one is more complex. This has made MAP algorithm unattractive for practical systems. However, the MAP scheme appeared again after the introduction of Turbo codes.

Assume that $\mathbf{u}$ is an information block of N bits, encoded by a convolutional code, modulated via BPSK and transmitted over an AWGN channel. The main objective of MAP algorithm is to estimate the transmitted block of information bits by observing the received sequence r. To this end, a trellis transition from a state s', at time instant $k - 1$, to a state s, at time instant k, is to be considered. After initialization, the *forward* and *backward recursions* are computed recursively through the trellis, as [112]

$$\alpha_k(s) = \sum_{s'} \alpha_{k-1}(s')\gamma_k(s', s) \qquad (4.35)$$

$$\beta_{k-1}(s') = \sum_{s} \beta_k(s)\gamma_k(s', s) \qquad (4.36)$$

where γ_k is the *branch metric* associated with the corresponding trellis transition. The decoder soft output (LLR) value of the transmitted bit can be computed as

$$L(\hat{u}_k) = \log \frac{P(u_k = +1|\mathbf{r})}{P(u_k = -1|\mathbf{r})} = \log \frac{\displaystyle\sum_{(s',s):u_k=+1} \alpha_{k-1}(s')\gamma_k(s', s)\beta_k(s)}{\displaystyle\sum_{(s',s):u_k=-1} \alpha_{k-1}(s')\gamma_k(s', s)\beta_k(s)}$$

$$(4.37)$$

4.3.2.4 Log-MAP and Max-Log-MAP decoding.

If the MAP algorithm operates in the *logarithmic domain* [113], then the Log-MAP algorithm is obtained. It makes a hardware decoder implementation easier by using additions instead of multiplications and a look-up table (LUT) for non-linear functions. Note that the error performance does not degrade, when these modifications are taken into account. The basic operation is the *Jacobian logarithm* (or *max** operation), which is defined as [114]

$$\max{}^*(a, b) = \max(a, b) + \log\left\{1 + \exp(-|a - b|)\right\} \qquad (4.38)$$

Therefore, the forward and backward recursions can be computed as

$$\widetilde{\alpha}_k(s) = \log \sum_{s'} \exp\left\{\widetilde{\alpha}_{k-1}(s') + \widetilde{\gamma}_k(s', s)\right\} = \max{}^*_{s'}\left\{\widetilde{\alpha}_{k-1}(s') + \widetilde{\gamma}_k(s', s)\right\}$$

$$(4.39)$$

$$\widetilde{\beta}_{k-1}(s') = \log \sum_{s} \exp\left\{\widetilde{\beta}_k(s) + \widetilde{\gamma}_k(s', s)\right\} = \max{}^*_{s}\left\{\widetilde{\beta}_k(s) + \widetilde{\gamma}_k(s', s)\right\}$$

$$(4.40)$$

where tilde denotes values in the logarithmic domain. Following this approach, the LLR (4.37) becomes

$$
\begin{aligned}
L(\hat{u}_k) &= \log \frac{P(u_k = +1|\mathbf{r})}{P(u_k = -1|\mathbf{r})} = \\
&= \log \frac{\displaystyle\sum_{(s',s):u_k=+1} \exp\left\{\widetilde{\alpha}_{k-1}(s') + \widetilde{\gamma}_k(s', s) + \widetilde{\beta}_k(s)\right\}}{\displaystyle\sum_{(s',s):u_k=-1} \exp\left\{\widetilde{\alpha}_{k-1}(s') + \widetilde{\gamma}_k(s', s) + \widetilde{\beta}_k(s)\right\}} = \\
&= \max{}^*_{(s',s):u_k=+1}\left\{\widetilde{\alpha}_{k-1}(s') + \widetilde{\gamma}_k(s', s) + \widetilde{\beta}_k(s)\right\} \\
&\quad - \max{}^*_{(s',s):u_k=-1}\left\{\widetilde{\alpha}_{k-1}(s') + \widetilde{\gamma}_k(s', s) + \widetilde{\beta}_k(s)\right\} \qquad (4.41)
\end{aligned}
$$

Omitting the LUT for $\log\{\cdot\}$ given in eq. (4.38), the Log-MAP algorithm converges to the Max-Log-MAP algorithm. That means the *max** operator in eq. (4.39), (4.40) and (4.41) is replaced by the *max* operator.

Note that the Max-Log-MAP algorithm can be regarded as a *dual-VA* by updating the LLR output after having processed the trellis both in the forward and the backward direction [114]. It is thus equivalent to the modified SOVA proposed by Battail [115].

From the implementation point of view, the Log-MAP algorithm is approximately three times more complex than SOVA, which is twice as complex as the standard VA [92].

4.3.2.5   Alternative decoding techniques.   The primary difficulty with Viterbi decoding is the fact that the memory order is limited to fairly small values. Furthermore, the algorithm requires a fixed amount of computations per trellis section, no matter whether the receiver samples are heavily distorted by noise or not.

Shortly after convolutional codes were introduced as an alternative to block codes, the sequential decoding technique was proposed. Threshold decoding as a further alternative was suggested a bit later, as a less powerful but computationally simple variant. Both algorithms lost their practical relevance with the advent of Viterbi decoding, mainly after Forney pointed out that it is in fact a maximum-likelihood method, which can be most efficiently implemented using a trellis [116] .

In the following, sequential as well as threshold decoding are presented in a nutshell for reasons of completeness.

4.3.2.5.1   Sequential decoding.   Two main solutions [87], [88] have been devised in the course of time. Both are search strategies applied to the code tree, which contains the same information like trellises or state diagrams:

1. Fano algorithm
2. Stack (Zigangirov-Jelinek) algorithm

In the Fano algorithm, the decoder examines a sequence of nodes in the code tree moving forward and backward to adjacent nodes. The decoder moves forward as long as the path metric continues to increase. When the latter dips below a given threshold, the decoder backs up and begins to examine other paths. If no path is found with a metric above the threshold, it is lowered and the decoder starts moving forward again.

In the stack algorithm, an ordered list of previously examined paths with different lengths is kept in a storage. Each entry contains a path along with its metric, the path with the largest metric placed on top and the others listed in order of the decreasing path metric. For each decoding step, the top path is extended by computing the metrics of the leaving branches, adding these to the path metric on top and rearranging the resulting path metrics in decreasing order.

It is to be observed that sequential decoding of convolutional codes requires a different metric compared to Viterbi decoding. The reason is that the paths being compared at any decoding step are all of different length, which has to be taken into account for an optimum metric design. Also, since powerful decoding calls for a rapid initial growth of the column distance, codes with optimum distance profiles are recommended.

4.3.2.5.2   Threshold decoding.   Threshold (majority-logic) algorithms differ from Viterbi or sequential devices insofar as the final decision covers only one constraint length, rather than on a multiple thereof or even the entire receiver sequence. This results in an inferior error performance, but the implementation is much simpler. With threshold decoding, any non-systematic code can easily be transformed in an equivalent systematic code, though only the latter one has been used in practice. The decoder principle follows the concept of orthogonal parity-check sums.

4.3.3 Turbo codes

Turbo codes are a new class of iterated decoding convolutional codes with small memory order, which closely approach the theoretical limits imposed by Shannon's theorem (see section 2.3.2). They have much less computational complexity than the Viterbi algorithm for decoding convolutional codes with large constraint lengths, which would be required for the same error performance. This sort of codes, being introduced by Berrou, Glavieux and Thitimajshima [117], is also characterized by an increased bandwidth and power efficiency, when compared to the classical (non-iterative) FEC solutions known before from the open literature.

4.3.3.1   The Turbo Principle.   Classical Turbo codes make use of parallel or serial concatenated RSC (see Section 4.2.1.3) component encoders separated by interleavers.

The Turbo code *interleaver* permutes the information bit sequence and the output is passed to the second RSC encoder. The role of the interleaver is to generate a long block code from convolutional encoders with small memory orders [92]. Also, it decorrelates the Turbo decoder inputs (extrinsic information and channel values) by spreading out the burst errors. Finally, it breaks the low-weight input sequences and, hence, increases the free distance of the code. The error performance is improved by increasing the interleaver size. For RSC codes, an increase of the interleaver by N reduces the bit error probability by a factor of N. This is called *interleaving performance gain*.

The presence of the feedback in RSC encoders makes the *trellis termination* more complicated than for non-recursive encoders. Normally, this can be done by allowing a tail bit sequence to terminate the encoder to the all-zero state. The number of tail bits needed for each component RSC encoder is equal to the memory order. *Circular trellis* or *tail biting* is another technique to cope with the trellis termination problem (see Section 4.3.3.3).

DEFINITION 4.39 *A Parallel Concatenated Convolutional Code (PCCC) consists of two RSC encoder components that are linked together in parallel through an interleaver (see Figure 4.4)*

In Figure 4.4 the basic block diagram of a PCCC Turbo encoder is shown. u_s is the *systematic* (information bit sequence) output of the first RSC encoder and $u_{p,1}/u_{p,2}$ is the *parity* (coded bit sequence) output of the first/second RSC encoder, respectively. The systematic output of the second RSC component is not transmitted.

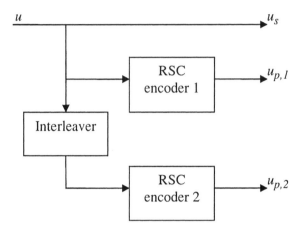

Figure 4.4. PCCC encoding scheme

DEFINITION 4.40 *A Serial Concatenated Convolutional Code (SCCC) consists of two RSC serially linked through an interleaver. For this reason, both the information bit sequence and the parity bit sequence of the first RSC component output (outer encoder) are interleaved before entering the second RSC component (inner encoder).*

If the outer and inner RSC components have a coding rate of $r_o = k/p$ and $r_i = p/n$, respectively, then the overall concatenated code has a coding rate of $r = r_o \cdot r_i = k/n$.

In general, the most used Turbo codes make use of binary RSC encoders. In addition to them, *non-binary* Turbo codes can be built from RSC component encoders with $m \geq 2$ inputs. Non-binary Turbo codes, in PCCC scheme, are supposed to perform better than binary Turbo codes, in particular at very low SNRs and higher coding rates [118].

PCCC and SCCC represent a class of Turbo coding schemes based on convolutional codes. Apart from such solutions, other forms of concatenation are amenable to iterative decoding.

For example, the Hybrid Concatenated Convolutional Code (HCCC) combines both PCCC and SCCC structures, while a Block Turbo code (BTC), frequently also called Turbo product code (TPC) [119], is a serial concatenation of two linear block codes $\mathcal{C}_1$ and $\mathcal{C}_2$, separated by a row-column permutation.

In the latter case, given two block codes $\mathcal{C}_1$ and $\mathcal{C}_2$ with parameters (n_1, k_1) and (n_2, k_2), respectively, the product code $\mathcal{C}$ has information block length $k = k_1 \times k_2$ and codeword length $n = n_1 \times n_2$. The encoder organizes the

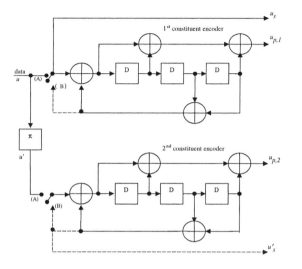

Figure 4.5.   3GPP Turbo encoder for UMTS and S-UMTS

information block in a $k_1 \times k_2$ two-dimensional array. Each row of the array is encoded with C_2 appending the $n_2 - k_2$ parity bits. Each column (including those composed by the parity bits obtained in the previous step) is then encoded with C_1, and the so-computed parity bits are appended to the related column. Coding rate and minimum distance of C are the product of the corresponding parameters of C_1 and C_2.

4.3.3.2   The 3GPP example.   The 3GPP standard for Universal Mobile Telecommunications System (UMTS) and Satellite UMTS (S-UMTS) recommends a Turbo encoder with two 8-states RSC components and coding rate of $1/3$, as illustrated in Figure 4.5. A rate of $1/2$ is also available by appropriate puncturing of the parity bits. The related *transfer function* is given by

$$G(D) = \left\{ 1, \frac{g_1(D)}{g_0(D)} \right\} = \left\{ 1, \frac{1 + D + D^3}{1 + D^2 + D^3} \right\} \qquad (4.42)$$

where $g_1(D)$ denotes the *feedforward polynomial* and $g_0(D)$ the *feedback polynomial*, both in octal form of the constituent RSC components.

4.3.3.3   The DVB-RCS example.   The Digital Video Broadcast - Return Channel via Satellite (DVB-RCS) standard [120], adopted by the European Telecommunications Standards Institute (ETSI), recommends for the return channel an 8-states duo-binary Turbo code, as illustrated in Figure 4.6.

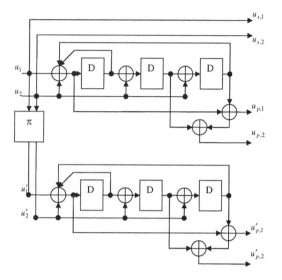

Figure 4.6. DVB-RCS Turbo Encoder

The need of puncturing is less crucial in case of duo-binary Turbo codes. This is because the RSC encoder components of rate 2/3 allow less redundant symbols to be discarded, in contrast to classical Turbo codes with RSC encoder components of rate 1/2.

Regarding the trellis termination of this code, it is considered that in satellite broadcasting applications the small Asynchronous Transfer Mode (ATM) packets (ranging from 12 to 216 bytes) require, apart from a very low frame error rate (around 10^{-7} and less), a trellis termination of the RSC encoder components without the need of tail bits, if possible, so as not to waste extra bandwidth on flushing bits. DVB-RCS has solved this difficulty via a technique called *circular trellis* or *tail biting* [121], i.e., the encoder retrieves the initial state at the end of the encoding process, such that the decoder can be initialized to any state and finish to this state in a circular way.

The other main characteristic of this code is that DVB-RCS has proposed to realize the internal interleaving in two steps. The first one performs *intra-symbol* permutations, which increases the minimum distance of the code. The second interleaving is as for classical binary Turbo codes and performs *inter-symbol* permutations, so as to reduce the correlation of symbols during the iterative decoding process.

The major disadvantage is that the 8-states duo-binary Turbo decoder is around 30% more complex than the 16-states binary variant. However, if the comparison is per bit, then the duo-binary Turbo decoder is simpler by 35%. We can summarize that duo-binary Turbo codes are *more attractive* than classical binary Turbo schemes, allowing also parallel architectures for high data throughput.

4.3.4   Turbo decoding

The principle of Turbo decoding relies on iterative process, which is found as solution to complex problems in many scientific domains (electronics, electromagnetism and mechanics). Concepts related to this type of process were not often used in digital communications, up to the advent of Turbo codes. The latter make use of the diversity effect of the extrinsic information, which ensures particular performance gains, if it remains uncorrelated with the channel noise.

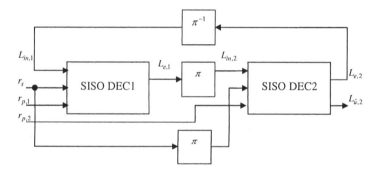

Figure 4.7.   Turbo decoder

4.3.4.1   The Turbo decoding principle.   As just mentioned, the Turbo decoding principle is based on the *iterative* process between two soft-input soft-output (SISO) decoders. In Figure 4.7, the input of a SISO decoder is fed by the *a priori* information of bits (L_{in}) and the received channel values corresponding to systematic and parity bits ($L_c r$). It produces a soft output ($L_{\hat{u}}$) value. The term $L_c = 2/\sigma^2$ is called *channel reliability* value depending on the noise variance of the channel denoted as σ^2. The *extrinsic* information (L_e) of information bits is calculated as

$$L_e = L_{\hat{u}} - (L_{in} + L_c r) \qquad (4.43)$$

with the same sign as the transmitted information bit sequence. $L_{e,1}$ is then used as *a priori* information for the subsequent SISO module. The extrinsic information of the latter is deinterleaved and looped back to the first SISO component, which finishes one round of the iterative process. After a certain number of iterations, hard decisions are taken from the second decoder output.

Applying a stopping criterion, the number of iterations can be reduced. By using semi-analytical methods (e.g. EXtrinsic Information Transfer (EXIT) charts, described in Section 4.3.4.3), the convergence behavior of the Turbo decoder can be predicted.

Nowadays, the Turbo principle, based on the iterative decoding process, rather than the parallel concatenation of two convolutional encoders, is a *de facto* algorithm that can be applied to many areas of digital communications systems [121], [122].

The MAP algorithm and its simplifications (Log-MAP and Max-Log-MAP) can be applied to Turbo decoding in an iterative process. Another algorithm is the iterative SOVA. The modification needed in both cases is based on the fact that the *a priori* information should be taken into account in branch metrics calculation. For non-binary Turbo codes the iterative decoding algorithms have to be extended appropriately, so as to handle symbols instead of bits. That introduces the symbol-by-symbol iterative MAP or SOVA decoder.

The iterative SOVA is *sub-optimum* in terms of BER performance. In particular, the degradation against the MAP Turbo decoder is approximately 0.7 dB at BER of 10^{-4}, if an AWGN channel is assumed [113]. In case of the iterative Max-Log-MAP, the performance degradation against the MAP Turbo decoder is approximately 0.4 dB at the same BER value [113]. Such performance comparison can be found in Section 4.3.5.3.

4.3.4.2   Stopping criteria.   Usually, the Turbo decoding process assumes a *fixed* number of iterations per frame. Using a *stopping rule*, the average number of decoding iterations can be reduced, providing an increase in the average decoding speed without significant degradation to the code performance.

In Turbo decoding there are four basic stopping criteria, which are summarized below [123]:

1. Cyclic Redundancy Check (CRC)

   A certain number of CRC bits are added to the Turbo encoding frame before transmitting to the channel. These extra bits are used at the decoder side to check for errors. The Turbo decoder makes hard decisions during the iterations.

2. Cross Entropy (CE)

   At each iteration i, the cross entropy $T(i)$ between the LLRs of the component decoders is measured. Iterations are discontinued if $T(i) < (0.0001 \sim 0.01)T(1)$.

3. Sign Change Ratio (SCR)

This method computes the number of sign changes $C(i)$ of the extrinsic information between iterations $i-1$ and i. Decoding is terminated when $C(i) < (0.005 \sim 0.03)N$, where N is the frame size.

4. Sign Difference Ratio (SDR)

This is similar to the SCR method, but the sign change $C(i)$ is replaced by the sign difference $D(j,i)$ between the *a priori* and the extrinsic information of each of the $j = 2$ component decoders. Decoding is terminated if $D(j,i) < (0.001 \sim 0.01)N$. The parameter $D'(j,i)$, which is the sign difference between the *a priori* plus the channel values and the extrinsic information, may also be employed.

The *CRC* method provides the smallest average number of decoding iterations but has the drawback of transmitting extra bits. The rest of the three methods behaves almost the same. However, the *CE* method requires more computations than the *SCR* and *SDR* methods. With respect to the latter, the *SDR* method is simpler than the *SCR* solution.

4.3.4.3   EXIT charts.   EXIT charts are a useful tool to analyze the *convergence behavior* of Turbo codes [124]. This is achieved by observing the *mutual information*, exchanged between the two component decoders, throughout the iterative decoding process. In order to do this, two assumptions are made. First, the *a priori* information is not correlated to the channel observations, which is approximately true for large interleaver sizes. Second, the extrinsic information follows a Gaussian-like distribution, which is also true as soon as the number of iterations increases.

Assuming a specific SNR value, the so-called EXIT chart shows for each component decoder the evolution of the *a priori* mutual information I_a against the *extrinsic* mutual information I_e. For identical components, the two curves are mirror-like images.

The *a priori* information at the component decoder input is modeled as $L_a = \mu_a s + n_a$, where s is the transmitted symbol value and n_a is a Gaussian random variable with zero mean and variance σ_a^2. After the decoding process, the extrinsic information of the decoder output L_e can be easily extracted. The *a priori* mutual information is defined as

$$I_a = \frac{1}{2} \sum_{s=-1,+1} \int_{-\infty}^{+\infty} p(L_a|S = s) \log_2 \frac{2p(L_a|S = s)}{p(L_a|S = -1) + p(L_a|S = +1)} dL_a$$

$$(4.44)$$

Similarly, the extrinsic mutual information is given by

$$I_e = \frac{1}{2} \sum_{s=-1,+1} \int_{-\infty}^{+\infty} p(L_e|S = s) \log_2 \frac{2p(L_e|S = s)}{p(L_e|S = -1) + p(L_e|S = +1)} dL_e$$

(4.45)

where the probability density functions of L_a and L_e are obtained from the corresponding histograms. Different pairs of the transfer function curve (I_a, I_e) are then related to different values of σ_a^2.

After the derivation of the transfer function curve, the *trajectory* of the mutual information can be observed during the iterative decoding process. This is done in a same way to the method described above, however, the *a priori* information of one component decoder input is replaced by the *extrinsic* information generated by the other component decoder in the previous decoding step.

4.3.4.4   SOVA improvements.   The standard SOVA output is too "optimistic" since only two trellis paths are considered – in contrast to a MAP device which runs through all the trellis paths. An attempt to improve the SOVA decoder is to *normalize* the extrinsic information passed from one component decoder to the other (see Figure 4.8). Two main algorithms have been proposed

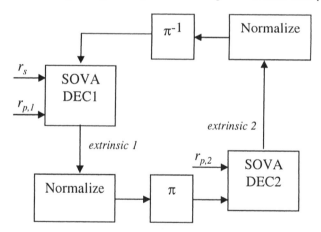

Figure 4.8.   Normalized SOVA Turbo decoder

in the past to improve the SOVA performance. In the first approach [125], the decoder soft output is normalized, based on the Gaussian assumption distribution. This normalization is done by multiplying with a correcting factor that depends on the decoder output variance. In the second approach [111], the decoder soft output values are limited to a smaller range, achieving an overall performance improvement.

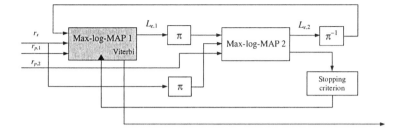

Figure 4.9. Hybrid Turbo-Viterbi decoder based on Max-Log-MAP's

Another method to improve the SOVA Turbo decoder was reported in [126]. It is based on an observation from [111], where both Hagenauer's and Battail's SOVA decodings are compared to each other. In more detail, the iterative SOVA decoder output can be improved, if the extrinsic information is increased in the last decoding iteration. It is thus a two step approach, as the extrinsic information is multiplied by a constant number, while in the last decoding iteration, it is multiplied by a greater number. Results obtained using this improvement are reported in Section 4.3.5.3.

4.3.4.5   Hybrid Turbo-Viterbi decoding.   Using Turbo decoders based on Max-Log-MAP components, the *error floor* may become a serious problem. The latter can be lowered most effectively by increasing the encoder memory and/or the block length. However, the memory length is limited by the available processing power and the block length can not be chosen arbitrarily depending on the application or service behind.

It is well known that the error floor is closely related to the number of low-weighted codewords, which may be reduced by an appropriate interleaver design; the S-random principle [92] is frequently recommended as a simple solution in this respect. In [127], the error floor problem is successfully tackled by concatenation with BCH outer codes, i. e., the error locations of the latter are identified and fed back to the Turbo decoder starting a second cycle of iterations.

A totally different solution [128] suggests the trellis of one of the Max-Log-MAP components to be used for Viterbi decoding after the iterating procedure has stopped, e. g., Max-Log-MAP1 as visualized in Figure 4.9. As soon as the stopping criterion terminates the Turbo process, the Viterbi algorithm is fed with the soft decisions $L_{e,2}$ of the Max-Log-MAP2 output.

A detailed analysis confirms that the error performance is improved by the fact that the Viterbi module follows a maximum-likelihood principle minimizing the sequence error probability, not the bit error probability as do MAP devices. Therefore, error events at the output of the Max-Log-MAP2 decoder

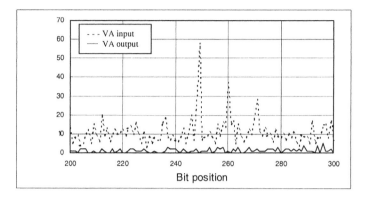

Figure 4.10.   Accumulated number of bit errors at the Viterbi input/output

can be corrected if they do not exceed the error correction capability of the
Viterbi algorithm. For a specific setup, Figure 4.10 exemplifies the accumulated number of bit errors at the VA input (Max-Log-MAP2 output) and the
VA output, respectively, for bit positions $l = 200 \ldots 300$. It can be seen that
the latter has definitely less errors at any position. Note also that the error
peaks at $l \in \{249, 260, 271\}$, probably due to imperfections of the S-random
interleaver, are significantly reduced.

Note that the Turbo-Viterbi hybrid must not be confused with any form of
concatenation such that no extra bandwidth is required. Furthermore, no additional resources are necessary insofar as both Turbo and Viterbi algorithm
use the same trellis. Finally, the additional complexity introduced by VA postprocessing is negligible since it is only applied once per packet and the VA
complexity is just a quarter of that consumed by a Max-Log-MAP device [92].

Results obtained using this improvement are also reported in Section 4.3.5.3.

4.3.5   Turbo code performance

In this section, an analytical characterization of the Turbo code performance
is sketched first and then computer simulation results are given. A brief complexity evaluation of all presented decoding algorithms is also provided.

4.3.5.1   Bounds and distance spectra for Turbo codes.   Since it has
not been possible to obtain analytical bounds on both packet and bit error rates
for Turbo codes with specified interleavers, the *uniform interleaver* concept has
been introduced in [129]. This approach allows an easy derivation of the weight
distribution of parallel and serial concatenated codes from the weight distributions of their component codes. The transfer function has been calculated in

order to obtain union bounds on the packet and bit error probabilities averaged over certain ensembles with random coding properties [130]. These bounding techniques are based on the union upper bound, which is widely used in the literature (see Section 2.4.1.3). The latter is an effective method of bounding a code performance provided its weight distribution is known. Since the weight distribution is calculated over the uniform interleaver that averages the weight function over the all possible interleaving schemes, then there must be at least one interleaver that performs better than the average. In [130], binary (k, n) Turbo codes with rate $r = k/n$ and minimum code distance d_{min} have been considered. Assuming coherent detection, ML decoding and channel state information (CSI), the conditional upper bound of the bit error probability can be written as:

$$P_b(E|\rho) \leq \sum_{i=1}^{k} \sum_{w=d_{min}}^{n} \frac{i}{k} A_{i,w}^{\mathcal{C}} Q \left(\sqrt{2r E_b/N_0 \sum_{j=1}^{w} \rho_j^2} \right) \qquad (4.46)$$

where E_b/N_0 is the signal-to-noise ratio per bit, $A_{i,w}^{\mathcal{C}}$ represents the number of codewords with output weight w associated to input weight i, the so-called Input-Output Weight Coefficient (IOWC), and ρ_j denotes the multiplicative fading coefficient. It shall be noted that this bound can be applied to codes on both AWGN and flat fading channels. The only unknown parameter in eq. (4.46) is the IOWC for Turbo codes. The unique method to tractably compute the IOWC is the use of the uniform interleaver concept. The latter is defined as a probabilistic device of length n which maps a given input word of weight i to all its distinct permutations $\binom{n}{i}$ with equal probability. Assuming such an interleaver as well as $A_{i,w_1}^{\mathcal{C}_1}$ and $A_{i,w_2}^{\mathcal{C}_2}$ as the IOWC's of the component codes, the IOWC for Turbo codes can be simply calculated as

$$A_{i,w}^{\mathcal{C}} = \sum_{w_1, w_2 : w_1 + w_2 = w} \frac{A_{i,w_1}^{\mathcal{C}_1} A_{i,w_2}^{\mathcal{C}_2}}{\binom{n}{i}} \qquad (4.47)$$

Hence, the error probability of Turbo codes is obtained from eq. (4.46) by taking into account the fading model according to

$$
\begin{aligned}
P_b(E) &= \int_{\rho} p(\rho) P_b(E|\rho) \mathrm{d}\rho \\
&\leq \sum_{w=d_{min}}^{n} B_w^{\mathcal{C}} \int_{\rho} p(\rho) Q \left(\sqrt{2r E_b/N_0 \sum_{j=1}^{w} \rho_j^2} \right) \mathrm{d}\rho
\end{aligned} \qquad (4.48)
$$

where $B_w^{\mathcal{C}} \triangleq \sum_{i=1}^{k} \frac{i}{k} A_{i,w}^{\mathcal{C}}$.

For non-binary Turbo codes, such as the coding scheme standardized for DVB-RCS, it was proved that for any interleaver the minimum distance grows at most logarithmically with the interleaver length [131], which implies that Turbo codes with two components cannot achieve the Gilbert-Varsharmov condition on the minimum distance. Moreover, as demonstrated in [132] in a portion of the low Hamming weights, the distance spectra of the ensembles of uniformly interleaved concatenated Turbo codes deviate notably from the binomial distribution, which characterizes the distance spectrum of the ensemble of fully random block codes.

4.3.5.2   Implementation complexity.   After many years of theoretical investigations aimed at understanding and explaining their impressive performance, Turbo codes have finally entered the field of practical applications. In this respect, they have been chosen by the Consultative Committee Space Data System (CCSDS) for the telemetry coding standard, which performance is depicted in section 4.3.5.3, and they are used for medium-high data rate transmissions both in the UMTS third generation mobile communications standards, as shown in section 4.3.3.2, and in the DVB-RCS standard, as shown in section 4.3.3.3.

When dealing with the implementation of a given algorithm, both the arithmetic and the memory complexity are the most important problems. In particular, one crucial design issue is the number of bits required to represent the quantities involved in the decoding algorithm. Therefore, the decoding complexity of PCCC or a SCCC can be summarized into two main parts: the memory complexity, which refers to the storage requirements of the algorithm and the arithmetic complexity, which only depends on the kind and number of operations that are performed inside the SISO modules.

For Turbo encoders with memory order M, the decoding complexity is estimated in [113], where it is assumed that one bit comparison costs the same as one addition. In Table 4.1 the SISO complexity has been examined, considering the most frequently used algorithms. In particular, the differences between the Max-Log-MAP, Log-MAP and SOVA have been highligthed. From the table, it can be concluded that the Max-Log-MAP Turbo decoder is around twice as complex as the SOVA Turbo decoder, while it requires around half the complexity compared to a Log-MAP Turbo decoder.

With respect to the arithmetic complexity, the implementation of an iterative decoder has to take into account problems related to the finite precision arithmetic of a digital implementation. In particular, these problems can be grouped in three main families; first of all, the integer representation of the LLRs, which are inherently real numbers; second, the internal precision of the

Table 4.1. Turbo decoding complexity

	max operations	additions	mult. by ± 1	compar.	look-ups
Max-Log-MAP	$5 \times 2^M - 2$	$10 \times 2^M + 11$	8		
Log-MAP	$5 \times 2^M - 2$	$15 \times 2^M + 9$	8		$5 \times 2^M - 2$
SOVA	$3(M + 1) + 2^M$	$2 \times 2^M + 8$	8	$6(M + 1)$	

decoder arithmetic operations; and third, the values of the quantities involved in the SISO operations that grow as long as the more iterations are performed.

These arithmetic problems can be solved finding the best design of the quantizer that operates on the LLRs. Being the number of bit used to represent the decimal part of the LLR the crucial parameter for the *max* operation performed by each SISO, it is really important to find the best trade-off among performance and quantization bits. This is reached considering 4 or 5 bits for the fixed-point LLR representation, where 1 or 2 bits are devoted to represent the decimal part of the LLR.

4.3.5.3   Simulation results.   A typical error rate curve for an iterative code is depicted in Figure 4.11. The error rate is plotted in logarithmic scale, while E_b/N_0 is expressed in dB. Two main regions can be identified on the chart: the so-called "waterfall region" and the "error floo". The former is usually referred to as the region of the error rate plot where the decoder begins to work effectively, characterized by a steep slope. In many cases, the slope of the curve changes significantly at higher signal-to-noise ratios, originating a phenomenon that is usually referred to as "error floor", i.e., the error rate begins to flatten as soon as the SNR increases. Thus, low error rates can be obtained only at the expense of very high signal-to-noise ratios. The error floor can be originated by different causes, usually dealing with the minimum distance [133] of the code or with particular error patterns which somehow "trap" the iterative decoder [134]. It should be noted that there exists a kind of trade-off between the performance in the waterfall and error floor domain: codes designed to exhibit their error correction capabilities at low SNR's usually show relatively high error floors, while codes achieving very low error rates without any sign of slope flattening are usually characterized by smaller coding gains in the waterfall region.

Figure 4.12 illustrates the error performance of two improved SOVA decoding schemes, referred to as *norm1* SOVA and *norm2* SOVA respectively. In the first scheme, the extrinsic information is multiplied by a constant number in every decoding iteration. This was proposed in [135] to improve the performance of the Max-Log-MAP Turbo decoder. The second scheme is based on the two step approach described in Section 4.3.4.4. For compari-

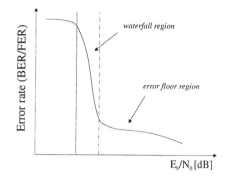

Figure 4.11.   Error rate curve of a typical iterative error correcting code

son purposes, the standard SOVA, Log-MAP and Max-Log-MAP results are also shown. The simulation parameters are 8-states 3GPP Turbo code with BPSK modulation/AWGN channel (1000 bits interleaver size, rate 1/3 and 8 iterations).

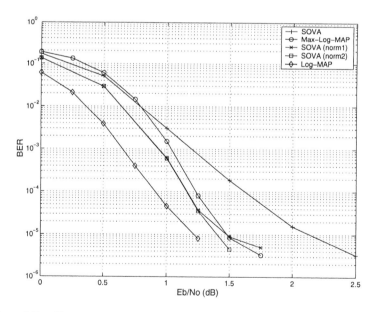

Figure 4.12.   SOVA, Max-Log-MAP, improved SOVA (*norm1/norm2*) and Log-MAP algorithms comparison for Turbo decoding (1000 bits interleaver size, rate 1/3, AWGN channel)

The first SOVA scheme (*norm1*) improves the performance with respect to the Max-Log-MAP Turbo decoder at high to medium BER values. In addition, the second scheme (*norm2*) improves the performance at low BER values. The performance degradation of both SOVA schemes against the Log-MAP Turbo decoder is 0.3 dB at BER of 10^{-4}. They both add very small extra decoding complexity since the extrinsic information is multiplied by a constant correcting factor. However, the best correcting factors are found by trial and error.

For an 8-state Turbo-Viterbi hybrid with S-random interleaving (TV-S), Figure 4.13 depicts the packet error rate (PER) as a function of E_b/N_0 (2040 bits interleaver size, rate $1/2$, AWGN channel). The iterative procedure stops as soon as no change of the bit signs between two Turbo cycles is detected (see Section 4.3.4.2 and Figure 4.9). For comparison purposes, the performance of a Turbo code with S-random (TC-S) and uniform (TC-U) interleaving is illustrated to show the improvement with respect to the error floor. Also, the error rate of TC-S concatenated with an outer $(255, 223)$ BCH code is shown (BCH+TC-S), which exhibits a loss of about 0.6 dB due to additional redundancy. For reasons of completeness, the classical concatenation of a $(255, 223)$ RS code with a rate $1/2$, constraint length 7 convolutional code with Viterbi decoding (RS+VA) is included.

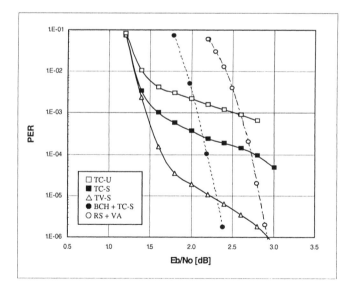

Figure 4.13. PER performance of an 8-state Turbo-Viterbi hybrid compared to different coding schemes (2040 bits interleaver size, rate $1/2$, AWGN channel, SCR stopping criterion)

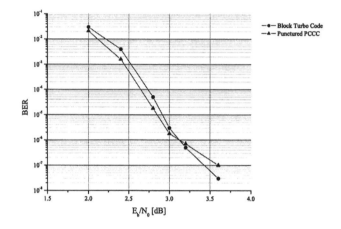

Figure 4.14. Bit error rates of a Block Turbo code and a PCCC (information block size $k \approx$ 1784, rate $3/4$, over the AWGN channel)

In Figures 4.14 and 4.15, bit and packet error rates are compared for a Block Turbo code and a PCCC [136], assuming an information block length of $k \approx 1784$ bits and code rate equal to $3/4$, considering transmission over the AWGN channel. The PCCC has been designed by puncturing the Turbo code with information block length $k \approx 1784$, as recommended by the Consultative Committee for Space Data Systems [100]. The Block Turbo code is a product code, where the two component codes are a shortened extended $(45, 38)$ Hamming code and a shortened extended $(54, 47)$ Hamming code, respectively. The product code has been decoded using the algorithm described in [137]. For both codes, the maximum number of iterations has been set to 15. It is evident that the PCCC outperforms the BTC by almost 0.1 dB in the waterfall region, whereas at higher SNR's the BTC is characterized by a lower error floor.

4.4   Graphs and LDPC codes

Robert Gallager introduced LDPC codes in the early sixties [41], but since technology was not ready for practical implementation, they were forgotten for many years up to the nineties, when they were rediscovered by MacKay [138]. The best way to study these codes is to use graph models and in particular the special class of factor graphs, called bipartite graphs.

In this section the basic notations on factor graphs will be provided in order to discuss how bipartite graphs can be used to design LDPC codes whose

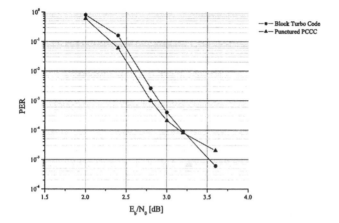

Figure 4.15. Packet error rates of a Block Turbo code and a PCCC (information block size $k \approx 1784$, rate $3/4$, over the AWGN channel)

performance approaches the Shannon limit. Several design techniques will be described as well as their decoding algorithms. For the performance evaluation of these codes, numerical simulations will be shown, trying also to do a fair comparison with Turbo code results.

4.4.1 Factor Graphs

DEFINITION 4.41 *A graph* $\mathcal{G} = (\mathcal{V}, \mathcal{E})$ *[139] is a pair of two finite sets: a set* $\mathcal{V} = \{v_1, v_2, \cdots \}$ *of points called vertices or nodes and a set* $\mathcal{E} = \{e_1, e_2, \ldots\}$ *of lines called edges, such that each edge connects two nodes. The number of nodes in* $\mathcal{G}$ *is* $|\mathcal{V}|$ *and the number edges in* $\mathcal{G}$ *is* $|\mathcal{E}|$.

Edges are denoted by their endpoints so that $e_k = (v_j, v_k)$, where $e_k \in \mathcal{E}$ and $v_j, v_k \in \mathcal{V}$. An edge e_k is called *incident* with a node v_i if e_k is connected to v_i. The number of edges that are incident to a node is called the (local) degree of the node. A k-regular graph is a graph in which all nodes have degree equal to k. A node that is not connected to any edge is called an isolated node.

DEFINITION 4.42 *A path in a graph is said* closed *if it ends at the node that it started from. A closed path is a* cycle.

The girth $g(\mathcal{G})$ is the length of the shortest cycle (if any) in a graph. The graph circumference is the length of the longest cycle in a graph. The girth g_v at node v is the length of the shortest cycle passing through v.

DEFINITION 4.43 *A bipartite graph [139] is a graph $\mathcal{G} = (\mathcal{V}, \mathcal{E})$ in which the node set $\mathcal{V}$ is partitioned into two sets $\mathcal{V}_c = \{c_1, c_2, \ldots, c_m\}$ and $\mathcal{V}_v = \{v_1, v_2, \ldots, v_n\}$ such that every edge in $\mathcal{G}$ has one endpoint in $\mathcal{V}_c$ and the other endpoint in $\mathcal{V}_v$. A bipartite graph can be referred as $\mathcal{G} = (\mathcal{V}_c, \mathcal{V}_v, \mathcal{E})$.*

Let $\mathcal{C}$ be a (n, k) binary block code with parity check matrix $\boldsymbol{H} = (h_{ij})$ of dimensions $(n - k, n)$. We denote by $\boldsymbol{h}_j$ the j-th column of $\boldsymbol{H}$. The code $\mathcal{C}$ can be represented as a bipartite graph $(\mathcal{V}_c, \mathcal{V}_v, \mathcal{E})$ where $\mathcal{V}_c$ is the set of check nodes, $\mathcal{V}_v$ is the set of variables nodes and $\mathcal{E}$ is the set of edges such that

$$(c_i, v_j) \in \mathcal{E} \quad \Longleftrightarrow \quad h_{ij} \neq 0$$

The bipartite graph associated to $\boldsymbol{H}$ is called a Tanner graph.

DEFINITION 4.44 *A stopping set is a set of variable nodes $\mathcal{S}$ such that all neighbors of $\mathcal{S}$ are connected to the nodes in $\mathcal{S}$ at least twice.*

From the parity check matrix perspective, a stopping set is a binary row vector $\boldsymbol{v}$ of length n that does not have exactly one 1 in common with any row of $\boldsymbol{H}$. A minimum stopping set is a non-zero stopping set with minimum weight. All codewords are also stopping sets.

4.4.2 LDPC codes

The last recent research years have shown that LDPC codes are able to provide performance comparable to, or in certain cases better than Turbo codes. This, jointly to the fact that LDPC codes can have a very flexible and high-speed decoder implementations, has generated a big interest in these codes. In particular, the second generation of the Digital Video Broadcasting - Satellite 2nd generation (DVB-S2) standard [101] has chosen LDPC codes as adopted forward error correction scheme, while since 2001 the Consultative Committee for Space Data Systems (CCSDS) has been investigating the use of LDPC codes in both deep-space and near-earth high data-rate missions [136].

4.4.2.1 Basic Properties.   LDPC codes are linear block codes that can be defined in many different ways. In the literature, the research attention is in general restricted to binary codes with block lengths of at least many thousand information bits. Looking at [41] and [138], an LDPC code can be defined as:

- the null-space of a parity-check matrix whose entries are almost all zero

- a linear block code that has a low density of non-zero entries in its parity check matrix

- a linear block code (k, n) such that the number of non-zero entries in its parity check matrix is proportional to n as the length of the code grows

As previously stated, it is very common to represent an LDPC code with the help of a bipartite graph. LDPC codes can be classified as regular or irregular. In particular, a regular LDPC code has parity-check matrix (or a Tanner graph) with a uniform column weight (or uniform variable-node degree) equal to d_v and a uniform row weight (or uniform check-node degree) equal to d_c. If the column weight or row weight is not constant throughout the matrix, then the code is called an irregular code. Typically, the average column weight is ≥ 3. For a regular code $n \cdot d_v = m \cdot d_c$ and the code rate can be defined in terms of the row and column weights.

The LDPC minimum distance is closely related to the structure of the parity check matrix H [138]. H is sparsely populated with non-zero terms that should not overlap more than once. Therefore, a large number of columns of H are involved to fulfil all parity check equations. For this reason the minimum distance can be large, which is a high desirable property for all codes, ensuring the correction of more errors.

4.4.2.2   LDPC matrix construction.   For LDPC codes, although both the encoding and the decoding can be made linear (or near-linear) in the block length, hardware implementation has not been easy due to their random structure. In fact, the research work has indicated that randomness is important for near-capacity performance, whereas codes with structure and regularity should be used in order to have simple hardware implementation [140].

Random LDPC codes were the first type of LDPC codes that was developed proving that they perform very well when the code is very long. For these codes the parity check matrix construction is really simple, being based on computer generation, which picks the binary entries in the $m \times n$ parity check matrix H at random. The most important disadvantage of this construction technique is that produces LDPC codes with quite complex encoding, due to the lack of structure as for structured construction. Two main families of random LDPC codes can be identified: the regular Gallager codes [41] and the irregular MacKay codes [138].

Although methods exist for creating random codes of any length it is not obvious that the code is good until it is tested in decoding simulations. There is a large variance in decoding performance for codes chosen from random ensembles of short block length codes. However, it is computationally expensive to find a good code from those ensembles based merely on trial and error and simulations.

Since it was shown by Mao [141] [142] that codes with high girth averages perform better than codes with low girth averages from the same ensemble, some efforts have been made to construct LDPC codes with elimination of small cycles and stopping sets. In particular, some heuristic algorithms have been proposed to improve the girth of random codes. The bit-filling algorithm developed by Campello [143] grows the parity check matrix column by column in a uniform way such that the new column does not introduce a cycle of length g. The algorithm is run using parameters (n, r, d_v, d_c), setting a low target value for the girth g. The algorithm may fail, however if it succeeds, then the process is rerun until the best girth is found. This algorithm has been also improved in [144] by varying the girth constraint during the algorithm instead of failing and rerunning.

Considering that random constructions of the parity-check matrix H presents several drawbacks concerning implementation issues on both encoder and decoder sides, some pseudo-random construction techniques have been introduced with the aim of simplifying the implementation of both encoding and decoding procedures. This approach consists in the partially-random design of parity-check matrices constrained to have basic structures. Examples of such techniques are random quasi-cyclic LDPC codes [145] and the so-called Repeat-Accumulate (RA) codes [146].

Random quasi-cyclic LDPC codes can be obtained starting from a $M \times N$ base matrix H_{base}, where each element of H_{base} [147][148] can be any integer number greater or equal to 0.

$$H_{base} = \begin{pmatrix} h_{0,0}^{base} & \cdots & h_{0,N-1}^{base} \\ \vdots & \ddots & \vdots \\ h_{M-1,0}^{base} & \cdots & h_{M-1,N-1}^{base} \end{pmatrix}$$

Usually the values M and N are much smaller in respect to the values m and n that we want to obtain for the final parity-check matrix H. The parity-check matrix is then organized as an $M \times N$ array of $p \times p$ square sub-matrices. Each sub-matrix is filled with a circulant matrix of column (row) weight equal to the value stored in the corresponding position of the base matrix. The choice of the circulant matrices is performed randomly, usually imposing constraints to the girth. H has therefore $m = M \cdot p$ rows and $n = N \cdot p$ columns. After a proper permutation of the columns, H can be recognized to be the parity-check matrix of a quasi-cyclic code [149]. A quasi-cyclic code can be efficiently encoded through simple shift-register-based procedures, while on the decoder side the structure of H can simplify the decoding algorithm's implementation.

A Repeat-Accumulate (RA) code is a simple serial concatenation of an outer repetition code, a random bit interleaver and an inner rate-1 recursive convolutional code with transfer function

$$T(D) = \frac{1}{1 \oplus D},$$

which can be simply implemented with an accumulator. The encoding scheme is depicted in Figure 4.16.

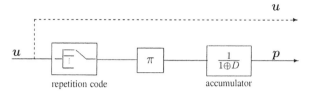

Figure 4.16. Encoder for Repeat-Accumulate codes

The code can be either systematic or non-systematic, and is characterized by a code rate lower or equal to $1/2$. Despite of their simple structure, RA codes perform reasonably well on AWGN channel, and they can be classified as a particular class of LDPC codes, thus they are amenable to message-passing decoding. Irregular Repeat-Accumulate [150] codes introduce variable (irregular) repetition rates in the outer repetition code, allowing to achieve performance comparable to the best designed irregular LDPC codes. A more general construction of efficient-encodable LDPC codes is related to the so-called extended Irregular Repeat-Accumulate codes (eIRA) [151]. Here, the outer repetition code is replaced by a sparse matrix multiplication. The encoding scheme is depicted in Figure 4.17.

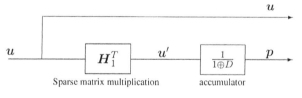

Figure 4.17. Encoder for eIRA codes

The information vector u (composed by k bits) is multiplied by a $k \times m$ low-density matrix H_1^T. The resulting m-bits long vector u' is then processed by an accumulator which produces the parity vector p of length m. The encoded codeword c is given by the concatenation $c = [u|p]$, and the final code rate is $\frac{k}{m+k}$. The parity-check matrix is given by

$$H = [H_1|H_2]$$

where H_2 is a double-diagonal (staircase) $m \times m$ square matrix related to the accumulator

$$H_2 = \begin{pmatrix} 1 & 0 & 0 & 0 & \cdots & 0 & 0 \\ 1 & 1 & 0 & 0 & \cdots & 0 & 0 \\ 0 & 1 & 1 & 0 & \cdots & 0 & 0 \\ \vdots & \vdots & \vdots & \vdots & \ddots & \vdots & \vdots \\ 0 & 0 & 0 & 0 & \cdots & 1 & 0 \\ 0 & 0 & 0 & 0 & \cdots & 1 & 1 \end{pmatrix}$$

Obviously only H_1 can be randomly designed, since the structure of H_2 is already defined. Other classes of RA-like schemes have been invented [152] [153], extending the efficient structure of RA codes in order to design codes suited for different application scenarios.

Another LDPC code design technique makes large use of algebra, geometry or non-random algorithm [154] [155], producing structured codes. The performance of these codes is generally not as good as random codes at long or very long block lengths. However, such codes are often designed so that hardware encoding or decoding is more efficient. The major advantage, from a coding theory perspective, is that the predictable well defined structure of these codes leads to an easier analysis than random codes. Various algebraic and geometric codes are also possible, even if the main interest is on the cyclic and quasi-cyclic LDPC codes [156] [157] [158], which girth and minimum distance property are completely under the designer control.

4.4.2.3   The DVB-S2 example.   The DVB-S2 standard [101] presents a FEC encoding which is performed in three steps, with LDPC codes as the core of this scheme. It is based on an outer systematic BCH encoding, an inner LDPC encoding and a block bit-interleaving to accommodate four possible constellations: QPSK, 8-PSK, 16-APSK, 32-APSK (see Section 5.2.4.1). The FEC scheme produces 64800 or 16200 bit-long codewords with eleven possible rate ranging from $1/4$ to $9/10$. The combination of the different code rates and the different constellations makes it possible to optimize spectral efficiency as transmission conditions evolve.

LDPC codes standardized by DVB-S2 belongs to the extended IRA family defined above where the sparse matrix H_1 is quasi-cyclic. Thanks to this structure, DVB-S2 LDPC codes can be encoded very simply.

4.4.3   LDPC decoding

The sparseness of the parity check matrix of LDPC codes is crucial to the successful implementation of the decoder which makes use of iterative algorithms [159]. Indeed, the information about each edge in the graph must be calculated on each iteration of the decoding algorithm. The information for a particular edge depends on the degree of the nodes that are connected to it. Therefore, the complexity of the decoder is reduced by reducing the number of edges in the graph and by reducing the degree of the nodes, which is the same as increasing the sparseness of $\boldsymbol{H}$.

LDPC decoding can be performed exploiting hard or soft algorithm. In case of hard decoding, it is also known as Bit-Flipping (BF), while in the other case the algorithm is commonly called as Sum-Product Algorithm (SPA), as well as Message Passing Algorithm (MPA) or Belief Propagation Algorithm (BPA) [160], depending on the context. Issues in sum-product decoding are high girth requirement, decoding complexity and convergence.

The most efficient hard decision type LDPC decoder was proposed by Gallager [41] and is based on flipping bits in the received codeword.

In general, the objective of decoding is to find the most likely codeword sent. To this end it must be verified that $\boldsymbol{x} \cdot \boldsymbol{H}^T = 0$, where $\boldsymbol{x}$ is the received word. If $\boldsymbol{x}$ does not satisfy this equation, then the aim of the decoder is to modify $\boldsymbol{x}$ so that it can satisfy the equation. Intuitively, the fewer of these alterations that are made the better, since the nearest codeword to $\boldsymbol{x}$ is to be found. By fracturing this problem into its m parity check equations, bits can be altered (or flipped) on a parity-by-parity basis. By flipping the fewest number of bits in each parity check the received word will move incrementally towards a codeword. This procedure is done iteratively until all checks are satisfied, or a maximum number of iterations is reached. This process is known as local decoding, since each parity check is concerned solely with its own received bits and it is the same procedure in essence that is done in standard sum-product decoding.

The LDPC sum-product decoding procedure is an iterative soft-decision algorithm similar, in principle, to the above hard decision decoder. Rather than flipping bits in the received word, the sum-product decoder alters the probability (or likelihood) that each bit is a symbol x_i on each iteration. This probability is sent through the Tanner graph of the code and is related to the code-bit corresponding to the variable-node from which the considered edge starts. In the following, the logarithmic version of this algorithm is briefly described. The log-likelihood ratio of the a priori probabilities is $\Lambda^0(x_i)$ and it is defined as

$$\Lambda^0(x_i) = \log \frac{Pr\{x_i = 0|y_i\}}{Pr\{x_i = 1|y_i\}} \tag{4.49}$$

The log-domain version of the SPA is preferable because it is more stable and foresees additions instead of multiplications, that are more complex to implement.

At the k-th iteration, it can be denoted by $\Lambda^{k,v \to c}$ a message sent from a variable node to a check node, and by $\Lambda^{k,c \to v}$ a message in the opposite direction. A variable node of degree d_v receives and processes the messages $\Lambda_i^{k-1,c \to v}$, $i = 1, \ldots, d_v$, and sends back to its j-th with $j = 1, \ldots d_v$ neighbor check nodes the message

$$\Lambda_j^{k,v\to c} = \Lambda^0(x_j) + \sum_{\substack{i=1\\i\neq j}}^{d_v} \Lambda_i^{k-1,c\to v} \qquad (4.50)$$

At the first iteration, each variable node simply propagates its initial received message. It can be noted that the message $\Lambda_j^{k,v\to c}$ in eq. (4.50) does not depend on the message previously received on the same edge.

A check node of degree d_c receives and processes the messages $\Lambda_i^{k,v\to c}$, $i = 1,\ldots,d_c$, and sends back to its j-th with $j=1,\ldots d_c$ neighbors variable node the message

$$\Lambda_j^{k,c\to v} = 2\cdot\tanh^{-1}\left\{\prod_{\substack{i=1\\i\neq j}}^{d_c}\tanh\left\{\frac{1}{2}\Lambda_i^{k,v\to c}\right\}\right\} \qquad (4.51)$$

The decoding algorithm proceeds iteratively until the code parity-check constraints are all verified or a maximum number of iterations is reached. Then, a decision on each element of the received vector is performed by summing all the messages incoming to the corresponding variable node (i.e., the message coming from the channel and the messages sent by its neighbors), and comparing the sum with a threshold usually set to zero. If at a particular variable node, the sum is greater than the threshold, the decoded bit will be a "0". Otherwise, the decoder decides for a "1". A simplified min-sum version of the updating law eq. (4.51) is described in [161][162].

An important instrument used to analytically evaluate the decoding performance of LDPC codes through the message passing algorithm is the Density Evolution (DE) tool [163], which is a technique that determines the performance of an iteratively decoded code in the limit of large block lengths. More precisely, density evolution determines the expected (over the ensemble) message distribution along each class of edge and the bit error probability at each decoding iteration. For practical applications the density evolution approach has the drawbacks that it applies only in the limit of large block-lengths and it concerns the behavior of the bit error probability only.

4.4.4 LDPC Performance

In this section a characterization of the distance spectrum for LDPC codes is firstly shown, then the results obtained through an extensive simulation campaign are presented. In particular a fair comparison among LDPC and Turbo performance is shown at the end of this section.

4.4.4.1 Distance spectrum for LDPC Codes. The main problem in calculation of any kind of LDPC bounds is the determination of its weight enumerating function. Hence, in this section the derivation of the asymptotic distance spectrum of various LDPC ensemble will be provided.

The first results in this field were determined by Gallager for regular ensembles [41], while for the irregular ensembles the first attempt is reported in [164], where LDPC average spectrum calculation was found for a related irregular ensemble, in which the parity check matrices have known row and column weight profiles. However the most interesting and innovative results in enumerating function determination are due to Burshtein and Miller [165], that derived the spectrum for irregular bipartite graph ensembles.

Considering regular LDPC, characterized by a regular (d_v, d_c)-regular ensemble with block-length n, the average spectrum $\overline{S}_i(0 \leq i \leq n)$ is defined as $E[S_i]$, where S_i is a random variable related to the number of codewords of weight i in the drawn code. According to [165], for any α multiple of $1/n$ with $0 < \alpha < 1$ it can be written:

$$\frac{\log_2 \overline{S}_{\alpha n}}{n} = h(\alpha) + \frac{P(X_{\alpha n} = 1)}{n} + o(1) \tag{4.52}$$

where $P(X_{\alpha n} = 1)$ is the probability that a specific αn weight word is a codeword. Without loss of generality, this is the word beginning with a sequence of αn ones, and ending with all zeros. This word is a codeword if and only if all check nodes are connected to the set of the first αn variable nodes by an even number of edges. Thus the average spectrum of the regular ensemble satisfies:

$$\lim_{n \to +\infty} \frac{\log_2 \overline{S}_{\alpha n}}{n} = (1 - r) \log_2 \inf_{x>0} \frac{(1+x)^d + (1-x)^d}{2x^{\alpha d}} - h(\alpha)(c - 1) \tag{4.53}$$

where r is the coding rate and $h(\alpha)$ is the binary entropy of α. A similar result was obtained in [41] and [164] using different approaches.

Considering irregular bipartite graph based ensembles, the average spectrum $\overline{S}_i$ can be written as [165]

$$\lim_{n \to +\infty} \frac{\log_2 \overline{S}_{\alpha n}}{n} = \max_{\beta>0} \log_2 \inf_{x>0,y>0} \frac{\prod_i (1 + xy^i)^{\tilde{\lambda}_i}}{x^\alpha y^\beta} +$$

$$+ (1 - r) \log_2 \inf_{x>0} \frac{\prod_i \left[(1+x)^i (1-x)^i \right]^{\tilde{\rho}_i}}{2x^{\beta/(1-r)}} - \gamma h(\beta/\gamma) \tag{4.54}$$

where $\tilde{\lambda}_i$ and $\tilde{\rho}_i$ denote the fraction of left and right nodes with the degree i and $\gamma = \sum_i i\tilde{\lambda}_i$.

Once the spectrum is computed, the bounds on the probability of decoding error for the given ensemble and for various propagation channels could be obtained.

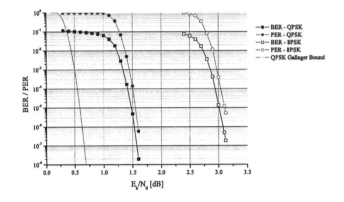

Figure 4.18.   Bit error rate and packet error rate for a quasi-cyclic a rate $1/2$ (3,6) regular LDPC code with $k = 8000$, QPSK and 8-PSK signalling, compared to the Gallager bound

4.4.4.2   Simulation results.   In Figures 4.18, 4.19 and 4.20 the performance of three regular LDPC codes with code rates ranging from $1/2$ to $7/8$ and information block size $k \sim 8000$ are presented. The bit error rate and packet error rate curves have been obtained with 50 iterations of the sum-product algorithm (SPA). The modulations used in the simulations are QPSK and 8-PSK. The performance with QPSK signalling are compared to the Gallager bound [166] for linear block codes. Bounds on error probability for codes with given code rate, allowing the code length going to infinity, are a well-known result of Shannon information theory. However, since we are interested in block codes, we must constraint the codeword length to be finite. In [166] Gallager derives an analytical formulation on performance bounds averaging over the set of all possible codes with the desired code rate and block length [133]. This ensemble bound gives us information about the performance of good codes. The three codes are random quasi-cyclic codes constructed with the technique presented in Section 4.4.2.2, with girth equal to 6. The regular design results to be very effective in order to achieve low error floors, while in the waterfall region the measured performance are within 1 dB from the theoretical bounds.

4.4.4.3   LDPC/Turbo codes comparison.   In the following, the comparison will be focused merely on the performance, recalling that some of the presented codes possess different characteristics which make them more or less appealing depending on the application scenario.

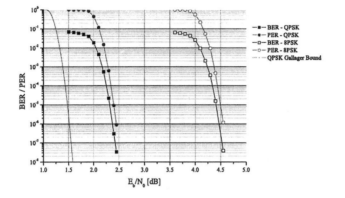

Figure 4.19. Bit error rate and packet error rate for a rate $2/3$ quasi-cyclic regular LDPC code with $k = 8000$, QPSK and 8-PSK signalling, compared to the Gallager bound

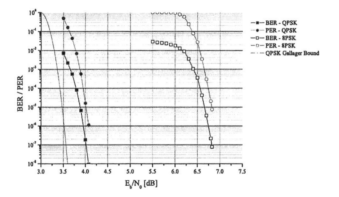

Figure 4.20. Bit error rate and packet error rate for a rate $7/8$ quasi-cyclic regular LDPC code with $k=8100$, QPSK and 8-PSK signalling, compared to the Gallager bound

In Figures 4.21 and 4.22 the BER and PER performance of two LDPC codes are compared with the ones of a punctured PCCC [136]. For the presented codes, the information block is approximatively 7136, and the code rate is $7/8$. One of the two LDPC codes is an irregular one [133], with a distribution optimized by density evolution. The other LDPC code is a regular code with variable

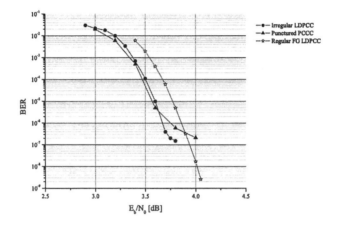

Figure 4.21.   Bit error rates of different LDPC codes and PCCC for a rate 7/8. Information
block size $k \approx 7136$

node degree $d_v = 4$, and is based on a finite geometry (FG) construction [154].
The presented PCCC has been obtained by puncturing the Turbo code with
information block length $k = 7136$ already recommended by the Consultative
Committee for Space Data Systems [100]. The irregular LDPC code and the
Turbo code exhibit almost the same performance in the waterfall region. Both
suffer for an error floor just below a PER $= 10^{-3}$, whereas the regular FG LDPC
code permits to reach low error rates without slope changes for the error rate
curves.

Although the purpose of this section is the performance comparison between
different coding solutions, it must be pointed out that the choice of a particular
coding scheme for a given application scenario is not so simple. In fact a
complete comparison between different coding schemes should be based on
the performance, but also on the complexity of encoders and decoders, on the
presence of error floors, on the flexibility of the coding scheme.

The complexity of the encoder for Turbo codes is relatively small compared
to the decoding. On the contrary, some cases of LDPC codes with non-invertible
parity check matrix H present an encoding complexity in the same order as
decoding complexity, which is the reason why such a structure is typically
avoided. On the other hand, there are LDPC codes based on a staircase structure
like in the DVB-S2 standard that have an encoding complexity which is linear
in regard to the block size.

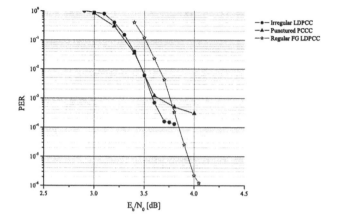

Figure 4.22. Packet error rates of different LDPC codes and PCCC for a rate 7/8. Information block size $k \approx 7136$

Moreover, for applications requiring low error rates it is desirable to design a channel codes with low error floors. On the contrary, for applications where error rates requirements are less strict or in communication links characterized by severe limitations in the transmitted power, more attention should be paid to the waterfall region performance.

4.5 Forward Error Correction at Higher Layers

Application of FEC is not limited to the Physical Layer (PL) only. For example, in mobile scenarios, the mean duration of fades can be in the range of some seconds, which results in a potential need for large interleaver sizes on physical layer. Even then, there is still a non-negligible probability that fades can last up to one or two minutes. Channel coding approaches on the physical layer with such long interleavers require a high memory and pose additional problems, especially concerning synchronization.

The introduction of coding at higher layer can be a solution for countering these periodic or very long deep fades. In fact, higher layer coding can be either introduced at network, transport or application layer to allow reliable transmission notwithstanding the impairments due to the propagation channel.

The basic idea is to specify a superstructure whose transmission takes longer than the fade events. The information inside this superstructure is subdivided into k individual units, which could be e.g. physical frames or Internet Protocol

Higher layer: $n = k + h$ packets

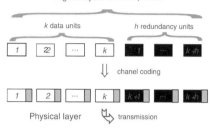

Figure 4.23.   Higher layer FEC

(IP) packets. Additionally, h redundancy units are created. The individual units are then transferred to the physical layer, which adds its independent channel coding to each unit (e.g. based on the DVB-S or S-UMTS standard). The principle is shown in Fig.4.23.

The approach results in a double structure, where the physical layer channel coding combats the effect of fast fading and white Gaussian noise for individual units. The additional redundancy units are then responsible to combat the effect of slow fading events.

Alternatively to the systematic encoding approach described before, it is of course also possible to build n non-systematic units out of the k source units, but a systematic approach is normally preferred.

Usage of MDS codes, like the Reed-Solomon code, for higher layer encoding would then allow the receiver to reconstruct the original information, if at least k out of n units of the superstructure are correctly received. The receiver can cope with erasures, as long as they result in a total loss not exceeding h unit, independently, where the erasures did happen. For achieving this, the decoder would need to be assured, that all units it gets have been correctly decoded. This could be either ensured via a Cyclic Redundancy Check (CRC) at unit level or using an outer block code already built in at the physical layer [167], [168].

One major advantage of higher layer FEC is that it can be added to an air-interface standard, without modifications, while at the same time it extends the capability of the standard to unforseen situations and environments, especially for mobile applications.

Overall, one can remark that in recent years there is an uptake of interest in higher layer coding activity, from areas encompassing DVB for file transfers, space telemetry applications at CCSDS, UMTS and even multicast IP networks, due to possible losses at packet switches or routers.

In the following two different approaches for higher layer coding are shortly presented. The first, called here, packet layer coding, is more closely related to the air interface and uses a constant size of the superstructure.

The second is more closely related to the IP based transport layer and uses a variable size superstructure. The coding is done on logical independent objects, which are normally files.

4.5.1   Packet Layer Coding

For packet layer coding , a fixed number of packets or frames of the physical layer are used to generate a common code group. These packets form the superstructure, which is encoded together and for which redundancy packets are generated.

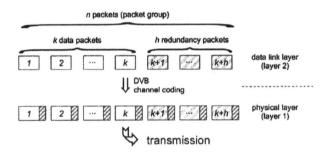

Figure 4.24.   Packet level FEC

Such an approach has been standardized for DVB-H ('Handheld') under the name of MPE-FEC [169] and it will be probably also introduced into the DVB mainstream.

The advantage of Packet Layer Coding is that the constant length of the encoding objects allows the usage of MDS codes, like the Reed-Solomon code, and the placement of the additional packets in an optimized time distance in regard to the physical layer effects.

Its drawback is that large file transfers have a higher probability to fail than small file transfers, since a file consists of different, independent packet groups and the failure of a single group, can lead to the failure of the full transfer.

4.5.2   Transport Layer Coding

The second approach therefore tries to encode full files, which can have a variable length. If a fixed percentage of coding is added, as in the previous paragraph, a long file will have a larger absolute number of redundancy packets,

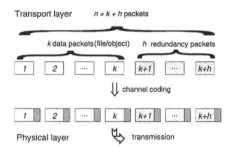

Figure 4.25. Transport layer FEC

that are not restricted to any specific part of the file and can therefore cope with longer bursts of error.

Since the encoding needs then to have access to the full file, it can be no longer done below the packet layer, but needs instead to be added to the transport layer.

This approach is used in the IETF Reliable Multicast Group, where the necessary Internet protocols for such a scheme are specified (RFC 3450 "Asynchronous Layered Coding Protocol" (ALC)). These protocols are especially well suited to be combine with the FILE Delivery over Unidirectional Transport (FLUTE) protocol, which can be used in different broadcast schemes.

A second advantage of the this approach is that the redundancy can be file specific, and therefore highly protected files can be mixed with best effort datagrams. Additionally, each file is encoded and decoded independently, allowing parallel execution and storage of files before decoding. Lastly, the possible multiplex between packets from different files increase the spreading of the file in time.

However, there are also disadvantages. First, short files consisting only of a low number of packets are not so well protected. Secondly, the signalling information of the coding scheme like e.g. packet position or the parameters of the coding have to be transferred. Lastly, the variable length and large size of the file make the usage of RS-codes either complex or sub-optimal. Especially to address this last point, alternative low complexity FEC scheme are of interest.

4.5.3   Erasure Decoding Algorithms

One especially interesting code family are the LDPC codes, since in case of erasure decoding, the decoding steps simplify to individual XOR calculations per row. Additionally, it is known when the decoding approach has reached its last step and can no longer improve.

In [170] it was proved that LDPC decoders can in principle reach the channel capacity for Binary Erasure Channel (BEC) channels with an arbitrary small deviation e, called also inefficiency.

In case there is only one erasure in a row, the sum of the XOR operations is equal to the value of the erased symbol. The symbol can be re-inserted in the codeword x and then can help to bring another parity check equation to the status of just having one erasure. This process is repeated until there is no longer any row with only one erasure in it (either because everything was successfully decoded or a decoding error occurred). A special approach is to try to decode as soon as one packet arrives and to fill all erasures as soon as possible.

Different approaches in a regard to which kind of LDPC codes to use are possible. A simple one step approach is based on the previously mentioned eIRA codes which have the same staircase structures as DVB-S2. They allow a systematic and linear encoding in time and are also very flexible in regard to the extension to different file sizes.

This due to the effect, that while short cycles are also in a erasure channel not optimal, since they waste the error correcting capacity, their influence on the overall code performance is small as long as there are only a few of them. Neglecting this minor effect, allows a fast on-line implementation of the random code generation.

Based on the previous staircase structure one can then propose a general approach for generating a variable code:

$$H = [H_1 | H_s]$$

The last part of H is build by the staircase structure, which allows a systematic and linear encoding. The first part of the matrix H_1 is randomly generated, based on the degree distribution of the check and variable nodes.

This approach also allows the extension of the code in case more redundancy is needed. For example for a type II hybrid ARQ schemes one could use:

$$H = \begin{bmatrix} H_1 & H_2 & 0 \\ H_3 & 0 & H_4 \end{bmatrix}$$

where H_3 is an additional random part, and H_4 is the extended part of the staircase structure. The distribution of the check node could be derived from the robust soliton distribution found by Digital Fountain and best explained in [171] or from code families with different rates like DVB-S2.

A step further in this direction are then the so-called rate-less codes, which do not a priori state the number of necessary redundancy packets, but instead create each redundancy packet independently, only based on the known systematic packets and a random seed.

The most advanced code of this kind is at the moment the so called Raptor Code from Digital Fountain, which has been selected in UMTS as standard, both for Transport Layer and Packet Layer coding [172], which uses a two step approach: first it transforms via a high rate LDPC code the systematic information in a suitable manner; then using the so called Luby-Transform (LT) code, it produces the redundancy packets.

4.6   Conclusions

In this Chapter, the state-of-the-art of channel coding has been reviewed, taking into account issues related to design, encoding and decoding. The most important techniques used for satellite applications have been presented, with particular emphasis on the developments over the last years, i.e., Turbo and LDPC codes.

It has been shown how the iterative principle revolutionized the design of communications systems. In fact, apart from FEC decoders, receiver modules like equalizers are already exploiting this novel approach. In this respect, it is to be noticed that, in future receiver designs, estimation and synchronization tasks will be merged with powerful decoding algorithms, as shown in Section 6.4 where techniques for code-aware synchronization are discussed in detail.

Chapter 5

MODULATION TECHNIQUES

P. T. Mathiopoulos[1], G. Albertazzi[2], P. Bithas[1], S. Cioni[2], G. E. Corazza[2], A. Duverdier[3], T. Javornik[4], S. Morosi[5], M. Neri[2], S. Papaharalabos[6], A. Ribes[3], N. C. Sagias[1]

[1] *Institute for Space Applications and Remote Sensing, National Observatory of Athens, Greece*

[2] *University of Bologna, Italy*

[3] *Centre National d'Etudes Spatiales, France*

[4] *Josef Stefan Institute, Slovenia*

[5] *University of Florence, Italy*

[6] *University of Surrey, U.K.*

5.1   Introduction

In this chapter, various modulation techniques , including those employing constant and non-constant envelope signaling formats are presented. Commonly used modulation schemes, such as Phase Shift Keying (PSK), Quadrature Amplitude Modulation (QAM) and Continuous Phase Modulation (CPM) and several coded modulation techniques are studied. Their performance in various channel environments, including Additive White Gaussian Noise (AWGN) and fading, is also presented.

5.2   Modulation schemes

This section reviews the characteristics and properties of the most commonly used modulation schemes. After classifying such modulation schemes, in terms of their envelope properties (i.e., constant or variable) a detailed description of the characteristics of PSK, QAM, and CPM signals is presented.

5.2.1 Basic principals

Digital modulation is a process of transforming a digital symbol to a signal suitable for transmission via a communication channel. According to this process the information is carried by a sinusoidal signal

$$s(t) = a(t)\sin[2\pi f(t)t + \phi(t)] \qquad (5.1)$$

with three parameters: amplitude $a(t)$, frequency $f(t)$ and phase $\phi(t)$. Thus, by encoding the information in one of the parameters, keeping the other two unchanged, three basic modulation schemes exist [173]:

- **Amplitude Shift Keying (ASK)** : The digital symbol is encoded in the amplitude of the signal. A simple example of ASK is the on/off modulation with digital 1,0 represented by signal with amplitude A and no signal, respectively.

- **Frequency Shift Keying (FSK)** : The signal with carrier frequency f_1 represents digital 1 and the signal with carrier frequency f_2 denotes digital 0. A simple FSK modulator consists of a switch and two oscillators with frequencies f_1 and f_2.

- **Phase Shift Keying (PSK)** : The information is encoded in the phase changes of the transmitted signal.

Advanced modulation schemes can be derived by combining basic modulation schemes to comply with transmission requirements, for example higher bandwidth efficiency. The waveforms of the three above basic schemes are shown in Figure 5.1. Two basic criteria exist to compare the performance of the modulation schemes, namely, power efficiency and bandwidth efficiency , [173].

- **Power efficiency** , which is defined as the ratio of the required bit energy to noise Power Spectral Density (PSD) determined for a certain Bit Error Probability (BEP) over a Gaussian channel. Normally, a Bit Error Rate (BER) equal to 10^{-5} is used to determine the power efficiency. Moreover, the symbol energy and symbol error rate are applied to define power efficiency.

- **Bandwidth efficiency** , which is defined as the number of bits per second that can be transmitted within one Hertz of the system bandwidth. The calculated values of bandwidth efficiency depend on the definition of the system bandwidth.

In practice, there are several ways of measuring system bandwidth , depending upon the employed filters of the telecommunication system under consideration. In order to fairly compare the bandwidth of different modulation schemes, the following definitions of system bandwidth can be used:

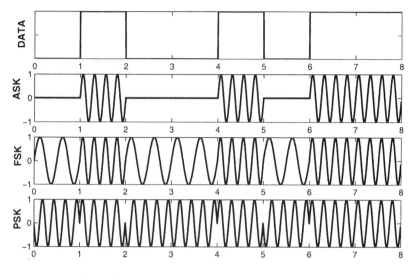

Figure 5.1. Waveforms for amplitude, frequency and phase modulations

- **Nyquist bandwidth:** In order to minimize the Inter-Symbol Interference (ISI) and spectral width of the modulated signal, the communication system should include a filter that satisfies Nyquist criterion for zero ISI [17]. The most commonly used Nyquist filter is the raised cosine α filter, $H(f)$, and is given by:

$$H(f) = \begin{cases} T_s & 0 \le |f| \le \frac{1-\alpha}{2T_s} \\ \frac{T_s}{2}\left[1 + \cos\left(\frac{\pi|f|-1/2T_s+\alpha}{2\alpha}\right)\right] & \frac{1-\alpha}{2T_s} < |f| \le \frac{1+\alpha}{2T_s} \\ 0 & |f| > \frac{1+\alpha}{2T_s} \end{cases}$$

(5.2)

where α is the roll-off factor, denoting excess bandwidth with respect to the minimum baseband Nyquist bandwidth, which is $W_N = 1/(2T_s)$, and T_s is the symbol duration. The corresponding impulse response, $h(t)$, of the Nyquist raised cosine α filter is:

$$h(t) = \frac{\cos(2\pi\alpha t)}{1 - (4\alpha t)^2}\frac{\sin\left(\pi t/T_s\right)}{\pi t/T_s}$$

(5.3)

The minimum bandwidth required to transmit digital signals free of ISI is $0.5R_s$, where $R_s = 1/T_s$ is the symbol rate. The minimum Nyquist bandwidth, W_N, of a modulated signal is twice as much as the baseband bandwidth, i.e., $W = R_s$. If the relation between the bit and the symbol rate is considered, the Nyquist bandwidth efficiency for M-ary modulation is defined as $\log_2 M$. The telecommunication systems have usually wider bandwidth, but for the theoretical comparison of modulation schemes this definition provides good results, because all modulations will suffer from the same degree of degradation at baseband.

- **null-to-null bandwidth:** The null-to-null bandwidth is defined as the width of the spectral main lobe when it reaches the first spectral null.

- **percentage bandwidth:** Percentage bandwidth is the portion of the bandwidth that contains a certain amount of signal energy. If the spectrum of the modulated signal does not have spectral nulls, percentage bandwidth is commonly used. For example the bandwidth where 99% of the signal energy is transmitted can be considered as the percentage bandwidth.

In satellite and mobile communication systems the power resources are limited and consequently all elements of the communication system should use the energy as efficient as possible. The high power amplifier consumes a high percentage of energy in a system; therefore it is forced to operate economically, i.e., close to its saturation point, and hence the non-linear distortion is introduced in the system.

The sensitivity of the modulation scheme to non-linear distortion depends on the Peak-to-Average Power Ratio (PAPR), which is defined as a ratio of the peak power of the modulated signal to the average power of the modulated signal. Signals with PAPR = 0 dB have no variation of the signal power and constant envelope and they are insensitive to nonlinear distortions. On the contrary, in signals with PAPR higher than 0 dB the non-linear amplifier introduces distortions. In general, an increase of the PAPR also increases the signal sensitivity to non-linear distortion. PAPR values for modulation schemes described in following sections are shown in Table 5.1. The PAPR of the filtered modulation schemes are obtained with Nyquist filter with roll off factor $\alpha = 1.0$. The CPM and unfiltered phase modulation have constant envelope, consequently they are rather insensitive to nonlinear distortion. For all amplitude modulation, multilevel quadrature amplitude modulation and filtered phase modulation schemes the envelope is variable. Furthermore, higher level QAM schemes have higher PAPR.

The envelope variation depends on the type (shape) of the filter and its passband. The PAPR of the Quadrature Phase Shift Keying (QPSK) signal filtered by Nyquist filter, with $\alpha = [0, 1]$, is plotted in Figure 5.2. Rectangular filter, $\alpha = 0$, introduces more signal envelope variation than filters with wider bandwidth, $\alpha = 1$, and smoother shape.

modulation scheme	unfilter	Nyquist $(\alpha = 1)$
CPM (different)	0.00 dB	0.00 dB
QPSK	0.00 dB	1.80 dB
8-PSK	0.00 dB	1.80 dB
16-PSK	0.00 dB	1.80 dB
16-QAM	2.55 dB	4.35 dB
64-QAM	3.68 dB	5.50 dB

Table 5.1.   Peak to average power ratio (PAPR) of several modulation schemes.

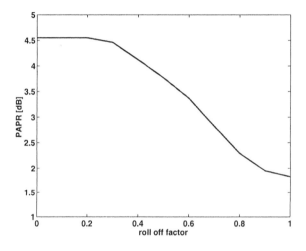

Figure 5.2.   Peak to average power ratio dependency on filter shape for QPSK signals

5.2.2   Modulation Schemes Classification

Many combinations and variations of basic modulation techniques exist. Some modulation schemes can be derived from others and some can be represented as successors of the basic modulation types. For satellite communication, the modulation schemes can be appropriately classified according to their sensitivity on non-linear distortion into two generic group of signals, i) constant, and ii) variable envelope modulation schemes. This classification is illustrated in Figure 5.3. The variable envelope modulation schemes include two basic types, phase (PSK) and amplitude (ASK) modulated signals.

The PSK modulation schemes, e.g., Binary Phase Shift Keying (BPSK) , QPSK , 8-PSK, and 16-PSK, have constant envelope without baseband filtering. On the contrary, baseband filtering, which is always used in practical communication systems to reduce the spectrum occupied by the transmitted

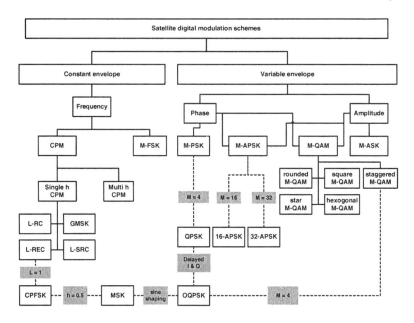

Figure 5.3. Modulation schemes classification

signals, introduces signal envelope variation. Therefore, it is appropriate to classify PSK as a variable envelope modulation scheme.

The multilevel QAM (M-QAM) , is a combination of amplitude and phase basic modulations. The in-phase (I) and quadrature phase (Q) channels of QAM signal can be superimposed either coherently, i.e., without delay between components, or with a delay of one half of the symbol interval between the components, which is known as an offset or staggered QAM. The different shapes of QAM constellation, e.g., square QAM, star QAM, hexagonal QAM, rounded QAM, have been found under different optimization criteria [174, 175], namely the average signal power, the minimum Euclidean distance, and finally the peak to average power ratio.

Special attention should be paid to the QPSK signal and its derivations. It can be represented as a "4-QAM" or "4-PSK" signal. If the Q channel of QAM signal is delayed by half of the symbol interval, regarding to the I channel, the Offset Quadrature Phase Shift Keying (OQPSK) is obtained. Furthermore, if I and Q phase channels are shaped by sine function, the Minimum Shift Keying (MSK) modulation scheme is obtained. The MSK signal can be also represented as a FSK signal with continuous phase and modulation index $h = 1/2$.

A conventional FSK signal is generated by shifting the carrier by a discrete amount of frequency to reflect the digital information that is being transmitted.

In general, this type of modulation does not take care of the signal phases, when the change of signal frequency occurs, and therefore does not introduce memory into the transmitted signal. Consequently, such abrupt phase switching results, in large side lobes outside the main spectral band, make the conventional FSK modulation unusable in modern communication systems. Assuring the smooth phase transition from symbol interval to symbol interval, significantly decreases the amount of spectrum outside the main spectrum lobe. This type of modulation is referred to as CPM . By choosing different pulse shape, varying modulation index h and size of the alphabet M, a great variety of CPM schemes can be obtained. Pulse shapes that have been widely used are: Rectangular Pulse Shaping (REC), pulse shaping Raised Cosine (RC) and Gaussian pulse shaping which is used in Gaussian Minimum Shift Keying (GMSK). Pulse shape function has smooth shape over finite time interval and is zero outside this interval. If the pulse shape interval is equal to one symbol interval, full response CPM is obtained and if pulse shape interval is greater than one symbol interval, partial response CPM is obtained. CPM scheme with $h = 0.5$, $M = 2$ and 1REC pulse shape is known as MSK modulation scheme [173, 17]. By varying the modulation index h, the CPM schemes can be grouped in two groups of modulated signals [173] known as single-h, and multi-h.

5.2.3   Phase shift keying

PSK is a modulation scheme whereby the phase of the transmitted signal is modulated by the data stream.

5.2.3.1   Binary phase shift keying .

In BPSK the phase of the carrier is shifted between two positions that are 180 degrees apart. On an I/Q diagram, the I channel has two different values as it is depicted in Figure 5.4. There are two possible locations in the state diagram, so a binary one or zero can be sent. The symbol rate is one bit per symbol, $E_s = E_b$, where E_s is the average signal energy and E_b is the average bit energy. Special attention should be paid to the subtle difference between the ASK and BPSK modulations. In BPSK the baseband signal takes the values of $+A$ and $-A$ instead of $+A$ and zero, as in ASK. Thereby, the phase of the output signal, for one and zero, is reversed rather than turned on or off.

The envelope of a BPSK signal is

$$s(t) = \sum_{n=-\infty}^{\infty} g_n\, p\,(t - n\,T_s) \tag{5.4}$$

where $p(t)$ is a unit amplitude shaping pulse of duration T_s and g_n is the BPSK symbol in the nth symbol interval $n\,T_s \le t < (n+1)\,T_s$ given by

$$g_n = A \exp\{\jmath\,\phi_n\} \tag{5.5}$$

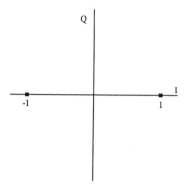

Figure 5.4.   BPSK constellation diagram

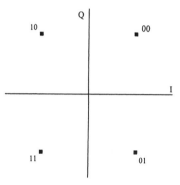

Figure 5.5.   QPSK constellation diagram

In the above equation, A is the constant BPSK symbol envelope and ϕ_n is the phase of the nth symbol obtained as

$$\phi_n = \begin{cases} 0, & \text{if } b_n = 0 \\ \pi, & \text{if } b_n = 1 \end{cases} \tag{5.6}$$

5.2.3.2   Quadrature phase shift keying .   In QPSK the phase of the carrier is shifted between four positions that are 90 degrees apart. On an I/Q diagram, the I channel has four different values, as it is depicted in Figure 5.5. There are four possible locations in the state diagram, so all possible binary combinations of one and zero can be sent. The symbol rate is two bit per symbol, $E_s = 2\,E_b$.

Usually, Gray mapping is being used for allocating bits to a symbol. In this way, adjacent symbols have the property that they differ by only one bit. This is obtained computing the phase ϕ_n of the nth symbol as

$$\phi_n = \begin{cases} \pi/4, & \text{if } b_{n-1} = 0 \cap b_n = 0 \\ 3\pi/4, & \text{if } b_{n-1} = 1 \cap b_n = 0 \\ 5\pi/4, & \text{if } b_{n-1} = 1 \cap b_n = 1 \\ 7\pi/4, & \text{if } b_{n-1} = 0 \cap b_n = 1 \end{cases} \tag{5.7}$$

5.2.3.3   Offset QPSK .   An equivalent modulation signaling, which is widely employed in satellite communications, is OQPSK (or staggered QPSK). OQPSK is a form of QPSK wherein the I and Q channels are misaligned with respect to one another by half a symbol time interval, $T_s/2 = T_b$. This signaling has the advantage over the QPSK that the abrupt phase change of 180 degrees is eliminated. Hence, the phase trajectories passing through the origin of the constellation diagram are avoided. This property makes the OQPSK signal less sensitive to nonlinear amplifier impairments as compared to QPSK. The complex OQPSK envelope is

$$s(t) = s_I(t) + j s_Q(t) \tag{5.8a}$$

$$s_I(t) = \sum_{n=-\infty}^{\infty} g_{In} \, p\,(t - n\,T_s) \tag{5.8b}$$

$$s_Q(t) = \sum_{n=-\infty}^{\infty} g_{Qn} \, p\,(t - n\,T_s - T_s/2) \tag{5.8c}$$

where g_{In} and g_{Qn} are the I and Q data symbols for the nth transmission interval that take on equiprobable ± 1 values.

5.2.3.4 M-ary PSK .   In M-ary PSK (M-PSK) the phase of the carrier is shifted between M positions that are

$$\varphi = (2i - 1)\frac{\pi}{M}, \quad i = 1, 2, \dots, M \tag{5.9}$$

radians apart. On an I/Q diagram the I channel has M different values. There are M possible locations in the state diagram, so all possible binary combinations of one and zero can be sent. In Figure 5.6, the constellation diagram of 8-PSK modulation scheme is depicted.

5.2.4   Quadrature amplitude modulation

The amplitude, phase and frequency of a transmitted signal can be demodulated independently of each other. Note that PSK and FSK receivers remove amplitude variations by limiting the incoming signal and that an ASK receiver pays no attention to the phase of the incoming signal. Therefore, it is possible

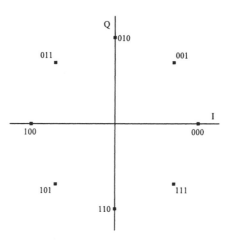

Figure 5.6. 8-PSK constellation diagram

to combine these schemes to produce a modulation scheme that conveys information both as amplitude and phase. Such a modulation scheme is known as QAM. QAM has the advantage of high bandwidth efficiency, however amplifier nonlinearities will degrade its performance due to the non-constant envelope.

The envelope of a M-ary QAM (M-QAM) signal is

$$s(t) = \sum_{n=-\infty}^{\infty} g_n \, p\,(t - n\,T_s) \tag{5.10}$$

where $p(t)$ is a unit amplitude shaping pulse of duration T_s. Moreover, g_n is the M-QAM symbol in the nth symbol interval $n\,T_s \leq t < (n+1)\,T_s$ given by

$$g_n = A\,(a_{In} + j\,a_{Qn}) \tag{5.11}$$

where A is the constant M-QAM symbol envelope, a_{In} and a_{Qn} are the I and Q information amplitudes of the nth symbol, respectively. For the case where M is a power of four, i.e., $M = 4^b$, square M-QAM constellations can be constructed and a_{In} and a_{Qn} take one of the equiprobable values ±1, $\pm3, \ldots, \pm(\sqrt{M}-1)$. When M is not a power of four, a variety of QAM signaling constellations can be constructed, as for example having the shape of a cross. In Figure 5.7, the constellation diagrams of several square M-QAM modulation schemes are depicted.

5.2.4.1   Non-rectangular modulations.   Satellite transmission, especially in the geostationary case, requires very high output powers leading the designer to push the RF amplifier's operating point toward saturation, causing

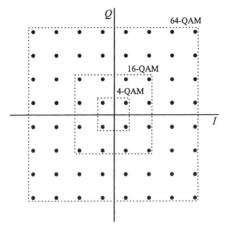

Figure 5.7. Square QAM constellation diagrams for $M = 4$, $M = 16$, and $M = 64$

strong non-linear distortion in the received signal. In order to mitigate this effect, which strongly affects QAM constellations, a new family of constellations has been introduced [176], aiming to reduce the PAPR. Amplitude Phase Shift Keying (APSK) constellations are made up of concentric PSK rings, with several possible combinations of points per ring, and have been proved to be more resilient to non-linear distortion than QAM constellations, [177]. In order to univocally characterize an APSK constellation two sets of parameters are defined:

- ϱ_i, corresponds to the i-th ring radius normalized by the inner ring radius;

- $\omega_i = \theta_1 - \theta_i$, identifies the displacement between the first and the i-th ring.

For instance, the constellation diagrams of several M-APSK modulation schemes are depicted in Figure 5.8. By considering 16-APSK, $\varrho_2 = R_2/R_1$, where R_1 and R_2 are the radii of the internal and external rings, respectively, and ω_2 is the relative phase shift between the outer and inner rings PSK constellations. Similarly, a 32-APSK modulation format can be derived with a triple ring $(4 - 12 - 16)$. In this case, $\varrho_3 = R_3/R_1$, and ω_3 is the relative phase shift between the third and first ring PSK constellations. Furthermore, a quaternary ring $(4 - 12 - 20 - 28)$ for 64-APSK modulation format is customarily introduced, defining $\varrho_4 = R_4/R_1$ and ω_4 as the relative phase shift between the fourth and first ring PSK constellations.

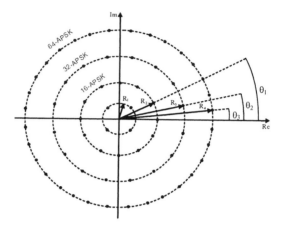

Figure 5.8. M-APSK constellation diagrams for $M = 16$, $M = 32$, and $M = 64$

5.2.5   Continuous phase modulation

Signals with constant envelope and information carrying phase, which are continuous in time, are known as CPM signals [178, 173]. In general the CPM signals can be expressed in complex form as:

$$s(t) = \sqrt{\frac{2E_s}{T_s}} \ \mathfrak{Re} \left\{ \exp\{j(2\pi f_c t + \Phi(t))\} \right\} \qquad (5.12)$$

where f_c is the carrier frequency. The information carrying phase, for the nth symbol interval, is

$$\Phi(t) = 2\pi h \sum_{i=n_0}^{n} \alpha_i q(t - iT) + \Phi_0 \qquad (5.13)$$

where h is the modulation index, Φ_0 denotes the arbitrary phase at the n_0th symbol interval. Data α_i are M-ary symbols taking equiprobable values [$\alpha_i \in \pm 1, \pm 3, \ldots, \pm(M-1)$], where M is the modulation level, which is usually a power of 2. The relationship between the phase function $q(t)$ and its derivative, frequency pulse $g(t)$, is given below

$$q(t) = \int_{-\infty}^{t} g(\tau) \, d\tau \qquad (5.14)$$

where the input α_i and h determine the phase changes of the CPM signal in time. The signal $g(t)$ is usually a pulse with a smooth shape over a finite interval, e.g. $0 \leq t \leq LT$, and zero outside. When the frequency pulse is limited to one symbol interval T ($L \leq 1$), the CPM signal depends on the initial phase at the symbol interval and on the transmitted data in current symbol interval. Such CPM signals are known as full response CPM. When $L > 1$, the frequency pulse is spread over more than one symbol intervals, and hence, overlapping of the frequency pulses occurs. The transmitted symbol has influence on L successive symbol intervals and the resulting CPM signal is called partial response. By applying continuous phase and partial response, signaling memory is introduced in modulated signal. Such an approach usually reduces the signal bandwidth at the expense of increased signal detection complexity.

The shape of frequency pulses determines the properties of CPM signals. The most widely used shapes are [178, 173]:

- rectangular frequency pulse of length L, L-REC:

$$g(t) = \tfrac{1}{2LT} \qquad \text{and} \qquad q(t) = \tfrac{t}{2LT} \qquad\qquad 0 \leq t \leq LT$$

- raised cosine pulse of length L, L-RC:

$$g(t) = \tfrac{1}{2LT}\left[1 - \cos\left(\tfrac{2\pi t}{LT}\right)\right] \text{ and } q(t) = \tfrac{t}{2LT} - \tfrac{1}{4\pi}\sin\left(\tfrac{2\pi t}{LT}\right) \ 0 \leq t \leq LT$$

- spectral raised cosine of length L, L-SRC:

$$g(t) = \frac{1}{LT}\frac{\sin[(2\pi t)/LT]}{(2\pi t)/(LT)}\frac{\cos[\beta(2\pi t)/(LT)]}{1 - [(4\beta t)/(LT)]^2} \text{ and } 0 \leq \beta \leq 1 \ \ 0 \leq t \leq LT$$

- Gaussian shaped MSK, GMSK:

$$g(t) = \tfrac{1}{2T}\left[Q\left(2\pi B_b \tfrac{t-T/2}{\sqrt{ln2}}\right) - Q\left(2\pi B_b \tfrac{t+T/2}{\sqrt{ln2}}\right)\right] \text{ and } 0 \leq B_b T \leq 1$$
$$Q(t) = \tfrac{1}{\sqrt{2\pi}}\int_t^\infty e^{-\tau^2/2}d\tau$$

where B_b is the -3 dB bandwidth of the filter.

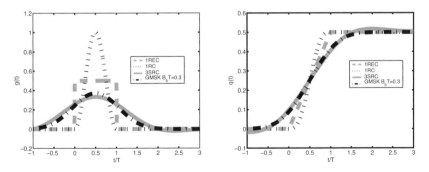

Figure 5.9. Frequency and phase pulses for various CPM signals

Commonly used frequency pulses, $g(t)$, and phase functions, $q(t)$, are plotted in Figure 5.9. All frequency pulses are normalized in order to give $q(t) = 1/2$,

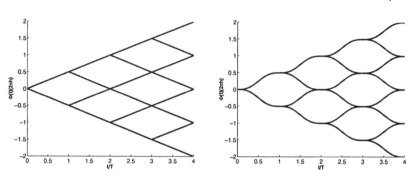

Figure 5.10. Phase trees for binary a) 1REC and b) 1RC CPM signals

for $t \geq LT$. This normalization limits the maximum phase changes in a symbol interval to $\pi h (M - 1)/L$.

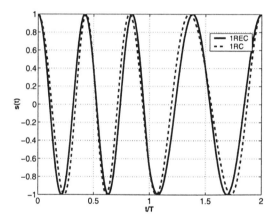

Figure 5.11. CPM signals for 1REC and 1RC frequency pulses

The signal phase is linear within the symbol interval, while a break of the signal phase may occur between symbol intervals. The phase tree of 1 REC CPM signals are shown in Figure 5.10a. The 1 REC CPM signals are also known as Continuous Phase Frequency Shift Keying (CPFSK) signals. Further on, for $M = 2$ and $h = 0.5$ the MSK signal is obtained. The smallest value is $h = 0.5$, where the waveforms of CPFSK signals are orthogonal. Other frequency pulse shapes give smoother changes of the signals frequency and consequently more compact spectrum. The phase tree of 1 RC signals is plotted in Figure 5.10b.

The waveforms of two CPM signals with different frequency pulses, namely 1 REC and 1 RC, and $h = 0.75$ are shown in Figure 5.11. This figure illustrates the properties of CPM signals discussed above: constant envelope, continuous phase, constant frequency within a symbol interval for the 1 REC frequency pulse and smooth changes of the signal frequency for the 1 RC frequency pulse. The phase values at the end of the symbol interval are equal to $h\pi\alpha$. When the rectangular frequency pulse is shaped by a Gaussian filter, then GMSK signals occur. Essentially, the MSK signals are GMSK signals with infinite normalized bandwidth $B_bT \to \infty$. The GMSK signals, with the normalized bandwidth between 0.25 and 0.5, are widely used in mobile communication systems, due to their excellence spectral properties, robustness against distortions and straightforward detection.

In general, h can be any real number and may even alternate between symbol intervals resulting the so-called multi$-h$ CPM signals. However, limiting h to rational numbers, $h = 2Q/P$, where P and Q are integers, limits the number of initial phase states at the transmitter to $S = PM^{L-1}$. Hence, this enables maximum likelihood signal detection algorithms, e.g., the Viterbi algorithm, for signal detection.

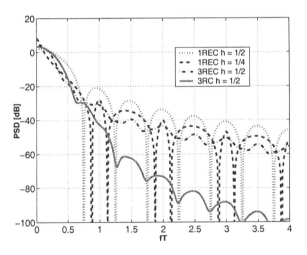

Figure 5.12. CPM signal power spectral density

The CPM signals PSD has been calculated by applying the autocorrelation method [178]. The PSD for various modulation indices, frequency pulse shapes and its duration have been collected in [173] and compared to MSK signals. In general, the following observations can be made (see Figure 5.12):

- for fixed h and M, the longer and smoother frequency pulses yields narrower spectrum

- a smaller h narrows the spectrum of the CPM signal, due to lower frequency deviation.

The waveform of CPM signals is determined by the signal state, which includes the influence of all previously transmitted signals and the current input symbol. Thus, for optimal signal detection, i.e., Maximum Likelihood Sequence Detection (MLSD), processing of an infinite long symbol sequence is required. However, for a practical receiver this processing is limited to N symbols. This approach will lead to a suboptimal receiver structure. The error performance of MLSD receiver in AWGN channel can be bounded by union bound as

$$P_{eb} = KQ(d_{min}\sqrt{E_b/N_0}) \tag{5.15}$$

where $Q(\cdot)$ is the Gaussian Q function, [179, eq. (26.2.3)], N_0 is the noise PSD and d_{min} is the normalized minimum Euclidean distance. The error performance of CPM signals mainly depends on d_{min}. Therefore, the analysis of error rate performance can be significantly simplified by finding d_{min}, which is usually calculated by thoroughly search of the complete CPM signals phase trellis. However, for CPFSK signals an analytical solution also exists [173]:

$$d_{min}^2 = \begin{cases} \log_2 M, & \text{integer } h \\ \min_\gamma \{2\log_2 M[1 - \text{sinc}(\gamma h)]\}, & Q \geq M \\ 2\log_2 M[1 - \text{sinc}(\gamma h)], & h < 0.3016773, Q < M \\ \log_2 M, & h < 0.3016773, Q < M \end{cases} \tag{5.16}$$

where the value for γ is taken from following set $\{2, 4, 6, ..., 2(M-1)\}$. A detailed analysis of how to obtain d_{min} and how it affects the overall performance for various CPM signals can be found in [178].

CPM signals can be generated by direct implementation of eq. (5.12). The data sequence is passed through the frequency pulse shaping filter and multiplied to form the input to the Voltage Controlled Oscillator (VCO), the output of which are CPM signals. The block diagram of such a modulator is shown in Figure 5.13. The required stability can not be achieved by free running VCOs; therefore in practice Phase Locked Loop (PLL) is used instead. CPM signals can also be generated by quadrature modulator and by different digital implementation, details can be found in [178].

5.3   Coded modulation

This Section reviews different coded modulation techniques, emphasizing in Trellis and Turbo coded modulations.

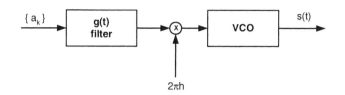

Figure 5.13. Block diagram of CPM modulator

5.3.1 Trellis coded modulation

In digital telecommunication systems the achievement of simultaneous power saving and spectral efficiency is generally a difficult problem. Reduction of bandwidth can be obtained with the use of M-ary modulation schemes, like for example M-PSK or M-QAM, while power reduction involves the use of Forward Error Correcting (FEC) codes. Unfortunately, the use of M-ary modulations, for $M > 4$, introduces a drastic reduction of the Euclidean distance, considered in a two dimensional signal space, and hence it is responsible for the loss of power efficiency. The use of FEC codes, as described in Chapter 4, counterbalance this effect but comes at the expense of bandwidth expansion. Hence, independently of the modulation and instead of maximizing the Hamming distance of a FEC code, coding could be used as an alternative method to maximize the Euclidean distance between coded sequences.

This idea, which was first proposed in [180], is at the basis of Trellis Coded Modulation (TCM) where coding and modulation are considered as a unique function with the following basic properties:

- the bandwidth is not expanded

- the same symbol rate, and consequently the same power spectrum, is obtained.

However, redundancy is introduced by using a constellation with twice the number of points that would be required without the use of coding.

5.3.1.1 TCM principle. The general principle of the coding function is described in Figure 5.14. The raw data are considered to be blocks of q bits, which are divided into two sub-blocks of k and $q - k$ bits. The k bit blocks are coded with rate $r = k/(k + 1)$, while the $q - k$ bits remain uncoded. The $k + 1$ bits at the output of the coder are used to select a sub-constellation that has 2^{q-k} points, while the $q - k$ uncoded bits are used to select a point in the selected sub-constellation.

The minimum Euclidean distance among all possible coded sequences observed for a given TCM, also known as the free distance d_{free} of the modulation,

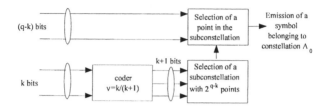

Figure 5.14. Principle of the TCM coder

is the key parameter to evaluate the BER performances. For an addition the bound of error event probability is roughly obtained, in AWGN channel, with PSD $\sigma^2 = N_0$, for high Signal to Noise Ratio (SNR) by

$$P_e \approx N_{free}\exp\left\{-d_{free}^2/2\sigma^2\right\} \qquad (5.17)$$

where N_{free} denotes the average number of nearest neighbor signal sequences with distance d_{free} from the transmitted signal sequence [181].

The asymptotic coding gain of a TCM [182] is defined as

$$G_{dB} \triangleq 10\log\left(d_{free}^2/d_M^2\right) \qquad (5.18)$$

in which d_M represents the minimum Euclidean distance of the M-ary modulation and has the same spectral efficiency.This is reached, with a good approximation, when a single additional binary element is added to each initial symbol representing the raw data. For a constellation with $M = 2^m$ symbols, the corresponding TCM will have $2M = 2^{m+1}$ symbols, e.g., for a QPSK modulation, where $M = 4$, the corresponding TCM has 8 symbols, yielding usually to a 8-PSK modulation.

Block codes, instead of convolutional, are mainly used for TCM, because the coder and modulator can easily be represented by a trellis diagram with a number of states NS depending on the constraint length L of the code, with $NS = 2(L-1)$. This type of coding, in association with a soft decision and maximum likelihood decoder of the received sequences, gives near optimum performance. The choice of the code and partitioning of the constellation are driven by three heuristic rules, first proposed in [181], usually referred to as *set partitioning*:

- in order to provide regularity and symmetry during exploration of the constellation, the $2M$ points of constellation A_0, see Figure 5.15, should occur with equal frequency

- if 2^{q-k} parallel transitions exist, they are labeled with signal points belonging to the same subset and having 2^{q-k} points

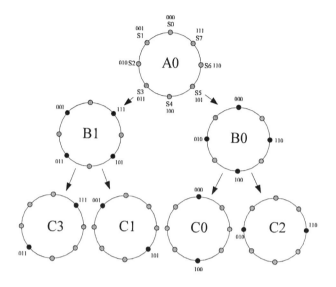

Figure 5.15. Partitioning of the 8-PSK circular constellation

- the 2^q transitions originating from or merging in one state, are labeled with signal points, belonging to the same subset and having 2^q points.

The two last rules guarantee that d_{free}, which is associated with all single and multiple signal error events, is always greater than d_M of the corresponding uncoded modulation, which has the same spectral efficiency and is used as the reference to calculate the coding gain. It should be noted that, when the number of states of the trellis is too small, then the third rule cannot be verified.

Coherent detection of M-ary modulation can suffer by $2\pi/M$ phase ambiguity at receiver during signal acquisition and possible symbol skipping due to high noisy conditions. The introduction of a differential coder at the transmitter side, dedicated to the choice of the transmitted symbol, allows to interleave information and redundancy between several symbols, and thus, removes this ambiguity. This procedure is the so-called multi-dimensional TCM.

5.3.1.2   TCM example.   According to the general method presented above, the principle of TCM is illustrated in Figure 5.15, with the following example, based on the 8-PSK circular constellation. The initial bi-dimensional constellation, A_0, characterized with $d_2 = 2\sqrt{E}\sin(\pi/8)$, where E is the signal energy per code word, is partitioned into two sub-constellations (or subsets) B_0 and B_1. Each of these subsets consists of a QPSK constellation with $d_1 = \sqrt{2E}$. The mapping of the 3 bit symbols is chosen in such a way that for B_0 the minimum significant bit is 0, while it is 1 for B_1. Each subset B_0 (or B_1)

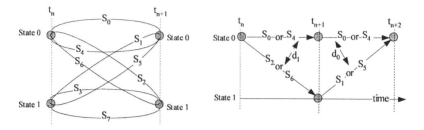

Figure 5.16.   Two states trellis and related minimum distance paths

can be divided into two new subsets C_0 and C_2 (or C_1 and C_3), these subsets have only a BPSK constellation with $d_0 = 2\sqrt{E}$; for subsets C_i the mapping of the 3 bit symbols is obtained with the two minimum significant bits, while the maximum significant bit is used to detect which of the two points is sent, as shown in Figure 5.15. This choice yields into a natural binary mapping for constellation A_0.

This modulation having one redundant bit per symbol and spectral efficiency of 2 bits/Hz must be compared, in terms of BER performance, with an uncoded QPSK modulation having the same efficiency and characterized with $d_1 = \sqrt{2E}$. Figure 5.16 illustrates a trellis diagram of a convolutional code with $k = 1$ and $q = 2$, with rate $r = 1/2$ and constraint length $L = 2$. It is noted that there are two parallel branches between states and the third rule is not satisfied. The minimum distance for the two possible first paths is such that $d_{free}^2 = d_1^2 + d_0^2 = 2E\left[1 + 2\sin^2\left(\pi/8\right)\right]$. The asymptotic gain in comparison to the QPSK modulation is $G_{dB} = 1.12$ dB.

For the four states trellis diagram shown in Figure 5.17, $k = 1$, $q = 2$, $r = 1/2$ and $L = 3$. There are also two parallel branches between states and the third Ungerboeck rule is now satisfied. The asymptotic gain, as compared to the QPSK modulation, is $G_{dB} = 3.01$ dB.

Table 5.2 gives the asymptotic coding gains as calculated in [180] for 8-PSK TCM with trellis from 2 up to 256 states. It should be noted that in practice the number of states is, usually, limited to 64 due to the increased complexity of the decoding function for larger number of states.

5.3.2   Bit interleaved coded modulation

In classical, TCM coding is combined with modulation in order to be seen as a single entity at the receiver. In a fading channel, a symbol interleaver is usually employed to improve the system performance. Unlike TCM, the Bit Interleaved Coded Modulation (BICM) employs a single bit interleaver at the encoder output and a soft demodulator at the receiver [183].

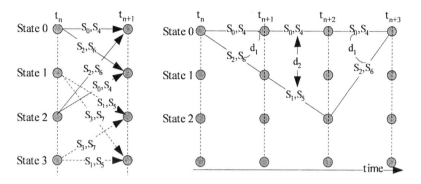

Figure 5.17. Four states trellis and related minimum distance paths calculation

Number of States	k	Asymptotic Gain (dB)
2	1	1.12
4	1	3.01
8	2	3.6
16	2	4.13
32	2	4.59
64	2	4.8
128	2	5.0
256	2	5.4

Table 5.2. Asymptotic coding gain for 8-PSK TCM with trellis from 2 up to 256 states

Let us assume uniform input distributions, ideal interleaving, C_{TCM} be the channel capacity of TCM and C_{BICM} be the channel capacity of BICM. It can be proved that $C_{BICM} \leq C_{TCM}$, which shows the sub-optimality of the BICM scheme. The maximization of the BICM channel capacity is achieved by Gray mapping, in contrast to the *set partitioning* rule by Ungerboeck.

When the comparison is with respect to the *cut-off rate*, TCM outperforms BICM in the AWGN channel. On the other hand, BICM outperforms TCM in a Rayleigh fading channel, especially for high rates ($R_o > 1$ bit/dim).

From both theoretical analysis and simulation results it can be shown that BICM improves the performance over a fading channel when it is compared to TCM. This is because the code diversity in BICM depends on minimum Hamming distances rather than on minimum Euclidean distances, as it does in TCM. However, BICM requires extra decoding complexity for the branch metrics computation compared to TCM.

In [183] the encoder consists of binary convolutional codes. It is possible that the encoder consists of turbo codes, introducing the concept of Bit Interleaved Turbo Coded Modulation (BITCM), [184], as will be presented next.

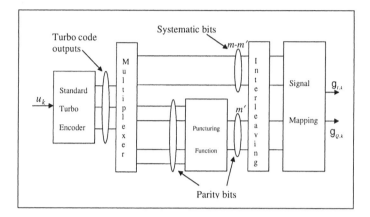

Figure 5.18.   BITCM transmitter [185] ©1994 IEEE

5.3.3   Bit interleaved turbo coded modulation

The simplest way to get various turbo coded modulation schemes is to use a pair of a standard turbo encoder/decoder and then modify a puncturing function and the signal constellation. This introduces the concept of BITCM or pragmatic binary turbo coded modulation [185]. Special care should be taken at the demodulator output, so that soft bit Log-Likelihood Ratio (LLR) values are produced associated with the transmitted symbol values.

An advantage of BITCM is the low decoding complexity compared to other turbo coded modulation schemes, e.g., turbo TCM and parallel concatenated TCM. This makes BITCM applicable to systems that are flexible to select the coding rate and the spectral efficiency according to the channel characteristics, without redesigning the encoder/decoder.

5.3.3.1   BITCM encoder/mapping.   A typical BITCM encoder and the signal mapping in case of 16-QAM are shown in Figures 5.18 and 5.19 respectively. At time k the information sequence u_k is encoded by a turbo encoder with coding rate $1/n$. The parity bits are punctured, so as m' coded bits and $(m - m')$ information bits are obtained. The signal is then mapped into $M = 2^m$ modulated symbols according to Gray coding. In a fading channel a bit interleaver is usually added between the puncturing function and the modulator. The overall coding rate is $R = 1/n \times n \times (m - m')/m = (m - m')/m$ and the spectral efficiency is $\Gamma = (m - m')$ bits/sec/Hz.

For example, a rate $1/2$ turbo coded 16-QAM scheme can be obtained, if $n = 3$, $m = 4$ and $m' = 2$. The mapping rule is illustrated in Figure 5.19, where $m' = 2$ parity bits are assigned to the most protected bits, i.e., $(u_{k,1}, u_{k,3})$,

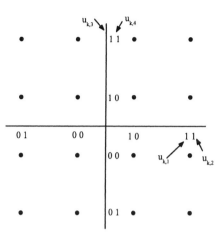

Figure 5.19. BITCM mapping for 16-QAM [185] ©1994 IEEE

of the signal constellation. Other combinations of bit mapping that affect the turbo code performance are possible, and will be discussed in Section 5.3.3.3.

5.3.3.2   BITCM receiver.   The basic component in the BITCM receiver, shown in Figure 5.20, is the LLR computation module. This converts the received symbol values to soft bit values, which are then passed at the standard turbo decoder input.

Assume that the in-phase and quadrature demodulator outputs of a coherent receiver, at an instant time k, are

$$y_{I,k} = g_{I,k} + n_{I,k}$$

$$y_{Q,k} = g_{Q,k} + n_{Q,k} \tag{5.19}$$

where $g_{I,k}$, $g_{Q,k}$ are the I and Q phase channels of the transmitted signal and $n_{I,k}$, $n_{Q,k}$ are two uncorrelated Gaussian processes with zero mean and variance σ_N^2 respectively.

The soft demodulator outputs are

$$\Lambda(u_{k,i}) = K \ln \left(\frac{P_r[u_{k,i} = 1/(y_{I,k}, y_{Q,k})]}{P_r[u_{k,i} = 0/(y_{I,k}, y_{Q,k})]} \right), \quad i = 1, 2, \dots, m \tag{5.20}$$

where K is a constant and m is the number of bits that correspond to a modulated symbol. Using *Bayes' rule* we have

$$\Lambda(u_{k,i}) = K \ln \left(\frac{P_r[(y_{I,k}, y_{Q,k})/u_{k,i} = 1]P(u_{k,i} = 1)}{P_r[(y_{I,k}, y_{Q,k})/u_{k,i} = 0]P(u_{k,i} = 0)} \right), \quad i = 1, 2, \dots, m \tag{5.21}$$

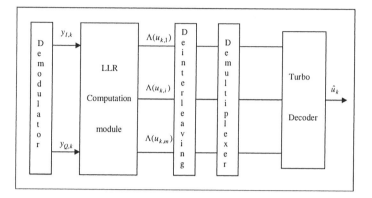

Figure 5.20. BITCM receiver [185] ©1994 IEEE

Considering M-QAM constellations with $M = 2^m$ and $m = 2p$, we can further approximate the above expression, by taking into account that the observation $(y_{I,k}, y_{Q,k})$ is affected by two independent Gaussian noises with identical variance σ_N^2. Thus, the LLRs associated to bits $u_{k,i}, (i = 1, 2, \ldots, p)$, depend only on the observation $y_{I,k}$. Similarly, the LLRs associated to bits $u_{k,i+p}, (i = 1, 2, \ldots, p)$, depend only on the observation $y_{Q,k}$.

In case of the LLRs associated to bits $u_{k,i}, (i = 1, 2, \ldots, p)$ and assuming equal transmitted bit probabilities, equation (5.21) becomes

$$\Lambda(u_{k,i}) = K \ln \left(\frac{\sum_{i=1}^{2^{p-1}} \exp\{-1/\left(2\sigma_N^2\right)(y_{I,k} - \alpha_{1,i})^2\}}{\sum_{i=1}^{2^{p-1}} \exp\{-1/\left(2\sigma_N^2\right)(y_{I,k} - \alpha_{0,i})^2\}} \right), i = 1, 2, \ldots, p$$

(5.22)

where $\alpha_{1,i}$ and $\alpha_{0,i}$ represent the realization of symbols conditionally on $u_{k,i} = 1$ and $u_{k,} = 0$. A simple approximation on above equation, assuming 16-QAM constellation as reported in [185], is

$$\Lambda(u_{k,1}) = y_{I,k} \text{ and } \Lambda(u_{k,2}) = |y_{I,k}| - 2$$

(5.23)

Similar equations hold for the LLRs associated to bits $u_{k,i+p}, (i = 1, 2, \ldots, p)$, by substituting $y_{Q,k}$ instead of $y_{I,k}$.

5.3.3.3   Simulation results.

The basic simulation block diagram is shown in Figure 5.21. The system parameters are 16-state Turbo encoder with generator polynomials $G(D) = [(1, 35/23)]_o$ in octal form, non-uniform interleaver of length 4096 bits, similar to [185]. At the receiver, three iterations are performed by using an improved SOVA turbo decoder with the *norm1* method, e. g. see Section 4.3.5.3.

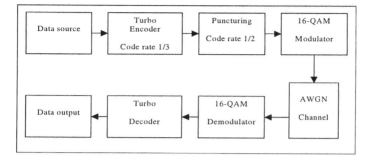

Figure 5.21. BITCM simulation block diagram

The effect of BITCM priority mapping is illustrated in Figure 5.22. Three types of priority mapping are considered. The fist type is to assign the *parity* bits as the most protected bits, denoted by $(p/s, p/s)$, similar to [185]. The second type is to assign the *systematic* bits as the most protected bits, denoted by $(s/p, s/p)$, similar to [186]. The third type is a *combination* of the former two types, assigning one parity bit and one systematic bit as the most protected bits, denoted by $(p/s, s/p)$ [187].

It can be noticed that the optimum performance is achieved when the second type of priority mapping, i.e., $(s/p, s/p)$ is used. This affects the distribution's received values of the systematic bits, to be more Gaussian-like and thus facilitates the turbo decoding process. At lower BER values this does not seem to happen, making all the three types of priority mapping to perform similarly.

5.4 Detection with ideal channel state information

In this section, the error performance of digital receivers operating in AWGN and fading channel is presented, assuming that ideal channel state information is available at the receiver.

5.4.1 Detection over the AWGN channel

Considering the AWGN channel, the received signal can be expressed as

$$y(t) = s(t) + n(t) \tag{5.24}$$

where $s(t)$ is the transmitted signal at timing instant t and $n(t)$ is the corresponding noise. Assuming perfect synchronization, after Nyquist filtering and ideal sampling, the received kth sample is

$$y_k = g_k + n_k \tag{5.25}$$

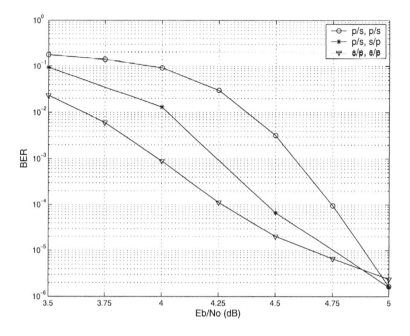

Figure 5.22.   BITCM BER performance with three types of priority mapping

where g_k and n_k are the received symbol and noise samples, respectively. They are in general complex numbers, represented as

$$g_k = g_{I,k} + jg_{Q,k}$$
$$n_k = n_{I,k} + jn_{Q,k} \qquad (5.26)$$

5.4.1.1   Phase shift keying .   Considering Gray encoding, the BEP of M-PSK signaling is presented, while the special cases of BPSK and QPSK are studied in details.

5.4.1.1.1   Coherent binary PSK .   In case the of BPSK where there are two real symbols with amplitude $\pm\sqrt{2E_b}$, the BEP is

$$P_{eb} = \frac{1}{2}\Pr\left(n_{I,k} > \sqrt{2E_b}\right) + \frac{1}{2}\Pr\left(n_{I,k} < -\sqrt{2E_b}\right) \qquad (5.27)$$

Since noise is Gaussian distributed, the above equation can be expressed as

$$P_{eb} = Q\left(\sqrt{2\gamma_b}\right) \qquad (5.28)$$

where γ_b is the SNR per bit

$$\gamma_b = \frac{E_b}{N_0} \tag{5.29}$$

5.4.1.1.2   Coherent quadrature PSK .   In the case of QPSK where there are four complex symbols $\sqrt{E_s}$ $(\pm 1 \pm \jmath)$, the Symbol Error Probability (SEP) is

$$P_{es} = 1 - \Pr\left(n_{Q,k} > \sqrt{E_s}, n_{I,k} > \sqrt{E_s}\right) \tag{5.30}$$

and since the I and Q channels of the noise samples are mutually independent, the above equation can be expressed as

$$
\begin{aligned}
P_{es} &= 1 - \Pr\left(n_{I,k} > -\sqrt{E_s}\right) \Pr\left(n_{Q,k} > -\sqrt{E_s}\right) \\
&= 1 - [1 - Q\left(\sqrt{\gamma_s}\right)]^2 \\
&= 2\,Q\left(\sqrt{\gamma_s}\right) - Q^2\left(\sqrt{\gamma_s}\right)
\end{aligned}
\tag{5.31}
$$

where γ_s is the SNR per symbol

$$\gamma_s = \frac{E_s}{N_0} \tag{5.32}$$

with $\gamma_s = 2\gamma_b$ and $E_s = 2E_b$. For high values of γ_s, $Q\left(\sqrt{\gamma_s}\right) \gg Q^2\left(\sqrt{\gamma_s}\right)$, using eq. (5.31) and considering Gray encoding, the BER performance of QPSK can be expresses as

$$P_{eb} \simeq Q\left(\sqrt{2\gamma_b}\right) \tag{5.33}$$

It is interesting to be noted, that the BER performance of QPSK is identical with BPSK, but QPSK has the advantage of double the spectral efficiency.

5.4.1.1.3   Coherent PSK .   For the general case of M-PSK, a closed form solution for the SEP is not available, however the SEP can be written in terms of an integral with finite limits as

$$P_{es} = \frac{1}{\pi} \int_0^{(M-1)\pi/M} \exp\left\{-\gamma_s \frac{\sin^2(\pi/M)}{\sin^2(\theta)}\right\} d\theta \tag{5.34}$$

It can be easily shown that for the special cases of $M = 2$ and $M = 4$, eq. (5.34) reduces to eq. (5.28) and (5.31), respectively. A well-known bound for eq. (5.34) is

$$P_{es} \simeq 2\,Q\left[\sqrt{2\gamma_s}\,\sin\left(\frac{\pi}{M}\right)\right] \tag{5.35}$$

which can be used in combination with

$$P_{eb} = \frac{P_{es}}{\log_2 M} \tag{5.36}$$

for Gray encoding, for determining the BER of M-PSK modulation format.

5.4.1.1.4   Coherent differentially encoded PSK .   A useful expression relating the SEP of coherently detected with differential encoding PSK (M-Differential Encoding Phase Shift Keying (DEPSK)) and coherent detected M-PSK is given by

$$P_{es} = 2\, P_{es}|_{M-\text{PSK}} - P_{es}^2|_{M-\text{PSK}} - \sum_{k=1}^{M-1} P_k^2 \tag{5.37}$$

where

$$P_k = \frac{1}{2\pi} \int_0^{\pi - \frac{(2k-1)\pi}{M}} \exp\left\{ -\gamma_s \frac{\sin^2\left[(2k-1)\frac{\pi}{M}\right]}{\sin^2(\theta)} \right\} d\theta$$
$$- \frac{1}{2\pi} \int_0^{\pi - \frac{(2k+1)\pi}{M}} \exp\left\{ -\gamma_s \frac{\sin^2\left[(2k+1)\frac{\pi}{M}\right]}{\sin^2(\theta)} \right\} d\theta \tag{5.38}$$

and $P_{es}|_{M-\text{PSK}}$ is given by eq. (5.34).

5.4.1.1.5   Differential PSK .   The BER of Differential Binary Phase Shift Keying (DBPSK) is given by

$$P_{eb} = \frac{1}{2} \exp\{-\gamma_b\} \tag{5.39}$$

while the SEP of differentially detected M-PSK (M-DPSK) is given by

$$P_{es} = \frac{1}{2\pi} \int_0^{(M-1)\pi/M} \exp\left\{ -\gamma_s \frac{\sin^2(\pi/M)}{\sin^2(\theta) + \sin^2(\theta + \pi/M)} \right\} d\theta \tag{5.40}$$

which for large values of M simplifies to

$$P_{es} \simeq \frac{1}{\pi} \int_0^{(M-1)\pi/M} \exp\left\{ -\frac{1}{2}\gamma_s \frac{\sin^2(\pi/M)}{\sin^2(\theta)} \right\} d\theta \tag{5.41}$$

By comparing eq. (5.41) with eq. (5.34) it can be easily concluded that for large values of M, the performance of M-PSK is 3 dB better than the equivalent performance of M-DPSK.

5.4.1.2   Quadrature amplitude modulation .   In case of coherent detection of QAM signals, the SEP of a coherently detected M-QAM modulation scheme is given by

$$P_{es} = 4\left(1 - \frac{1}{\sqrt{M}}\right) Q\left(\sqrt{\frac{3\gamma_s}{M-1}}\right) - 4\left(1 - \frac{1}{\sqrt{M}}\right)^2 Q^2\left(\sqrt{\frac{3\gamma_s}{M-1}}\right) \tag{5.42}$$

For $M = 4$, the above equation reduces to eq. (5.31). For large values of γ_s, with the aid of eq. (5.36), the BER of Gray encoded M-QAM can be simplified as

$$P_{eb} \simeq \frac{4}{\log_2 M} \left(1 - \frac{1}{\sqrt{M}}\right) Q\left[\sqrt{\frac{3\gamma_b \log_2 M}{M - 1}}\right] \tag{5.43}$$

It is noted that, typically M-QAM is detected using coherent demodulation methods. However, techniques using noncoherent detection for M-QAM have been also proposed in [188][189][190].

5.4.1.3   AWGN performance bounds .   The aim of this section is the analytic determination of the probability of error, P_{es}, for an arbitrary two-dimensional constellation C.

In general, and for arbitrary modulation schemes, the derivation of a closed form expression for the bit or word error probabilities is a very difficult task. As a result, there are two particular reasons that analytical bounding techniques are proved to be an insuperable engineering tool for the design and performance evaluation of codes and modulation schemes:

- to gain insight on the effect of system parameters on the performance of such schemes and hence, facilitate the design of better schemes

- to provide assessments of performance for the existing systems without time-consuming for very long simulations.

Considering a signal point set s_i, $i = 1, 2, \ldots, M$, with M denoting the number of signal points, it is important to know how constellation points are distributed. In the two-dimensional space E_s can be computed as

$$E_s = \sum_{i=1}^{M} |s_i|^2 \pi_i \tag{5.44}$$

where π_i is the i-th symbol a-priori probability. To compute the error probability, for a given signal point s_i, it is necessary to define its decision region D_i. This will allow the computation of the error probability $P_{es}(i)$ for the i-th symbol by using the following integral

$$P_{es}(i) = \iint_{D_i} f(x; y) \, \mathrm{d}x \, \mathrm{d}y \tag{5.45}$$

where $f(x; y)$ is the bivariate Gaussian probability density function (pdf) whose integral over the decision region provides the exact expression for $P_{es}(i)$. In this way the global error probability of the given constellation can easily work out as

$$P_{es} = \sum_{i=1}^{M} P_{es}(i) \pi_i \tag{5.46}$$

Since the boundaries of the decision region are often described by complex polygons, the exact P_{es} expression contains integrals which are difficult to be evaluated. The only reasonable way to calculate them is rewriting and bounding every double integrals as the sum of a finite number of more simple one-dimensional integrals. In addition, each of these integrals will be rearranged in order to contain the same elementary function as integrand function, over a finite integration range.

It was shown in eq. (2.186) that the classical union bound is the most used and simple bound. Due to the fact that a union bound is not very tight at low SNR, a first significant improvement was provided only for sphere constellations, by the Tangential Bound of Berlekamp. This bound combines the Gallager First Bounding Technique (GFBT), eq. (2.188), with the union bound, separating radial and tangential components of the Gaussian noise in a half-space referred as Gallager region $\mathcal{R}$.

Tangential Bound (TB). Given a transmitted signal point s_0, denoting the received sample as r, the radial component of the noise as z_1 and its tangential component as z_2, the Gallager region is defined as

$$\mathcal{R} = \{r | z_1 \leq \gamma\} \qquad (5.47)$$

where γ is used to tighten the bound. Applying first the GFBT and then the union bound to (5.47), the BEP can be bounded by

$$P_{eb} \leq \sum_{k=1}^{L} A_k Q\left(\frac{d_k}{2\sigma}\right) Q\left(\frac{d_k^2/4 - \sqrt{E_k}\gamma}{\sigma\sqrt{E_k - d_k^2/4}}\right) + Q\left(\frac{\gamma_{opt}}{\sigma}\right) \qquad (5.48)$$

where $\sqrt{E_k}$ is the square root of the transmitted codeword energy, d_k is the distance among the received vector and the considered codeword, $A_k = 1, 2, \ldots, L$ is the number of codewords at a distance d_k from the transmitted codeword and γ_{opt} is the root of the following equation

$$\sum_{k=1}^{L} A_k Q\left(\frac{\sqrt{E_k} - \gamma_{opt}}{\sqrt{E_k - d_k^2/4}} \frac{d_k}{2\sigma}\right) = 1 \qquad (5.49)$$

For $\gamma \to \infty$, the tangential bound becomes identical to the simple union bound, which proves that the union bound is never tighter than the tangential bound. Similar to the TB of Berlekamp, the Sphere Bound of Herzberg and Poltyrev has been the subsequent significant improvement to the union bound for ML-decoded PSK constellations. In this case the Gallager region is a sphere centered at the transmitted signal point whose radius r must be optimized in order to tighten the bound

$$\mathcal{R} = \{r | \|r - s_0\| \leq r\} \qquad (5.50)$$

Sphere Bound . For an ML-decoded PSK sequence, considering the above decision regions and applying GFBT (see section 2.4.1.3) the word-error probability can be written as

$$P_{es} \leq Pr\{\mathbb{E}, \|r - s_0\| \leq r\} + Pr\{\|r - s_0\| > r\} =$$
$$= Pr\{\mathbb{E}, y \leq r^2\} + Pr\{y > r^2\} \tag{5.51}$$

where $\mathbb{E}$ denotes the error event, and $y = \sum_{i=1}^{N} n_i^2$ is a random variable with Chi-square distribution with N degree of freedom. With $M = 2^k$ and applying the union bound to equation (5.51) the decoding probability error can be written as

$$P_{es} \leq \sum_{k:\, d_k < r} A_k Pr\{\mathbb{E}_k, y \leq r^2\} + Pr\{y > r^2\} \tag{5.52}$$

where $\mathbb{E}_k$ is the error event that the received vector r is closer to s_j than the transmitted signal s_0 and $d_k = \|s_0 - s_k\|^2$ is the distance from the transmitted codeword.

Assuming $z_1 \sim \mathcal{N}(0, \sigma^2)$ the noise component directed from the transmitted codeword towards the origin, and y_1 a Chi-square distribution with $N - 1$ degree of freedom, the overall ML decoding error probability is bounded by:

$$P_{es} \leq \min_r \left\{ \sum_{k:\, d_k^2/4 < r^2} \left(A_k \int_{d_k/2}^{r} f_{z_1}(z_1) \int_{0}^{r^2 - z_1^2} f_{y_1}(y_1) \mathrm{d}y_1 \mathrm{d}z_1 \right) \right.$$
$$\left. + \int_{r^2}^{\infty} f_y(y) \mathrm{d}y \mathrm{d}z_1 \right\} \tag{5.53}$$

Another commonly used bound to evaluate the code performance in the noisy region was proposed by Herzberg and Poltyrev in [191], where *radial* component of the noise z_1 is separated by the rest of the noise vector.

Tangential sphere bound for an AWGN channel. Given two constellation points c_i and c_j, the word-error probability for an ML-decoded PSK sequence can be written as

$$P_{es} = Pr\{\mathbb{E}|n \in C_N(\theta)\}\, Pr\{n \in C_N(\theta)\} + Pr\{\mathbb{E}|n \notin C_N(\theta)\}\, Pr\{n \notin C_N(\theta)\} \tag{5.54}$$

where $C_N(\theta)$ is a bi-dimensional cone with half-angle θ centered at the origin, and n is the noise that affects the received codeword. This probability of error can be upper bounded by

$$P_{es} \leq \min_{\theta} Pr\{\mathbb{E}|n \in C_N(\theta)\}\, Pr\{n \in C_N(\theta)\} + Pr\{n \notin C_N(\theta)\} \tag{5.55}$$

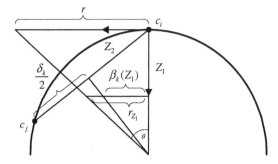

Figure 5.23.   Graphical illustration of the tangential sphere bound

Let

$$r_k = \sqrt{E_k}\tan(\theta) \tag{5.56}$$

$$r_k(z_1) = \tan(\theta)(\sqrt{E_k} - z_1) = r_k - \tan(\theta) \tag{5.57}$$

$$\beta_k(z_1) = \frac{\sqrt{E_k} - z_1}{\sqrt{E_k - d_k^2/4}}\frac{d_k}{2} \tag{5.58}$$

where $\beta_k(z_1)$, as seen in figure 5.23, is the projection of the perpendicular bisector hyper-plane among s_0 and s_k. The probability of error for a given value of the noise component z_1 is upper bounded by

$$P_{es}(z_1) \leq \min_{r_k}\left\{\sum_{k:\beta_k(z_1)<|r_k(z_1)|} A_k Pr\{\mathbb{E}_k|\, z_1, y \leq r_k^2(z_1)\} + Pr\{y > r_k^2(z_1)\}\right\} \tag{5.59}$$

where $\mathbb{E}_k$ is the error event with a given z_1, $y = \sum_{i=2}^{N} n_i^2$ is a random variable with Chi-square distribution with $(N-1)$ degree of freedom.

Thus, separating the tangential component of the noise z_2, with $z_2 \perp z_1$, the tangential sphere bound is given by

$$P_{es} \leq \int_{-\infty}^{\infty}\left\{\sum_{k:\beta_k(z_1)<r_k(z_1)}\left[A_k \int_{\beta_k(z_1)}^{r_k(z_1)} f_z(z_2)\int_0^{r_k(z_1)^2 - z_2^2} f_{y_1}(y_1)\mathrm{d}y_1\mathrm{d}z_2\right]\right.$$

$$\left. + \int_{r_k(z_1)^2}^{+\infty} f_y(y)\mathrm{d}y\right\}\mathrm{d}z_1 \tag{5.60}$$

where z_2 as well as z_1 is a zero mean Gaussian variable with variance σ^2. The optimum value r_{opt} of r_k can be found iteratively through the following

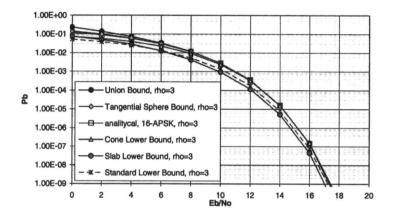

Figure 5.24. Bounds on the bit error probability versus E_b/N_0 for 16-APSK and $\varrho = 3$

equation

$$\sum_{k:\beta_k(z_1)<r_k(z_1)} A_k \frac{\Gamma(N-1)}{2\sqrt{\pi}\Gamma(\frac{N}{2}-1)} \int_0^{\theta_k} \sin^{(N-3)}(u)\mathrm{d}u = 1 \qquad (5.61)$$

where $\theta_k = \arccos\left(\dfrac{d_k/2r_{opt}}{\sqrt{1-d_k^2/2N}}\right)$, and $\Gamma(\cdot)$ is the Gamma function.

Then, this r_{opt} value shall be substituted in eq. (5.56) and (5.57) to finally compute the bound (5.60).

5.4.1.4   APSK analytical performance.   In this section the classical problem of detecting one of M-APSK uncoded signals in the AWGN channel is considered. Since the region of integration is not trivial and the numerical evaluation of $P_e(i)$ is quite difficult, it has been preferred to calculate both an upper and lower bound of $P_e(i)$.

According to Section 2.4.1.3, the classical union bound is the first approach that has been taken into account and it is the most common approach to upper-bounding the conditional probabilities

$$P_e(i) = \sum_{i=1,j\neq i}^{M} Q\left(\frac{d_{i,j}}{2\sigma}\right) \qquad (5.62)$$

where $d_{i,j} = \|s_i - s_j\|$ is the Euclidean distance between the signal points s_i and s_j. The BER obtained by numerical integration shows that this bound is

Figure 5.25. Partitioning of integration regions in equal width slabs centered at minimum distance

asymptotically tight and it attains the true error probability for high SNR, see Figure 5.24. This difference is quite large and the time required to compute the union bound is comparable with that needed for the exact calculation, and hence, it has been chosen to improve this bound with other well-known bounds, as well as, the tangential sphere bound described in Section 5.4.1.3. Since it is very important to examine the tightness of the upper bounds for small SNR values, some lower bounds have been considered looking for both the tightest and the simplest lower bound. One commonly employed lower bound to $P_e(i)$ uses only the closest adjacent signal

$$P_e(i) \geq Q \left[\frac{d_{min}(i)}{2\sigma} \right] \qquad (5.63)$$

where $d_{min}(i)$ is the minimum distance for the transmitted s_i. Since Figure 5.24 shows that unfortunately this "standard" lower bound is not asymptotically tight, both the slab and the cone lower bounds [192] have been studied.

The first bound restricts the integration region to the rectangular areas (*slabs*) centered at the minimum distance and set to the same width for a generic bidimensional constellation, as shown in Figure 5.25. The second approach extends the integration region by converting the slabs into cones. In fact it can be seen that the slabs miss the triangular area on the plane between two adjacent slabs. One way to include part of this area is to widen each slab and angling the parallel sides, as shown in Figure 5.26. By integrating each of these angled regions, tighter lower bound can be obtained than the slab one.

In Figure 5.24, the BER for all lower and upper bounds, for the $\varrho = 3$ case, is presented. Tangential sphere upper bound and cone lower bound yield results that, besides realizing an optimum trade-off between the calculation elapsed time ensures the tightness with the true error probability. These results

Figure 5.26. Partitioning of integration regions in equal width cones centered at minimum distance

are interesting because they allow to determine which constellation parameters provide the best performance, without performing a waste time simulations.

5.4.2 Detection over fading channels

Let R_k be the kth channel envelope of a flat fading channel, after filtering and sampling, the baseband received complex signal can be expressed as

$$y_k = R_k\, g_k + n_k \tag{5.64}$$

In case of fading channels and alleviating index k, γ_s is a varying RV according to R^2, i.e.,

$$\gamma_s = R^2 \frac{E_s}{N_0} \tag{5.65}$$

and since $P_{es}(\gamma_s)$ is a function of γ_s, $P_{es}(\gamma_s)$ is also a RV hereafter called as Conditional Symbol Error Probability (CSEP). Following the pdf-based approach, the average value of $P_{es}(\gamma_s)$ can be derived averaging the CSEP over the pdf of γ_s, i.e.,

$$\overline{P}_{es}(\overline{\gamma}_s) = \int_0^\infty P_{es}(\overline{\gamma}_s|\gamma)\, f_{\gamma_s}(\gamma)\, d\gamma \tag{5.66}$$

An alternative and efficient method to study the average Average Symbol Error Probability (ASEP) is to use the moments generating function (mgf) based approach . By following this approach and using an available formula for the γ_s mgf, of the fading channel under consideration, the ASEP can be derived using Table 5.3. For example, a well-known expression for the average

Modulation scheme	ASEP $\overline{P}_{es}$
BPSK	$\frac{1}{\pi} \int_0^{\pi/2} \mathcal{M}_{\gamma_s} \left[\frac{1}{\sin^2(\varphi)} \right] d\varphi$
BDPSK	$\frac{1}{2} \mathcal{M}_{\gamma_s}(1)$
M-PSK	$\frac{1}{\pi} \int_0^{\pi-\pi/M} \mathcal{M}_{\gamma_s} \left[\frac{g_{psk}}{\sin^2(\varphi)} \right] d\varphi, \quad g_{psk} = \sin^2 \left(\frac{\pi}{M} \right)$
M-DPSK	$\frac{1}{\pi} \int_0^{\pi-\pi/M} \mathcal{M}_{\gamma_s} \left[\frac{g_{psk}}{1 + \cos(\varphi) \cos(\pi/M)} \right] d\varphi$
Square M-QAM	$\frac{4}{\pi} \left\{ \left(1 - \frac{1}{\sqrt{M}} \right) \int_0^{\pi/2} \mathcal{M}_{\gamma_s} \left[\frac{g_{qam}}{\sin^2(\varphi)} \right] d\varphi \right.$ $\left. - \left(1 - \frac{1}{\sqrt{M}} \right)^2 \int_0^{\pi/4} \mathcal{M}_{\gamma_s} \left[\frac{g_{qam}}{\sin^2(\varphi)} \right] d\varphi \right\},$ $g_{qam} = \frac{3}{2(M-1)}$

Table 5.3.   ASEP formulae as a function of the mgf of γ_s.

BER of coherent BPSK is[1]

$$\overline{P}_{eb} = \frac{1}{2} \mathcal{M}_{\gamma_b}(1) \tag{5.67}$$

Next, the ASEP of various modulations schemes, i.e., M-PSK and M-QAM, for Rayleigh, Rice, Nakagami-m and Weibull fading channels is presented.

5.4.2.1   Detection over Rayleigh fading channel .   One of the most widely used distributions to model fading channels is the Rayleigh with its pdf given by eq. (2.68). Hence, replacing eq. (2.75) and eq. (5.34) in eq. (5.66), the ASEP performance of coherent M-PSK is

$$\overline{P}_{es}(\overline{\gamma}_s) = \frac{M-1}{M} \left\{ 1 - \sqrt{\frac{\sin^2(\pi/M)\,\overline{\gamma}_s}{1 + \sin^2(\pi/M)\,\overline{\gamma}_s}} \frac{M}{(M-1)\,\pi} \right.$$
$$\left. \times \left[\frac{\pi}{2} + \tan^{-1} \left(\sqrt{\frac{\sin^2(\pi/M)\,\overline{\gamma}_s}{1 + \sin^2(\pi/M)\,\overline{\gamma}_s}} \cot \left(\frac{\pi}{M} \right) \right) \right] \right\} \tag{5.68}$$

---

[1]It is noted that the mgf of γ_s is considered with negative sign, i.e., $\mathcal{M}_{\gamma_s}(s) = \mathcal{E} \langle \exp\{-s\,\gamma_s\} \rangle$.

Using Table 5.3 and eq. (2.77) the ASEP of M-QAM can be obtained as

$$\overline{P}_{es}\left(\overline{\gamma}_s\right) = 2\frac{\sqrt{M}-1}{\sqrt{M}}\left(1-\mu_1\right) - \left(\frac{\sqrt{M}-1}{\sqrt{M}}\right)^2\left[1 - \frac{4\mu_1}{\pi}\tan^{-1}\left(\frac{1}{\mu_1}\right)\right]$$

(5.69)

with

$$c_1 = \frac{1.5\overline{\gamma}_s}{M-1} \quad \text{and} \quad \mu_1 = \sqrt{\frac{c_1}{c_1+1}}$$

(5.70)

5.4.2.2   Detection over Rice fading channel.   In several wireless channels, there is a specular or a Line of Sight (LOS) component added to the multipath. These kind of channels can be modeled by the Rice distribution, given by eq. (2.96), where $0 \leq K = n^2 < \infty$ is the Ricean factor. In this channel the ASEP of digital receivers for M-PSK modulation is given by

$$\overline{P}_{es}\left(\overline{\gamma}_s\right) = \frac{1}{\pi}\int_0^{\pi-\pi/M} \frac{(1+K)\sin^2(\theta)}{(1+K)\sin^2(\theta) + \overline{\gamma}_s\, g_{psk}}$$

$$\times \exp\left\{\frac{K\,\overline{\gamma}_s\, g_{psk}}{(1+K)\sin^2(\theta) + \overline{\gamma}_s\, g_{psk}}\right\} d\theta$$

(5.71)

while for M-QAM is given by

$$\overline{P}_{es}\left(\overline{\gamma}_s\right) = \frac{4}{\pi}\left\{\left(1 - \frac{1}{\sqrt{M}}\right)\int_0^{\pi/2}\frac{(1+K)\sin^2(\theta)}{(1+K)\sin^2(\theta) + g_{qam}\overline{\gamma}_s}\right.$$

$$\times \exp\left\{\frac{Kg_{qam}\overline{\gamma}_s}{(1+K)g_{qam} + g_{qam}\overline{\gamma}_s}\right\} d\theta - \left(1 - \frac{1}{\sqrt{M}}\right)^2$$

$$\times \int_0^{\pi/4}\frac{(1+K)\sin^2(\theta)}{(1+K)\sin^2(\theta) + g_{qam}\overline{\gamma}_s}\exp\left\{\frac{Kg_{qam}\overline{\gamma}_s}{(1+K)g_{qam} + g_{qam}\overline{\gamma}_s}\right\} d\theta\right\}$$

(5.72)

It is worth mentioning that the optimal maximum likelihood detector for Ricean fading channels has been derived for the first time in [193] and later in [194]. This work has shown for the first time that the optimal receiver structure consists of bank of Multiple Differential Detectors (MDD). Such MDD structure involves more than one distinct differential detector, with delay elements progressively increasing multiples of the symbol duration. This detection technique has been acknowledged as a breakthrough in the field of differential detection and is widely known as "Multiple differential detection", [195].

5.4.2.3   Detection over Nakagami-m fading channel.   The Nakagami-m distribution is a very flexible statistical model which spans a range of use from "worst" than Rayleigh channel conditions to no fading. With the aid of Table 5.3 and eq. (2.92), the ASEP of digital receivers for M-PSK modulation and integer values of m is given by

$$
\overline{P}_{es}\left(\overline{\gamma}_s\right) = \frac{M-1}{M} - \frac{1}{\pi}\sqrt{\frac{g_{psk}\,\overline{\gamma}_s/m}{1+g_{psk}\,\overline{\gamma}_s/m}}
$$

$$
\times \left\{ \left[\frac{\pi}{2}+\tan^{-1}(\alpha)\right]\sum_{k=0}^{m-1}\binom{2k}{k}\frac{1}{\left[4\left(1+g_{psk}\,\overline{\gamma}_s/m\right)\right]^k} \right.
$$
$$
\left. +\sin\left[\tan^{-1}(\alpha)\right]\sum_{k=1}^{m-1}\sum_{i=1}^{k}T_{i,k}\frac{\left[\cos\left(\tan^{-1}(\alpha)\right)\right]^{2(k-i)+1}}{\left(1+g_{psk}\,\overline{\gamma}_s/m\right)^k} \right\} \tag{5.73}
$$

with

$$
\alpha = \sqrt{\frac{g_{psk}\,\overline{\gamma}_s/m}{1+g_{psk}\,\overline{\gamma}_s/m}}\,\cot\left(\frac{\pi}{M}\right) \tag{5.74}
$$

and

$$
T_{i,k} = \binom{2k}{k}\left\{\binom{2(k-i)}{k-i}4^i\left[2(k-i)+1\right]\right\} \tag{5.75}
$$

Moreover, the ASEP for M-QAM can be expressed as

$$
\overline{P}_{es}\left(\overline{\gamma}_s\right) = \frac{2\sqrt{M}}{\sqrt{M}-1}\left[1-\mu_m\sum_{k=0}^{m-1}\binom{2k}{k}\left(\frac{1-\mu_m^2}{4}\right)^k\right] - \frac{M}{\left(\sqrt{M}-1\right)^2}
$$

$$
\times \left\{1-\frac{4\mu_m}{\pi}\left[\left[\frac{\pi}{2}-\tan^{-1}(\mu_m)\right]\sum_{k=0}^{m-1}\binom{2k}{k}\frac{1}{\left[4\left(1+c_m\right)\right]^k}\right.\right.
$$
$$
\left.\left. -\sin\left[\tan^{-1}(\mu_m)\right]\sum_{k=1}^{m-1}\sum_{i=1}^{k}T_{i,k}\frac{\left[\cos\left(\tan^{-1}(\mu_m)\right)\right]^{2(k-i)+1}}{\left(1+c_m\right)^k}\right]\right\} \tag{5.76}
$$

with

$$
c_m = \frac{1.5\overline{\gamma}_s}{m(M-1)} \quad \text{and} \quad \mu_m = \sqrt{\frac{c_m}{c_m+1}} \tag{5.77}
$$

5.4.2.4   Detection over Weibull fading channel .   Weibull distribution is another flexible statistical model for describing multipath fading channels in different radio propagation environments. Based on Table 5.3 and eq. (2.87), the ASEP of any modulation scheme under consideration can be derived, in

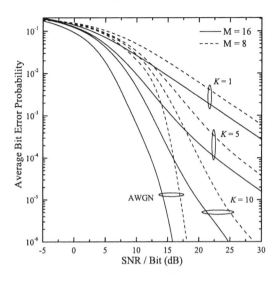

Figure 5.27. ABEP performance of M-PSK signaling in Rice fading

conjunction with the following equation for the mgf of the SNR

$$\mathcal{M}_\gamma(s) = \frac{\beta}{2\left(a\overline{\gamma}_s\right)^{\beta/2}} \frac{\sqrt{k/l}\,(l/s)^{\beta/2}}{\left(\sqrt{2\pi}\right)^{k+l-2}}$$
$$\times G_{l,k}^{k,l}\left[\frac{\left(a\overline{\gamma}_s\right)^{-k\beta/2}}{s^l}\frac{l^l}{k^k}\left|\begin{array}{c}\frac{1-\beta/2}{l},\frac{2-\beta/2}{l},\dots,\frac{l-\beta/2}{l}\\[4pt]0,\frac{1}{k},\frac{2}{k},\dots,\frac{k-1}{k}\end{array}\right.\right] \tag{5.78}$$

with

$$\alpha = \frac{1}{\Gamma\left(1+2/\beta\right)} \tag{5.79a}$$

$$\frac{l}{k} = \frac{\beta}{2} \tag{5.79b}$$

where k and l are positive integers. Depending upon the value of β, a set with minimum values of k and l can be properly chosen in order (5.79) to be valid (e.g., for $\beta = 2.5$, $k = 4$ and $l = 5$). For instance, the ASEP for M−PSK modulation scheme is given by

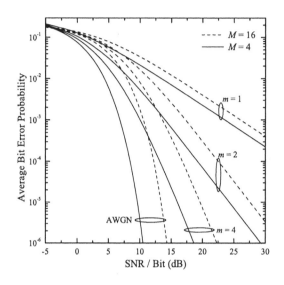

Figure 5.28.   ABEP performance of M-QAM signaling in Nakagami-m fading

$$\mathcal{M}_\gamma(s) = \frac{\beta}{2\pi \, (a\overline{\gamma}_s)^{\beta/2}} \frac{\sqrt{k/l}\, l^{\beta/2}}{\sin^2(\pi/M) \left(\sqrt{2\pi}\right)^{k+l-2}}$$

$$\times \int_0^{\pi-\pi/M} \sin^2(\theta) G_{l,k}^{k,l}\left[\left.\frac{(a\overline{\gamma}_s)^{-k\beta/2}\sin^{2l}(\theta)}{\sin^{2l}(\pi/M)} \frac{l^l}{k^k}\, \right| \begin{array}{l} \frac{1-\beta/2}{l}, \frac{2-\beta/2}{l}, \ldots, \frac{l-\beta/2}{l} \\ 0, \frac{1}{k}, \frac{2}{k}, \ldots, \frac{k-1}{k} \end{array}\right] d\theta$$

$$(5.80)$$

The application of the Weibull fading model to satellite communications has been investigated in [196].

5.4.2.5   Performance analysis.   Next, various performance evaluation results obtained by analytical and bounding techniques are presented.

5.4.2.5.1   Analytical evaluation.   Using the previously derived expressions, the average BEP (ABEP) can be derived for a great variety of modulation schemes and channel models. For instance the ABEP in Figures 5.27, 5.28, and 5.29 is obtained for Rice (5.71), Nakagami-m (5.76), and Weibull (5.80) fading channels, respectively.

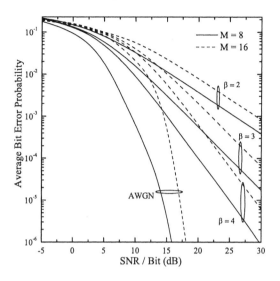

Figure 5.29. ABEP performance of M-PSK signaling in Weibull fading

5.4.2.6 Performance evaluation through bounds .

In this section some bounds for M-PSK and TCM error probability over Rayleigh- and Rice-Lognormal channels are respectively provided.

5.4.2.6.1 M-PSK signaling over Rayleigh-Lognormal channel .

Considering ideal interleaving and constant fading over a symbol duration, the SEP in the Rayleigh-Lognormal channel, which is described in 3.2.2, can be written as

$$P_{es} = \int_0^\infty P(e|r)p_r(r)\mathrm{d}r = \mathrm{E}_S\left[\mathrm{E}_R\left[P(e|r)\right]\right] \tag{5.81}$$

where $P(e|r)$ is the SEP conditioned on r and $p_r(r)$ is the Rayleigh-Lognormal pdf. The inner expectation provides the average error probability in the presence of Rice fading only (small-scale average) while the outer expectation is the average over lognormal shadowing (large-scale average). Equation (5.81) can be used for both coherent and noncoherent demodulation with perfect phase recovery.

Considering M-PSK signaling with coherent demodulation, the SEP conditioned on r can be written as

$$P(e|r) = \frac{1}{2}\overline{N}_a\,\mathrm{erfc}\left(r\sqrt{\rho}\right) \tag{5.82}$$

where [4, eq. (8.250.4)], $\rho = d^2/(8\sigma_\xi^2)$, σ_ξ^2 is the variance of the additive disturbance (e.g., noise, interference) acting with the channel, $\overline{N}_a$ and d are the average number of adjacent points and the minimum Euclidean distance within the signal constellation, respectively. It is important to note that eq. (5.82) is rigorous for binary transmission ($M = 2$) and is a good approximation for M-PSK transmission with high SNR.

Substituting eq. (5.82) into eq. (5.81), the ASEP in the Rayleigh-Lognormal channel may be shown to be [197]:

$$P_{es} = \overline{N}_a E_S \left[Q(U, V) - \frac{1}{2} \left[1 + \sqrt{\frac{p}{1+p}} \right] \exp\left\{ -\frac{U^2 + V^2}{2} \right\} I_0(UV) \right]$$

(5.83)

where $Q(U, V)$ is the Marcum Q-function, and

$$U = \sqrt{K} \left[\frac{1 + 2p}{2(1+p)} - \sqrt{\frac{p}{1+p}} \right]$$

(5.84)

$$V = \sqrt{K} \left[\frac{1 + 2p}{2(1+p)} + \sqrt{\frac{p}{1+p}} \right]$$

(5.85)

$$p = \frac{p_0}{K+1} \quad , \quad p_0 = S^2 \rho$$

(5.86)

Since the evaluation of Marcum Q-function requires high computational complexity for its exact expression, recently some novel tight bounds have been developed [18][198]. These bounds are noticeably good for large values of Marcum Q-function parameters and are stable in regions where exact computation becomes critical due to numerical problems.

In case of noncoherent reception the conditional error probability can be written as [197]

$$P(e|\mathbf{r}) = \frac{1}{M} e^{-2r^2\rho} \sum_{i=2}^{M} (-1)^i \binom{M}{i} e^{(2/i)r^2\rho}$$

(5.87)

Hence, the large scale ASEP in the Rayleigh-Lognormal channel can be shown to be

$$P_{es} = \frac{K+1}{M} e^{-K} \sum_{i=2}^{M} (-1)^i \binom{M}{i} E_S \left[\frac{\exp\left\{ \dfrac{K(K+1)}{K+1+2p_0\frac{i-1}{i}} \right\}}{K+1+2p_0\frac{i-1}{i}} \right]$$

(5.88)

It is important to note that eq. (5.88) is also valid for the special case of DPSK transmission, $M = 2$ and $d = 2\sqrt{E_b}$.

5.4.2.6.2 TCM signaling over Rice-Lognormal channel. An analytic bound can be also found in case of TCM [199], where codewords are mapped onto $L \times M$-PSK constellation points. Let $\boldsymbol{x}_N = (x_1, \ldots, x_N)$ be a sequence of $2L - D$ transmitted signals affected by the fading sequence $\boldsymbol{r}_N = (r_1, \ldots, r_N)$ and by AWGN, the classical union bound on ABEP becomes

$$P_{eb} \leq \sum_{N=1}^{\infty} \frac{1}{2^{mN}} \sum_{\boldsymbol{x}_N} \sum_{\hat{\boldsymbol{x}}_N} n(\boldsymbol{x}_N, \hat{\boldsymbol{x}}_N) P(\boldsymbol{x}_N \rightarrow \hat{\boldsymbol{x}}_N) \qquad (5.89)$$

where $n(\boldsymbol{x}_N, \hat{\boldsymbol{x}}_N)$ is the number of bit errors occurring when $\hat{\boldsymbol{x}}_N$ is detected, and $P(\boldsymbol{x}_N \rightarrow \hat{\boldsymbol{x}}_N)$ is the pairwise error probability associated to this error event of length N. The sum is extended over all possible error events. The pairwise error probability conditioned on fading is

$$P(\boldsymbol{x}_N \rightarrow \hat{\boldsymbol{x}}_N | \boldsymbol{r}_N) = \frac{1}{2} \operatorname{erfc} \left(\sqrt{\sum_{j=1}^{N} \sum_{k=1}^{N} L r_{jk}^2 \alpha_{jk}^2} \right) \qquad (5.90)$$

where $\alpha_{jk}^2 = \dfrac{E_s}{4N_0} d_{jk}^2$ and d_{jk}^2 is the Euclidean distance between symbols x_{jk} and $\hat{x}_{jk}$. The unconditional pairwise error probability is therefore

$$P(\boldsymbol{x}_N \rightarrow \hat{\boldsymbol{x}}_N) = \int_0^\infty \cdots \int_0^\infty \frac{1}{2} \operatorname{erfc} \left(\sqrt{\sum_{j=1}^{N} \sum_{k=1}^{N} L r_{jk}^2 \alpha_{jk}^2} \right)$$
$$\times p_r(r_{11}) \cdots p_r(r_{NL}) dr_{11} \cdots dr_{NL} \qquad (5.91)$$

The above integral must be solved numerically, except when shadowing is absent and all the d_{jk}^2 are the same. To avoid numerical multiple integration the upper-bound, explained in [199], can be used. This minimization corresponds to evaluating the $\operatorname{erfc}(\cdot)$ for the maximum α_{uv} to improve the tightness of the bound. Hence, an upper-bound to pairwise error probability may be written as

$$P(\boldsymbol{x}_N \rightarrow \hat{\boldsymbol{x}}_N) \leq \frac{1}{2} \mathrm{E}_{r_{uv}} \left[\operatorname{erfc} \left(r_{uv} \alpha_{uv} \right) \right]$$
$$\times \prod_{\substack{j=1 \\ (j,k) \neq (u,v)}}^{N} \prod_{k=1}^{L} \mathrm{E}_{r_{jk}} \left[\exp \left\{ -\frac{E_s}{4N_0} d_{jk}^2 r_{jk}^2 \right\} \right] \qquad (5.92)$$

This is very useful since results over the Rice channel can be found analytically. Averaging over the lognormal pdf can be accurately performed by Gauss-Hermite quadrature, as shown in [199].

Using the Chernoff bound, eq. (5.92) can be simplified obtaining a pairwise error probability in the form of a product of terms

$$P(\boldsymbol{x}_N \to \hat{\boldsymbol{x}}_N) \leq \frac{1}{2} \prod_{j=1}^{N} \prod_{k=1}^{L} \mathrm{E}_{r_{jk}} \left[\exp\left\{ -\frac{E_s}{4N_0} d_{jk}^2 r_{jk}^2 \right\} \right] \qquad (5.93)$$

To compute BEP like shown in eq. (5.89), it is necessary to consider all error events or, alternatively, the set of more dominant errors.

5.5   Conclusions

In this chapter the most commonly used coded and uncoded modulation schemes have been reviewed. Emphasis is given to PSK, QAM, and CPM signalling formats, as well as Trellis and Turbo-Trellis coded modulation techniques. Performance evaluation results, obtained by analysis and using bounds, for these schemes in AWGN and fading environments have been also presented and discussed.

Chapter 6

PARAMETER ESTIMATION AND SYNCHRONIZATION

C. Mosquera[1], M.-L. Boucheret[2], M. Bousquet[3], S. Cioni[4], W. Gappmair[5], R. Pedone[4], S. Scalise[6], P. Skoutaridis[7], M. Villanti[4]

[1] *University of Vigo, Spain*

[2] *GET/ENST, France*

[3] *ONERA/TeSA, France*

[4] *University of Bologna, Italy*

[5] *Graz University of Technology, Austria*

[6] *DLR (German Aerospace Center), Germany*

[7] *The University of Surrey, U.K.*

6.1   Introduction

Every communications receiver must perform some operations on the incoming signal to extract the necessary information for a feasible decoding. A number of parameters must be estimated, such as the channel response, the clock and carrier (frequency/phase) references, and the time epochs corresponding to the start of frames or code words. All these ancillary tasks are commonly considered as synchronization issues, and must be especially designed to minimize their impact on the final performance of the decoding.

The choice of synchronization algorithms largely depends on the constraints imposed by their expected performance and the frame formats. The performance of an estimator is usually given in terms of its average value and variance. For those estimators which are unbiased, i.e., with an average value equal to the estimated parameter, the Cramér-Rao bound (CRB) establishes a performance bound for the estimate variance, as detailed in Chapter 2. In the case of biased estimators, the CRB can be conveniently modified provided that the bias is characterized [3]. This bound will be used in this chapter, together with the Modified Cramér-Rao bound (MCRB), easier to compute in many practical cases (see Chapter 2). When it comes to parameter estimation in communication receivers, the behavior with time of the respective synchronization algorithms is of paramount importance. As an example, let us consider bursty transmissions, as those found in Time Division Multiple Access (TDMA) systems: short bursts impose necessarily short acquisition periods, i.e., the time elapsed since data reception starts till estimated parameters can be used to assist the decoding process must be as short as possible, so as to avoid a significant detriment of the useful bandwidth. Short acquisition time operation requires feed-forward synchronization schemes. In fact, there are different types of structures attending to the following criteria:

- Feed-Forward (FF): correction of the affected parameter is performed after the error parameter detection.

- Feed-Backward (FB): correction is performed prior to the parameter estimation.

- Data Aided (DA): Training symbols assist the extraction of the synchronization parameters.

- Decision Directed (DD): Estimated symbols are used for the synchronization.

- Non-Data Aided (NDA): Synchronization is symbol blind.

Equally important is the tracking of the synchronization parameters: once acquired, the synchronization references can change with time, making it necessary to keep the estimation algorithms in continuous operation. As a result, the estimated values will exhibit fluctuations, known as jitter, about an equilibrium point. The amplitude of these fluctuations is usually measured in terms of their variance, which is commonly used as the reference to study the sensitivity of the bit error rate performance to synchronization errors. The fluctuations get larger as the Signal to Noise Ratio (SNR) decreases. If very low SNR values are expected, as usually occurs in satellite systems with strong coding protection, cycle slips become an issue. When cycle slips occur, the estimated values are pushed away from their current stable point to a neighboring stable point which is also a valid solution, causing a loss of data. This is a nonlinear phenomenon difficult to analyze and whose probability of occurrence must be negligible.

The following sections will detail the ideas exposed in this Introduction. The final section on receiver architectures is especially significant; it will show how to combine and put together the different synchronization structures in practical systems.

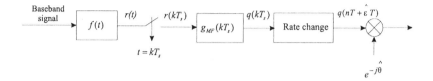

Figure 6.1. Maximum Likelihood receiver.

6.2 Synchronization Principles

Let us consider the detection of a sequence of N symbols $c_n, n = 0, \ldots, N\text{–}1$. The received signal $r(t)$ is written as

$$r(t) = \sum_{n=0}^{N-1} c_n g_r(t - nT - \epsilon_0 T) e^{j(\theta_0 + 2\pi \Delta f t)} + n(t) \qquad (6.1)$$

where $g_r(t)$ is the convolution of the transmission pulse $g(t)$ with the channel response $h_{eq}(t)$:

$$g_r(t) = g(t) * h_{eq}(t) \qquad (6.2)$$

The channel response $h_{eq}(t)$ includes the channel effects and the analog front-end $f(t)$ at the receiver. The parameters ϵ_0 and θ_0, which are assumed constant in this approach for simplicity, express the uncertainty about the beginning of the symbol interval and the carrier phase, respectively. We consider that the possible frequency misadjustment Δf is small enough such that $T\Delta f \ll 1$, with T the symbol period; we will give some details later as to how to proceed without this assumption. The noise $n(t)$ has a Power Spectral Density (PSD) given by

$$S_n(f) = \sigma_n^2 |F(f)|^2 \qquad (6.3)$$

with $\sigma_n^2 = N_0/2$ and $\sigma_n^2 = N_0$ for real and complex modulations respectively, and $F(f)$ the Fourier transform of the analog front-end $f(t)$. A complete digital receiver processes the samples $r(kT_s)$ of the signal $r(t)$, with T_s the sampling period such that $T_s \leq T/2$. These samples are included in the vector $\boldsymbol{r}$. The Maximum Likelihood (ML) criterion maximizes the probability of the observed data, referred to as the likelihood function:

$$\ell(\boldsymbol{r}; \boldsymbol{c}, \epsilon, \theta) \doteq f_{\boldsymbol{r}|\boldsymbol{c},\epsilon,\theta}(\boldsymbol{r}|\boldsymbol{c}, \epsilon, \theta) \qquad (6.4)$$

with the symbols $c_n, n = 0, \ldots, N - 1$ included in the vector $\boldsymbol{c}$. The derivation of the optimum strategy is quite complex (see [200]), although the final result is astonishing illustrative, as shown in Figure 6.1. The function to be minimized can be written as

$$L(\boldsymbol{c}, \epsilon, \theta) = \left[\boldsymbol{q}(\epsilon)e^{-j\theta} - \boldsymbol{H}^{-1}\boldsymbol{c}\right]^H \boldsymbol{H} \left[\boldsymbol{q}(\epsilon)e^{-j\theta} - \boldsymbol{H}^{-1}\boldsymbol{c}\right] \qquad (6.5)$$

for which the elements of the square $N \times N$ matrix $\boldsymbol{H}$ are given by

$$h_{m,n} = [\boldsymbol{H}]_{m,n} = \sum_{k=-\infty}^{\infty} g_r(kT_s - mT)g_{MF}(-kT_s + nT) \qquad (6.6)$$

$g_{MF}(t) = g_r^*(-t)$ represents the matched filter corresponding to the received pulse $g_r(t)$ in equation (6.2). The vector $\boldsymbol{q}(\epsilon)$ contains the samples at symbol rate of $q(t) = r(t) * g_{MF}(t)$ at a given sampling offset ϵT.

PROPOSITION 6.2.1 *If $g_r(t) * g_{MF}(t)$ is a Nyquist pulse then the matrix $\boldsymbol{H}$ boils down to the identity matrix, avoiding the Inter-Symbol Interference (ISI). In such a case, the previous function reads as*

$$L(\boldsymbol{c}, \epsilon, \theta) = \|\boldsymbol{q}(\epsilon)e^{-j\theta} - \boldsymbol{c}\|_2^2 \qquad (6.7)$$

In consequence, the optimum receiver must minimize the distance between all possible received sequences for different phase and timing estimations and all possible symbol sequences $\boldsymbol{c}$.

We conclude then that it is necessary to generate the following sequence of values included in the vector $\boldsymbol{q}(\epsilon)$ and shown in Figure 6.1:

$$q(nT + \epsilon T) = r(t) * g_{MF}(t)|_{nT+\epsilon T}$$
$$= T_s \sum_{k=-\infty}^{\infty} r(kT_s)g_{MF}(nT + \epsilon T - kT_s) \qquad (6.8)$$

This solution is obviously too complex, since it is necessary to evaluate the derived metric for all possible symbol sequences and to scan the whole space of parameters (ϵ, θ). As a result, practical implementations tackle these tasks in successive steps, with different options that we detail next.

A first solution, known as DA estimation, sends a known preamble $\boldsymbol{c}_0$ prior to sending the information symbols:

$$(\hat{\epsilon}, \hat{\theta}) = \arg\left[\max_{(\epsilon,\theta)} \ell(\boldsymbol{r}; \boldsymbol{c} = \boldsymbol{c}_0, \epsilon, \theta)\right] \qquad (6.9)$$

In order to track the slow variations of the synchronization parameters, the decoded symbols $\boldsymbol{c}_D$ can be used, thus avoiding the systematic use of training sequences:

$$(\hat{\epsilon}, \hat{\theta}) = \arg\left[\max_{(\epsilon,\theta)} \ell(\boldsymbol{r}; \boldsymbol{c} = \boldsymbol{c}_D, \epsilon, \theta)\right] \qquad (6.10)$$

The algorithms under this category are referred to as DD, and their correct operation depends on a good initial estimation of the parameters (ϵ_0, θ_0). It is also possible to estimate the parameters without using the transmitted symbols:

$$(\hat{\epsilon}, \hat{\theta}) = \arg \left[\max_{(\epsilon, \theta)} \ell(\boldsymbol{r}; \epsilon, \theta) \right] \qquad (6.11)$$

with

$$\ell(\boldsymbol{r}; \epsilon, \theta)) = \sum_{\substack{\text{set of sequences} \\ \boldsymbol{c}}} \ell(\boldsymbol{r}; \boldsymbol{c}, \epsilon, \theta) p(\boldsymbol{c}) \qquad (6.12)$$

for a given symbol sequence probability distribution $p(\boldsymbol{c})$. These algorithms are referred to as NDA.

Once discarded the joint estimation and detection due to the involved complexity for most practical cases, we must analyze how to proceed to the estimation of the parameters (ϵ, θ). It is usual to perform the estimation of the phase θ_0 after the recovery of the sampling instant ϵ_0, since one sample per symbol period taken at the correct sampling instant is enough for the phase recovery.

The exposed derivation of the ML receiver, developed in more detail in [200], uses an important assumption, namely, that the carrier frequency error is negligible. However, received signals in satellite links usually offer a certain degree of frequency shift (see Chapter 3), which may pose difficulties especially for the lowest symbol rates. This shift must be corrected together with the other parameters affecting the decoding such as the offset in the sampling frequency. In fact, frequency errors can be incorporated to the derivation of an NDA low-SNR Maximum Likelihood receiver. This study was done in [201], where it was shown that for a high degree of uncertainty in the carrier frequency, timing estimation must be performed prior to the matched filtering. In such a case, the matched filter is not really matched to the received signalling pulse, and the potential improvement in the SNR is overcome by the misadjustment due to the frequency shifted pulse. If important frequency errors are expected and cannot be decreased before the timing adjustment, then some classical timing recovery algorithms should be modified, see, e.g., [201] or [202].

Before presenting some synchronization schemes in detail, it is important to note that the choice of a particular algorithm and its design parameters will depend, among other factors, on the impact of the unavoidable estimation errors in the overall system performance. The sensitivity of the BER with respect to the synchronization errors will determine the specifications to be met by the corresponding synchronization algorithms. As an illustration, Table 6.1 shows the performance loss of an M-PSK modulated signal caused by errors in the

	Δf	σ_ϕ^2	σ_ϵ^2
D_{dB}	$\frac{10}{\ln 3}\frac{1}{3}\left(\pi\frac{\Delta f}{R}\right)^2$	$\frac{10}{\ln 10}\left[1 + 2\frac{E_s}{N_0}\cos^2\left(\frac{\pi}{M}\right)\right]\sigma_\phi^2$	$\frac{10}{\ln 10}\left[A + 2\frac{E_s}{N_0}B\right]\sigma_\epsilon^2$

Table 6.1. Degradation [dB] in the performance of an M-PSK modulated signal due to a frequency offset Δf [Hz], a phase jitter σ_ϕ^2 [rad^2] and a normalized timing jitter σ_ϵ^2. R is the symbol rate, and the parameters A and B depend on the raised-cosine roll-off factor.

different synchronization parameters, based on [203] and [204]. The frequency error Δf is assumed to be corrected for after the matched filter for a symbol rate R, and E_s denotes the energy per symbol. The analysis of the impact of synchronization errors in coded systems is more tedious. As a general rule, we can say that coding increases the sensitivity to synchronization errors. In addition, coding allows to operate at lower signal to noise ratios, for which synchronization becomes harder.

6.3 Code Unaware Synchronization

A detailed taxonomy of parameter estimation and synchronization in digital receivers is a tedious task, mainly due to the large number of variants and the considerations behind each of them. Throughout this chapter, we will outline the most relevant algorithms for the *satellite scenario* in terms of the role that coding plays in this context. Thus, we distinguish between *code unaware* and *code aware* algorithms. The former are employed for both coded and uncoded systems; as such they do not exploit any potential information coming from the decoding process. The latter are believed to become a relevant element for future systems working at low signal to noise ratios; they will be presented in Section 6.4.

6.3.1 Frequency Recovery

Following [205], next we present the most common estimators for FF recovery of the carrier frequency. It is assumed that perfect symbol timing has been achieved (see Section 6.3.3) and that the independent and identically distributed (i.i.d.) M-PSK symbols are solely distorted by Additive White Gaussian Noise (AWGN). FB recovery and the application to modulation schemes other than M-PSK will be discussed later.

In the complex baseband model with timing index k, let the M-PSK symbols be denoted by c_k. Furthermore, let the samples q_k at the output of the matched filter be rotated by $2\pi k\Delta f T + \theta$, i.e.,

$$q_k \triangleq q(kT) = c_k e^{j(2\pi k\Delta f T + \theta)} + n_k \qquad (6.13)$$

where θ is the carrier phase, uniformly distributed over $[0, 2\pi)$, and Δf denotes the carrier frequency offset to be estimated. The real and imaginary parts of the complex zero-mean AWGN samples n_k are assumed to be independent, each with the same power spectral density equal to $N_0/2$. Note that the signal model holds only true for rather limited values of Δf (in general, on the order of a few percent of the symbol rate $1/T$), because equation (6.13) does not contain the mismatch between the incoming signal and the receiver filter due to frequency errors, namely, the intersymbol interference and the reduction of the available signal energy. If a DA approach is adopted, the data ambiguity is easily removed by

$$z_k = c_k^* \cdot q_k \tag{6.14}$$

while for NDA algorithms

$$z_k = q_k^M \tag{6.15}$$

In the following, N denotes the observation length in symbol intervals for all frequency recovery techniques. Note also that M describes the cardinality of the M-PSK alphabet in case of NDA, whereas $M = 1$ for DA solutions.

Rife and Boorstyn (R&B) derived the ML solution, with frequency estimates computed as [206]

$$\Delta \hat{f} = \frac{1}{MT} \arg \left[\max_f |Z(f)| \right] \tag{6.16}$$

where $Z(f)$ is the discrete Fourier transform applied to z_k, $k = 0, 1, \ldots, N-1$, i.e.,

$$Z(f) \triangleq \mathcal{F}\{z_k\} = \sum_{k=0}^{N-1} z_k e^{-j2\pi k f} \tag{6.17}$$

In other words, a *coarse* search is performed over a discrete set of f-values, returning the one which maximizes $|Z(f)|$. In order to enhance the frequency resolution, zero-padding by a pruning factor K is suggested, i.e.,

$$z_k' \triangleq \begin{cases} z_k & 0 \leq k < N \\ 0 & N \leq k < KN \end{cases} \tag{6.18}$$

In practice, the coarse search will be implemented as a Fast Fourier Transform (FFT), followed by a *fine* search carried out by interpolation of the samples next to the maximum of $|Z(f)|$. R&B estimator is very powerful in terms of both jitter variance and operational range, but it is fairly complex from a computational point of view (see Table 6.2).

Tretter and *Kay* algorithms, which are equivalent as shown in [205], follow a least squares approach. Compared to R&B synchronizer, they are simpler but not very powerful for smaller SNR's in particular, which makes them less attractive for advanced satellite communications.

Subsequently, two algorithms will be presented which are based on the autocorrelation of the z_k sequence, henceforth defined as

$$R(m) \triangleq \frac{1}{N-m} \sum_{k=m}^{N-1} z_k z_{k-m}^* \tag{6.19}$$

Derived from the ML principle, the first method has been proposed by Luise and Reggiannini (L&R). Via a design parameter $L, 1 \leq L < N$, the frequency estimate develops as [207]

$$\Delta \hat{f} = \frac{1}{\pi M(L+1)T} \arg \left[\sum_{m=1}^{L} R(m) \right] \tag{6.20}$$

In [207], it is shown that the jitter variance achieves a minimum for $L = N/2$. By inspection of eq. (6.20), it is obvious that the operational range is limited to $\pm 1/[M(L+1)T]$. An alternative solution, proposed by *Fitz* in [208], has an estimation range half of L&R's for the same values of M and N.

For both L&R and Fitz estimators, the operational range might be too narrow for practical applications, where the initial frequency offset could be on the order of $\pm 0.1/T$. A possible solution to this problem has been investigated by Mengali and Morelli (M&M). Their algorithm provides a frequency estimate [209]

$$\Delta \hat{f} = \frac{1}{2\pi MT} \sum_{m=1}^{L} w_m \arg \left[R(m+1)R^*(m) \right] \tag{6.21}$$

which is theoretically unbiased up to $\pm 1/2MT$. The related weighting coefficients are given by

$$w_m = \frac{3[(N-m)(N-m+1) - L(N-L)]}{L(4L^2 - 6LN + 3N^2 - 1)} \tag{6.22}$$

with $L = N/2$ as the optimum choice.

In order to conclude this brief overview of FF frequency estimation techniques, the most essential aspects are summarized in Table 6.2. In particular, the *accuracy* stands for the closeness to the MCRB

$$\text{MCRB}(\Delta f) = \frac{3}{2\pi^2 N(N^2-1)T^2 E_s/N_0} \tag{6.23}$$

whereas the *frequency offset sensitivity* addresses the accuracy degradation as a function of the initial frequency offset. Finally, the *normalized estimator range* reports the pull-in range of the corresponding FF algorithm, i.e., the relative frequency offsets that can be estimated, while the *complexity* assesses the associated computational load.

Algorithm	Accuracy	Frequency off-set sensitivity	Normalized estimator range ΔfT	Complexity
R&B	High	Low	$\pm \frac{1}{2M}$	High
Tretter/Kay	Low	Medium	$\pm \frac{1}{2M}$	Low
L&R	High	High	$\pm \frac{1}{ML}$	Low/Med.
Fitz	High	High	$\pm \frac{1}{2ML}$	Low/Med.
M&M	High	Medium	$\pm \frac{1}{2M}$	Medium

Table 6.2. Comparison of Feedforward frequency estimators

EXAMPLE 6.1 For 4-PSK ($N = 128, L = 16$) and 8-PSK ($N = 256, L = 16$), the jitter performance of the L&R estimator is exemplified in Figure 6.2. With decreasing values of E_s/N_0, the jitter variance begins to flatten since arg $[\cdot]$ in eq. (6.20) becomes more and more uniformly distributed between $]-\pi, \pi]$. On the other hand, if $E_s/N_0 \gg 1$, the NDA variance approaches asymptotically the performance of the DA estimator [207]. Note also that the jitter variance can never be located below the MCRB.

For NDA carrier frequency recovery of M-PSK signals, the idea of Viterbi and Viterbi (V&V) [210] – originally developed for NDA recovery of the carrier phase – can be applied as well. Therefore, the L&R estimate (6.20) may be improved if q_k^M is replaced by $|q_k|^\mu e^{jM \arg[q_k]}$, $0 \leq \mu \leq M$. In [211], it is verified that the modified estimator exhibits no bias. Furthermore, a closed-form solution of the jitter variance is derived, which agrees very accurately with the simulation results in the medium-to-high SNR range. The analytical results in [211] have been used to optimize the L&R variant with regard to μ such that the variance exhibits a minimum. For medium-to-high SNR's and smaller values of L, $\mu_{opt} = 0$ seems to be an appropriate solution, whereas $\mu_{opt} = 1$ might be chosen for, say, $L \geq N/4$. The jitter performance of the improved L&R estimator is illustrated in Figure 6.2 for 4-PSK ($N = 128, L = 16, \mu = 0$) and 8-PSK ($N = 256, L = 16, \mu = 2$). Note that the improvement of the variance is particularly pronounced in the transition range between flat top and asymptotic evolution.

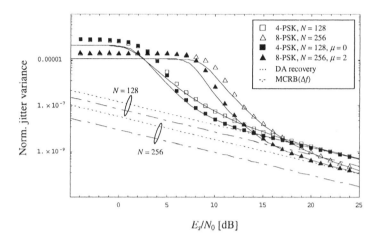

Figure 6.2.   NDA jitter performance for L&R algorithms $(L = 16)$

6.3.1.1   Application to M-QAM Schemes.

For M-QAM schemes, since they are quadrature-symmetric, it has been suggested to apply a fourth-order power nonlinearity which, unfortunately, results in a jitter floor for increasing SNR values [212]. The reason for this phenomenon is found in the data modulation caused by symbols, which are not part of the diagonals in a QAM constellation. Increasing the observation length is a simple but computationally expensive solution. Nevertheless, with $q_k^4 \rightarrow |q_k|^\mu e^{j4\arg[q_k]}$ applied to square-QAM schemes, the jitter floor decreases considerably for increasing values of μ. On the other hand, with cross-QAM schemes, the manipulation of μ does not help very much. A different solution for this problem is presented in [213] which avoids the jitter floor by circular partitioning of the signal constellation such that only the elements along the diagonals are selected. However, the procedure excludes powerful algorithms based on the correlation principle, e.g., L&R and M&M estimators.

6.3.1.2   Feedback Recovery.

Assuming that the symbol timing has been established by appropriate means (see Section 6.3.3), the carrier frequency of an M-PSK signal can as well be recovered by a FB solution according to Figure 6.3. In this case, the frequency error detector (FED) is fed with the synchronized samples at the output of the matched receiver filter, generating an error signal (delay-and-multiply method)

$$u_k = \Im\left\{(q_k q_{k-1}^*)^M\right\} \qquad (6.24)$$

If noise is negligible, it is easily shown that $u_k \sim \sin[2\pi M(\Delta f - \Delta \hat{f})T]$. For slowly varying phase drifts, the loop filter may be chosen as a constant. The Voltage Controlled Oscillator (VCO) is usually a simple integration unit.

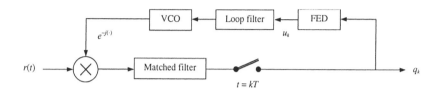

Figure 6.3. Baseband model for closed-loop frequency recovery.

If the symbol timing is not available, oversampling is mandatory to guarantee to resolve the time axis. In this case, $u_k = \Im\{q_k y_k^*\}$, where q_k and y_k denote the oversampled outputs of the matched filter and derivative matched filter, respectively [213]. Replacing the latter by suitably designed filters gives the quadricorrelator solution reported in [214]. As an alternative, the dual-filter detector can be designed so as to be equivalent to the quadricorrelator; the implementation issues will determine the choice [213].

It is important to mention that clock-aided recovery of the carrier frequency exhibits a definitely better synchronization performance, i.e., the jitter variance is much closer to the CRB compared to the case where no timing information is available.

6.3.2   Phase Recovery

For DA recovery of the carrier phase θ, the data symbols c_k are known in advance. If an AWGN channel is assumed [213], the log-likelihood function is given by $\Lambda(\boldsymbol{q};\theta) \sim \sum_{k=1}^{N}|q_k - c_k e^{j\theta}|^2$. Setting the derivative of $\Lambda(\boldsymbol{q};\theta)$ with respect to θ to zero provides the ML estimate. For a DD synchronizer, c_k has to be replaced by $\hat{c}_k$. However, decisions on symbols are not reliable as long as the carrier phase is not recovered [215]. On the other hand, for NDA recovery of M-PSK modulated signals, the symbol ambiguity can be eliminated via an M-th order power nonlinearity. In this respect, Viterbi and Viterbi [210] suggested an appropriate modification using a nonlinear function $f(q_k)$.

All in all, FF estimation of the carrier phase can be summarized as

$$
\hat{\theta} = \begin{cases}
\arg\left[\sum_{k=1}^{N} c_k^* q_k\right] & \text{DA} \\[2mm]
\arg\left[\sum_{k=1}^{N} \hat{c}_k^* q_k\right] & \text{DD} \\[2mm]
\frac{1}{M}\arg\left[\sum_{k=1}^{N} f(q_k)e^{jM\arg[q_k]}\right] & \text{NDA}
\end{cases}
\tag{6.25}
$$

For V&V synchronizers, most times a monomial with additional design parameter μ, $0 \leq \mu \leq M$, is employed, i. e., $f(q_k) = |q_k|^{\mu}$.

In any case, the jitter variance is limited by the modified CRB

$$
\mathrm{MCRB}(\theta) = \frac{1}{2NE_s/N_0}
\tag{6.26}
$$

FF schemes are frequently used for acquisition purposes in burst as well as in continuous modes. For parameter tracking, however, FB synchronizers are frequently suggested because of their simplicity. In order to follow a phase drift, FB synchronizers are normally established as second-order loops such that bias problems (loop stress) are avoided. The error signal u_k at the output of the detector module is straightforwardly computed as [213]

$$
u_k = \begin{cases}
\Im\left\{c_k^* q_k\right\} & \text{DA} \\[2mm]
\Im\left\{\hat{c}_k^* q_k\right\} & \text{DD} \\[2mm]
\frac{1}{M}\Im\left\{f(q_k)e^{jM\arg[q_k]}\right\} & \text{NDA}
\end{cases}
\tag{6.27}
$$

The V&V synchronizer in eq. (6.27) is considered to be applied to M-PSK modulated signals. As with FF algorithms, $f(q_k) = |q_k|^{\mu}$ is employed in most practical cases. Also, $1/2N$ has to be replaced in eq. (6.26) by the equivalent noise bandwidth[1]. The performance of a tracking loop is determined by the detector characteristic (S-curve) defined as the averaged output of the open loop. For linear analysis of the loop, the slope in the stable equilibrium point is required; with respect to M-PSK, closed-form solutions are given in [216] for DD and in [217] for NDA recovery. As a consequence, the design parameter μ of the V&V tracker has been optimized such that the jitter variance achieves a minimum. Figure 6.4 illustrates the result for μ_{opt} and different M-PSK schemes as a function of E_s/N_0. With E_s/N_0 decreasing, it is obvious that $\mu_{\mathrm{opt}} \to M$, whereas in the opposite case $\mu_{\mathrm{opt}} \to 1$.

For ML-oriented NDA synchronization of M-QAM signals, a fourth-order power nonlinearity has to be used as shown in [218]. The self-noise problem may be circumvented in the same manner as already suggested for frequency

---

[1] The equivalent noise bandwidth times the constant noise power spectral density gives the noise variance at the output of the loop.

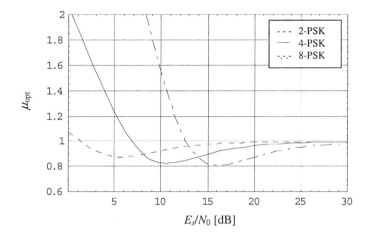

Figure 6.4. Evolution of the optimized design parameter for V&V trackers.

recovery: (i) lowering the jitter floor via a generalized nonlinearity, i. e., $q_k^4 \rightarrow |q_k|^\mu e^{j4\arg[q_k]}$ and increasing μ's for higher SNR values, (ii) circular partitioning of the signal constellation such that only the elements along the diagonals are selected.

With FB algorithms, parameter deviations are usually located near the stable equilibrium point achieved after successful initial acquisition. Due to the noise, however, tracking loops can lose lock, run into one of the nonstable equilibrium points (*hang-up*), and lock again with respect to a stable equilibrium point in the neighborhood (*cycle slip*). A detailed analysis is extremely difficult (Fokker-Planck theory); in [200], however, it is shown that the mean cycle-slip period, as the main figure of merit, depends on both the area under half the period of the detector characteristic and the jitter variance.

The phase recovery of higher order modulation formats such as Amplitude Phase Shift Keying (APSK) requires some extra processing. Thus, following [202], if a 16-APSK constellation is raised to the 3rd power a new constellation with most of the energy in an outer QPSK ring is obtained. In similar terms, raising a 32-PSK to the 4th power yields a new distribution with most of the energy in an outer QPSK constellation[2]. A feed-back DD scheme is suggested in [202] to process the outcome of the raising operation, handled as a true QPSK.

---

[2]A $\pi/4$ phase shift must be compensated for.

The most remarkable features of this ingenious procedure are the simplicity and its relative insensitivity to amplitude errors.

6.3.3   Timing Recovery

As stated earlier, carrier recovery algorithms usually require the symbol timing to be established in advance. Therefore, NDA synchronization procedures are of primary interest for burst as well as continuous transmission of data. In this respect, Oerder and Meyr (O&M) introduced an NDA square-law estimator [219], where the normalized timing offset ϵ is estimated as

$$\hat{\epsilon} = -\frac{1}{2\pi}\arg\left[\sum_{k=0}^{KN-1}|q_k|^2 e^{-j2\pi k/K}\right] \tag{6.28}$$

where q_k denotes the output of the matched receiver filter, sampled K times per symbol period T with $K = 4$ frequently adopted in practice. Alternatives may be realized via absolute value, logarithmic or fourth-order nonlinearities. Most interestingly in this context, in [220] an ad-hoc procedure is introduced for $K = 2$, which has been analyzed as well as optimized in [221].

Powerful FF estimators for symbol timing recovery are more or less mandatory for efficient acquisition in burst (TDMA) modems. However, FB solutions are often much less complex from the computational point of view. The latter can be an attractive alternative for continuous transmission of data, where rapid acquisition is not that important as with TDMA. Symbol timing recovery via FB loops, based on the ML principle and as such related to the first derivative of the matched receiver filter, is in many cases too complex for implementation without simplifications. The problem is usually circumvented by simple ad-hoc algorithms like DD zero-crossing (ZC) or NDA Gardner (GA) synchronizers as most prominent examples, generating an error signal [213]

$$u_k = \begin{cases} \Re\left\{(\hat{c}_{k-1} - \hat{c}_k)q_{k-1/2}^*\right\} & \text{ZC} \\ \Re\left\{(q_{k-1} - q_k)q_{k-1/2}^*\right\} & \text{GA} \end{cases} \tag{6.29}$$

at the output of the detector.

Note that both ZC and GA synchronizers are operated with two samples per symbol. However, Mueller and Müller detectors [213] need only one sample per symbol; compared to ZC or GA devices, no self noise is observed with Mueller and Müller algorithms although the increased jitter variance in the medium-to-low SNR range makes them less attractive.

The GA detector may be modified appropriately if q_k is expressed in generalized polar form, i. e., $q_k \rightarrow \rho_k^\mu e^{j\arg[q_k]}$, $\rho_k \triangleq |q_k|$, such that u_k develops as

$$u_k = \Re\left\{(\rho_{k-1}^\mu e^{j\arg[q_{k-1}]} - \rho_k^\mu e^{j\arg[q_k]})q_{k-1/2}^*\right\} \tag{6.30}$$

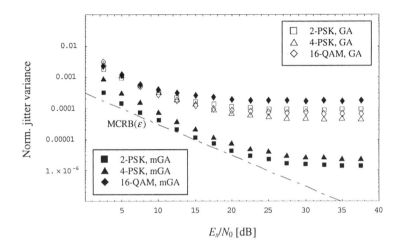

Figure 6.5. Evolution of the normalized jitter variance for GA and mGA ($\mu = 0$) trackers ($\alpha = 0.1$, $B_L T = 10^{-3}$).

Of course, with $\mu = 1$, the error signal converges to its initial shape. In [222], it is shown that the jitter performance may be improved considerably if the modified Gardner (mGA) detector is used with M-PSK signals. Assuming root-raised cosine filters for baseband shaping, this turns out to be particularly true for small values of roll-off factor α.

EXAMPLE 6.2 Assuming 2-PSK, 4-PSK and 16-QAM as modulation schemes, Figure 6.5 illustrates the normalized jitter variance of a Gardner synchronizer for a roll-off $\alpha = 0.1$ and an equivalent noise bandwidth $B_L T = 10^{-3}$. The figure depicts also the jitter variance of the mGA tracker ($\mu = 0$). It is checked easily that the variance decreases significantly. For values of $\mu > 0$, a degradation of the jitter performance is observed which is simply explained, because a radial noise component is introduced in addition to the angular jitter. Unfortunately, the mGA detector does not work satisfactorily for modulation schemes with non-constant envelope as depicted in Figure 6.5 for 16-QAM. For comparison purposes, the MCRB is included in Figure 6.5. Normalized to T, the latter is for root-raised cosine filters given by

$$\text{MCRB}(\epsilon) = \frac{B_L T}{[\frac{1}{3}\pi^2(1 + 3\alpha^2) - 8\alpha^2]E_s/N_0} \qquad (6.31)$$

As already mentioned previously with carrier phase recovery, the mean cycle-slip period depends on both the area under half the period of the detector characteristic and the jitter variance [200].

6.3.4   Automatic Gain Control

The Automatic Gain Control (AGC) represents a key component in any demodulator. The amplitude of the incoming symbols must be normalized, especially when the information is contained in the magnitude as well as in the phase of the symbols, as it is the case with QAM or APSK formats. Together with the phase recovery subsytem, the AGC can be considered as a sort of one-tap channel equalizer, i.e., adjustment of both magnitude and phase of the signal in order to meet the desired values. Assuming carrier frequency and symbol timing to be perfectly recovered, the output of the matched filter is expressed as

$$q_k = \lambda e^{j\theta} c_k + n_k \qquad (6.32)$$

The AGC must compensate for the unknown value of λ. An ML estimate of λ has been derived in [223] for large SNR values such that we have the recursive adaptation

$$G_{k+1} = G_k - \mu \left(G_k |q_k| - E[|c_k|] \right) \qquad (6.33)$$

where $G \triangleq 1/\hat{\lambda}$ and μ is the step-size parameter. For low SNR's, the performance of this NDA AGC degrades and the adjustment of the gain based on the pilot symbols is preferred; the AGC gain is then frozen while receiving data symbols and adjusted only at those instants where pilot slots are available [202]. Therefore,

$$G_{k+1} = G_k - \mu \left(G_k q_k - c_k \right) c_k^* \qquad (6.34)$$

if c_k is a pilot symbol; otherwise, $G_{k+1} = G_k$. The magnitude $|G_k|$ must be used to multiply q_k, so as to avoid undesired interaction effects with the phase recovery subsystem.

6.4   Code Aware Synchronization

In the following, the receiver architecture is designed to fully exploit the synergy between channel parameter estimation and data decoding techniques. The presented synchronization algorithms are embedded into the decoding process to overcome poor estimation performance of stand-alone techniques when they have to operate at low signal to noise ratios.

6.4.1   Per-Survivor Processing

The principal goal of this section is the description of a recent receiver design methodology, identified as Per-Survivor Processing (PSP) [224] [225], and its application to advanced digital satellite communications.

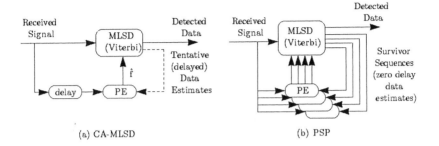

Figure 6.6. The structure of: (a) CA-MLSD, and (b) PSP.

DEFINITION 6.3 *According to the PSP concept, the estimation of all the relevant unknown parameters can be embedded into the structure of the search algorithm itself, and thus joined with the fundamental task of data detection.*

In general, PSP can offer superior and robust performance versus the classical segregated-task architectures in many challenging environments. Clearly, this advantage comes at the price of a complexity increase, which could be significant.

Traditional receiver structures operate the Parameter Estimation (PE) in a *segregated* way [226], as in Figure 6.6(a), which allows for two options: in the NDA approach, PE is based on the observed data only, and its output estimated vector is fed to the decoder. In the second option, the alternative DD structure is depicted, which includes a FB link (the dashed line) from the data detector to the PE box. This link provides *tentative*[3] data decisions out of the detector, to be used by the external estimator with the purpose of removing the data uncertainty from the observation sequence. Basically, the quality of the estimated parameter will depend on the quality of the estimated data fed to it (i.e., its error rate); thus, a low input SNR (or other equivalent measure) will create a vicious cycle of poor tentative decisions driving the PE, whose deteriorated estimator worsens further the detector, etc. This error propagation, explicit or implicit, is the Achilles heel of any such DD estimator or Conventional Adaptive-MLSD (CA-MLSD) structure. Note that these two structures in Figure 6.6(a) tend to perform similarly in terms of delivered BER.

---

[3] *Tentative* means allowing the decisions destined for the PE box to be extracted from the Viterbi Algorithm (VA) at a smaller delay than those destined for the deliverable output, a necessary compromise in rapidly varying environments.

On the other hand, the essence of PSP can be graphically explained in Figure 6.6(b), where each retained path in the search (a *survivor*) keeps and updates its own individual vector of estimated parameters, based on its own data sequence, namely its associated data history. It follows that PE is performed in a *decentralized* manner, as opposed to the *global* manner of conventional receivers, which can also be interpreted as a zero-delay, localized decision FB.

Next we present an application of the PSP paradigm and analyze its performance. The focus of this application is placed on coherent reception of a convolutional encoded PSK signal in the presence of phase/frequency uncertainty for future multimedia broadband satellite networks [227]. In fact, frequency offsets and phase noise are common link impairments that must be faced by these digital satellite systems due to the Doppler effect and oscillator imperfections. In the following, the detection performance of CA-MLSD and PSP are compared.

As in [227], the information bit sequence $\{b_k\}$ is encoded by way of a rate $1/2$, constraint length $L = 7$ convolutional code, and then fed to a QPSK mapper. Letting the encoder state at time k be $\mu_k = (b_{k-1}, b_{k-2}, \ldots, b_{k-L+1})$ and denoting the state transition as $\mu_k \rightarrow \mu_{k+1}$, the QPSK symbol corresponding to the state transition $\mu_k \rightarrow \mu_{k+1}$ is represented as $c_k = c(\mu_k \rightarrow \mu_{k+1})$. The transmitted QPSK symbol sequence $\{c_k\}$ is formatted into packets with $N_{symb} = 150$ symbols, and with a preamble of $N_{pre} = 12$ known symbols to ease initial parameter acquisition in the receiver section. Assuming negligible filter distortion, and perfect symbol timing recovery, the complex samples at the output of the matched filter can be expressed as:

$$q_k = c_k e^{j\theta_k} + n_k \tag{6.35}$$

where n_k is the complex AWGN process whose independent quadrature component samples have identical variance equal to $N_0/2$ and θ_k is the phase rotation

$$\theta_k = 2\pi \Delta f T k + \phi_k \tag{6.36}$$

In the above equation $\Delta f T$ represents the residual carrier frequency offset normalized to the symbol rate and ϕ_k accounts for the phase noise introduced by the fluctuations in signal source and local oscillators within the transmitter, satellite transponder and receiver. The frequency offset $\Delta f T$ is assumed to be deterministic and constant during the transmission of each packet, whereas the phase sequence $\{\phi_k\}$ is modelled as a Wiener random process characterized by zero mean Gaussian independent increments with standard deviation σ_ϕ.

At each discrete time instant k the VA updates the path metrics by performing the classical Add-Compare-Select (ACS) operation. In particular, let

$$\lambda(\mu_k \rightarrow \mu_{k+1}) = F\left[c_k, q_k, \hat{\theta}_k\right] \tag{6.37}$$

be the branch metric pertaining to the transition $\mu_k \rightarrow \mu_{k+1}$ at discrete time
k. $F[\cdot]$ represents the metric functional dependence on the QPSK symbol
corresponding to the state transition, on the received sample, and on the channel
phase estimate. The phase estimation is performed by a second order loop
(SOL) tracker [213], used either in the single external (CA-MLSD) or in the
distributed (PSP) configuration. The SOL equations are:

$$\hat{\theta}_{k+1} = \hat{\theta}_k + \xi_k \tag{6.38}$$

and

$$\xi_k = \xi_{k-1} + \gamma \left(\beta e_k - e_{k-1} \right) \tag{6.39}$$

where γ and β are the loop parameters, whereas ξ_k and e_k are the radial frequency error and the error signal at time k, respectively. In general, the SOL
error signal would be computed as

$$e_k = \Im \left\{ q_k c_k^* e^{-j\hat{\theta}_k} \right\} \tag{6.40}$$

but, in the following, this equation is personalized according to the considered
estimation technique.

In CA-MLSD the data symbol c_k in the error signal computation is replaced
with a low delay tentative data decision. In other words, instead of tracking
back the currently best path many steps in search of a very reliable data estimate,
only few steps are traced back in search for a more recent but less reliable data
estimate. Letting d denote the number of steps traced back, then, at time k, the
most recent phase and radial frequency estimates available are $\hat{\theta}_{k-d}$ and ξ_{k-d}.
Thus, the branch metric in the CA-MLSD must be computed as

$$\lambda(\mu_k \rightarrow \mu_{k+1}) = \left| q_k e^{-j(\hat{\theta}_{k-d} + \xi_{k-d}d)} - c(\mu_k \rightarrow \mu_{k+1}) \right|^2 \tag{6.41}$$

where the term $\xi_{k-d}d$ provides phase unwrapping, i.e. the projection of the
radian frequency error estimate to the present time instant. The value for d
drives the trade-off between the reliability of data aiding and loop response to
rapid phase changes.

In PSP each survived data sequence in the trellis produces its own parameter
estimate, i.e., as many SOLs as survivors in the trellis are run in parallel. Note
that in PSP no tentative decision must be taken since every hypothesized data
sequence generates an estimate, and thus each estimate is perfectly time aligned
with the channel samples. Letting $\hat{\theta}(\mu_k)$ be the phase estimate associated to the
survivor path ending up at state μ_k, the branch metric computation becomes

$$\lambda(\mu_k \rightarrow \mu_{k+1}) = \left| q_k e^{-j\hat{\theta}(\mu_k)} - c(\mu_k \rightarrow \mu_{k+1}) \right|^2 \tag{6.42}$$

The dependence on the survivor path modifies the SOL equations. Thus, if $\bar{\mu}_k$ denotes the survivor state, we have

$$\hat{\theta}(\mu_{k+1}) = \hat{\theta}(\bar{\mu}_k) + \xi(\bar{\mu}_k) \tag{6.43}$$

$$\xi(\bar{\mu}_k) = \xi(\bar{\mu}_{k-1}) + \gamma \left[\beta e(\bar{\mu}_k) - e(\bar{\mu}_{k-1}) \right] \tag{6.44}$$

whereas the error signal is

$$e(\bar{\mu}_k) = \Im \left\{ q_k c^* (\bar{\mu}_k \to \mu_{k+1}) e^{-j\hat{\theta}(\bar{\mu}_k)} \right\} \tag{6.45}$$

Note that the two branch metrics associated to the transitions leading to state μ_{k+1} are computed using not only different data symbols but also different phase estimates. In fact, $\hat{\theta}(\mu_k)$ in eq. (6.42) depends on the starting state of the transition $\mu_k \to \mu_{k+1}$. In other words, SOL equations are updated only after the ACS algorithm has determined the survivor state.

Finally, it is worthwhile underlining that if the correct data path is present among the survivors then there will be a tracker that will enjoy zero-delay correct data, thus performing as an ideal DA scheme would. This advantage must be traded off against the implementation complexity increase.

Figure 6.7 shows the BER performance for the CA-MLSD[4] and PSP algorithms as a function of the frequency offset $\Delta f T$, for phase noise standard deviation $\sigma_\phi \in \{0, 0.3, 1.0, 3.0, 5.0\}$ degrees and $E_b/N_0 = 3$ dB [227]. Each PSP curve (solid) outperforms the corresponding CA-MLSD one (dashed). PSP curves keep close to the *known channel* bound longer than the CA-MLSD ones and, therefore, performance degradation is slower. This is well evidenced considering $\Delta f T = 10^{-3}$, and $\sigma_\phi \leq 1$ degree: BER for PSP is in the order of 9×10^{-4} and crosses 10^{-2} only for $\Delta f T > 3 \times 10^{-2}$ for any phase noise standard deviation. It is worthwhile to note that PSP is more robust against phase noise. In fact, PSP curves for different σ_ϕ are less spread than those for CA-MLSD.

6.4.2 Turbo Embedded Estimation

In the literature, the term *joint iterative estimation* usually refers to the external iterative loop between the MAP decoder (Sec. 4.3.2.3) and the channel parameter estimator; e.g., see [228][229][230]. Therefore, the channel estimation is implemented iteratively and recursively using the a-Posteriori Probability (APP) given by the turbo decoder. After PSP, joint data detection and parameter estimation have been also applied in Turbo coded systems [231][232] [233] [234]. In the following we outline the joint procedure called Turbo Embedded Estimation (TEE), first presented in [233][234].

---

[4]For the CA-MLSD approach, the delay d for the tentative decisions has been set to one QPSK symbol.

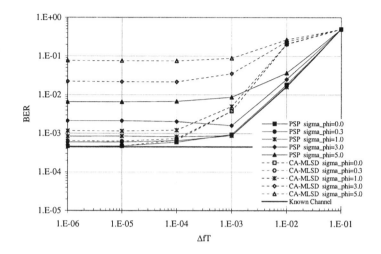

Figure 6.7. BER vs. ΔfT for CA-MLSD and PSP. $\sigma_\phi = 0, 0.3, 1, 3, 5$ degrees and $E_b/N_0 = 3$ dB. ([227] ©2001 IEEE)

As in PSP, the parameter estimator is integrated into the Maximum-A-Posteriori (MAP) algorithm. The aim is to exploit the advantages posed by the joint parameter estimation and turbo decoding process. Nevertheless, as opposed to PSP, the VA is not usually adopted in Turbo coded systems[5], but the BCJR algorithm is preferred. The former is a Maximum Likelihood Sequence Detection (MLSD) method which minimizes the probability of word error for convolutional codes, while the latter algorithm is a symbol-by-symbol process derived to minimize the bit error probability [112].

The TEE technique is integrated in the *forward recursion* (eq. (4.35)), and the most probable received sequence is created. Let the state of the encoder at time k be S_k, which can take on values in $m = 0, 1, ..., M_s = 2^\nu - 1$, where ν is the encoder constraint length. If the observed sequence is denoted as $Q_1^N = \{q_1, ..., q_k, ..., q_N\}$, the probability associated with the state $S_k = m$ and conditioned by the first k symbols is expressed as

$$\alpha_k(m) = Pr\left\{ S_k = m | Q_1^k \right\} \qquad (6.46)$$

There is not a real path linking the different states inside the MAP decoder as a function of the time index k but, from $\alpha_k(m)$, the most probable state can

---

[5]The survivor sequence of the VA is used in PSP to lock a channel tracker.

be detected at any time instant. Since the TEE algorithm can be applied to high order modulation schemes [234], it shall be noted that demodulation and depuncturing must be performed jointly with the decoding process. The TEE must provide for the channel compensation of the received symbols, which are subsequently demodulated. Finally, the systematic information is separated from the parity checks.

The main steps of the TEE algorithm are detailed next.

1 **Pre-processing.** This is the fundamental step to take into account the presence of high order modulation. The received symbols must be converted into their constituent bits. The turbo decoder, just before a new step of the forward recursion, performs a two fold task at time k:

   - *Symbol correction:* the current received symbol is extracted from the integrated estimation block. For instance, in case of channel phase estimation, the received symbol is derotated according to the last phase estimate.

   - *Soft demodulation and depuncturing:* based on the corrected symbol, the soft information of the constituent bits is computed and according to the puncturing pattern, the systematic and the parity check bits are discriminated.

2 **Most probable state detection.** The state with the maximum value of probability among the M_s possible states is searched at time k in the forward recursion updating $\alpha_{k-1}(m) \rightarrow \alpha_k(m)$:

$$\hat{S}_k \triangleq \max_m \{ \alpha_k(m) \} \quad m \in M_s$$

3 **Most probable transition detection.** After the most probable state at time k has been computed, the most probable trellis transition can be chosen. Letting M_k^p be the set of possible predecessors of $\hat{S}_k$, this yields

$$\hat{S}_k' \triangleq \max_m \{ \alpha_{k-1}(m) \} \quad m \in M_k^p$$

4 **Estimate updating.** The decided symbol may be rebuilt from the bits, systematic and parity checks, pertaining to the most probable transition $\hat{S}_k' \rightarrow \hat{S}_k$ used for the sole purpose of updating the estimation block.

As an example of TEE application and performance, let us consider a Turbo coded system in presence of carrier frequency and phase uncertainty. The TEE algorithm is benchmarked with an external iterative technique combining both the V&V estimates and the APP values from the Turbo decoder [233, 234]. The turbo encoder, consisting of two Recursive Systematic Convolutional (RSC) encoders and an interleaver, is a R_c=1/3 code rate with constraint length ν. The

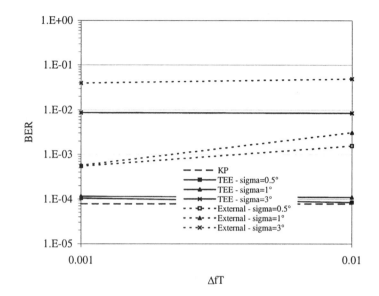

Figure 6.8. BER vs. ΔfT with $\sigma_\phi = 0.5, 1, 3$ degrees and $E_b/N_0 = 2$ dB. $R_t = 1/2$ and QPSK modulation.

information bits sequence $\{b_k\}$ is organized in packets of length $N = 1024$. In the turbo encoder, a puncturing block is introduced with puncturing rate R_p to reduce the effective encoder rate. Two possible values are considered: $R_p=1$, for which all parity bits are transmitted, and $R_p=2/3$, which amounts to the alternate transmission of the parity check bits. The overall coding rate is $R_t = \frac{R_c}{R_p}$. Then, an M-PSK modulator produces the symbol sequence α_k. At the beginning of each packet a preamble of $N_{pre} = 32$ known symbols is inserted to ease external phase estimation or to get the initial phase to start the embedded algorithm. The received sequence can be expressed as in eq. (6.35), where the channel model is the same reported in eq. (6.36). In all the simulations we used two identical RSC encoders with constraint length $\nu = 4$, and generators $g_0(D) = (37)_8$ and $g_1(D) = (21)_8$. The normalized carrier frequency offset $\Delta fT \in \{0.001, 0.01\}$ and the phase noise standard deviation $\sigma_\phi \in \{0.5, 1.0, 3.0\}$ degrees are considered. The BER results will be contrasted against the known phase (KP) curve, corresponding to the performance ideally achievable when the channel distortion is perfectly known by the receiver. In Figure 6.8 the detection performance for $R_t = 1/2$ and QPSK modulation is

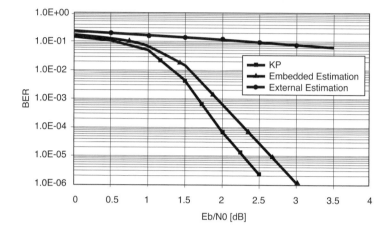

Figure 6.9. BER vs. E_b/N_0 with $\Delta fT = 0$ and $\sigma_\phi = 1$ degrees. $R_t = 1/3$ and 8-PSK modulation.

shown as a function of the normalized frequency offset and the phase noise standard deviation, assuming $E_b/N_0 = 2$ dB. Several interesting points can be observed. For example, for $\Delta fT \leq 0.01$ and $\sigma_\phi \leq 1$ degree, the embedded proposal is insensitive to channel variations and it keeps very close to the ideal estimation case. On the other hand, the iterative external approach achieves worse BER performance than the TEE algorithm and, moreover, it is sensitive to the residual normalized frequency offset from 0.001 up to 0.01. Then, when $\sigma_\phi = 3$ degrees is simulated, both approaches fail to obtain good performance, but again, the TEE outperforms significantly the conventional scheme.

In Figure 6.9 BER is reported for R_t=1/3 and 8-PSK modulation. No normalized frequency offset is assumed, and $\sigma_\phi = 1$ degrees is considered. In this condition, the external approach completely fails to obtain reliable results. This is due to the NDA algorithm ambiguity and the higher phase sensitivity of 8-PSK, which rules an important factor in decreasing drastically the estimation and detection performance during the first iteration. On the other hand, the TEE algorithm demonstrates again its robustness and good phase tracking performance. At BER=10^{-5}, the distance between the TEE and KP curves is less than 0.4 dB.

6.5   Channel Estimation

Satellite channels fall within the category of fading channels, i.e., channels evolving with time with responses fluctuating in such a way that the receiver needs to keep track of those changes. Channel tracking can be used for power control purposes and adaptive coding and modulation, and the channel estimates can be used for equalization in frequency-selective channels, in such a way that the quality of the channel estimates has an important effect on the overall receiver performance [200]. As usual, channel estimation schemes can be divided according to the use of data, i.e., DA or Pilot Aided (PA), NDA or blind, and DD. For narrowband communications for which the channel can be considered frequency non-selective, the receiver must track the channel gain. For frequency-selective channels more parameters must be estimated and tracked. The rate of change of the channel properties will determine how estimates are updated. For DA schemes, training symbols must be inserted in the transmitted signal, with important engineering decisions to take ragarding the positions and number of pilot symbols. Provided that the number and position of the training symbols is fixed, it is necessary to devise how to fill the gaps for which there are no known symbols. Thus, interpolation is also necessary for proper channel tracking. Next we describe the main features of channel estimation algorithms, with a special subsection devoted to the estimation of the SNR.

6.5.1 Pilot Aided Channel Estimation

If the transmitter inserts known symbols, referred to as pilot symbols, in the data stream, then the receiver can derive an amplitude and phase reference. The insertion pattern of the pilot symbols is not unique, and will depend on the rate of change of the channel fading, the available bandwidth and the allowed error in the channel estimation. For the case of single carrier transmissions, the interlacing of known symbols in the time domain offers some advantages with respect to the insertion of symbols in the frequency domain, because it is not necessary an in-band filtering and the signal amplitude characteristics are not changed. We will assume that the statistical channel information is available and that the correct frame synchronization has been achieved (see Section 6.7). In its simplest form, pilot-assisted channel estimation is achieved through re-modulation of the received signal at the reference positions with the inverse pilot symbols. Let the received signal after matched filtering at time k be denoted by:

$$q_k = h_k c_k + n_k \qquad (6.47)$$

where q_k is the output of the matched filter, h_k is the channel gain at time k and n_k is AWGN. Assuming that pilot symbols are inserted periodically at $k = nM, n = 0, 1, \ldots$, then a pilot-based channel estimate, denoted by $\hat{h}_{nM}$ is also available to the receiver at the same rate. These primary channel estimates are then used to derive estimates of the channel gain at positions occupied by data symbols.

A first simple approach would estimate the channel gain as

$$\hat{h}_{nM} = \frac{q_{nM}}{c_{nM}} \qquad (6.48)$$

and would interpolate for the remaining values $\hat{h}_k, k \neq nM$. The choice of the interpolation kernels will depend on the allowed margin for the interpolation error and the relation of the channel dynamics with the pilot symbol rate. The application of this basic scheme does not require the knowledge of the channel statistics. The optimum estimation filter in the minimum square error sense is the *Wiener filter*, which can be computed if the statistics of the channel and the noise are known. Thus, the filter input-output relation can be written as [235]

$$\hat{h}_k = \sum_{n=-N}^{N} w_{n,k} q_{nM} \qquad (6.49)$$

where $\hat{h}_k$ is the estimated channel gain at time k and $w_{n,k}$ are the coefficients used. Assuming stationarity, the knowledge of $E[h_{k+n}h_k^*]$ makes it possible to derive the optimum coefficients $w_{n,k}$, which play the role of the interpolation weights for $k \neq nM$.

Adaptive algorithms are usually adopted to track channel coefficients as a practical alternative to the Wiener filter. Thus, a simple Least Mean Square (LMS) algorithm can be used, or a more complex Recursive Least Square (RLS). If a state-space model of the channel coefficients evolution is available, Kalman filtering is also an option [200].

6.5.2   Blind Channel Estimation

All the channel estimation techniques mentioned so far rely on the existence of known training sequences used by the receiver to derive the channel response. However, the use of training sequences is either impractical or not feasible in some cases, and the process of estimating the channel in the absence of pilot sequences is termed *blind channel estimation/identification*.

For blind channel estimation to be feasible, some knowledge of the transmitted signal and/or the channel statistics is required. The most commonly utilized characteristics of the transmitted signal and/or the channel that are employed for blind channel estimation are [236]: the *distribution of the transmitted signal*, the *relative durations* of the signal and the channel impulse response and the *stationary or time-varying characteristics* of the signal and the channel. According to [236], *Conditional Mean*, ML and MAP are all techniques that can be employed in the context of blind channel estimation.

In addition, second and higher order statistics of the signal and/or channel can be employed in an effort to blindly identify the channel response. As mentioned in [237], the cyclostationarity of digitally modulated signals can be taken advantage of for this purpose. Examples of methods using second order signal statistics are presented in [238], [239] and [240]. In most cases some form of diversity (spacial or temporal) is required for the application of such estimation techniques [241]. If this diversity is not readily available in the form of multiple receive antennas or delayed tap propagation, it can be generated through over-sampling of the received signal as in [242]. Finally, alternative methods need to be employed in case the cyclostationary property of the signal is somehow destroyed, as occurs with long code WCDMA [243]. Obviously, higher or-der statistics can also be employed for blind channel estimation. Nonetheless, this comes at the expense of computational cost and the resulting increase in processing delay, which limit the range of techniques that can be employed in practical systems.

A key point to note is that blind estimation techniques are usually more com-plicated and computationally demanding than PA estimation schemes. There-fore, they are only employed in practical systems if the use of training sequences is either impractical or impossible.

6.5.3   Signal to Noise Ratio Estimation

An accurate knowledge of the SNR is necessary for the implementation of Fade Mitigation Techniques, and in particular, of Adaptive Coding and Modu-lation (Chapter 8). The MAP procedure for iterative soft decoding is another illustration of the need to know the SNR, as shown in Section (Chapter 4.3).

Several methods for the SNR estimation are presented and compared in [244], namely, Split-Symbol Moments estimator (SSME), ML estimator, Squared Sig-nal to Noise Variance estimator (SNVE), second and four order moments (M2, M4) estimator, and Signal to Variation Ratio (SVR) estimator. These methods assume that the channel fading is a slowly time-varying process compared to the SNR estimation time interval, and that interference contributions are mod-elled as a white Gaussian random process and added to the thermal noise. The authors in [244] show that the ML estimation algorithm performs better than the others. A particular case of the ML estimation is the Signal to Noise Ra-tio Estimation (SNORE) algorithm that has been proposed initially by the Jet Propulsion Laboratory (JPL) for BPSK modulated signals and has been further developed in [245]. In the DA version, the SNORE algorithm and the ML al-gorithm are equivalent. In the NDA approach, the SNR estimate is obtained by using the modulus operator to remove the data modulation uncertainty. In this form, the SNORE shows a strong estimation bias for SNR values below 3 dB. To extend the SNORE algorithm to QAM and APSK modulation schemes, it is necessary to know the transmitted symbols.

For the purpose of illustration, the SNORE estimation formulas at the output of the matched filter are reported assuming that carrier and timing recovery has already been established with a proper algorithm (see Sections 6.3.1 - 6.3.3). For complex signals distorted by AWGN, $\rho \triangleq$ SNR can be effectively estimated as:

$$\hat{\rho} = \begin{cases} \frac{M_0^2}{M_2 - M_0^2} & \text{DA-SNORE algorithm} \\ \frac{M_1^2}{M_2 - M_1^2} & \text{NDA-SNORE algorithm} \end{cases} \tag{6.50}$$

where

$$\begin{aligned} M_0 &= \frac{1}{N} \sum_{k=1}^{N} \Re\left\{c_k^* q_k\right\} \\ M_1 &= \frac{1}{N} \sum_{k=1}^{N} |\Re\left\{q_k\right\}| + |\Im\left\{q_k\right\}| \\ M_2 &= \frac{1}{N} \sum_{k=1}^{N} |q_k|^2 \end{aligned} \tag{6.51}$$

and N denotes the observation length.

In [246], the authors derive an analytical characterization of the DA-SNORE estimation performance. In particular, the useful signal and the noise power are modelled by a χ^2-distribution, thus the resulting SNR estimate is described in terms of a non-central F-distribution. The analytical form of $\hat{\rho}$ is modelled by:

$$\hat{\rho} \sim \frac{1}{2N - 1} \cdot \Phi_{1,\,2N-1}(2N\rho) \tag{6.52}$$

where Φ is a non-central F-distribution with $n_1 = 1$ and $n_2 = 2N - 1$ degrees of freedom, and non-centrality parameter $\lambda = 2N\rho$. Then, from eq. (6.52), the average SNORE estimate and its variance are respectively:

$$\mu(\hat{\rho}) = \frac{2N}{2N - 3}\left(\rho + \frac{1}{2N}\right) \tag{6.53}$$

$$\sigma^2(\hat{\rho}) = \frac{N^2\left(8\rho^2 + 16\rho\right) + N\left(4 - 16\rho\right) - 4}{8N^3 - 44N^2 + 78N - 45} \tag{6.54}$$

Finally, and in order to assess the estimator efficiency, we need the CRB for *biased* estimators, which normalized to ρ^2 is given by [246]:

$$\text{CRB}(\rho) = \frac{8N}{\rho(2N - 3)^2} + \frac{4N}{(2N - 3)^2} \tag{6.55}$$

It can be checked that the variance of the estimator is almost identical to the CRB, thus asserting the efficiency of the DA-SNORE estimator.

6.6   Code Synchronization for Spread Spectrum Systems

DEFINITION 6.4 *Code synchronization for Spread Spectrum (SS) systems is the problem of estimating the code epoch between the Pseudo Noise (PN) spreading sequence and its replica generated by the receiver for despreading.*

Code synchronization is the necessary requisite for correct signal reception in SS systems, because an untrained receiver cannot distinguish the useful received signal from the background noise. In particular, the most critical step is the initial synchronization, i.e., code acquisition, which provides the coarse code epoch estimate subsequently refined by the following code tracking stage. In fact, the SNR before despreading is very low (-20 dB, for example) and this fact makes accurate parameter estimation typically unfeasible. Thus, code acquisition must be performed at low SNR and in the presence of phase uncertainty, frequency errors, and with no information on channel propagation conditions. Also, for a Code Division Multiple Access (CDMA) system, multiple access interference has to be faced.

Code acquisition usually discretizes the uncertainty region [6] into a finite number of *cells* or *hypotheses*, so transforming the estimation problem into a detection problem. As a consequence, acquisition can be formulated as a classical binary testing problem, which consists of deciding in favor of a H_1 hypothesis (*correct detection*), while discarding all incorrect H_0 hypotheses (*correct rejection*), and trying to avoid *missed detection* and *false alarm* events. The discretized uncertainty region can be scanned in a serial or parallel way, in order to search for the H_1 hypotheses. The selection of a parallel or serial search strategy determines a specific trade-off between the issues of complexity and delay. In fact, a parallel search requires a dedicated hardware per cell (maximum complexity), but is able to scan the entire region in the time period required to perform a single test. Dually, a serial search optimizes complexity requiring only a single dedicated circuit that is repeatedly employed to scan the entire uncertainty region. Also, hybrid serial/parallel solutions are possible.

The usual complexity limitations in terminals determine the fact that for forward link transmission a serial search (or hybrid with low/moderate parallelism) is normally selected. Different choices can be in general adopted for the return link synchronization.

The associated incremented delay of serial search can be limited adopting an appropriate scanning strategy. The most commonly used is the so called *straight line*, which explores the uncertainty region starting from a random point and straightly proceeding cell by cell. Possible alternatives are represented by *expanding window* and by the *broken* Z [247], which basically exploit a-priori information to perform a larger number of observations in the most likely uncertainty sub-region. These techniques are usually adopted for post-initial acquisition, for example to resolve several paths in a frequency selective channel.

---

[6]The *uncertainty region* is the domain of the unknown code epoch.

In the following, straight line serial search is considered. In this case, the correct hypothesis can be selected according to different criteria, such as MAX, TC and MAX/TC. The MAX criterion decides in favor of the cell with the largest detection variable. This strategy has the drawback of taking the decision after having scanned the entire uncertainty region. This aspect can be penalizing when the correct hypothesis is well distinguishable from the misaligned cells. In these cases, the decision could be anticipated through the employment of a Threshold Crossing (TC) criterion, which compares the decision variable of each cell with a threshold and declares the acquisition in correspondence of a threshold crossing. Finally, a general criterion identified as MAX/TC [248] foresees to divide the uncertainty region into sectors, and to perform MAX search within each sector, and TC criterion between sectors. The MAX and TC criteria can be seen as particular cases of MAX/TC.

The threshold design is usually pursued adopting the Constant False Alarm Rate (CFAR) criterion. The CFAR threshold setting allows to have a constant false alarm probability, P_{fa}, by varying the SNR. This is an essential feature because the detection problem is usually strongly unbalanced, having to discriminate one (or a few) H_1 cell from a large number of H_0 cells. Thus, being the P_{fa} specification typically a very critical constraint, the CFAR criterion enables an optimized detector design, independently of the SNR.

Code acquisition performance strongly depends on the specific detection scheme employed to extract the decision variable. The optimal solution under the ideal assumption of perfect carrier recovery in AWGN channel is the employment of a coherent correlator that performs correlation between the received signal and the locally generated PN sequence. If a phase uncertainty affects the received signal, the optimal solution consists in the energy detector. For satellite communications, frequency uncertainty is also present due to the TX/RX oscillator mismatch and the relative satellite/terminal motion. To limit the degradation introduced by the frequency offset, code acquisition can be performed according to a windowing technique [249], limiting coherent integration onto a sub-section of the transmitted PN sequence. The residual integration is performed after non linear detection, and is identified as Post Detection Integration (PDI). PDI improves the SNR notwithstanding the noise enhancement introduced by possible non linearities. In the literature, several different PDI schemes have been proposed, among which the classic Non Coherent PDI (NCPDI) [249], [250] and the Differential PDI (DPDI) [251] can be found. NCPDI and DPDI block diagrams are reported in Figure 6.10, where M is the coherent correlation length, L is the PDI length, and c_i denotes the i-th chip of the PN sequence. Note that $T = T_c$ for a spread spectrum system, being T_c the chip period.

Theoretical approaches to solve the frequency uncertainty provide two alternative PDI schemes: the Generalized PDI (GPDI) [252] and the Average

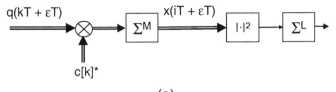

(a)

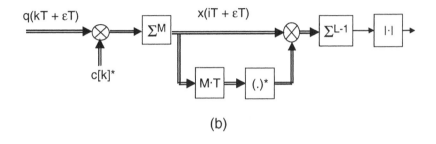

(b)

Figure 6.10. Non Coherent Post Detection Integration (NCPDI) (a) and Differential Coherent Post Detection Integration (DPDI) (b) block diagrams.

PDI (APDI) [253], which are based on the Generalized Likelihood Ratio Test (GLRT) and Average Likelihood Ratio test (ALRT) criteria described in Section 2.4.

DEFINITION 6.5 *GPDI is the sum of NCPDI, DPDI and L-2 terms identified as n-Span DPDI, which are structurally similar to DPDI, but takes the phase reference on the n-th past sample instead on the previous predecessor.*

Mathematically, the decision variable for GPDI is

$$\Lambda^G = \Lambda_0 + 2 \sum_{n=1}^{L-1} \Lambda_n^G \tag{6.56}$$

where $\Lambda_0 = \sum_{i=1}^{L} |x[i]|^2$ is the classic NCPDI, and $\Lambda_1^G = \left| \sum_{i=2}^{L} x[i]x^*[i-1] \right|$ corresponds to the DPDI, being $x[i] = x(iT + \epsilon T)$ the received samples after coherent accumulation. Finally, the additional $L-2$ terms are

$$\Lambda_n^G = \left| \sum_{i=n+1}^{L} x[i]x^*[i-n] \right| \tag{6.57}$$

identified as n-Span DPDI. For GPDI, a good performance-complexity trade-off is achieved by introducing several truncated schemes as alternative to the total case (GPDI[Λ]). An appealing truncated case is GPDI[$\Lambda(2)$] achieved by summing NCPDI and DPDI only.

DEFINITION 6.6 *APDI is the sum of NCPDI and L-1 terms identified as n-Span DPDI-Real, which takes the phase reference on the n-th past sample and computes the real part instead of the module.*

More formally, the detection variable for APDI is

$$\Lambda^{A} = \Lambda_0 + 2\sum_{n=1}^{L-1} \text{sinc}(2n\Delta f_{max}MT)\,\Lambda_n^{A} \qquad (6.58)$$

where, Λ_0 is again the classical NCPDI, and

$$\Lambda_n^{A} = \sum_{i=n+1}^{L} \Re\left\{x[i]x^*[i-n]\right\} \qquad (6.59)$$

is the n-Span DPDI-Real term. Δf_{max} is the maximum frequency error affecting the received signal. Similarly to GPDI, also for APDI the complexity increase can be limited by introducing several truncated schemes as alternative to the total APDI[Λ]. We introduce APDI[$\Lambda(2)$] that sums NCPDI and DPDI-Real only. In AWGN channels, GPDI outperforms other PDI schemes in the presence of large frequency errors. An example is provided in Figure 6.11, where performance in terms of Receiver Operating Characteristics (ROC), i.e. probability of missed detection P_{md} vs. probability of false alarm P_{fa} is illustrated for NCPDI, DPDI, GPDI/APDI[Λ], and GPDI/APDI[$\Lambda(2)$] with M=10, L=20, E_c/N_0= -8 dB, ΔfT_c=0.02, and $\Delta f_{max}T_c$=0.02. In AWGN channels, APDI outperforms other PDI schemes in the presence of relatively small frequency errors. An example is provided in Figure 6.12, where performance is presented for NCPDI, DPDI, GPDI/APDI[Λ], and GPDI/APDI[$\Lambda(2)$], with M=10, L=20, E_c/N_0=-8 dB, ΔfT_c=0.0001, and $\Delta f_{max}T_c$=0.0001.

Multi-dwell procedures can be adopted in order to reduce the false alarm probability, introducing a verification stage in cascade with the search mode.

There are different strategies for verification. With immediate rejection, the procedure stops as soon as a single test fails and the search mode is re-initiated. In case of no immediate rejection, a common approach is to confirm the acquisition if A tests out of B are above threshold.

False alarm and missed detection probabilities, P_{fa} and P_{md}, contain information on the performance related to the exam of a single cell and do not depend on the number of tests to be performed. The overall performance is instrumental to accomplish a complete code acquisition design dependent on

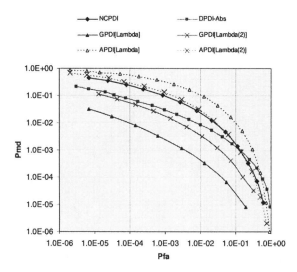

Figure 6.11. Performance comparison between NCPDI, DPDI, GPDI, and APDI with M=10, L=20, E_c/N_0=-8 dB, $\Delta f T_c$=0.02, $\Delta f_{max} T_c$=0.02.

the specific scenario at hand. Different overall metric figures can be introduced according to the transmission type. With continuous transmission, a typical performance measure is given by the acquisition time, the pdf of which can be derived through a direct approach that enumerates all possible events. However, the drawback of direct approach is that the computation of the acquisition time pdf can be very lengthy, especially with a large number of cells. A faster way to evaluate the overall performance is thus provided by the *flow graph* approach [254], which leads to closed form expressions for the moments of the acquisition time, such as the mean value and the variance. This approach exploits the fact that the acquisition procedure can be modelled as a discrete Markov chain; with this approach, the states of the chain become the nodes of the graph, and nodes are interconnected through branches with appropriate gains. The flow graph approach relies on the assumption that different cells provide statistically independent variables. In general, an overall false alarm state is present in the flow graph, which corresponds to the case of false acquisition after the entire single/multidwell procedure. This state is classified as *absorbing* when the procedure does not restart, while it is identified as *non absorbing* when the procedure restarts after a penalty time T_p dependent on the employed tracking circuit. The mean value and the variance of the acquisition time for a N-dwell procedure with the MAX/TC criterion and non absorbing false alarm is provided for example in [255].

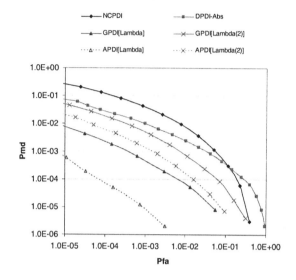

Figure 6.12.   Performance comparison between NCPDI, DPDI, GPDI, and APDI with $M{=}10$, $L{=}20$, $E_c/N_0{=}{-}8$ dB, $\Delta fT_c{=}0.0001$, $\Delta f_{max}T_c{=}0.0001$.

6.7   Frame Synchronization

DEFINITION 6.7 *Frame Synchronization is the procedure that performs frame epoch recovery for Time Division Multiplexing (TDM)/TDMA systems.*

Frame synchronization is the necessary pre-requisite for TDM/TDMA systems to correct data decoding. For example, to avoid that frame synchronization becomes the system bottle-neck with packet transmission, the related performance requirements are constrained by the achievable Packet Error Rate (PER). This constraint typically amounts to an overall erroneous synchronization probability at least one order of magnitude below the PER.

Frame synchronization is typically performed employing a DA approach that introduces a known preamble identified as Unique Word (UW) marking the start of the frame.

Frame synchronization is based on the UW location inside the frame. In general, the longer the UW, the better the synchronization performance. However, a long UW translates into an increased overhead, calling for the identification of a convenient performance/overhead trade-off. Besides, the UW pattern is another important parameter to be carefully designed, to aid the identification of the UW epoch.

Similarly to code synchronization, frame synchronization is typically divided into two phases in cascade: acquisition, which provides a coarse epoch estimate, and tracking, which refines the outcome of acquisition improving the synchronization accuracy. The domain of the unknown frame epoch, identified as the *uncertainty region*, is typically very large for initial frame synchronization (i.e. terminal startup), and usually equals the frame length. This fact makes acquisition the most critical phase. To cope with the large uncertainty, acquisition is performed discretizing the uncertainty region into a finite number of *cells* or *hypotheses*, thus transforming the estimation procedure into a discrete delay detection problem. A single hypothesis per symbol can be adopted if symbol timing recovery precedes frame synchronization, because in this case most of the useful energy is collected and limited ISI is present. However, symbol timing estimation prior to frame synchronization is possible only with blind estimators, which could not provide satisfactory performance if the SNR is very low. In addition, this approach does not apply to burst transmissions where the frame (packet) must be detected in a single attempt (one shot acquisition).

If symbol timing estimation cannot be performed prior to frame synchronization, two or more hypotheses per symbol have to be considered. In this case, one or more correct hypotheses can be identified, according to the estimate accuracy goal. A design based on a single synchronous cell allows a precision of $T/2h$, where h is the number of hypotheses per symbol, and T the symbol duration. On the other hand, h correct hypotheses lead to a lower accuracy (equal to $T/2$), but the robustness against timing shifts increases. The selection of the synchronization accuracy practically depends on the pull-in range of the tracking circuit operating in cascade.

The problem of discriminating between U_R hypotheses, where U_R is the number of cells in the uncertainty region, can be optimally solved applying the ML criterion described in Chapter 2. This requires the formulation of the likelihood function that strongly depends on the working assumptions in actual systems. As a consequence, very different solutions have been presented in the literature. Massey presented in [256] one of the first instances of ML approach in which the presence of interferer data is taken into account, neglecting the presence of timing, frequency, and phase uncertainty. The resulting solution foresees that the decision variable, z, for the hypothesis μ is given by the correlation between the received signal and the locally generated UW with a correction term dependent on the received samples, as

$$z(\mu) = \sum_{i=0}^{L_{UW}-1} q[i+\mu]c_i - \sum_{i=0}^{L_{UW}-1} f(q[i+\mu]) \qquad (6.60)$$

where $q[i+\mu] \equiv q((i+\mu)T)$ is the received sample, c_i a known UW symbol, the correction function $f(x)$ is given by $f(x) =$

$N_0/(2\sqrt{E_s})\ln\cosh(2\sqrt{E_s}\,x/N_0)$, $N_0/2$ is the two-sided noise power spectral density, L_{UW} is the UW length, and E_s is the symbol energy. Notably, it holds $\ln\cosh(x) \simeq x^2/2$ for small SNR values, so that eq. (6.60) becomes

$$z(\mu) = \sum_{i=0}^{L_{UW}-1} q[i+\mu]c_i - \frac{\sqrt{E_s}}{N_0} \sum_{i=0}^{L_{UW}-1} |q[i+\mu]|^2 \qquad (6.61)$$

providing a more suitable low complexity detector implementation for actual systems operating at low SNR.

Starting from [256], several alternatives have been proposed, taking into account also the presence of phase and frequency uncertainties. For example, [257] and [258] present various solutions obtained by modelling the unknown phase and frequency offset as uniform random variables. Also pragmatic approaches can provide satisfactory performance, as shown for example in [258, 259]. In particular, to perform detection in the presence of frequency errors, PDI techniques can be employed, as discussed in Section 6.6. In [259], NCPDI and DPDI, depicted in Figure 6.10, are considered as low complexity robust solutions to limit the energy degradation induced by the frequency error.

Frame synchronization can be considered as a particular case of code synchronization. This analogy allows to apply to frame synchronization all techniques, criteria and procedures discussed in Section 6.6 for code acquisition. However, the distinctive SNR and symbol duration require a specific optimization that can lead to different results with respect to code acquisition.

Identifying the aligned cells as H_1 hypotheses, and all non-synchronous cells as H_0 hypotheses, the frame acquisition problem has to locate the H_1 cells in the uncertainty region. This is in general a very unbalanced detection problem since there is typically one or a few H_1 against a lot of H_0 (in the order of thousands).

As discussed in Section 6.6, having selected the detector structure that provides the decision variable, different strategies can be identified in order to take the decision, namely MAX, TC, or MAX/TC. In the following, MAX and TC strategies will be considered.

DEFINITION 6.8 *Adopting the TC criterion, the basic performance evaluation is in terms of single cell performance, i.e. missed detection probability, P_{md}, and false alarm probability, P_{fa}, as defined in Section 2.4.*

Starting from single cell performance, overall metric figures can be introduced to characterize the frame acquisition procedure. With continuous transmission, the most common performance measure is the *acquisition time*, defined as the time interval between the procedure start and the correct acquisition. The acquisition time is a random variable that can be statistically characterized.

However, it is usual practice to simply refer to the *mean acquisition time*. Differently, with burst transmission, performance is in terms of overall correct detection probability, P_D, or false alarm probability, P_{FA}. In fact, a discontinuous operation requires a one-shot procedure to achieve frame acquisition, while a continuous transmission allows to take the final decision after a number of frames, introducing a verification step as discussed in Section 6.6 for code acquisition. For example, with the TC criterion, the introduction of a verification phase is useful when the single cell performance does not allow to jointly have limited P_{md} and P_{fa}.

EXAMPLE 6.9 *Frame Synchronization for Downlink Continuous Transmission.* Downlink continuous transmission allows timing recovery before frame synchronization; thus a single hypothesis per symbol ($h = 1$) and no fractional timing displacement, $\epsilon = 0$, can be considered. To provide an application example, a frame duration of 32 ms and a baud-rate of 90 Mbaud are considered. The corresponding uncertainty region results to be very large and the TC criterion is adopted because it can provide the best performance for very large uncertainty region. A single-dwell procedure (1TC) and a multi-dwell procedure are investigated, and in particular a verification phase with 2 steps is considered, identified in the following as 3TC. Notably, overall false alarm events are not absorbing, meaning that, in these cases, the synchronization procedure is restarted after a penalty time T_p that has been assumed equal to $2T_F$. In Figure 6.13, the mean acquisition time is reported as a function of E_s/N_0 for 1TC and 3TC with a frequency error of $\Delta f = 5$MHz at 90 MBaud. To cope with the large frequency error, the detection of the whole UW, of length 96, has been divided into a coherent integration length $M = 6$ and a PDI length $L = 16$. It can be noted that 3TC and 1TC performance are very tight for NCPDI, with a slight improvement of 3TC at the lowest SNR, and the opposite behavior otherwise. Differently, considering DPDI, 1TC is able to outperform 3TC for all the considered SNR.

In the example at hand, 1TC seems to be the most suitable solution. However, this is not a general trend and different results can be obtained varying the signal to noise ratio. A useful observation is that a verification phase is in general useful at low SNR while at large SNR, the asymptotic mean acquisition time is $T_F/2$ for 1TC and $2.5T_F$ for 3TC. Thus, a crossing point between the two approaches is always present.

EXAMPLE 6.10 *Frame Synchronization for Downlink Discontinuous Transmission.* When considering discontinuous transmission, a single dwell procedure must be performed. This is the case of beam hopping, for example, where the transmission is periodically switched between different beam subsets, in order to cope with non uniform traffic conditions. No accurate timing recovery is possible in this case before frame synchronization and multiple hypotheses

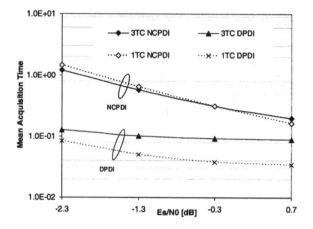

Figure 6.13. Mean Acquisition Time vs. SNR for continuous transmission. 1TC and 3TC are considered with $T_F = 32$ms, 90 MBaud, $\Delta f = 5$MHz, $h = 1$, $\epsilon = 0$, $L_{UW} = 96$, $M = 6$, $L = 16$.

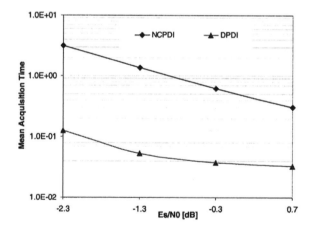

Figure 6.14. Mean Acquisition Time vs. SNR for discontinuous transmission. 1TC is considered with $T_F = 32$ms, 90 Mbaud, $\Delta f = 5$MHz, $h = 2$, $\epsilon = 0.25$, $L_{UW} = 96$, $M = 6$, $L = 16$.

per symbol and a residual timing shift ϵ have to be considered. In Figure 6.14, the mean acquisition time is reported as a function of E_s/N_0 for $T_F = 32$ ms at 90 MBaud, $\Delta f = 5$ MHz, $h = 2$, $\epsilon = 0.25$, with $T_p = 2T_F$, considering $L_{UW} = 96$ with $M = 6$ and $L = 16$, as done in the example above. Both

NCPDI and DPDI are reported. Due to its robustness against large frequency errors, as in the example at hand, DPDI provides a consistent performance improvement.

EXAMPLE 6.11 *Frame Synchronization for Uplink Burst Transmission.* Uplink burst transmission is characterized by the presence of a guard period, N_g, preceding the UW, and frame acquisition reduces to a packet acquisition problem. In this case, performance is typically in terms of overall probabilities, P_{FA} and P_D. Neglecting the terminal startup, it is reasonable to assume that a packet is present; thus, $P_{FA} + P_D = 1$ and the uncertainty region equals $N_g + 1$ symbols. Also in this case, timing recovery cannot be achieved before frame synchronization, so the design must take into account the residual timing shift, ϵ, and introduce $h > 1$ cells per symbol. Notably, in this case, a false alarm event is one-to-one with a packet loss; thus, as already observed, in order to prevent that frame synchronization becomes the system bottleneck, P_{FA} has usually to be one order of magnitude below the achievable packet error rate. Figure 6.15 shows P_{FA} versus E_s/N_0 with $N_g = 8$, $\Delta f = 20$ kHz, 512 kBaud, $h = 2$, $\epsilon = 0.25$, $L_{UW} = 80$, $M = 8$, $L = 10$. Also in this case DPDI outperforms NCPDI.

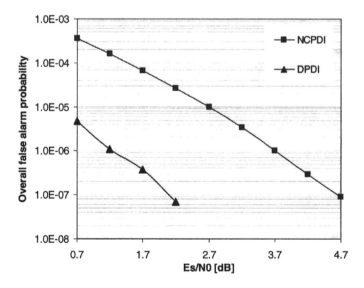

Figure 6.15. Overall false alarm probability, P_{FA}, vs. SNR for uplink burst transmission. A guard interval of 8 symbols is considered with $\Delta f = 20$ kHz, 512 kBaud, $h = 2$, $\epsilon = 0.25$, $L_{UW} = 80$, $M = 8$, $L = 10$.

The detector structure based on PDI has been shown to be robust with respect to large frequency errors. This characteristic makes at the same time the detector robust against significant phase noise effects.

Furthermore, the nonlinear effects usually associated to High Power Amplifier (HPA) on-board the satellite can be limited by employing a constant envelope modulation scheme for the UW, for example QPSK. This translates into a slight energy degradation and constant phase rotation that do not significantly affect the detection performance.

6.8   Receiver Architectures

This section provides some exemplary receiver architectural schemes applicable to different standardized air-interfaces for satellite communications. The goal is to show how the different synchronization algorithms presented throughout this chapter may be combined together in specific practical cases, taking into account the constrains introduced by the adopted frame format and the requirements deriving from the type of service and application.

EXAMPLE 6.12 *DVB-S Receiver Architecture [99]*. Digital Video Broadcast - Satellite (DVB-S) is an established air-interface to broadcast digital TV over satellite using Ku-Band (10-12 GHz). A block diagram of a generic DVB-S receiver is shown in Figure 6.16. Since the frame structure at physical layer relies entirely on the MPEG Transport Stream (MPEG-TS) data format, only NDA synchronization schemes can be employed. In fact, as can be seen in Figure 6.16, the sync-byte indicating the beginning of each MPEG-TS can be only detected after Viterbi decoding. Consequently, also the ambiguity present at the output of the phase detector can only be solved after sync-byte detection. Although traditional solutions are based on FB schemes, such as the Gardner detector (see Section 6.3.3) for timing estimation and Costas loop for carrier recovery [260], the results in [261] shows that hybrid FB-FF solutions, capable of achieving fast synchronization (below 50 ms) for SNR as low as 1 dB can be devised. For instance, the scheme proposed in [261] makes use of R&B and V&V algorithms to respectively estimate and correct the carrier frequency and phase offset after Gardner timing recovery. The overall acquisition time can be seen in Table 6.3. In practical cases, a coarse frequency acquisition prior to timing recovery may be performed for initial synchronization only, since the frequency offsets, mainly due to the Low Noise Block-converter (LNB) and typically in the order of some MHz, can prejudice the timing recovery operation. As seen in Table 6.2, the main drawback of the R&B algorithm is the high computational complexity. On the one hand, performing a first coarse frequency correction could permit to replace the R&B algorithm with another FF scheme having lower complexity and smaller estimation range. On the other hand, the R&B algorithm guarantees the robustness of the scheme under

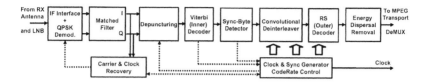

Figure 6.16. Exemplary DVB-S Receiver.

	Algorithm	Norm. $B_L T / L_0$ in symb.	T_{acq} in symb.
Timing Rec.	Gardner	$1.00 \cdot 10^{-4}$	$2.50 \cdot 10^4$
Freq. Est.	R&B	8192	8192
Phase Est.	V&V	1001	1001
		Overall T_{acq} in symb.	$3.42 \cdot 10^4$
		Symbol Rate in Mbaud	27.5
		Overall T_{acq} in ms	1.24

Table 6.3. Summary of Acquisition Time for DVB-S after Initial Synchronization with SNR $\geq$ 1 dB following the approach in [261].

discussion in regards to low SNR operating points, since, in contrast to many other algorithms like Kay's, the R&B algorithm does not suffer from threshold phenomenon (i.e. frequent occurrence of outliers if the SNR is below a given threshold) provided that the observation length is long enough.

EXAMPLE 6.13 *DVB-S2 Receiver Architecture [101].* Digital Video Broadcasting - Satellite 2nd generation (DVB-S2) is the new standard for broadcast, multicast and unicast services over satellite using Ku and Ka (20-30 GHz) bands. With respect to the previous version, more powerful Forward Error Correcting (FEC) schemes are employed, together with a large range of possible modulations, from traditional QPSK to 32-APSK, thus improving the spectral efficiency up to 30%. For unicast profiles, Adaptive Coding and Modulation (ACM) may be used to compensate for unfavorable channel conditions, such as rain fading: higher order modulations and code rates are used at high SNR, more robust modulations and lower code rates are used at lower SNR. To support ACM on a frame basis and to facilitate the synchronization, a more elaborated framing structure with respect to the first version of DVB-S, with longer preamble and distributed pilot symbols has been adopted in the standard. As depicted in Figure 6.17, mainly taken from [202], typical demodulator architectures are based on PA solutions, avoiding when possible FB schemes in order to speed-up the overall acquisition time. As in the case of DVB-S version 1, coarse frequency acquisition may be required only during the initial synchronization, where large frequency offsets up to some MHz may be present. For subsequent

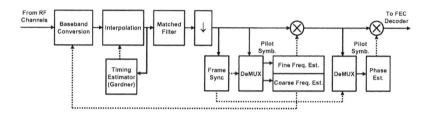

Figure 6.17.   Exemplary DVB-S2 Demodulator.

	Algorithm	Norm. $B_L T / L_0$ in symb.	T_{acq} in symb.
Timing Rec.	Gardner	$5.00 \cdot 10^{-5}$	$1.00 \cdot 10^5$
Frame Sync.			$1.00 \cdot 10^5$
Freq. Est.	L&R	36000	$1.51 \cdot 10^6$
Phase Est.	PA Lin. Interp.	72	1548
		Overall T_{acq} in symb.	$1.71 \cdot 10^6$
		Symbol Rate in Mbaud	25.0
		Overall T_{acq} in ms	68.54

Table 6.4.   Summary of Acquisition Time for DVB-S2 after Initial Synchronization with $SNR \geq 1$ dB following the approach in [202].

resynchronizations, e.g. in case of channel zapping, the residual frequency offset is typically in the order of hundreds of kHz. For low symbol rates the frequency offset becomes too large with respect to the symbol rate, and a dual timing recovery circuit may be necessary, working with two frequency shifted versions of the input so that one of them falls within the frequency range of the timing recovery circuit. Frame synchronization, in contrast to DVB-S version 1, plays here a key role, since it permits to identify the location of the pilot symbols in order to perform carrier fine acquisition and tracking. It has to be noticed that, if ACM is employed, the frame configuration field containing the information about the employed modulation and code rate can be decoded only after that the complete synchronization has been finalized. Results in [202], summarized in Table 6.4 show that, although phase tracking for modulations higher than 8-PSK may require additional effort with respect to the values reported in the table, all modulation schemes can be successfully used without excessive penalties.

EXAMPLE 6.14 *DVB-RCS Receiver Architecture [120].* Digital Video Broadcast - Return Channel via Satellite (DVB-RCS) is the complement of DVB-S: DVB-RCS terminals use a DVB-S-based TDM forward link and a Multi Frequency TDMA (MF-TDMA) return link in order to support interactive services. After having acquired the synchronization in the forward link, each terminal

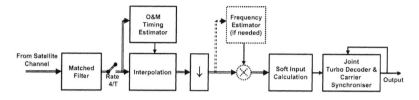

Figure 6.18. DVB-RCS Gateway Receiver with Joint Turbo Decoding and Carrier Synchronization.

initiates its internal clock, by tracking the Network Clock Reference (NCR) which is transmitted by the Network Control Center (NCC) on the forward link. The NCR distribution follows the mechanism as defined in [262]. After NCR synchronization, the terminal will perform a network synchronization procedure to access the MF-TDMA return channel. The requirements for the terminal synchronization are the following [120]:

- the burst synchronization accuracy shall be within 50% of the symbol period with a resolution equal to 1 NCR count interval;

- clock accuracy shall be within 20 ppm from the nominal symbol rate;

- the symbol clock rate shall have a short-term stability that limits the time error of any symbol within a burst to 1/20 of the symbol duration;

- Root Mean Square (RMS) normalized carrier frequency error shall be lower than 10^{-8}.

At this point the terminal is allowed to transmit traffic bursts, provided that the synchronization maintenance procedure does not fail. The receiver located in the gateway shall hence employ synchronization algorithms capable of acquiring the very short logon burst as well as the longer traffic bursts, being the first one normally affected by larger delay and frequency offset. The burst length according to [120] can vary from 96 to 36096 information bits (typical values are 128 bits for synchronization and signalling bursts and suitable multiples of 424 or 1504 bits for traffic bursts), whereas the preamble has a maximum length of 256 symbols, although in practical implementation the preamble overhead does normally not exceed 10% and 40% for traffic and signalling bursts respectively. Timing recovery in the gateway is hence normally achieved using O&M timing estimator presented in Section 6.3.3. For what concerns carrier frequency and phase, FF schemes are preferable, given the bursty nature of the receive signal. Since DVB-RCS uses Turbo Codes in the return channel, their high sensitivity to phase noise may become a critical issue: currently deployed

DVB-RCS systems seem to be limited by the performance of the carrier synchronizer at low code rates and thus cannot fully exploit the potential gain of the Turbo Codes. For this reason, the investigation of joint turbo decoding and carrier synchronization techniques (see Section 6.4) depicted in Figure 6.18 is for the time being very promising. In addition, possible extensions to 8-PSK constellations can be beneficial for future systems (i.e. evolution of DVB-RCS towards ACM).

6.9   Conclusions

The constant demand for more efficient synchronization schemes working in more adverse scenarios means that further advances are expected in the upcoming years. In particular, the joint acquisition of carrier frequency/phase *and* symbol timing is especially interesting for TDMA systems, to reduce the preamble overhead of short bursts. New iterative decoding schemes are foreseen; the embedding of the carrier and timing recovery in the decoding process makes it possible to decrease the SNR level, and increase the spectral efficiency, as shown in this chapter. Finally, the quest for bandwidth opens the floor for discussion on optimal pilot positioning, including the use of superimposed pilot sequences in satellite scenarios.

Chapter 7

DISTORTION COUNTERMEASURES

A. A. Rontogiannis[1], M. Alvarez-Diaz[2], M. Casadei[3], V. Dalakas[1], A. Duverdier[4], F.-J. Gonzalez-Serrano[5], M. Iubatti[3], T. Javornik[6], L. Lapierre[4], M. Neri[3], P. Salmi[3]

[1] *National Observatory of Athens, Greece*

[2] *University of Vigo, Spain*

[3] *University of Bologna, Italy*

[4] *Centre National d'Etudes Spatiales, France*

[5] *University Carlos III of Madrid, Spain*

[6] *"Jožef Stefan" Institute, Slovenia*

7.1   Introduction

As already mentioned in previous chapters, the role of a satellite is to receive a signal from an earth station or another satellite (uplink) and, acting as a simple payload, to transmit it to another earth station or satellite (downlink) [72]. The satellite channel introduces linear distortion to the transmitted signal due to linear filtering, shadowing and multipath fading. Moreover, the need to maximally exploit on-board resources often imposes driving a High Power Amplifier (HPA) at or near its saturation point, resulting in nonlinear distortion of the signal, and rendering the overall link nonlinear. For this reason, modulated signals with a constant envelope have been widely used in practice until now [263]. However, contemporary and future mobile satellite communication systems require high bandwidth efficiency. Thus, bandwidth efficient modulation schemes, with large signal constellations, must be also considered and evaluated [264].

In this chapter, the distortion present in satellite links is described, modeled and analyzed with emphasis given on nonlinear effects, which are mainly due to the HPA operation. Then, the two basic distortion compensation approaches for reliable symbol detection in satellite systems, are presented. The first approach calls for pre-processing of the nonlinear effects at the transmitter side and refers mainly to predistortion methods. The second approach deals with equalization methods, i.e., processing of the signal at the receiver side to recover the transmitted information sequence, by post-cancelling satellite link's linear and nonlinear distortion. Based on various models of the satellite channel, a number of predistortion and equalization techniques are analytically described and their performance is evaluated.

7.2   Distortion in Satellite Systems

In this section, after presenting the general model of a satellite communications system, we proceed with the analysis of the system's nonlinear effects for several modulation schemes. Then, various memoryless HPA models are described and HPA as a nonlinear device with memory is also considered.

7.2.1   System Model

A satellite system processes the uplink signals to ensure - after re-transmission (downlink) - that they are detectable by the earth station receiver. This is achieved by a chain of components called transponder. Figure 7.1 depicts a

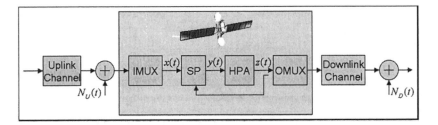

Figure 7.1.   Model of a transparent satellite transponder

block diagram of a transparent satellite transponder . The transparent transponder merely amplifies and frequency shifts the input signals. Its main elements are the Input MUltipleXer (IMUX), the HPA, and the Output MUltipleXer (OMUX). The IMUX separates the individual signals $x(t)$ from the down-converted uplink beam into narrow band channels. Typical IMUX designs comprise amplitude and phase equalization to enhance passband performance. The HPA is a high gain, high power, broadband amplifier that provides the RF power required for the downlink equivalent isotropic radiated power. Owing to the nonlinear distortion introduced by the HPA as explained below, HPAs are usually driven by a Signal Predistorter (SP), which properly modifies the phase and amplitude of the input signal in order to reduce these nonlinear effects.

The OMUX combines the channelized, amplified signals $z(t)$ and directs them to the transmit antenna input port. OMUX typically comprises input isolators, lowpass or harmonic reject filters, low order bandpass filters, a waveguide manifold and switches.

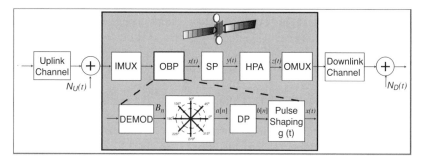

Figure 7.2. Model of a regenerative satellite transponder

Figure 7.2 depicts a block diagram of a regenerative satellite transponder . The regenerative transponder recovers the original symbols from the modulated signals received from the ground, performing, among others, demodulation, data perdistortion of the information symbols and remodulation. These functions are usually, but not necessarily, implemented digitally by the On-Board Processor (OBP). More specifically, the OBP maps the input bit stream into a constellation of M complex symbols $\{a[n]\}$. The complex symbols pass through a pulse shaping filter, $g(t)$, that limits the bandwidth of the transmitted spectrum. The resulting baseband complex signal can be written as :

$$x(t) = \sum_{n=-\infty}^{\infty} a[n]g(t - nT_s) , \qquad (7.1)$$

where T_s is the symbol period. The signal $x(t)$ drives a quadrature modulator, which generates a Radio Frequency (RF) signal $y(t)$ that is amplified by the HPA. If the OBP includes a Data Predistorter (DP), the original complex symbols $\{a[n]\}$ are transformed into a set of predistorted complex symbols $\{b[n]\}$.

In Figures 7.1 and 7.2, $N_U(t)$ and $N_D(t)$ stand for the additive uplink and downlink noise respectively. Due to the high Signal to Noise Ratio (SNR) in the uplink channel, uplink noise is often omitted from the system model. The uplink and downlink channels as well as the transponder's IMUX and OMUX filters introduce linear distortion to the system. Especially the downlink channel linear distortion comes also from shadowing of the satellite signal and multipath fading.

As already mentioned, nonlinear distortion is mainly due to the on-board HPA. The energy resources are limited in satellite communication systems at the satellite and the customer side. The HPA consumes up to 70% of the available system energy, and for this reason it has to be utilized as efficiently as possible. The common way to increase HPA efficiency is to force it to work close to the saturation point, which introduces nonlinear distortion in the transmitted signal. The HPA amplifies higher magnitudes less than smaller ones, thus causing a squeezing of the signal constellation. Moreover, higher signal magnitudes are phase shifted more than small ones, and consequently the signal constellation is also twisted.

To limit distorsion introduced by the amplifier, the average input power of the latter may be reduced. This reduction is called Input Back-Off (IBO) , and is defined as the ratio of the input power that causes saturation of the amplifier (maximum output power) to the actual input power (see Figure 7.3). Similarly, Output Back-Off (OBO) can be defined.

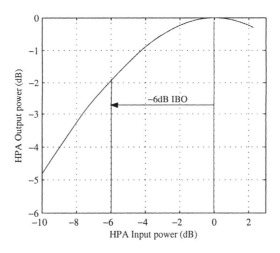

Figure 7.3. Input back off

7.2.2   Nonlinear effects

The nonlinear effects of the system mostly depend on the modulation schemes applied, which can be classified into two basic categories (see Chapter 5):

- high bandwidth efficient multilevel modulation schemes with variable signal envelope, i.e., M-QAM, and M-ASK signals

- low bandwidth efficient, high power efficient constant or near constant envelope modulation schemes, i.e., CPM, D-M-PSK, M-PSK, etc.

7.2.2.1 **Nonlinear effects for M-QAM and DxPSK.** For the first class of modulation schemes simulation results [265] show high correlation between Peak-to-Average Power Ratio (PAPR) and spectrum spreading in neighboring frequency bands. Higher level M-QAM modulation schemes are more sensitive to nonlinear distortions.

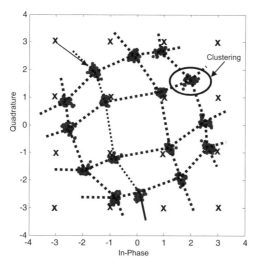

Figure 7.4. Original and received symbol constellation after nonlinear distortion (OBO = 0dB)

Due to the memoryless nonlinear behavior of the HPA, the received constellation is no longer lying on the original lattice ("warping effect"). Additionally, the inclusion of the HPA between linear transmission and receiving filters leads to nonlinear Inter-Symbol Interference (ISI) that produces spreading of the received constellation point in small clusters ("clustering effect"). The last two effects, sketched in Figure 7.4, can be described in terms of the following discrete-time dynamic equation

$$r[n] = F(\ldots, a[n-1], a[n], a[n+1], \ldots) , \qquad (7.2)$$

where $r[n]$ is the received symbol, $\{a[n]\}$ is the original symbol sequence, and $F(\bullet)$ is the multi-variable nonlinear mapping, which describes the behavior of the channel[1].

---

[1] We consider the channel including all the elements and devices between the modulator and the detector.

When a signal with time-varying envelope is nonlinearly amplified, inter-modulation distortion produces a spectral widening of the transmitted pulses. This effect, also known as spectrum regrowth, restores the out-of-band side-lobes, which might cause severe Adjacent Channel Interference (ACI). The spectrum regrowth phenomenon is illustrated in the Power Spectral Density (PSD) curves of Figure 7.5.

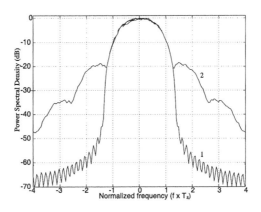

Figure 7.5. Transmitted PSD for a nonlinearly amplified 64-QAM signal. Curve 1: Ideal (linear) PSD; curve 2: Transmitted spectrum when the amplifier is at the saturation

Figure 7.6 depicts, for various values of IBO, the BER performance versus SNR per bit ($\frac{E_b}{N_0}$) for a 64-QAM system using square-root raised cosine filters. It is shown that BER performance can be improved by increasing $\frac{E_b}{N_0}$. However, for 64-QAM, there is an irreducible error floor where BER is almost a constant with respect to $\frac{E_b}{N_0}$. Moreover, the smaller the IBO is, the larger the error floor becomes.

On the other side, spectrum spreading is highly dependent on PAPR for near constant envelope modulation schemes, like differential DxPSK signals [265]. Due to lower signal envelope variation, introduced mainly by the effects of filtering, spectrum spreading in adjacent frequency bands is smaller. When differential phase detection is used only slight increase of BER is observed due to nonlinear distortion.

7.2.3   Amplifier Models

There exist two basic technologies for HPA design on-board satellites, i.e., the Travelling Wave Tube Amplifier (TWTA) and the Solid State Power Amplifier (SSPA) . In recent years, there has been an ongoing discussion about the relative merits of tube-based versus solid state-based amplifiers. TWTA was

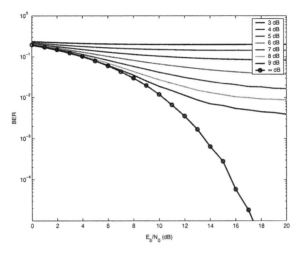

Figure 7.6. BER vs. $\frac{E_b}{N_0}$ after nonlinear distortion. The nonlinear behavior is controlled by the IBO. Notice that $IBO = \infty$ dB is equivalent to linear transmission and when $IBO = 0$ dB the amplifier operates at the saturation

invented in 1940s by Rudolph Kompfner and has been used extensively in a variety of applications. Advances in TWTA technology have given considerable boost to satellite communications. The large bandwidth and high power levels they can provide [266] made TWTAs the dominant power amplifier technology on-board for satellite communications. Recent improvements in solid-state material and amplifier design have pushed the output power level of a single microwave monolithic integrated circuit to a few watts level. Although solid-state electronics are generally more desirable in terms of size, weight, reliability and manufacturability, it is still difficult and costly at the present time to realize significant RF output power at a single device level. Thus, power combining techniques have been adopted [267]. That makes the SSPAs qualified candidates to replace TWTAs for applications in satellite communications by providing high power over a broad bandwidth.

HPA models are required in order to study the effect of the HPA on the over-all system performance. Two approaches exist in modeling a HPA, namely, physical modeling and behavioral modeling. The first approach considers specific electronic elements to model the amplifier, while the second define the HPA model based on the device response to various input excitations. Behavioral models can be further classified to memoryless models and models with memory.

7.2.3.1   Memoryless models.   Let V_{In} and V_{Out} be the input and output voltage of an HPA such that:

$$V_{\text{In}} = \rho_{\text{In}} e^{j\Theta_{\text{In}}} \tag{7.3}$$

$$V_{\text{Out}} = \rho_{\text{Out}} e^{j\Theta_{\text{Out}}}. \tag{7.4}$$

The output voltage V_{Out} can be expressed as follows:

$$V_{\text{Out}} = G(\rho_{\text{In}}) V_{\text{In}}, \tag{7.5}$$

where the complex gain $G(\rho_{\text{In}})$ is defined by:

$$G(\rho_{\text{In}}) = \frac{A(\rho_{\text{In}})}{\rho_{\text{In}}} e^{j\Phi(\rho_{\text{In}})}. \tag{7.6}$$

$|G(\rho_{\text{In}})|$ and $\Phi(\rho_{\text{In}})$ correspond to the Amplitude-to-Amplitude conversion (AM/AM) and Amplitude-to-Phase conversion (AM/PM) characteristics of the amplifier respectively. Due to their simplicity, these models are frequently used to model the on-board HPA response. Based on a narrowband hypothesis for the input signal, it turns out that the only intermodulation products in the output signal bandwidth are due to even degree terms of $G(\rho_{\text{In}})$, i.e., $G(\rho_{\text{In}})$ depends only on ρ_{In}^2 [268]. This is equivalent to have both $A(\rho_{\text{In}})/\rho_{\text{In}}$ and $\Phi(\rho_{\text{In}})$ being functions of ρ_{In}^2. In general, the AM/AM curve is also normalized in power, i.e.,

$$[G(\rho_{\text{In}})]_{\rho_{\text{In}}=1} = 1. \tag{7.7}$$

Moreover, for a TWTA, the saturation is obtained at normalized power:

$$\left[\frac{dG(\rho_{\text{In}})}{d\rho_{\text{In}}}\right]_{\rho_{\text{In}}=1} = 0. \tag{7.8}$$

In general, TWTAs are accurately modeled using this approach. Nevertheless, this is not the case for SSPAs due to non-negligible memory effects.

The most commonly used model for TWTAs is the Saleh model [269], whose AM/AM and AM/PM characteristics are given by the following expressions:

$$A(\rho) = \frac{\alpha_a \rho}{\beta_a \rho^2 + 1} \tag{7.9}$$

$$\Phi(\rho) = \frac{\alpha_p \rho^2}{\beta_p \rho^2 + 1}, \tag{7.10}$$

where $\alpha_a, \alpha_p, \beta_a, \beta_p$ are constant coefficients, and subscripts a and p denote amplitude and phase. The Saleh model is illustrated in Figure 7.7 with parameters α_a, β_a, α_p and β_p taking typical values, i.e., $\alpha_a = 2.1587$, $b_a = 1.1517$, $\alpha_p = 12.5767$ and $b_p = 9.1040$.

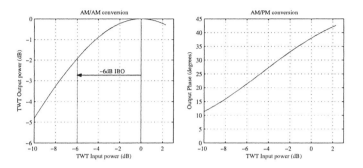

Figure 7.7. (a) AM/AM and (b) AM/PM conversions for Saleh model

Other models can be devised in order to improve the match between analytic and measured characteristics. For example, a more accurate model, involves the use of 8 parameters, i.e.,:

$$A(\rho) = \frac{\alpha_{a1}\rho + \alpha_{a2}\rho^3}{1 + \beta_{a1}\rho^2 + \beta_{a2}\rho^4} \tag{7.11}$$

$$\Phi(\rho) = \frac{\alpha_{p1}\rho^2 + \alpha_{p2}\rho^4}{1 + \beta_{p1}\rho^2 + \beta_{p2}\rho^4}. \tag{7.12}$$

Regarding the modeling of the SSPA, a simple model based on measurements of the TriQuint Semiconductors TGA1135B amplifier operating in Ka band is proposed in [270]. The AM/AM characteristic is modeled as an exponential function and the AM/PM characteristic as a polynomial function, i.e.,:

$$A(\rho) = a(1 - e^{-b\rho}) + c\rho\, e^{-d\rho^2} \tag{7.13}$$

$$\Phi(\rho) = g_5\rho^5 + g_4\rho^4 + g_3\rho^3 + g_2\rho^2 + g_1\rho^1 + g_0, \tag{7.14}$$

where $a, b, c, d, g_5, g_4, g_3, g_2, g_1, g_0$ are constant coefficients. Another model encountered is the following [270]:

$$A(\rho) = \frac{\gamma\rho}{[(\gamma\rho)^{2\delta} + 1]^{1/2\delta}} \tag{7.15}$$

$$\Phi(\rho) = \frac{\alpha_p\rho^2}{\beta_p\rho^2 + 1}, \tag{7.16}$$

where $\gamma, \delta, \alpha_p, \beta_p$ are constant coefficients, and the subscript p denotes phase. The problem of such a model is that it considers $G(\rho)$ to be dependent not only on even but also on odd powers of ρ.

Starting from the measured AM/AM and AM/PM characteristics, the model parameters can be extracted through Minimum Mean Square Error (MMSE) optimization. Given N the number of points of the discrete characteristics, $A_{\text{emp}}[\rho(i)]$ the ith empiric value of the output amplitude and $\Phi_{\text{emp}}[\rho(i)]$ the ith empiric value of the output phase distortion, the Mean Square Error (MSE) for the AM/AM curve can be expressed as:

$$MSE_{AM} = \frac{1}{N} \sum_{i=1}^{N} |A_{\text{emp}}[\rho(i)] - A[\rho(i)]|^2 \qquad (7.17)$$

and for the AM/PM curve as:

$$MSE_{PM} = \frac{1}{N} \sum_{i=1}^{N} |\Phi_{\text{emp}}[\rho(i)] - \Phi[\rho(i)]|^2. \qquad (7.18)$$

As measurements are expressed in dB, it is optimal to realize this minimization problem on dB values, regularly distributed in dB. An example of the application of the MMSE method to the generalized Saleh model of (7.11) and (7.12) is illustrated in Figure 7.8. The resulting parameters are computed as $\alpha_{a1} = 2.2446$, $\alpha_{a2} = 0.3859$, $\beta_{a1} = 1.5404$, $\beta_{a2} = 0.0867$, $\alpha_{p1} = 2.2111$, $\alpha_{p2} = 0.0532$, $\beta_{p1} = 2.1755$, and $\beta_{p2} = -0.0330$.

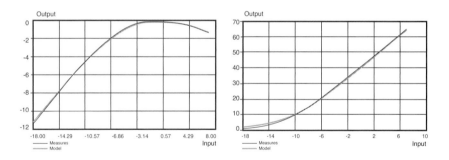

Figure 7.8. (a) AM/AM and (b) AM/PM conversions: model and measurement curves

7.2.3.2 Models with Memory. Memory effects in electrical circuits come from delays and reactive elements like inductors and capacitors. Memory is a critical phenomenon for nonlinear devices and the accuracy of its modeling impacts drastically on the quality of the system simulation. However, sometimes memory can be omitted without any significant impact.

Due to unacceptable computation time, simple behavioral models are used instead of complete electrical circuit models. The amplifiers are often known only by their characteristics measured at input and output ports and then are considered as black-boxes, as described in the previous section. It is not very difficult to model nonlinear devices with memory, if it is possible to split them in two separate blocks. One would represent the nonlinearity without memory and the other would correspond to the memory under linear conditions.

With a high degree of accuracy a TWTA can be considered as a memoryless nonlinear device. On the contrary, the situation is more complicated for SSPAs because nonlinearities are mixed with memory elements and it is not easy to separate them. Specific and more complex models have to be developed. It is also interesting to note that there are two main types of 'memories':

- High frequency memory, which is due to reactive elements that can be found in the high frequency electrical models of the transistors used in SSPAs and in the matching circuits around these transistors.

- Low frequency memory, which is due to three kinds of phenomena:

  - Temperature variations of the transistors with respect to the power of the amplified RF signal. The characteristics of the transistors change with temperature and these variations are limited by thermal time constants that are around a few milliseconds. Then, the signal is distorted mainly on the low frequency components of the complex envelope spectrum.

  - A behavior, similar to the thermal one, which is due to relaxation times of traps in the transistor semiconductor materials.

  - Time constants of the power supply and its associated filters are also an important cause of signal distortion. Careful design of power supply circuits reduces this contribution.

Thus, SSPA AM/AM and AM/PM characteristics measured at a single frequency give only a first approximation of the behavior of this kind of amplifier. The high frequency memory can be modeled by AM/AM and AM/PM characteristics measured at several frequencies [271], [272], but this kind of model is still limited since the low frequency memory is an important factor not taken into account. Moreover, when AM/AM and AM/PM characteristics are measured, the input power changes generally at a lower rate compared to signal amplitude variations. For each input signal power level the temperature is nearly stabilized at a different value. Thus, in case of fast signal variations the temperature has no time to change. This clearly shows the impact of low frequency memory on SSPA's behavior.

In conclusion, SSPAs require specific models with associated measurement techniques. The basic idea of a recently proposed behavioral model with memory is presented next [273]. The model is based on the fact that the behavior of an SSPA depends on the slope of the input RF signal variations. This calls for the use of the first order derivative of the signal. The signal derivative takes into account the present time as well as the near past time and thus a non-negligible part of the memory is taken into account. The complex envelope signal $z(t)$ at the output of the SSPA due to an input complex envelope signal $y(t)$ can be approximated by the following expression:

$$z(t) = f\left(y(t), \frac{dy(t)}{dt}\right).$$ (7.19)

If we use a complex gain the last expression becomes:

$$z(t) = g\left(A(t), \frac{\partial A(t)}{\partial t}, \frac{\partial \Phi(t)}{\partial t}\right).$$ (7.20)

The modeling procedure consists on determining the dynamic complex gain $g(\cdot, \cdot, \cdot)$. This is a three variable function that can be described by a neural network with the advantage of being easily extracted from measurements. A good learning signal is a Gaussian noise sequence. The above model is well adapted for telecommunication applications since it provides a significant improvement on the quality of SSPA modeling. Many on going activities take place to develop more powerful models.

In the literature, an SSPA with memory can also be modeled by a memoryless nonlinearity followed by a linear Finite Impulse Response (FIR) filter [274].

7.3   Transmitter techniques: Predistortion

The objective of the predistorter is to invert the nonlinear function of the HPA so that the response of the cascade of these two devices is linear. Thus, the generic predistortion complex function $y = H(x)$ has to be designed in such a way that the following relation holds:

$$z = \Gamma(H(x)) = \Gamma_0 x,$$ (7.21)

where z and x are the complex envelopes of the amplified signal and the modulated signal, and Γ_0 is the nominal gain of the amplifier (see Figure 7.9). The exact inverse can be obtained for input values in the range where the amplifier characteristic is a one-to-one function. As an example, if the Saleh model is used the analytic inversion of the amplifier is possible whenever $\rho \leq \frac{\alpha_a}{2\sqrt{\beta_a}}$. When $\rho > \frac{\alpha_a}{2\sqrt{\beta_a}}$, the amplitude of the predistortion device output is taken equal to $\frac{1}{\sqrt{\beta_a}}$. Regarding phase distortion compensation, when $\rho \leq \frac{\alpha_a}{2\sqrt{\beta_a}}$, $\Psi[\rho(t)] = -\Psi[R[\rho(t)]]$, and $\Psi[\rho(t)] = -\frac{\pi \alpha_p}{\beta_a + \beta_p}$, otherwise. The resulting

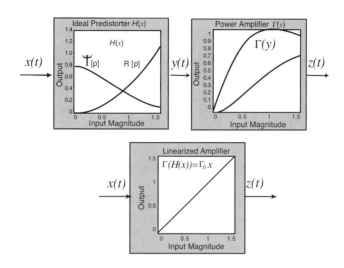

Figure 7.9. The ideal predistorter $y = H(x)$ in series with the power amplifier $z = \Gamma(y)$ results in a linear response

AM/AM and AM/PM characteristics of the predistorter are:

$$
R[\rho(t)] = \begin{cases} \frac{\alpha_a - \sqrt{\alpha_a^2 - \beta_a(2\rho)^2}}{2\beta_a \rho} & \rho \leq \frac{\alpha_a}{2\sqrt{\beta_a}} \\ \frac{1}{\sqrt{\beta_a}} & \rho > \frac{\alpha_a}{2\sqrt{\beta_a}} \end{cases}
\tag{7.22}
$$

$$
\Psi[\rho(t)] = \begin{cases} \frac{\pi\alpha_p(\alpha_a - \sqrt{\alpha_a^2 - \beta_a(2\rho)^2})^2}{2\left(-\alpha_a^2\beta_p + 2\beta_a(-\beta_a+\beta_p)\rho^2 + \alpha_a\beta_p\sqrt{\alpha_a^2 - \beta_a(2\rho)^2}\right)} & \rho \leq \frac{\alpha_a}{2\sqrt{\beta_a}} \\ \frac{-\pi\alpha_p}{\beta_a+\beta_p} & \rho > \frac{\alpha_a}{2\sqrt{\beta_a}} \end{cases}
\tag{7.23}
$$

Since the HPA behaves linearly for a large range of input amplitudes, in practice the predistortion function is modeled as follows:

$$
H(x(t)) = x(t) + \hat{H}(x(t)) ,
\tag{7.24}
$$

where $\hat{H}(\cdot)$ represents the deviation from linearity. This way, the predistorter just compensates for the residual nonlinear distortion.

The advantage of predistortion lies in the fact that only a single system (on-board the satellite) for canceling the HPA nonlinearity is needed (compared to using an equalizer in each terminal). On the other hand, its main drawback is that processing takes place on-board, so it cannot be applied to the satellite payloads already on orbit and needs ground control. Moreover, in case multipath is present, which is the realistic case, an equalizer at the terminal side is still needed to compensate for distortions such ISI or ACI.

7.3.1   Classes of predistorters

Predistorters can be grouped based on two criteria, i.e., a) the way they perform the modification on the transmitted signal (signal and data predistorters) and b) the applied technology (analog and digital predistorters)

7.3.1.1   Signal Predistortion.   A SP generates a signal that compensates for the nonlinearities introduced by the RF module without accessing to the underlying original data symbol sequence. SPs are placed after the baseband pulse shaping filter and are usually implemented with adjustable nonlinear devices. If they work on baseband or intermediate frequency signals, adaptive and digital implementations are feasible. In this case, the most usual design criterion to adjust the parameters of the SP is the minimization of the mean squared error between the original signal and the received one[2] (time domain optimization). SPs working on radio frequency signals are implemented with analog technology and are usually designed in order to minimize the intermodulation products and the out-of-band PSD (frequency domain optimization).

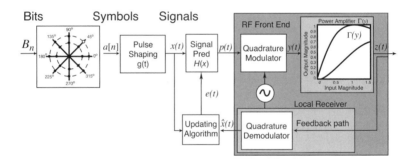

Figure 7.10.   Signal predistorter at the baseband

Figure 7.10 illustrates the hardware configuration for an adaptive signal predistorter at the baseband. The data modem generates the desired complex signal $x(t)$ and the predistorter generates a complex signal $p(t) = H(x(t))$, that corrects for the nonlinearities introduced by the RF module.

7.3.1.2   Data Predistortion.   A DP works on the baseband data symbols. It modifies the transmitted constellation in such a way that, after linear filtering and nonlinear processing in the downlink, the constellation of the received samples on average will match (or approximate) the desired signal constellation.

---

[2] Adaptive implementations require the inclusion of a local receiver at the satellite.

Obviously, data predistortion can only be applied to linear modulation schemes.

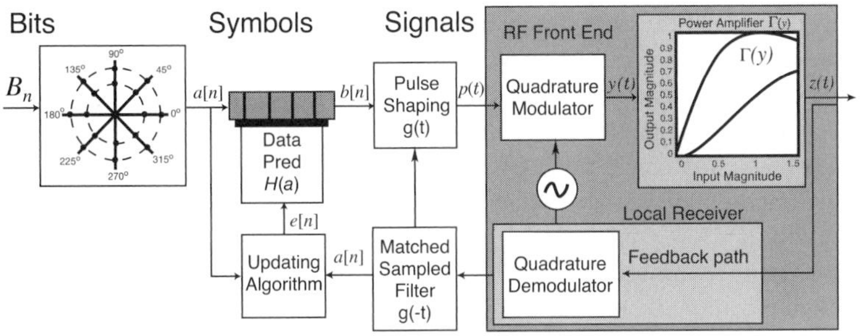

Figure 7.11. Model of the transmission system with data predistortion

Figure 7.11 illustrates the hardware configuration for an adaptive DP. As the predistorter is placed before the pulse shaping filter, memory is needed to learn its response. Thus, data predistorter transforms a finite sequence of P symbols and produces one predistorted symbol $b[n]$:

$$
\begin{aligned}
b[n] &= H(a[n - Q], \ldots, a[n], \ldots, a[n + R]) = \\
&= H(\mathbf{a}_P[n]) ,
\end{aligned}
\tag{7.25}
$$

where $P = Q + R + 1$ is the memory length of the predistorter.

7.3.1.3   Analog Predistortion.   Analog predistortion is provided by an analog device, usually placed just before the HPA, that modifies the transmitted waveform (signal predistortion).   Typically, analog SPs invert the third- or fifth-order polynomial approximation (only the odd terms are present given the bandpass nature of the transmitter) of the amplifier characteristic's model. For example, the cubic distortion generator, one of the most frequently used predistorters, is composed of a pair of diodes connected in antiparallel and a linear impedance, combined with a 180° hybrid. A wide-band analog multiplier, that is available as a commercial IC device, can be used to get the fifth-order distortion component.

7.3.1.4   Digital Predistortion.   Unlike the analog approach, a digital implementation of the predistorter lends itself to the design of adaptive schemes, capable of tracking changes in the HPA response, such as temperature and aging drifts, variations in the operating point, etc. Obviously, digital implementation is mandatory for data predistortion.

In digital signal predistortion , the signal at baseband is usually A/D converted, then it is fed to a processor providing sample-by-sample predistortion[3] and finally converted back to analog form prior to frequency translation and power amplification. The digital implementation of a SP requires specification of the sampling rate and the levels of the quantizer. The sampling rate must be selected according to both the bandwidth of the signals to be processed and generated, and the power consumption in the digital processor. The number of levels of the quantizer is selected to keep the performance of the SP between the desired margins of quality. Nonuniform quantization may be used to compress areas with larger signal variations and enlarge those with smaller variations.

In order to summarize, Figure 7.12 represents both technology (analog, digital) and position along the transmission chain (baseband, intermediate frequency, radio frequency) of signal and data predistorters.

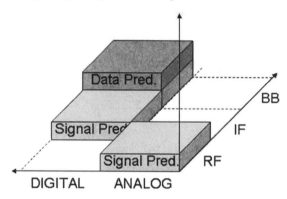

Figure 7.12.   Technology (analog, digital) and position along the transmission chain (baseband, intermediate frequency, radio frequency) of signal and data predistorters

7.3.2   Predistortion Techniques

In this section a number of predistortion techniques are described namely the Look-Up Table (LUT) approach, Volterra model based predistortion, neural networks based predistortion and the clustering approach.

7.3.2.1   LUT approach.   Predistortion can be interpreted as a function approximation. The smoothness property of the ideal predistortion functions is the key element to design simple, but effective, predistorters. For this reason, architectures with a few parameters, such as memoryless polynomial networks

---

[3]The memory effects of the nonlinear RF front-end are negligible.

or LUTs, are often used, especially in signal predistortion. Rather than computing in real time the inverse characteristic, requiring a considerable amount of computational power, it is possible to compute the coefficients off-line and place them into a LUT, so as to need only one memory access for each processed value.

The use of LUTs is preferable when adaptation is required, as they can be updated by means of simple algorithms so as to track HPA characteristic variations. The main drawback of this method is high memory occupation, which grows exponentially with the number of precision bits in the input variable. For this reason, accuracy of representation has to be chosen carefully, as quantization noise increases for decreasing precision. In the following, two LUT approaches are described, i.e., the mapping predistorter and the gain-based predistorter.

7.3.2.1.1   Mapping Predistorter.   This strategy consists in indexing the LUT with two entries, the real and the imaginary part of the input signal. The outputs of the corresponding predistorted values are stored in two distinct LUTs, one for the real and one for the imaginary part of the predistorted signal. This method, proposed by Nagata [275], has the advantage of presenting a low computational complexity, as no coordinate conversion is required. Morever, the output of the predistorter can be input to the HPA without further elaboration, since the desired predistorter output values are stored in the two LUTs. One of the disadvantages of this technique is that the Signal to Quantization Noise Ratio (SQNR) is proportional to the number of table entries, leading to high memory occupation [276]. Another issue related to this predistortion technique is the need for a phase shifter in the feedback loop.

7.3.2.1.2   Gain Based Predistorter.   While the mapping technique, can be applied to predistort any nonlinear function, in gain-based predistorters the complex gain is assumed to be a function of the input amplitude only, independent of the input phase. Mapping predistorters have the complex gain implicitly embedded into the ratio between the output and the input complex values, and do not take advantage of the phase-independent gain property. A smart optimization thus consists in indexing the LUT considering only the module of the input signal, neglecting the phase and storing in the LUT the complex gain instead of the output values. In this way, one indexing dimension is removed with considerable reduction of memory requirements. In addition to this, it can be proved that the error due to quantization is now relative and not absolute like it is for the mapping predistorter, resulting in better performance for low signal levels. The indexing strategy is not determined by this method, and it can be chosen so as to trade off complexity, speed, and accuracy. For example, to ensure low complexity achieving good performance, linear in power indexing can be applied by addressing the LUT with the square module of the input signal. Notably, this approach does not require the presence of a phase shifter in the feedback loop [276]. Moreover, the SQNR achievable with this technique is proportional to the square of the table size, compared to the mere proportionality to the size of the LUT of the former approach.

Smart LUT indexing techniques, in addition to the classical ones (linear in amplitude or in power) can be devised in order to optimize memory occupation for a given complexity.

7.3.2.1.3   Classical indexing strategies.   In the following, an overview of classical indexing techniques is presented:

- **Linear in amplitude.** This method exhibits good performance by uniformly spacing the table entries along the input signal range. There are some issues related to the complexity of the computation of the input signal amplitude, which requires either the availability of a rectangular to polar coordinates converter or the presence of a square-root block.

- **Linear in power**. This method requires only a square calculation module resulting in simple implementations. It exhibits a good performance if the HPA characteristic is linear for low amplitudes, where the table entries spacing is coarser. This is not the case for class AB amplifiers, whose characteristic exhibits strong nonlinearities for low amplitudes. For this class of amplifiers, this indexing technique leads to poor performance due to the high quantization error affecting small input values. If instead nonlinear behavior is exhibited for high amplitudes only, this approach is both effective and simple to implement.

7.3.2.1.4   Advanced Indexing Strategy.   Besides linear in amplitude and linear in power indexing techniques, other approaches can be applied. Cavers, starting from the definition of companding functions, obtained the optimum LUT mapping [277], which is back-off, modulation, and access scheme dependent. Unfortunately, this approach requires large computational power. A possible way to reduce complexity is to keep the optimum mapping obtained for a single IBO value (e.g. 3 dB) for all IBOs. In this case, however, the gain with respect to the "linear in amplitude" mapping is only 1 dB, compared to the 1 to 4 dB improvement achievable when the optimum mapping is computed dynamically. Hassani and Kamarei [278] investigated the possibility to introduce an advanced LUT mapping technique, which envisages a denser mapping where the gradient of the AM/AM gain characteristic is larger. In particular, they devised a low complexity mapping, which employs a pre-indexing unit that maps the entries of a 'virtual' linear and equispaced LUT, onto the real LUT. This technique allows to reach the same performance achievable using a LUT with 256 entries with 'linear in amplitude' mapping using a LUT with

64 entries only, thus reducing the LUT dimension by a factor 4. Firstly, input amplitude range is partitioned in a number of regions, keeping the same level of nonlinearity in each region. If for a given precision with amplitude method 2^n entries are needed, the considered LUT has 2^m entries, with $m < n$. Both LUTs are segmented, but in the real LUT the smaller the step is, the more nonlinear the characteristic is in that region. Proper mapping translation functions can be devised to keep the complexity at low levels, for example envisaging a ratio of 2 between the number of entries in the two LUTs. This approach is particularly beneficial compared to the linear in power indexing when working with AB class amplifiers, which present large nonlinearity for low amplitudes. In the following, an overview of advanced indexing techniques is presented:

- **PCM Companding law (A-law, μ-law).** Its main advantage is the availability of commercial devices used in PCM applications, but exhibit bad behavior, and are thus not suitable for predistorters [277].

- **Proportional to the gradient of the AM/AM gain characteristic.** This indexing method yields good results without requiring a significant complexity increase. With respect to the aforementioned indexing methods, this approach takes into account the actual shape of the gain characteristic, increasing the samples density as the nonlinearity of the characteristic increases. In this way an improved flexibility can be achieved, using denser mapping only where required by the gain characteristic [278].

- **Optimum indexing.** This approach, introduced by Cavers [277], can yield valuable predistortion efficiency improvement, but has the drawback of presenting a high complexity, as it must be re-calculated dynamically. To reduce complexity it is possible to keep the optimum predistortion computed for a given IBO also for other values of IBO, but in this case performance improvement is very small.

7.3.2.2   Volterra model-based predistortion.   Most existing systems assume that the power amplifier has a memoryless nonlinearity and thus employ a memoryless predistorter to compensate for this nonlinearity. In practice, however, the combination of baseband (or Intermediate Frequency (IF)) pulse shaping at the transmitter with the HPA leads to an overall nonlinear channel with memory. Moreover, amplifiers (especially SSPA) exhibit themselves memory effects, whose cause may be electrical or electrothermal, as described in section 7.2.3.2. As a result, the conventional memoryless predistortion appears to be ineffective in linearizing the HPA response. More specifically, as explained in [279], memoryless predistortion is able to compensate for constellation warping, but fails to reduce clustering, which is mainly due to intersymbol interference.

Improvement over conventional memoryless predistortion can be achieved by modeling the HPA as a nolinear system with memory and designing a memory predistorter accordingly. A general well-studied nonlinear model with memory is the Volterra series model . The input-output expression of a finite order, finite support Volterra model, which can be used for HPA modeling, can be represented as follows

$$z[n] = \sum_{k_1=0}^{K} h_1[k_1]y[n-k_1] + \cdots$$

$$+ \sum_{k_1=0}^{K} \cdots \sum_{k_L=0}^{K} h_L[k_1,\ldots,k_L]y[n-k_1]\cdots y[n-k_L], \quad (7.26)$$

where $h_i[k_1,k_2,\ldots,k_i]$, $i = 1,\ldots,L$ is called kernel of order i. Predistortion of the Volterra model is usually implemented using the pth-order inverse technique [280], [279], [281]. According to this technique, the predistorter is also modeled as a Volterra system of order L and is designed in such a way that the Volterra kernels of orders 1 to L of the overall system are zero. The method is difficult to implement and results in an approximation of the inverse system, which may by unacceptable in practice.

Special cases of the Volterra model for the HPA have been recently proposed, including the Wiener model [282],[283], and the memory polynomial model [284],[285]. The Wiener model consists of a linear time invariant system followed by a memoryless nonlinearity and under certain conditions accepts an exact inverse (predistorter), in the form of a Hammerstein system. However, the identification of a Hammerstein system is in general a complicated task. The memory polynomial model retains only the 'diagonal' terms of the Volterra model and is thus expressed as follows:

$$z[n] = \sum_{l=1}^{L} \sum_{k=0}^{K} h_l(k,\ldots,k)y^k(n-k). \quad (7.27)$$

As in the Volterra model case, the inverse of a memory polynomial model is difficult to obtain even as an approximation. Instead of modeling the HPA and attempting to implement the predistorter as the inverse system, the emphasis can be given directly to the identification of the predistorter itself, irrespective of the model of the system to be compensated. The problem with this approach is that the desired output of the predistorter is not known beforehand. To circumvent this problem, one possibility is to use the architecture shown in Figure 7.13, which is known as the indirect learning architecture. This configuration comprises two identical blocks, one for predistortion and the other for training. If we assume that the HPA nonlinearity is invertible, then as the error $e(n)$ approaches

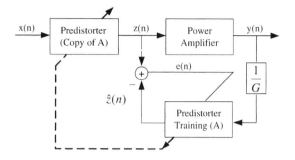

Figure 7.13. The indirect learning architecture [285] © 2004 IEEE

zero, the output of the overall system tends to $Gx(n)$, where G is the gain of the linearized amplifier and $x(n)$ is the input of the system. After convergence of the training phase, the setup of Figure 7.13 can switch to the open loop mode, until an update of the predistorter settings is required due to possible changes of the HPA characteristics. The indirect learning architecture has been introduced in [286] assuming a general Volterra predistorter and adopted in [285] for a memory polynomial predistorter with reduced computational complexity.

7.3.2.3 Neural networks based predistortion. A reasonable alternative for approximating the ideal data predistortion function is to use a Neural Network (NN) . NNs are parallel distributed information processing systems that are capable of learning and self-organizing. They are composed of a large number of simple processing units, called neurons, which are interconnected to form a network that performs complex computational tasks. Neural networks perform two major functions: learning and generalization. Learning is the process of adapting the connection parameters in a neural network in order to minimize a loss function given an input vector. Generalization (or recall) is the process of accepting a new input vector and producing an output response. In this case, the (fixed) network weights are used in order to compute the output.

NN predistortion is well suited for satellites with regenerative payloads. Relevant works commonly resort to nonlinear predistorters [287, 288] based on Multi-Layer Perceptrons (MLP) NN structures [289].

7.3.2.3.1 Multi-layer NNs. The MLP is one of the most popular neural network architectures used in digital communications. Its basic unit is the neuron, shown in Figure 7.14. The MLP network consists of an input layer, one or more so-called hidden layers and an output layer. Hidden and output layers include several neurons.

The classical neuron computes the weighted sum of its inputs and feeds it to a nonlinear function called activation function . Assuming that the layer index is denoted by i, x_{ik} is the output of neuron k of layer i, w_{ijk} is the weight that links the output $x_{i-1,j}$ to neuron k of layer i and $N(i)$ is the number of neurons in layer i, the output y_{ik}, of neuron (i,k) is given by

$$y_{ik} = f\left(\sum_{j=1}^{N(i-1)} w_{ijk}x_{i-1,k} + b_{ik}\right). \tag{7.28}$$

$\{w_{ijk}\}$ and $\{b_{ik}\}$ form the free parameters of the neuron. Thus the behavior of the neural network depends on the choice of the activation function. Different classes of nonlinear activation functions, depending on some free parameters, have been widely studied and applied. The most commonly used activation functions are sigmoids (see Figure 7.14). It has been shown that an MLP with

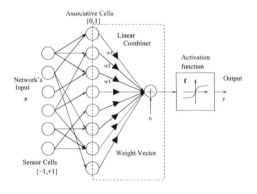

Figure 7.14.   The neuron

one hidden layer is able to uniformly approximate any function defined on a compact space, if it has enough neurons in the hidden layer [289]. This is an interesting property of neural networks, which is known as the 'universal approximation theorem'. In digital communication applications, experience showed that some tenths of neurons in the hidden layer can be sufficient to deal with most problems.

To perform predistortion, the basic architecture comprises two MLPs, one for amplitude and the other for phase predistortion. Moreover, there are two basic methods that have been investigated for an HPA predistorter implemented by MLPs. The first one, uses an identification model of the HPA (via a mimic model) for updating the NN predistorter. The second one attempts to build the predistorter parts (i.e., the two sub-NNs) from the set of input-output patterns of the HPA. Two different algorithms have been tested for training these networks, i.e., the classical ordinary gradient backpropagation algorithm and the natural gradient descent algorithm. Simulations showed that the gradient decent algorithm converges faster and attains a lower mean squared error compared to the gradient backpropagation method [290].

Alternative network structures that can be used for predistortion are local basis functions networks and the Generalized Cerebellar Model Articulation Controller (GCMAC) network, which are described below.

7.3.2.3.2   Local Basis Functions Networks.   The Local Basis Functions (LBF) networks are two-layer feedforward networks, whose input layer activation function $\phi(x)$ is a continuous function. In contrast to MLP-NN, the activation function is a compact one, and no longer a sigmoid. The general structure of an LBF network is shown in Figure 7.15. A special case of an

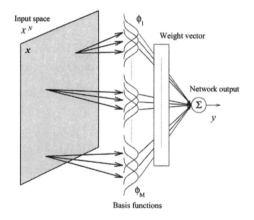

Figure 7.15.   The LBF network: two-layer architecture

LBF network is the Radial Basis Functions (RBF) network, whose activation function is Gaussian shaped, i.e.,:

$$\phi(t) = exp\left(-\frac{t^2}{2\sigma^2}\right). \tag{7.29}$$

The outputs of the first layer neurons are written as:

$$x_{1,k} = \phi_k\left(||\boldsymbol{x} - \boldsymbol{c}_k||\right), \tag{7.30}$$

where x is the input vector and c_k is the weight vector (or center) associated to neuron k. The network outputs are written as:

$$y_j = \sum_{k=1}^{M} w_{k,j}\phi_k\left(||x - c_k||\right), \qquad (7.31)$$

where M is the number of neurons in the first layer, and $w_{k,j}, j = 1,\ldots, N$, are the weights associated with the output layer. The free parameters are therefore the centers $\{c_k\}$ and the weights $\{w_{k,j}\}$.

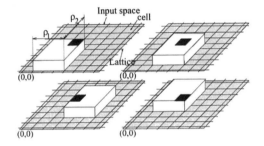

Figure 7.16. The GCMAC network with constant LBFs

7.3.2.3.3 The GCMAC network.

The GCMAC network divides the input space into cells using a lattice[4]. This division normalizes the input space (see Figure 7.16). In the GCMAC approach, the LBFs are supported on rectangular domains that are evenly distributed on the normalized input space. The vector $\rho = [\rho_1,\ldots,\rho_P]^T$ specifies the size in cells of the LBF along each input axis[5]. Unlike RBF networks, the size and location of these LBFs are predefined at the initialization step. The free parameters are the weights $\{w_k\}$ of the LBFs.

7.3.2.4 Clustering approach.

In the following, our attention is focused on data predistortion with memory. Let a_n be the input data symbol entering the predistorter, taking values α_i, $i = 1, 2,\ldots, M$. In addition, let r_n be the received signal sample at time n. Then, the received signal constellation is a set of M clusters identified as c_i, $i = 1, 2,\ldots, M$, where $\bar{c}_i = E[r_n|\alpha_i]$ is their center of gravity.

Let b_n be the symbol at the output of the predistorter. Assuming that the predistorter memory spans P symbols at a given instant n, as in (7.25), b_n is

---

[4]This preliminary quantification is not required when the input space is digital (data predistortion).

[5]When the size of the domain of the LBFs is one cell, the GCMAC becomes a LUT.

a signal point of an M^P-point multidimensional signal constellation. In order to optimize the predistorter, the minimization of the mean square error D' is pursued:

$$D' = E\left\{|E[r_n|b_n] - a_n|^2\right\}. \tag{7.32}$$

In other words, this predistortion technique, acts on an equivalent signal constellation of M^P points. Hence, each a_n symbol will have M^{P-1} possible predistorted values, identified as $b_{n,j}, j = 1, 2, \ldots, M^{P-1}$, depending on the P preceding and subsequent symbols. At the receiver side, each cluster can be split into M^{P-1} overlapped sub-clusters, referred to as $c_{i,j}, j = 1, 2, \ldots, M^{P-1}$, that are significantly smaller than the cluster as a whole. Based on the MMSE approach, the aim of the predistorter is to make the center of gravity of each sub-cluster overlapped to the nominal point of the signal constellation, by making an appropriate correction on the basis of the estimate of the center of gravity of each sub-cluster, performed by averaging over a sequence of sufficient length.

The predistorter with memory based on the clustering approach can be implemented using a RAM addressed by the bits of a set of consecutive symbols that are within the span of the predistorter memory. The required memory size depends on both the modulation scheme and the memory span of the predistorter. For example, if a M-QAM modulation scheme and a predistorter with a memory of P symbols are used, a memory with M^P locations, i.e., $P \log_2 M$ address bits, is needed. Actually, the number of address bits can be reduced by 2 if the quadrant symmetry of QAM signal constellation is exploited.

The optimum predistorter with memory based on the clustering approach can be implemented in the presence of a small number of signal points and small memory spans due to the complexity introduced by the required memory size. Different trade-offs between performance and complexity can be proposed by designing a number of suboptimum predistorters for a given modulation scheme and memory size. A suboptimum predistorter is obtained by partitioning the signal set into M/Y subsets of Y points.

7.3.2.5 Performance evaluation of predistortion techniques.

The performance of predistortion techniques is quantified using the equivalent SNR degradation caused by the residual nonlinear distortion at a specified BER. For this purpose, the channel is assumed to have a flat frequency response with additive, white, circularly symmetric, Gaussian noise. If $[SNR]_C$ is the SNR, expressed in dB, required by the compensated system to obtain the specified BER at a given OBO, and $[SNR]_G$ is the required SNR to obtain the same BER on the Gaussian channel, then the Total Degradation (TD) is defined as:

$$[TD] = [SNR]_C - [SNR]_G + [OBO]. \tag{7.33}$$

The total degradation results in a convex function of the OBO, taking the minimum value at the optimum OBO. This function can be obtained by following the quasi-analytical procedure described in [291].

A 64-QAM transmitter was simulated including a raised-cosine pulse shaping filter with roll-off factors $\alpha = 0.5$ and $\alpha = 0.25$. The desired BER is 10^{-4}. In Figure 7.17, total degradation versus OBO is plotted for the following compensation methods (schemes are listed below in the order they are labeled in Figure 7.17):

1 Automatic Control Gain (no predistortion).

2 Memoryless Data Predistortion: 1-length LUT ($Q = R = 0$, $P = 1$).

3 Data predistortion: 3-length LUT ($Q = R = 1$, $P = 3$).

4 Data predistortion: MLP ($P = 3$, i.e., three layers).

5 Data Predistortion: GCMAC ($P = 3$, $\rho = [63, 1, 63]^T$).

6 Complex signal predistortion; 6 bit ADC; sampling frequency $= 8 \times \dfrac{1}{T_s}$.

7 Fractionally-spaced Data Predistorter: oversampling factor $\lambda = 4; P = 3$.

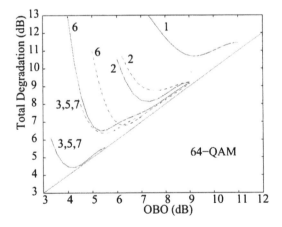

Figure 7.17. TD vs. OBO for a 64-QAM system for a BER $= 10^{-4}$. Solid line: $\alpha = 0.5$. Dashed line: $\alpha = 0.25$

Although speed of convergence is not a hard restriction in predistortion (for slow varying amplifiers, learning can be done off-line), it can be measured for comparison purposes. The following predistorters have been simulated for a

nonlinear 16-QAM transmitter: 1. 3-length LUT; 2. 5-th order Volterra filter;
3. GCMAC ($\rho = [15, 1, 15]$). The IBO is 5 dB. The Signal-to-Distortion Ratio
(SDR)[6] is depicted in Figure 7.18. It is observed that the 3-length LUT has the
slowest convergence: for each input sample only one (of 16^3) weight is updated.
The Volterra filter has a faster convergence, but the final SDR is about 10 dB
lower than that provided by the LUT. Finally, the GCMAC provides almost
the same final SDR than the LUT but its speed of convergence is considerably
higher than that of the Volterra filter and the LUT.

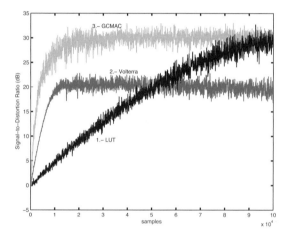

Figure 7.18. SDR for a nonlinear 16-QAM transmitter: 1. LUT; 2. 5-th order Volterra filter;
3. GCMAC ($\rho = [15, 1, 15]$)

Regarding the MLP, we have used a batch learning algorithm to adjust the
weights. In particular, the Levenberg-Marquardt algorithm [292] was used for
training a 3-layer perceptron with 15 neurons in the hidden layer. Results for
a nonlinear 16-QAM transmitter (IBO 5 dB) are depicted in Figure 7.19. It is
observed that after 50 epochs (i.e., 5×10^5 samples) the final SDR is about 23
dB (consider the 30 dB provided by the 3-length LUT or the GCMAC).

Better results are provided by the RBF network (see Figure 7.20). We have
used a RBF with 21 basis functions. The location of the centers were determined
using the batch mode k-means clustering algorithm, and the orthogonal least
squares algorithm was used for determining the weights [293].

All the results provided in this section are summarized in Table 7.1. The
main conclusion is that local approaches, which approximate the ideal predis-

---

[6]The SDR is the ratio between the transmitted output power and the power of the distortion measured as the
difference between the ideal constellation and the actually transmitted one.

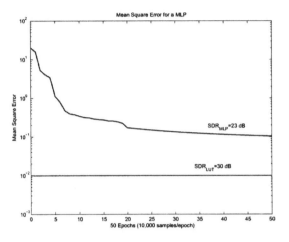

Figure 7.19.   SDR for a 16-QAM transmitter with a MLP-based (15 neurons in the hidden layer) predistorter

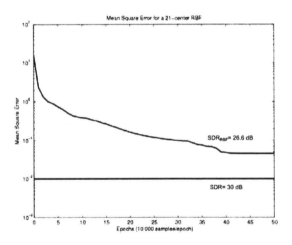

Figure 7.20.   SDR for a 16-QAM transmitter with a 21 center RBF-based predistorter

tortion function using basis functions with compact support, perform better than global ones (the Volterra filter uses polynomials and the MLP sigmoid-functions, whose domain is the whole input space). The reason is that local basis functions are able to capture fine details of the ideal predistortion function, and thus result in a better performance.

	SDR (dB)	Convergence (samples)	FLOPS per sample
LUT	30	10^5	4
GCMAC	30	2×10^4	64
5th-order Volterra	20	2×10^4	3255
MLP	23	5×10^5	22920
RBF	26	5×10^5	22540

Table 7.1. Performance of different nonlinear networks in a predistortion problem. We have considered a nonlinear 16-QAM transmitter (OBO 5 dB). The networks are driven by three input samples (data predistortion with memory)

Figure 7.21. Channel equalization

7.4 Receiver techniques: Equalization

The most important distortion compensation technique at the receiver side is channel equalization. Channel equalization can be viewed as a classification problem. The optimal solution to this classification problem is inherently nonlinear. Various nonlinear equalizer design techniques have been proposed that enhance the performance of conventional equalizer techniques presented in [17] and more detailed in [294]. In this section a brief overview of classical equalization techniques for linear channels is first provided. Then nonlinear methods appropriate for equalization of digital satellite channels are described.

7.4.1 Linear Channel Equalization

The task of the equalizer is to reconstruct the transmitted symbols as accurately as possible based on noisy channel output observations. In Figure 7.21 the discrete baseband equivalent of a conventional digital communication system is shown. The channel, which is the convolution of the transmitter filter, the transmission medium and the receiver filter, is modeled as an FIR filter with a transfer function $H(f)$ and L denotes the length of the channel impulse response. The transfer function of the equalizer is denoted by $C(f)$. Equalizers may be classified into two main categories, namely, symbol-decision equalizers and sequence-estimation equalizers. These can be further classified according to their structure, the optimization criterion and the algorithm used to adapt their coefficients.

On the basis of its structure, a symbol-decision equalizer can be a Linear Equalizer (LE) or a Decision Feedback Equalizer (DFE).The LE is using the information present in the observed channel output vector

$$\boldsymbol{r}_k = [r_k \cdots r_{k-n_f+1}]^{\mathrm{T}} \tag{7.34}$$

to produce an estimate $\hat{a}_{k-d}$ of the transmitted symbol a_{k-d}. The integers n_f and d are known as the equalizer Feed-Forward (FF) order and decision delay, respectively. Traditionally this is viewed as an inverse filtering procedure, in which the equalizer is designed to approximate the inverse of the distortion channel [295]. Thus, the LE is a linear filter with output

$$u_k = f(\boldsymbol{r}_k) = \boldsymbol{w}^{\mathrm{T}} \boldsymbol{r}_k, \tag{7.35}$$

where $\boldsymbol{w} = [w_1 \cdots w_{n_f}]^{\mathrm{T}}$ are the FF filter coefficients. The output of the equalizer u_k is then quantized into one of the symbol values with the aid of a decision slicer and an estimate $\hat{a}_{k-d}$ is produced.

By expanding the inputs of the LE to include past detected symbols, i.e.,

$$\hat{\boldsymbol{a}}_{k-d} = [\hat{a}_{k-d-1} \cdots \hat{a}_{k-d-n_b}]^{\mathrm{T}}, \tag{7.36}$$

we obtain a DFE where n_b is the equalizer FB order. The conventional DFE is based on linear filtering of this expanded input vector and its output is expressed as

$$u_k = f(\boldsymbol{r}_k, \hat{\boldsymbol{a}}_{k-d}) = \boldsymbol{w}^{\mathrm{T}} \boldsymbol{r}_k + \boldsymbol{b}^{\mathrm{T}} \hat{\boldsymbol{a}}_{k-d}, \tag{7.37}$$

where $\boldsymbol{b} = [b_1 \cdots b_{n_b}]^{\mathrm{T}}$ is the vector of FB filter coefficients.

Equalizers can also be distinguished according to the criterion used to optimize their coefficients. The so-called Zero-Forcing (ZF) criterion implies ISI elimination and assuming zero delay is defined as [295]

$$H(f)C(f) = 1. \tag{7.38}$$

It is not difficult to understand that the operation of the ZF equalizer is to provide gain at frequencies where the channel transfer function experiences attenuation and vice versa, removing completely the ISI. However, noise enhancement occurs with this process rendering ZF equalizers ineffective when the channel transfer function exhibits spectral nulls.

In contrast to the ZF criterion, the MMSE criterion allows joint minimization of both ISI and noise. The criterion is based on the minimization of the following MSE:

$$MSE = \mathrm{E} \left[\|a_k - u_k\|^2 \right]. \tag{7.39}$$

Various adaptive algorithms can be invoked, in order to adapt the coefficients of an equalizer, especially when the channel is time varying. Such algorithms are the Least Mean Square (LMS) algorithm or the Recursive Least Square (RLS) algorithm [294]. The interested reader may resort to a number of books that treat this subject more thoroughly, e.g. [294].

7.4.2 Nonlinear Channel Equalization

In the previous section equalization of linear channels was considered. As already seen, in satellite systems the communication channel has nonlinear characteristics and thus nonlinear equalizers need to be designed. An overview of nonlinear channel equalization methods is given below.

7.4.2.1 Maximum Likelihood Sequence Estimation.

Two different approaches have been developed for the design of the optimum receiving filter for bandlimited nonlinear channels. The approaches differ in the optimization criterion adopted for the derivation of the optimum filter . Specifically:

- The minimization of the MSE between the receiver output and the transmitted symbol, under the constraint that the receiving filter is linear, is considered in [296], [297], [298].

- The ML criterion is applied in [299], [300].

In [297], [298] the optimum MSE filter is derived for specific modulation formats. A general approach is followed in [296], which encompasses a broad class of modulation schemes and nonlinearities. It is shown that the optimum linear receiver consists of a bank of filters matched to specific waveforms, which are related to the nonlinear channel, in cascade with infinite-length transversal filters.

Using an analytical model for the bandpass nonlinearity, the ML receiver is derived for binary PSK symbols in [300]. The ML receiver for a nonlinear bandlimited satellite channel under a general formulation is presented in [299]. Both approaches result in a similar structure for the optimum receiver: a bank of matched filters followed by a maximum likelihood detector. More specifically, let $\{a_n\}$ be the information sequence and let the noiseless channel comprise both the linear and nonlinear parts of the transmission channel. We assume that the channel has a finite memory, say L. This means that the channel output $y(t)$ depends on the $L - 1$ previous symbols and the symbol transmitted at time t. Following [299], the channel can be modeled as a finite state machine, whose noiseless output $y(t)$ is expressed as:

$$y(t) = \sum_n h(t - nT, a_n, \sigma_n), \tag{7.40}$$

where $\sigma_n = (a_{n-1}, a_{n-2}, \ldots, a_{n-L})$ is the state of the channel and $h(t - nT, a_n, \sigma_n)$ is a waveform of duration T defined in the interval $[(n - 1)T, nT]$. Obviously the number of possible waveforms equals M^{L+1}, where M is the size of the input signal constellation.

The objective of the ML receiver is the detection of a sequence of transmitted symbols $\{a_0, a_1, \ldots, a_{N-1}\}$ based on the noisy observation

$$r(t) = y_N(t) + \eta(t) = \sum_{n=0}^{N} h(t - nT, a_n, \sigma_n) + \eta(t), \qquad (7.41)$$

where N is assumed to be much greater than L so that boundary effects can be ignored and $\eta(t)$ is AWGN with one-sided power spectral density N_0. The log-likelihood function can be expressed as

$$\lambda_N = -\frac{2}{N_0} \mathfrak{Re} \left\{ \int_0^{NT} y_N^*(t) u(t) dt \right\} + \frac{1}{N_0} \int_0^{NT} |y_N(t)|^2 dt, \qquad (7.42)$$

where $\mathfrak{Re}\{\cdot\}$ denotes real part and * stands for the conjugate transpose operation. Due to the finite duration of the waveforms $h(t - nT, a_n, \sigma_n)$, after some mathematical manipulations λ_N is written as

$$\lambda_N = -\frac{2}{N_0} \mathfrak{Re} \left\{ \sum_{n=0}^{N} Z_n(a_n, \sigma_n) \right\} + \frac{1}{N_0} \sum_{n=0}^{N} E_n(a_n, \sigma_n). \qquad (7.43)$$

$E_n(a_n, \sigma_n)$ is the energy of the waveform $h(t, a_n, \sigma_n)$ defined as

$$E_n(a_n, \sigma_n) = \int_0^T |h(t, a_n, \sigma_n)|^2 dt. \qquad (7.44)$$

$Z_n(a_n, \sigma_n)$ can be obtained as the response to $r(t)$ sampled at time $(n+1)T$ of a filter matched to $h(t, a_n, \sigma_n)$, i.e.,

$$Z_n(a_n, \sigma_n) = \int_{nT}^{(n+1)T} h^*(t - nT, a_n, \sigma_n) r(t) dt. \qquad (7.45)$$

Therefore a number of M^{L+1} matched filters are required corresponding to all different waveforms $h(t, a_n, \sigma_n)$. Since the symbols of the information sequence are related through a finite state machine, the Viterbi decoder can be employed to provide the ML estimate of the sequence.

7.4.2.2   Volterra model-based equalization methods.   In [301] the nonlinear satellite channel is modeled using a Volterra series representation. The Volterra model is a polynomial approximation for nonlinear systems, which is mathematically tractable, since it constitutes a natural extension of the conventional convolution scheme. A number of different methods have been developed for the equalization of Volterra channels. These methods comprise nonlinear Wiener-type filters [302], [303], nonlinear decision feedback equalizers [304], fixed-point schemes [305] and root methods [306].

In [302] the equalizer is modeled as a Volterra system. Specifically, if a finite support Volterra model is assumed, the output of the equalizer is expressed as

$$u[n] = \sum_{k_1=0}^{N_1} h_1[k_1] r[n - k_1] + \cdots$$

$$+ \sum_{k_1=0}^{N_m} \cdots \sum_{k_m=0}^{N_m} h_m[k_1, \ldots, k_m] r[n - k_1] \cdots r[n - k_m], \quad (7.46)$$

where the kernels $h_i[k_1, k_2, \ldots, k_i]$, $i = 1, \ldots, m$ correspond to the taps of the equalizer. By defining the vectors

$$\boldsymbol{r}[n] = [r[n], r[n-1], \ldots, r[n-N_1], r[0]r[0], \ldots, r[n-N_m] \cdots r[n-N_m]] \quad (7.47)$$

and

$$\boldsymbol{h} = [h_1[0], h_1[1], \cdots, h_1[N_1], h_2[0,0], \cdots, h_m[N_m, N_m, \cdots, N_m]], \quad (7.48)$$

the output of the equalizer can be rewritten as

$$u[n] = \boldsymbol{r}^{\mathrm{T}}[n]\boldsymbol{h}. \quad (7.49)$$

A stochastic gradient method can be applied for the adaptation of the Volterra equalizer tap weights. The weight update equation becomes:

$$\boldsymbol{h}[n + 1] = \boldsymbol{h}[n] + \mu e[n]\boldsymbol{r}[n], \quad (7.50)$$

where μ is the step size parameter and $e[n]$ the error between $u[n]$ and the desired symbol.

7.4.2.3 Blind equalization.

In a satellite link, even if we assume that the system nonlinearity is time invariant, the uplink and downlink channels change with time, especially in high data rate mobile applications. Under time varying conditions a periodically transmitted training sequence is necessary to update the equalizer coefficients, otherwise the system performance might significantly deteriorate. However, the periodic transmission of training symbols reduces the effective data rate leading to wasteful use of the available bandwidth. Thus, blind equalization techniques, which do not require a training sequence, are very attractive for mobile satellite communications.

Blind identification and equalization of linear channels has been studied extensively during the last twenty years. Of particular interest are methods that resort to Second Order Statistics (SOS) of the channel output, since previous higher order statistics methods were computationally demanding and usually required a larger data record. The basic idea behind SOS blind methods is that the sequence, which results after oversampling the output signal or after receiving the output signal by a sensor array, is cyclostationary with period equal to the oversampling factor or the number of antennas in the array. Based on this property, blind SOS-based channel identification and equalization methods have been designed. Recently, extending the derivation for linear channels, blind equalization of nonlinear channels has been made possible [307],[308], [309].

In [307] a deterministic approach is described for equalization of nonlinear channels that can be represented by finite order, finite memory Volterra filters. By oversampling the channel output a single-input multiple-output nonlinear Volterra-type system is obtained. Since each Volterra filter can be represented by a multiple-input single-output linear in the parameters model, the whole system has multiple-input multiple-output form. Based on this structure, a number of linear FIR equalizers are designed, which satisfy the ZF criterion. Conditions on the existence and uniqueness of the solution are also derived.

In [308] a SOS-based method is proposed for a more general class of nonlinear channels and specifically those that can be linearized by finite order, finite memory Volterra filters. Oversampling of the channel output by a factor equal to K leads to K nonlinear channels with common input. By assuming that there exist finite order finite memory Volterra filters that linearize these channels, a least squares blind technique, proposed for linear channels, is applied to obtain the Volterra equalizers.

Zero-forcing linear equalizers based on the SOS of the nonlinear channel output, are presented in [309]. More general channel models than previous techniques are considered, of which polynomial approximations such as Volterra models is a special case. The proposed method is a generalization of a known SOS-based blind technique for linear channels to the nonlinear channels case.

7.4.2.4   Neural network methods.   Various NN architectures have been investigated for nonlinear channel equalization. These include MLP [310] [311], RBF [312][313][314], and Self Organizing Maps (SOM) [315][316] [317] equalizers. The interested reader can find in [318][290] a description and performance comparison of MLP, RBF and SOM equalizers. Hybrid neural equalizers have also been developed. These equalizers combine neural networks with conventional equalization techniques such as the LE and the DFE [319].

The characteristic that distinguishes these techniques from Volterra model-based techniques or linear equalization techniques is their capability of learning the underlying relationship between the input and output of a system with data provision. The network's architecture defines the neurons' arrangement in the network. We can identify three different classes of network architectures. The so-called *layered feedforward networks* that exhibit a layered structure, where all connection paths are directed from input to output, with no feedback. Various algorithms, such as the gradient descent method can be used to train such networks. The so-called *recurrent neural network* distinguishes itself from a layered feedforward network by having at least one FB loop. Finally, the *lattice neural networks* are networks in which neurons are organized into one or higher dimensional lattice structures. One of the main problems in the application of NN equalization remains the very long data sequence required in the learning phase that leads to slow convergence, making them inappropriate when dealing with time-varying channels. Note that all NN structures, which have been described for predistortion in section 7.3.2.3, can also be used for equalization. Two such examples are given below.

MLP equalization. The MLP network has been introduced in the early 1960's. However, its application to channel equalization has not been considered since the beginning of the 1990's. The MLP equalizer for satellite channels consist of two separate MLP NNs, one for amplitude adaptation and the other for phase adaptation. There are several practical difficulties in the application of an MLP equalizer, such as the absence of a rule to determine the required number of hidden layers and nodes per layer or the slow convergence of the backpropagation algorithm. Algorithms which attempt to trade off between complexity and convergence speed have been recently developed, e.g. [289].

RBF equalization. We will refer here mainly to the symbol-by-symbol Bayesian M-ary RBF-DFE [314][295]. The structure of an RBF-DFE is specified by the equalizer decision delay d, FF order n_f and FB order n_b. These parameters were chosen in [314][295] as follows:

$$
\begin{aligned}
d &= L - 1 \\
n_f &= d + 1 = L \\
n_b &= L + n_f - 2 - d = L - 1,
\end{aligned}
\tag{7.51}
$$

where L is the channel order.

The RBF-DFE comprises M RBF networks. The output of each RBF network gives a conditional Bayesian decision variable defined as:

$$
\zeta_i(k) = \sum_{j=1}^{N_{h,l}^i} w_{l,j}^i \exp\left\{ -\frac{\left\| \boldsymbol{y}_k - \boldsymbol{c}_{l,j}^i \right\|^2}{2\sigma_j^2} \right\}, \qquad \begin{cases} 1 \leq i \leq M \\ 1 \leq l \leq M^{n_b} \end{cases}
\tag{7.52}
$$

where $\boldsymbol{c}_{l,j}$ is the jth center formed by the lth FB vector and a transmitted symbol x_{k-d}. The centers for the RBF networks, can be computed via channel estimation, while $w_{l,j}^i$ is the *a priori* probability of occurence for each center $\boldsymbol{c}_{l,j}^i$ respectively. In contrast to conventional DFEs this equalizer has a FB section that serves to select a subset of centers for a particular decision, assisting the operation of the feedforward section. Thus, the number of hidden RBF nodes, which is represented in eq. (7.52) by $N_{h,l}^i$, is reduced to M^{n_f+L-1}/M^{n_b} and from eq. (7.50) we have $N_{h,l}^i = M^L$. In order to achieve minimum error-probability we select as $\hat{x}_{k-d}$ the QAM symbol that maximizes the output $\zeta_i(k)$ of these M RBF networks.

7.4.2.5   Performance evaluation of nonlinear equalization methods.

In order to compare the performance of nonlinear equalization methods, we have simulated a 16-QAM system (symbol rate 10 Msymbols per second) with root-raised cosine filters having a roll-off factor equal to 0.22 over a nonlinear satellite channel. The satellite channel consists of two cascaded processes: the 'satellite process', and the 'terrestrial process'. The 'satellite process' is typically described in terms of a memoryless nonlinear function. The 'terrestrial process' includes the effects of multipath fading and shadowing. The nonlinear behavior of the on-board 'satellite process' is described by the Saleh model [269]. For the 'terrestrial process', we have used a discrete multipath model with N rays: $h_{ch}[k] = \sum_{n=0}^{N-1} h_n \delta[k-n]$.

Four multipath models, numbered 0 to 3, are shown in Table 7.2, corresponding respectively to channels with no multipath, 'good' multipath, 'medium' multipath, and 'bad' multipath. They are all normalized to have unit energy. These models are intended to be useful for comparison purposes. They are not necessarily typical responses, although their echo delays and amplitudes are similar to those reported for satellite channels. Instead, they are intended to provide varying degrees of 'stress' for evaluation of the equalizers performance. Model # 1 has a small (-20 dB) echo at 100 ns. delay. Model # 2 has -10.5 dB and 20 dB echoes at 0 ns. and 500 ns., respectively. This model exhibits a non-causal characteristic, since the first echo, located at 300 ns., is a precursor. Precursors represent situations where the shortest radio path is attenuated relative to some slightly longer paths, for example due to partial attenuation or blocking of the LOS path by heavy localized rain, or to misalignment of the subscriber's antenna. Because its second echo, at 200 ns. from the main pulse, is negative, Model # 3 will cause severe attenuation (a manifestation of a multipath fade) of signals.

In the first test, we focus on the equalizers ability of compensating for channel nonlinearities. We have computed the equivalent SNR degradation caused by the residual nonlinear distortion at a specified BER. The SNR degradation, expressed in dB, is defined as the difference between the SNR required by the equalized system to reach the specified BER at a given OBO, and the SNR required to obtain the same BER on the Gaussian flat channel. We obtained this function using the quasi-analytical procedure described in [291]. The results for a target BER of 10^{-4} on model # 0 channel are shown in Figure 7.22.(a). All equalizers have the same order. The equalizers we compare are a linear FIR, a Volterra Filter (VF) and a GCMAC-based equalizer in transversal and decision-feedback configurations. As the backoff decreases, the equalizers operation,

Tap Delay (ns.)	Model # 0 Tap Ampl.	Model # 1 Tap Ampl.	Model # 2 Tap Ampl.	Model # 3 Tap Ampl.
0	1	0.995	$0.286e^{-j\frac{3}{4}\pi}$	0.804
100	0	$0.0995\, e^{-j\frac{3}{4}\pi}$	0	0
200	0	0	0	-0.581
300	0	0	0.953	0
400	0	0	0	-0.123
500	0	0	-0.095	0

Table 7.2. Multipath models

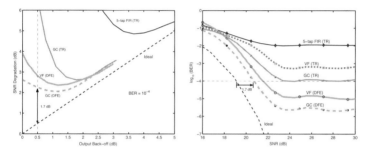

Figure 7.22. (a) SNR degradation vs. OBO for a BER $= 10^{-4}$. (b) BER vs. SNR for an OBO of 0.5 dB

and especially that of the transversal versions, gets worse. For the case of a saturated amplifier (0 dB backoff), the Volterra DFE requires about 1.5 dB more signal power than the GCMAC equalizer to achieve the same BER.

Figure 7.22.(b) is the complement of the previous figure, showing the BER evolution against SNR for 0.5 dB OBO. Again, the transversal filters barely reach the 10^{-4} BER limit, even for extremely high SNR.

The second test is a comparison between the convergence curves of these equalizers for a time-varying channel, shown in Figure 7.23. Again, the amplifier operates at 0.5 dB OBO and the SNR is 20 dB. In the first 15000 sample periods, the equalizers are trained with the actual symbols. Then they switch to the decision-directed mode. After 5000 sample periods, a sudden change in the channel impulse response occurs (Model #0 → Model #1), while the equalizers remain in the decision-directed mode. To rank the equalizers behavior, all of them were trained using the learning step that provides the lowest MSE.

Figure 7.23 represents the smoothed (we have applied an exponentially weighted time average over the ensemble average of 15 independent runs) BER

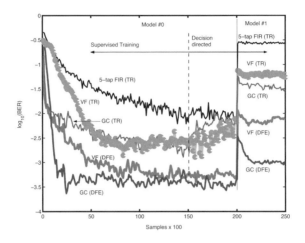

Figure 7.23. BER for transversal and DFE equalizers on a nonlinear time-varying channel. A SNR = 20 dB is assumed

evolution. It is observed that in both training and decision-directed regions, the DFE equalizers provide a lower BER than their transversal counterparts.

In the third test, the SDR at the equalizers output is analyzed for the multipath scenarios given by Models #2 and #3. This time, the amplifier operates at 2 dB OBO and the SNR is 30 dB. In the test, all equalizers have the same order and configuration as in the previous test. A sudden change in the channel impulse response takes place after 25000 samples (Model # 2→ Model #3) Again, the DFE equalizers outperform the transversal ones, although the difference in performance is reduced as the multipath gets worse.

7.4.3   Turbo Equalization

The turbo equalization method, first proposed by Douillard et al. [320], is based on soft information exchange between the equalizer and the decoder in a closed loop. This technique represents a suboptimal alternative to the traditional MLSD optimal receiver, providing comparable performance with significantly reduced computational complexity.

The block diagram of the data transmission system is described in Figure 7.25. The data bits a_i, $i = 1, 2, \ldots, D$ are processed by a convolutional encoder and modulated to generate the symbols c_n, $n = 1, 2, \ldots, C$. The coded symbols are interleaved in order to reduce the impact of burst errors before transmission over the channel, which is assumed to be an ISI, white gaussian additive noise channel. The receiver input r_n is given by

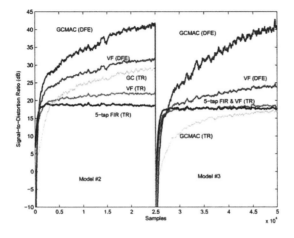

Figure 7.24. SDR at the equalizers output for Model # 2 channel ("medium" multipath) and Model # 3 channel ("bad" multipath)

Figure 7.25. Serial turbo data transmission system

$$r_n = \sum_{l=0}^{L-1} h_l x_{n-l} + \eta_n, \tag{7.53}$$

where L denotes the channel memory and h_l is the sampled impulse response of the cascade of the transmit filter, the channel, and the receive filter. The coefficients h_l are assumed to be known at the receiver as an output of a channel parameter estimation circuit.

The iterative turbo equalizer structure is represented in Figure 7.26. The terms $\Lambda_E(x_n)$ and $\Lambda_D(c_n)$ correspond to the Log-Likelihood Ratio (LLR) of the extrinsic information provided by the equalizer and the decoder, respectively. This soft information, after the interleaving process, is treated as *a priori* information by the following soft-input soft-output block. In this section, two turbo equalization techniques are described, respectively for MAP and MMSE equalization. In both cases, for simplicity, we assume a BPSK modulation scheme ($c_n \in \{+1, -1\}$) and MAP is used for decoding due to its optimal

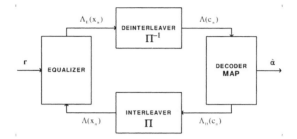

Figure 7.26. Turbo equalization receiver

BER performance [321].

7.4.3.1 Turbo equalization using a MAP equalizer.

The MAP equalizer computes the a-Posteriori Probabilitys (APPs) $Pr\{x_n = x \mid r_1, \ldots, r_C\}$ and outputs the difference $\Lambda_E(x_n)$ between the *a posteriori* LLR and the *a priori* LLR $\Lambda(x_n)$ provided by the decoder in the previous step

$$\Lambda_E(x_n) = \ln \frac{Pr\{x_n = +1 \mid r_1, \ldots, r_C\}}{Pr\{x_n = -1 \mid r_1, \ldots, r_C\}} - \ln \frac{Pr\{x_n = +1\}}{Pr\{x_n = -1\}}. \qquad (7.54)$$

The initial value of $\Lambda(x_n)$ is 0 $\forall n$, due to the absence of *a priori* information in the first equalization step. The MAP decoder computes the APPs $Pr\{c_n = x \mid \Lambda(c_1), \ldots, \Lambda(c_C)\}$ using the de-interleaved information $\Lambda(c_n) = \Pi^{-1}(\Lambda_E(x_n))$ and outputs the quantities

$$\Lambda_D(c_n) = \ln \frac{Pr\{c_n = +1 \mid \Lambda(c_1), \ldots, \Lambda(c_C)\}}{Pr\{c_n = -1 \mid \Lambda(c_1), \ldots, \Lambda(c_C)\}} - \ln \frac{Pr\{c_n = +1\}}{Pr\{c_n = -1\}}, \qquad (7.55)$$

where the last term is the LLR $\Lambda(c_n)$.

7.4.3.2 Turbo equalization using an MMSE-LE.

The MMSE-LE calculates the estimate $\hat{x}_n$ using the *a priori* information $\Lambda(x_n)$ by minimizing the mean square error $E\left[\mid x_n - \hat{x}_n \mid^2\right]$, and outputs

$$\Lambda_E(x_n) = \ln \frac{Pr\{x_n = +1 \mid \hat{x}_n\}}{Pr\{x_n = -1 \mid \hat{x}_n\}} - \ln \frac{Pr\{x_n = +1\}}{Pr\{x_n = -1\}} \qquad (7.56)$$

$$\overset{\text{Bayes}}{=} \ln \frac{f(\hat{x}_n \mid x_n = +1)}{f(\hat{x}_n \mid x_n = -1)}, \qquad (7.57)$$

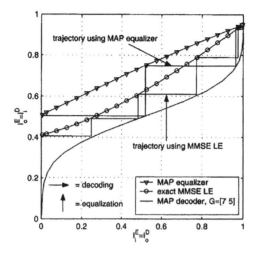

Figure 7.27. Receiver EXIT charts with traces of turbo equalization algorithms [321] © 2002 IEEE

which requires knowledge of the distribution $f(\hat{x}_n \mid x_n = x)$ where $x \in \{-1, +1\}$.

7.4.3.3   Turbo equalization using an MMSE-DFE.   The MMSE-DFE equalizer computes the estimate $\hat{x}_n$ using n_f FF and n_b FB time-varying filter coefficients $a_{n,k}^f$ and $a_{n,k}^b$, respectively as follows

$$\hat{x}_n = \bar{x}_n + \left(\sum_{k=-(n_f-1)}^{0} a_{n,k}^f \left(r_{n-k} - \mathrm{E}\left[r_{n-k} \right] \right) \right) - \left(\sum_{k=1}^{n_b} a_{n,k}^b \left(\hat{x}_{n-k}^d - \mathrm{E}\left[\hat{x}_{n-k}^d \right] \right) \right), \qquad (7.58)$$

where $\bar{x}_n = \mathrm{E}\left[x_n \right]$ and $\hat{x}_{n-k}^d$ are past estimates.

As shown in Figure 7.27 linear MMSE turbo equalization performs very closely to the BER-optimal turbo MAP equalization scheme. It requires a few more iterations to achieve similar performance in terms of BER, but has a lower computational complexity [321]. The performance of the turbo MMSE-DFE, at the same computational complexity is inferior compared to turbo MMSE-LE.

7.5   Emerging techniques

In previous sections of this chapter, different transmitter and receiver distortion countermeasures have been presented, focusing attention on the most relevant for satellite systems. Apart from these, other techniques have also been described in the literature, that have never been directly applied to satellite systems, due to the presence of nonlinear distortion in these systems. The latter has effectively hindered further investigation and development of new methods capable of reducing sensitivity to nonlinear effects.

In this section, we present a transmitter technique for counteracting linear distortion, called precoding . This technique is used mainly in cable communications (xDSL), whereas its application has not been foreseen within the satellite field. Since precoding is only valid against linear distortion, nonlinear distortion must be counteracted with appropriate techniques. We will first present the basics of precoding and then analyze the joint performance of precoding and fractionally-spaced predistortion[7] (both transmitter techniques) against the distortion caused by an HPA and a dispersive channel. Comparisons will be made with respect to a reference system composed of fractionally-spaced predistortion at the transmitter and standard DF equalization at the receiver.

The results of this study will then be instrumental to drive further investigations and development, with the aim to enable exploitation of the full benefits of precoding.

7.5.1   Introduction to Precoding

DF equalization is a classical equalization technique, able to cope with linear distortion introduced by a dispersive channel. However, DF equalization presents two main disadvantages: (i) it suffers from error propagation at low signal-to noise ratios and (ii) coded modulation is not directly applicable, since DFE requires zero-delay decisions to work.

These two problems can be avoided if the equalization procedure is moved to the transmitter side. This is known as pre-equalization and of course can only be applied when the channel impulse response is —at least up to a certain point— known to the transmitter. This can be accomplished by a setup procedure at the beginning of the transmission, through which channel information is estimated at the receiver and passed back to the transmitter. In Time Division Duplex (TDD) systems, the channel impulse response can be estimated during the reception phase and then used to pre-equalize the signal during the transmission phase. Nevertheless, linear pre-equalization poses a new disadvantage: in general, the transmitted power is increased.

---

[7]The input to the predistorter is an oversampled version of the original signal.

At this point, precoding arrives as a solution which maintains the advantages of linear pre-equalization while limiting transmit power. This is achieved using nonlinearities at a certain point. There exist different ways to perform precoding, but we will focus on one of them known as Tomlinson-Harashima Precoding (THP). The use of precoding techniques has been up to now restricted to channels which introduce linear distortion, and low or negligible nonlinear distortion (for instance xDSL technology). This restriction is due to the dynamic range of the precoded signal, which stimulates the non linear effects of power amplifiers. For this reason, the application of precoding to satellite systems has not been explored up to now.

7.5.1.1   Tomlinson-Harashima Precoding.   The precoding technique known as Tomlinson-Harashima Precoding (THP) [322], [323], [324], [325], [326, chap. 3] makes use of modulo arithmetic to achieve both pre-equalization and transmit power limitation. Its principle lies upon lattice theory: the use of modulo arithmetic allows us to represent symbols not only by those lying in the original constellation set, but also by their correspondents in other constellation replicas, as long as these replicas form a well-constructed lattice.

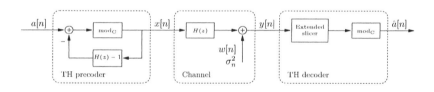

Figure 7.28.   System using Zero-Forcing Tomlinson-Harashima precoding

The basic implementation of ZF-THP for square Quadrature Amplitude Modulation (QAM) constellations is shown in Figure 7.28. The original symbol sequence $\{a[n]\}$, which is chosen from a given square QAM constellation, is precoded at the transmitter yielding the precoded sequence $\{x[n]\}$. The precoding operation consists of the typical inverse channel filtering – performed by a feedback $[H(z) - 1]$ filter – with the addition of a complex modulo operation – denoted as mod_C – which performs the modulo on both real and imaginary parts of its input signal. The $\mathbb{R} \rightarrow \mathbb{R}$ modulo operations are performed in such a way that their results lie in the interval $[-M, M)$. M is chosen so that the result of the modulo operation lies in the same square where the QAM constellation is defined. Mathematically, let $r, y, M \in \mathbb{R}$ and $v, x \in \mathbb{C}$; then the real and complex modulo operations are defined as follows:

$$\mathrm{mod}\,\mathbb{R} \to \mathbb{R}, \quad r \bmod 2M = y \in [-M, M) \tag{7.59}$$
$$\mathrm{mod}_C\,\mathbb{C} \to \mathbb{C}, \quad v \bmod_C 2M = x, \quad \text{where} \tag{7.60}$$
$$\mathfrak{Re}\,\{x\} = \mathfrak{Re}\,\{v\} \bmod 2M \quad \text{and} \tag{7.61}$$
$$\mathfrak{Im}\,\{x\} = \mathfrak{Im}\,\{v\} \bmod 2M \tag{7.62}$$

The precoded sequence $\{x[n]\}$ enters the channel, which introduces ISI and complex AWGN noise, yielding the received sequence $\{y[n]\}$. The input signal $\{x[n]\}$ is contained inside the original constellation region, but due to ISI the output signal set is actually expanded. Finally, the TH decoder estimates the original sequence of symbols, giving $\{\hat{a}[n]\}$ as a result. In the example of Figure 7.28 hard decoding is used by means of an extended slicer and a mod_C block. The extended slicer is a quantizer, which hard-decodes each received sample. Extended stands for the fact that it does not only operate inside the original constellation set, but also considers all replicas of the original constellation, which may occur after ISI has been introduced by the channel.

Some notes of interest are listed below:

- The discrete channel response $H(z)$ is assumed monic and minimum-phase, and represents the symbol-rate equivalent response of the cascade of the transmit, channel and receive filters.

- THP can be combined with channel coding in a straightforward manner.

- THP presents an SNR penalty with respect to normal QAM transmission, due to a slight increase in the transmitted power. The so-called precoding loss can be calculated as [326] $\gamma_{\mathrm{p}}^2 = \mathrm{E}\left[|x[n]|^2\right]/\mathrm{E}\left[|a[n]|^2\right]$, and gets smaller the bigger the original constellation is.

- If the channel introduces a considerable degree of ISI, then $\{x[n]\}$ is (approximately) uniformly distributed inside the square region $[-M, M) + j[-M, M)$.

- We have presented ZF-THP, but THP can be designed following the MMSE criterion. See [326] for details on MMSE-THP.

7.5.2 Performance of Joint Precoding and Predistortion

In this section we study the performance of a precoded system when transmitting over a satellite channel, which introduces both linear and nonlinear distortion. The case study comprises the use of fractionally-spaced predistortion techniques, as a means of reducing the impact of nonlinear distortion on the precoded signal. The reference system used as a benchmark for the performance curves makes use of DF equalization, with and without fractional predistortion techniques. In the following, the two configurations are presented.

7.5.2.1   Case study.   Figure 7.29 shows the block diagram of the proposed transmission scheme, combining precoding and predistortion. The source bit sequence enters the mapper, which maps the bits into complex symbols belonging to the square QAM family. The QAM symbols are then precoded by the TH precoder, and converted into a waveform by a square-root raised cosine filter. Then, the waveform is predistorted prior to the HPA stage, whose output is finally sent to the channel. The channel is a tapped delay line which introduces linear distortion and AWGN of power density N_0. The received signal is filtered by a Whitened-Matched-Filter (WMF), which is the optimum receive filter yielding a signal composed of ISI distorted symbols plus AWGN. After symbol-rate sampling, the TH decoder performs hard decisions on the complex-modulo processed symbols.

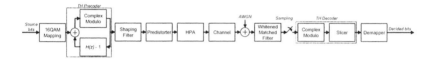

Figure 7.29.   Block diagram of the system using ZF-THP and fractionally-spaced predistortion

The reference system, which considers the use of a DFE at the receiver, is depicted in Figure 7.30. In this case, in the transmitter the symbols are directly passed to the pulse shaping filter. At the receiver, a ZF-DFE, with FB filtering only, takes the place of the TH decoder.

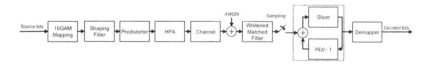

Figure 7.30.   Block diagram of the system using ZF-DFE and fractionally-spaced predistortion

The simulations have been performed for source packets of 500000 bits length. A 16-QAM mapping is considered. The shaping filter is a square-root raised cosine with a two-sided finite length of 40 symbol periods and a roll-off factor of 0.3, yielding an output waveform which is oversampled 4 times per symbol period. The symbol rate was 3.84×10^6 symbols per second. The fractionally-spaced predistorter is implemented through a LUT of size 1024 compensating the AM/AM and AM/PM characteristics of the TWT amplifier. The amplifier is operated at different IBOs, ranging from 0.5 to 10 dB, including the linear HPA case. Finally, as channel model, an urban environment with

Intermediate Module Repeater (IMR)s has been considered, whose power delay is depicted in Figure 7.31.

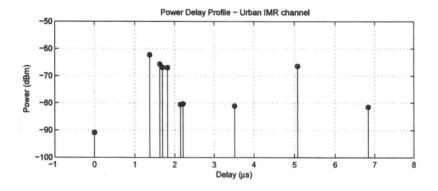

7.5.2.2   Evaluation and discussion.   The two systems introduced above have been simulated, reporting system performance in terms of Symbol Error Ratio (SER). Both ZF and MMSE cases were simulated, but we focus on the results obtained for the MMSE case.

Figures 7.32 and 7.33 summarize the simulation results for the THP and the DFE systems, respectively. The main conclusions that can be extracted are:

- When predistortion is not used, both systems are strongly affected by the presence of the HPA. The presence of the HPA causes error floors in the performance of both systems. For example, the error floors for IBO = 10 dB are located around SER values of $3 \cdot 10^{-2}$ for THP and $1 \cdot 10^{-2}$ for DFE.

- The introduction of predistortion improves greatly the performance of both systems. For example, for IBO = 5 dB and $E_b/N_0 = 12$ dB, the SER is reduced from $2 \cdot 10^{-1}$ to $2 \cdot 10^{-2}$ for THP and from $2 \cdot 10^{-1}$ to $1 \cdot 10^{-2}$ for DFE. For IBO = 10 dB, the performance with predistortion is indistinguishable from that of the linear case.

- The system with THP is, however, more affected by the HPA than the system with DFE. For example, using predistortion for IBO = 5 dB and $E_b/N_0 = 16$ dB, the system with THP offers a SER of $5 \cdot 10^{-2}$ while this value is reduced to $7 \cdot 10^{-3}$ in the DFE case.

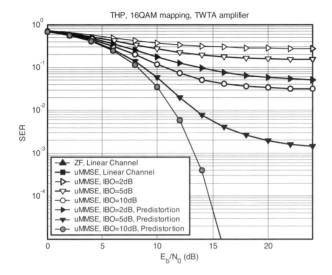

Figure 7.32. Performance of joint THP and fractionally-spaced predistortion in terms of symbol error ratio [327] © 2005 IEEE

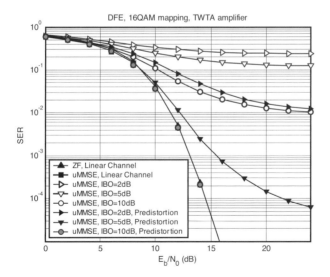

Figure 7.33. Performance of fractionally-spaced predistortion and DFE in terms of symbol error ratio [327] © 2005 IEEE

Our last conclusion can be explained recaling the nature of the TH-precoded signal. As stated in subsection 7.5.1, under strong ISI conditions, as is our case, the precoded signal tends to be uniform within the square defined by the modulo operation. Thus, the dynamic range of the signal entering the transmit filter is higher than that of a QAM signal (DFE case). After pulse shaping filtering, this property remains, so, at the end, the signal entering the HPA has a higher dynamic range in the THP case. This is clearly a handicap for the THP system, since the HPA introduces more distortion for signals with larger dynamic range.

This is illustrated in Figure 7.34. Scatter plots of the HPA output (oversampled waveform) and of the slicer input (symbol rate) are shown. In general, it can be seen that the signal entering the slicer is less distorted in the DFE case than in the THP case (compare columns (2) and (4)). Also, the benefits of predistortion are clear if we compare slicer's inputs between rows (b) and (c) or between rows (d) and (e). As explained above, the signal entering the HPA spans along a wider amplitude range in the THP case, so that higher amplitude values are more likely to appear. Thus, a greater portion of the waveform is affected by the HPA in the THP case, as can be seen comparing (b1) and (b3) or (c1) and (c3) (the surrounding circle of points is denser in the THP case).

7.6   Conclusions

In satellite communication systems, linear distortion comes from filtering along the satellite link, as well as from multipath fading and shadowing of the uplink and downlink signals. However, the basic source of distortion is the on-board HPA, which is usually forced to work at the saturation in order to maximize its performance. This introduces nonlinear distortion to the transmitted signal and renders the overall satellite link nonlinear. There exist two approaches to compensate for linear and nonlinear distortions in satellite communications, namely predistortion and equalization. Predistortion refers to pre-processing of the signal entering the HPA so that the response of the overall system is linear. Equalization, on the other side, attempts to counteract the distortion introduced by the channel by processing the signal at the receiver side. Combination of these two techniques or the use of a precoder at the transmitter side could offer significant performance gains over existing conventional methods.

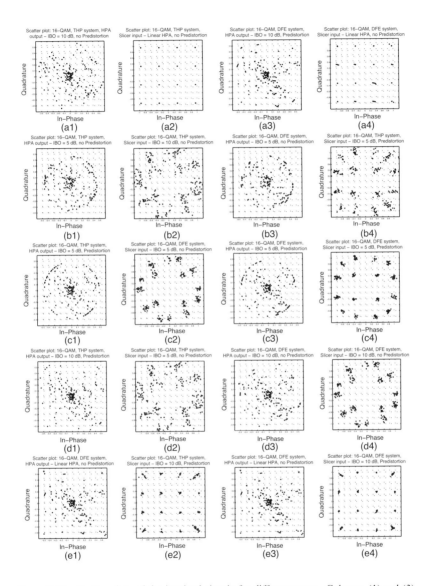

Figure 7.34.   Scatter plots of the involved signals for different cases. Columns (1) and (3) correspond with the HPA output, and columns (2) and (4) with the input of the slicer in the DFE and in the TH decoder, respectively. Different IBO cases are shown in each row. Row (a) shows scatterplots for the linear HPA case; rows (b) and (c) show IBO = 5 dB respectively without and with predistorion; rows (d) and (e) show IBO = 10 dB respectively without and with predistortion.

Chapter 8

DIVERSITY TECHNIQUES AND FADE MITIGATION

G. K. Karagiannidis[1], M. Bousquet[6], C. Caini[4], L. Castanet[6], M. A. Vázquez Castro[2], S. Cioni[4], I. Frigyes[3], P. Horvath[3], T. Javornik[7], G. Kandus[7], M. Luglio[5], P. Salmi[4], D. A. Zogas[1]

[1]*Aristotle University of Thessaloniki, Greece*

[2]*Autonomic University of Barcelona, Spain*

[3]*Budapest University of Technology and Economics,*

[4]*University of Bologna, Italy*

[5]*University of Rome "Tor Vergata", Italy*

[6]*ONERA/TeSA, France*

[7]*"Jožef Stefan" Institute, Slovenia*

8.1   Introduction

The purpose of any communication system is to reliably transfer information between the source and destination. As a signal propagates through a wireless channel, it experiences random fluctuations in time, due to changes in reflections and attenuations. Thus, the channel characteristic of the channels change randomly with time. The average signal strength received by an antenna element over a local area in the propagation environment can be quite large, but during some time intervals it is not uncommon for the instantaneous signal level in a multipath environment to fall 30 dB or more below its mean level.

A substantial decrease in Signal to Noise Ratio (SNR) occurs in a flat fading channel when all arriving multipath components add destructively at the receiver antenna. In this case, the receiver is essentially experiencing in a deep fade or signal null. To cope with these results, during these time periods the receiver requires an alternate signal path to the transmitted signal with a sufficiently large SNR in order to reliably decipher the desired signal. Accordingly, *diversity* is achieved by using the information on the different branches available to the receiver in order to increase the SNR at the decoding stage. The additional branches increase the probability that at least one branch, or the combined branch outputs, produce a sufficiently high SNR to permit reliable decoding of the useful message at the receiver.

There are several domains to produce additional diversity branches; the main ones are antenna, time, and frequency domains. Space or site diversity refers to the method of transmission or reception, or both, in which the effects of fading are minimized by the simultaneous use of two or more physically separated antennas (or sites). Antenna diversity requires multiple antennas at the receiver and is therefore usually bulkier. However, operating at high frequency bands allows for the size reduction of antenna elements, and it becomes feasible to have multiple antennas not only at the base stations, but also on the mobile handset. Time diversity takes advantage of the dynamics of the channel; at some point in time the received signal might be in a deep fade, while at a later time the channel has changed significantly such that the received SNR is at an acceptable value. Frequency diversity is implemented by transmitting information on more than one carrier frequency. The rationale behind this technique is that frequencies separated by more than the coherence bandwidth of the channel are uncorrelated and thus do not experience the same fade.

Generally, this chapter focuses on all diversity domains and mitigation techniques in general against the blockage/multipath and atmospheric effects for satellite communication systems. More specifically, the satellite diversity concept is introduced and some results are presented for Code Division Multiple Access (CDMA) based systems. Furthermore, several other issues of the diversity concept applied in satellite communication systems, as Multiple-Input Multiple-Output (MIMO) and space time coding, Power Control (PC), Adaptive Coding and Modulation (ACM), are analyzed. Concerning the blockage/multipath mitigation, recent results for joint fading mitigation, Intermediate Module Repeater (IMR) multipath diversity and combining techniques are presented.

8.2   Diversity domains and mitigation techniques

Various methods exist to counteract propagation effects exploiting diversity domains with of the satellite communications systems. The design of these methods has to take into account operating frequency bands, performance objectives of the system, and geometry of the network (system architecture, multiple access schemes, etc). In particular, here the focus is on:

- Satellite, Site and Frequency Diversity : the objective is to re-route information in the network in order to avoid impairments due to an atmospheric perturbation

- Multiple-Input Multiple-Output and Space Time Coding: space diversity in transmitter and receivers is applied

- Power Control: the output power is matched to the link impairment

- Polarization diversity : two orthogonally polarized receive antennas are used

- Adaptive Coding and Modulation: a variable coding rate is used to match to impairments originating from propagation conditions.

8.2.1   Satellite Diversity

A satellite communication system exploits satellite diversity when it is designed to provide coverage of the same area from more than a single satellite. By means of satellite diversity, a user can potentially exploit different satellites inside its field of view in order to reduce the probability of paths to the satellites being blocked by natural or artificial obstacles. Although this is the main motivation for satellite diversity, there is also an impact of satellite diversity on system capacity , which cannot be neglected in the evaluation of the advantages and disadvantages of this diversity technique.

8.2.1.1   Improving service availability in urban and sub-urban areas.
Link obstruction represents a major limitation in achieving a satisfactory satellite coverage of urban or suburban areas, due to the low elevation angle at which satellites are observed for most of the time, especially for Low Earth Orbits (LEOs). With the aim to assess the benefits provided by satellite diversity to overcome this problem, we assume a satellite system designed to provide N_{sat} as the maximum diversity order. However, due to the presence of obstacles, some of the paths from a user to the N_{sat} satellites may be blocked, so that the actual number L of satellites in view for the generic user is generally lower and can reach zero, in which case the user is completely blocked and cannot receive the service.

To quantitatively assess this phenomenon, we start by observing that the likelihood of link obstruction increases for low satellite elevation angles. Let $P_B(i,k)$ be the blockage probability in the link from the i-th satellite to the k-th user. Following [51], an empirical model for path blockage probability can be written as

$$P_B(i,k) = \frac{1}{a}(90 - \theta_{i,k})^2 \qquad (8.1)$$

where $\theta_{i,k}$ is the elevation angle for the i-th satellite in view from the k-th user ($\theta_{i,k} \in [10° - 90°]$), and a is a normalization factor fitted on measured data ($a = 7000$ in urban areas, $a = 16600$ in suburban areas) [51]. Note that $10°$ is a realistic value for the minimum useful elevation angle of real systems because

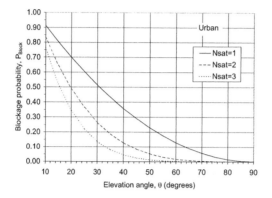

Figure 8.1.   Blockage probability in urban areas versus the number of satellites above the minimum elevation angle [328] ©2001 IEEE

below $10°$ blockage is extremely frequent. Under the simplifying assumption that blockage is independent in different links, the probability for a user to be inactive (completely blocked) can be computed as

$$P_{Block}(k) = \prod_{i=1}^{N_{sat}} P_B(i,k) = \frac{1}{a^{N_{sat}}} \prod_{i=1}^{N_{sat}} (90 - \theta_{i,k})^2 \qquad (8.2)$$

In order to avoid the consideration of a large number of particular cases, we also assume that $\theta_{i,k} = \theta$ (i.e. all the angles are the same, hence N_{sat} satellites "surround" each user) and therefore $P_B(i,k) = P_B$, for all i, k. The resulting probability P_{Block} is reported as a function of N_{sat} in Figures 8.1 and 8.2, for urban and suburban areas, respectively. The results are quite interesting. In an urban areas the blockage probability is extremely high. For example, with $\theta = 30°$, blockage probability is higher than 50% for $N_{sat} = 1$. With two satellites potentially in view, P_{Block} drops to about 25%, while $N_{sat} = 3$ corresponds to $P_{Block} < 15\%$. The situation is less dramatic in suburban areas, where for $\theta = 30°$ we have $P_{Block} = 22\%$ for $N_{sat} = 1$, $P_{Block} = 5\%$ for $N_{sat} = 2$, and $P_{Block} = 1\%$ for $N_{sat} = 3$. These results support the following remark: *satellite diversity is essential in improving service availability in urban and suburban areas.*

8.2.1.2   Exploiting satellite diversity using CDMA .   Assuming that satellite diversity is adopted, the problem is then to exploit the available satellites in the best possible way to minimize the probability of call dropping. To this aim, it is necessary to have an efficient handoff algorithm for passing control from

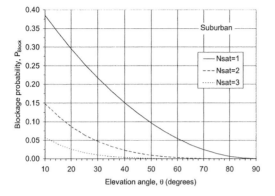

Figure 8.2. Blockage probability in suburban areas versus the number of satellites above the minimum elevation angle [328] ©2001 IEEE

one satellite to another, and more frequently from one beam to another. In this respect, the use of CDMA with universal frequency reuse and a Rake receiver is a very efficient solution, because it easily provides the soft handoff capability, whereby multiple signals from different satellites are linearly combined. In addition to improving the received signal quality, this technique is much better in terms of probability of call dropping than hard handoff from one frequency channel to another, and additionally it simplifies the Radio Frequency (RF) interface.

While the advantages of satellite diversity in terms of system availability are evident, the impact of this technique on system capacity is far from obvious. On the one hand, it is clear that using more than one satellite per user terminal corresponds to less resources (codes) available to other users. Moreover, the larger the number of satellites in view, the higher the interference level that the user terminal must face. On the other hand, these facts may not lead to a capacity penalty if, for example, satellite diversity is used to counteract heavy fading conditions. This is particularly true for slow moving terminals, where the interleaving depth is small compared to the channel coherence time. It should also be noted that the capacity of these satellite systems is generally limited by the satellite power budget, and therefore the limitation on capacity due to the number of available codes is hardly ever reached.

In [328] the capacity of a CDMA system is defined as the ratio between the total bit rate carried in a beam over the total available bandwidth. In formula,

$$C = \frac{N_{user} R_b}{MW} \quad (8.3)$$

where N_{user} is the average number of active users per beam, R_b is the user information bit rate, M is the number of orthogonal spreading codes (e.g. Walsh Hadamard) available per beam, and finally W is the bandwidth of the unspread signal. After introducing some necessary symmetry assumptions, in [328] the following analytical expression is derived for capacity:

$$C = \frac{\exp\left\{-h\left(\sigma\sqrt{2}\,\mathrm{erfc}\,(2\gamma)^{-1} + \frac{h\sigma^2}{2}\right)\right\} - \overline{Q(L)}\left(\frac{E_{b,tot}}{N_0}\right)^{-1}}{\frac{1}{N_{sat}}\left[\overline{Q(L)L}(1+\varsigma) - \overline{Q(L)}\right]} \qquad (8.4)$$

where $E_{b,tot}/N_0$ is the ratio between the average received energy per bit (after despreading and from all satellites) and the thermal noise one-sided power spectral density. N_{sat} is the maximum number of satellite in view (i.e. the maximum diversity order), and L the actual number of satellite in view (i.e. the actual diversity order, $L \leq N_{sat}$). $Q = Q(L)$ is the Signal to Interference plus Noise Ratio (SINR) threshold value. In fact, to achieve a predefined error probability, the overall SINR of a generic user must be greater or equal to a specified value Q. Note that Q differs from user to user according to the actual diversity order L, therefore we write $Q = Q(L)$. Further, Q depends upon channel characteristics (including the fading correlation in time, i.e. the user speed) and on the adopted modulation/coding/interleaving scheme. The over bar notation $\overline{Q}$ denotes its average value. $Q(L)L$ is the product of the SINR threshold value and the actual diversity for a given user. The over bar notation $\overline{Q(L)L}$ denotes its average value. ς is a parameter, identified as *normalized inter-beam interference*. It represents the ratio between the average inter-beam interference (i.e. interference coming from sidelobes of antennas that cover other cells) per satellite and the *received power per beam* (i.e. the average power a user receives through the main lobe covering its beam, before despreading). It depends on the antenna radiation pattern and it is zero if the antenna provides a perfect angle separation. σ(dB) is the standard deviation of the PC error when expressed in dB assuming that the power control error in dB follows a Gaussian distribution. The case of ideal PC is handled simply by setting $\sigma = 0$. Finally, γ is the outage probability, determined by the failure of providing a user with a satisfactory SINR due to power control errors, and $h = \frac{ln(10)}{10}$.

Note that C can only assume values inside the range $[0, N_{sat}R_b/W]$, so the value of the right hand side of the expression above must be correspondingly limited. Exploiting the Satellite diversity capacity formula, it can be shown than the introduction of satellite diversity may result in an increase or a decrease of system capacity basically depending on the fading correlation time, i.e. on user speed. For this reason most results will be provided in the section devoted to applications. Other results will be reported in the section on power control.

Note that system capacity has been defined considering active users, i.e. conditioning on the availability of at least one satellite per user, thus neglecting

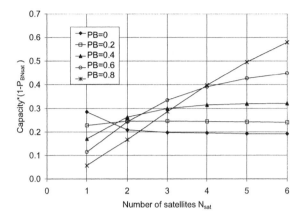

Figure 8.3. Product of capacity and probability of at least one clear link. Fast Rice channel ($K = 10$ dB, 10 ms interleaving delay, speed=100 km/h), $E_{b,tot}/N_0 = 8$ dB; normalized inter-beam interference $\zeta = 0.5$; PC error standard deviation $\sigma = 0.5$; outage due to PC errors $\gamma = 0.01$ [328] ©2001 IEEE

users that cannot get into the system altogether. To show in a single picture the impact of satellite diversity on both blockage and capacity, we consider a capacity-availability figure of merit, i.e. we multiply system capacity by the probability of having at least one satellite in view, $1\text{-}P_B^{Nsat}$. The result is shown in Figure 8.3, where the curves are reported as a function of the diversity order, N_{sat}, for several values of single satellite blockage probability P_B in the fast Rice fading channel, with $E_{b,tot}/N_0 = 8$ dB, PC error standard deviation $\sigma = 0.5$ dB, and outage due to PC $\gamma = 0.01$ and $\zeta = 0.5$.

It is interesting to observe that for P_B less than about 0.2, the capacity-availability product decreases for increasing diversity. This is due to the fact that with fast fading satellite diversity yields hardly any capacity gain (it may indeed be harmful) and the availability improvement brought about by diversity is also not very significant. However, for larger values of blockage probability the availability improvement becomes very significant and the capacity-availability product increases with increasing diversity. Evidently, the results would be much more favorable to diversity if "slow fading" were considered. The system designer should then make a decision on the exploitation of diversity depending also on the foreseen channel conditions. It should also be noted that for LEO constellations of reasonable size (less than 80 satellites), the probability that the satellites are seen at low elevation may be very large, depending on the latitude. Therefore, in view of a global Earth coverage, there are certainly areas where satellite diversity is beneficial, even in fast fading conditions, as well as areas where it is probably unnecessary.

8.2.1.3   Correlation aspects of satellite diversity links .   It is a general assumption for diversity studies that the links to the different satellites are uncorrelated. As a matter of fact, diversity gain can only be achieved if the different links behave in an uncorrelated manner or, even better, if they present negative correlation values.

Although most available Land Mobile Satellite (LMS) propagation models (see Chapter 3) can reproduce with fairly good accuracy the various channel effects (fading, time dispersion, Doppler, etc.) and can be used in the evaluation of physical layer techniques, only those models based on a physical approach can be directly used to simultaneously analyze various satellite-mobile links, and thus, evaluate the benefits of satellite diversity. A specially relevant scenario for diversity systems is the urban environment since other scenarios likely present Line of Sight (LOS) and there may be less need of diversity gain.

A number of relevant studies on path correlation related to satellite diversity can be found in the literature. In [329] an extension of the two-state model proposed in [45] for a single satellite-mobile link to two angle-spaced links was proposed. The approach followed to evaluate the correlation coefficient was to use circular scans within a given environment to obtain numerical landscape pictures in which a "0" or "1" would represent obstruction or visibility respectively, as shown in Section 3.2.2. A campaign in rural, suburban and urban environments using a video-camera to record the landscape was carried out. The outcome of this study was the formulation of an empirical model for the correlation coefficient for a number of environments.

The two-state model proposed in [45] was extended by the same author to model two correlated links [50]. Lutz proposed a four-state Markov model to describe the possible combinations of good and bad states in two different links. Equilibrium state and transition probabilities for a four-state model were computed for the correlated and the uncorrelated cases in terms of the individual two-state model probabilities and of the correlation coefficient, r. This model allows the simultaneous study of two satellite links with a given constant correlation behavior. In his paper, Lutz does not provide numerical values for the correlation coefficient.

Further studies [330] have provided correlation coefficient values extracted from experimental data. A method based on fisheye pictures for the analysis of path diversity for LEO Satellite-PCS networks in urban environments was presented in [331] and [332]. The method consisted of the following steps: 1) taking fisheye photos at potential user locations; 2) extracting from the images path-state information (clear/shadowed/blocked) as a function of look angles, and 3) combining each path-state for single or multiple satellites in a specific constellation using appropriate statistical fade models. Each of the three possible path states of the mobile satellite link is associated with a given fade distribution. The clear state is described using a Rice distribution, the shadowed and the blocked states are modelled using the Loo model. The derived urban three-state model was

$$f(r, \alpha) = C(\alpha) f_{Rice}(r) + S(\alpha) f_{Loo}(r) + B(\alpha) f_{Loo}(r). \qquad (8.5)$$

where α is the elevation angle. The Rice and Loo distribution parameters were extracted from measurements. The values of C, S and B were estimated from fisheye pictures [333].

In [334], the availability of the ICO and Globalstar systems was analyzed using the correlated four-state model developed by Lutz. From fisheye pictures taken in Guilford, Southampton, London and Los Angeles, blockage and correlation statistics were extracted. In this study it was observed that for azimuth separations smaller than 30°, satellite channels tend to be correlated.

In [335], two different approaches to shadowing correlation characterization/quantification are presented for the urban environment. One is based on the use of a physical-statistical model and the other is based on a purely deterministic model. These two approaches produce identical results of correlation coefficient values. These can be used to furnish correlation coefficient data to statistical models.

Average values of the correlation coefficient are given for urban canyon-like environments, as a function of the azimuth offset, $\Delta\phi$, for satellites at different elevations, θ_1 and θ_2, which are given by a reference elevation and an elevation increment $\Delta\theta$. It was found that for azimuth separations in the vicinity of both 0° and 180° high correlation values are observed in the street canyon geometry, both when considering same satellite elevations, $\theta_1 = \theta_2$, and different satellite elevations $\Delta\theta$. For street intersections, a further peak at 90° is also present. The correlation coefficient as a function of the azimuth increment, $\Delta\phi$, has circular symmetry about $\Delta\phi=0$, and for this reason only azimuth increments between 0° and 180° are meaningful. The shape and values of these correlation curves are in very good agreement with results presented by other Researchers using measured data [330]. General results are shown in Table 8.1.

8.2.2 MIMO and Space-Time Coding

A new approach, which promises a significant increase in system capacity , known as MIMO [336] is attracting the interest of the telecommunication research community over the last decade. The MIMO approach is an extension of the space diversity technique, often used in microwave communications. Transmitter diversity complements receiver diversity in mobile communication systems, and finally, both approaches merge into the MIMO approach [337]. MIMO increases the link data rate and/or the robustness of the system [338].

Table 8.1. Summary of geometrical conditions for which correlation can be expected in street canyon scenarios.

Reference Elevation	Elevation Increment	Azimuth Increment	Existence of Correlation
$10^\circ \leq \theta_{ref} < 50^\circ$	$\Delta\theta = 0^\circ$	$\|\Delta\phi\| < 30^\circ$ $170^\circ < \|\Delta\phi\| < 180^\circ$	Yes
$10^\circ \leq \theta_{ref} < 50^\circ$	$\Delta\theta = 0^\circ$	$30^\circ < \|\Delta\phi\| < 170^\circ$	No
$10^\circ \leq \theta_{ref} < 50^\circ$	$0^\circ < \Delta\theta = 0^\circ \leq 20^\circ$	$\|\Delta\phi\| < 15^\circ$ $170^\circ < \|\Delta\phi\| < 180^\circ$	Yes
$10^\circ \leq \theta_{ref} < 50^\circ$	$\Delta\theta > 20^\circ$	any	No
$\theta_{ref} \geq 50^\circ$	any	any	LOS

A direct increase in data rate can be achieved by spatial multiplexing, which means exploiting the rich scattering multipath propagation channel by transmitting independently modulated signals simultaneously by each antenna [339]. The other aim of MIMO techniques is to provide increased diversity involving the space domain. The robustness of the communication system is achieved by introducing the redundancy to transmitting antennas. This technique is conventionally known as space-time coding [340].

In a MIMO system with N transmit and M receive antennas, data are transmitted *concurrently* from all the N transmit antennas. The received signals in all of the M receive antennas exhibit inter-channel interference.

In this section, the channel is characterized by its $M \times N$ discrete-time channel matrix H which is assumed to model flat fading. Its entries α_{ij} correspond to the channel between transmit antenna i and receive antenna j. It was shown in [341] and [342], that, assuming ideal channel state information only at the receiver side, (i.e. equal power allocation is made across transmit antennas), the capacity amounts to (see Section 2.3.5)

$$C = \log_2 \det\left[I + \frac{\rho}{N} HH^H \right] \quad \text{bits/channel use} \quad (8.6)$$

where ρ is the SNR at one receive antenna. Denoting the nonzero eigenvalues of the matrix W defined as

$$W = \begin{cases} HH^H & \text{if } M \leq N, \\ H^H H & \text{if } N < M \end{cases} \quad (8.7)$$

as $\lambda_1, \lambda_2, \ldots, \lambda_m$, where $m = \min(M, N)$ is the rank of the channel matrix, we can formulate (8.6) as

$$C = \sum_{i=1}^{m} \log_2\left|1 + \frac{\rho}{N}\lambda_i\right| \quad \text{bits/channel use} \quad (8.8)$$

This equation shows that the MIMO channel can be thought as m orthogonal Soft-Input Soft-Output (SISO) subchannels (spatial modes) with different attenuations.

If we can assume feedback from the receiver to the transmitter, a better power allocation is possible between the different subchannels, corresponding to the classical waterfilling allocation [341].

In the first part of this section, basic space-time coding techniques will be presented and analyzed. Them, feedback methods will be shown. In these schemes, Adaptive Coding (AC) is introduced based on the distribution of the spatial modes of the channel.

8.2.2.1   Space-time coding schemes.   Space Time Coding (STC) techniques allow to exploit the domain of both antenna and time diversity that are joined to yield performance gains without increasing bandwidth requirements.

The gains can obviously be traded off for an increase in data rate, fixing power/interference requirements [343].

Space-Time Block Coding. The transmitter block-diagram of a Space-Time Block Coding (STBC) scheme is shown in Figure 8.4(A). Let b be the number of bits per modulated symbol, a sequence of bH information bits feeds the STBC encoder, which implements the orthogonal design matrix $\mathcal{O}$ [344], returning a sequence of bP coded bits at the input of each transmit branch. For convenience, we will indicate either input or output b-tuple of data bits with the corresponding modulated symbols $x_1, \ldots, x_h, \ldots, x_H$, which belong to the signal constellation $\mathcal{A}$ of 2^b elements. $\mathcal{O}$ is a $P \times N$ matrix whose elements are the indeterminates y_h, y_h^* and their linear combinations. P is the codeword length and $R_c = \frac{H}{P}$ is the coding rate. The encoder replaces y_h with the corresponding input symbol x_h, then the i-th column is transmitted by the i-th TX antenna during P symbol periods.

The information symbol, $x(k)$, is encoded at the time instant k by the Space-Time (ST) encoder into N coded symbols, $b_1(k), b_2(k), \ldots, b_N(k)$, mapped in turn by the modulator into $c_1(k), c_2(k), \ldots, c_N(k)$ that are transmitted simultaneously from different antennas. The encoder selects the N coded symbols so as to achieve a diversity order equal to NM, which is obviously the most relevant parameter for ST codes performance. The symbols, $c_i(k)$, are normalized so that the constellation average energy is 1. Letting $p(t)$ be the pulse shape waveform with unitary energy, the base-band equivalent signal transmitted from the i-th TX antenna is

$$S_i(t) = \sqrt{\frac{E_s}{N}} \sum_{k=-\infty}^{+\infty} c_i(k)p(t - kT) \quad i = 1, \ldots, N \qquad (8.9)$$

where E_s is the total average energy radiated by all N antennas in the symbol period T.

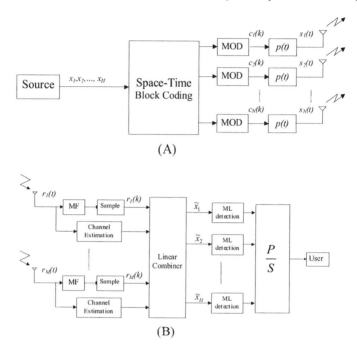

(A)

(B)

Figure 8.4. Space-Time Block Code Transmit and Receive Schemes. A: transmitter; B: receiver.

Orthogonal design matrices, $\mathcal{O}_2$ (with $R_c = 1$), $\mathcal{O}_3$ and $\mathcal{O}_4$ (with $R_c = 3/4$), presented in [344] and [345], will be adopted for two, three and four transmit antennas.

The decoding procedure for STBC follows the Maximum Likelihood (ML) criterion (an ideal constant envelope modulation is assumed). The receive block-diagram is reported in Figure 8.4(B). The signal received at the j-th RX antenna is matched-filtered and sampled, and in parallel the channel state information is recovered. Then, both data samples and channel estimates, indicated as $\hat{S}_j$ and $\hat{\alpha}_{ij}$, are linear combined to yield $\tilde{x}_1, \ldots, \tilde{x}_h, \ldots, \tilde{x}_H$, as follows:

$$\tilde{x}_h = \sum_{j=1}^{M} \sum_{k=1}^{P} \hat{S}_j(k) \sum_{i=1}^{N} \left[\delta_{ki}^R(h) \, \mathfrak{Re} \left\{ r_j(k) \hat{\alpha}_{ij}^* \right\} + j \delta_{ki}^I(h) \, \mathfrak{Im} \left\{ r_j(k) \hat{\alpha}_{ij}^* \right\} \right]$$

(8.10)

where $\delta_{ki}^R(h)$ and $\delta_{ki}^I(h)$ denote the coefficient of $\mathfrak{Re} \{y_h\}$ and $\mathfrak{Im} \{y_h\}$, respectively, in the (k, i)-th position of the matrix $\mathcal{O}$ and they are equal to zero whenever y_h does not appear in the (k, i)-th position. Equation (8.10) can be considered as an extension of (7) in [344] for complex orthogonal designs. Finally, the H symbols $\bar{x}_1, \ldots, \bar{x}_h, \ldots, \bar{x}_H$ are chosen following the decision metric:

$$\bar{x}_h = \arg \min_{x^h \in \mathcal{A}} \left| \tilde{x}_h - x^h \right|^2,$$

where x^h is the hypothesized symbol.

Space-Time Trellis Coding. The transmitter block-diagram of Space-Time Trellis Coding (STTC) is shown in Figure 8.5(A). The information bits are encoded by the STTC and split into N parallel streams of coded bits, subsequently mapped into the constellation $\mathcal{A}$. The modulated symbols enter into a block interleaver, then they are filtered by a pulse shape waveform filter to avoid intersymbol interference and transmitted over the fading and noisy channel. To describe STTC we refer to trellis diagrams. The input bit stream is divided into groups of b bits. At the beginning of the transmission the encoder is supposed to be in state A, then a transition branch is chosen on the basis of the b input bits and their ν predecessors. On each output a modulator maps coded bits into the constellation symbols in accordance with the selected branch label (q_i). Examples of STTC for Quadrature Phase Shift Keying (QPSK) modulation are shown in [340].

Figure 8.5(B) shows the STTC receiver scheme. The received signal $r_j(t)$ on the j-th RX antenna, is matched-filtered and sampled; in parallel the channel state information is recovered. Then both data samples and channel estimates are deinterleaved. A ML soft Viterbi decoder follows, and, under the hypothesis of perfect Channel State Information (CSI), the branch metric at time instant k is equal to:

$$\lambda(k) = \sum_{j=1}^{M} \left| r_j(k) - \sqrt{\frac{E_s}{N}} \sum_{i=1}^{N} S_j(k)\alpha_{ij}(k)q_i \right|^2 \tag{8.11}$$

Note that the case of channel estimation inaccuracies is considered in [346].
Performance analysis of STBC systems . The analytical average Bit Error Probability (BEP) for STBC system with coherent M-PSK data modulation, under the assumption of N transmit and one receive antennas and perfect channel estimation, can be provided in the presence of Rice lognormal fading channel constant during P symbol periods [347], [348]. Let's consider the well-known expression:

$$P_{b|S,\alpha} \simeq \frac{\overline{N}_a}{2b} \text{erfc} \left\{ \sqrt{\frac{b}{N} \frac{E_b}{N_0} \sin^2 \left(\frac{\pi}{2^b} \right) S^2 \sum_{i=1}^{N} |\alpha_i|^2} \right\} \tag{8.12}$$

where S is the shadowing associated to the received antennas following a lognormal distribution with parameters μ_S and σ_S (see Sec. 2.2.7, often referred to

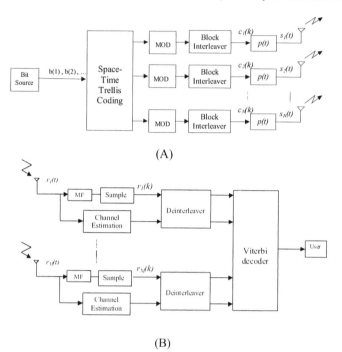

(A)

(B)

Figure 8.5. Space-Time Trellis Code Transmit and Receive Schemes. A: transmitter; B: receiver.

as the *dB-spread*; α_i is the i-th path gain modeled as a Gaussian fading process; $\overline{N}_a$ is the average number of adjacent points in the signal space and $E_s = bE_b$. To find the average bit error probability P_b, (8.12) has to be averaged over Rice and lognormal probability density functions (pdfs). The average over the N Rice random variables (rvs), α_i, can be expressed with a closed-form solution valid for $N \geq 2$, here reported:

$$
P_{b|S} = \frac{\overline{N}_a}{b}\left\{ Q(U,V) - I_0(UV)e^{-\frac{1}{2}(U^2+V^2)} + \frac{I_0(UV)e^{-\frac{1}{2}(U^2+V^2)}}{[2/(1-\xi)]^{(2N-1)}} \cdot \right.
$$
$$
\cdot \sum_{k=0}^{N-1}\binom{2N-1}{k}\left(\frac{1+\xi}{1-\xi}\right)^k + \frac{e^{-\frac{1}{2}(U^2+V^2)}}{[2/(1-\xi)]^{(2N-1)}} \cdot
$$
$$
\left. \cdot \sum_{n=1}^{N-1} I_n(UV) \sum_{k=0}^{N-1-n}\binom{2N-1}{k}\cdot\left[\left(\frac{V}{U}\right)^n\left(\frac{1+\xi}{1-\xi}\right)^k - \left(\frac{U}{V}\right)^n\left(\frac{1+\xi}{1-\xi}\right)^{2N-1-k}\right]\right\}
$$

(8.13)

where $I_n(\cdot)$ is the modified Bessel function of first kind and n-th order; $Q(U, V)$ is the Marcum Q function, and U, V, ξ are given by

$$\left.\begin{array}{c} U \\ V \end{array}\right\} = \sqrt{NK}\left[\frac{N(K+1)+2b\sin^2(\pi/2^b)S^2\gamma}{2N(K+1)+2b\sin^2(\pi/2^b)S^2\gamma} \mp \sqrt{\frac{b\sin^2(\pi/2^b)S^2\gamma}{N(K+1)+b\sin^2(\pi/2^b)S^2\gamma}}\right]^{\frac{1}{2}}$$
(8.14)

$$\xi = \sqrt{\frac{b\sin^2(\pi/2^b)S^2\gamma}{N(1+K)+b\sin^2(\pi/2^b)S^2\gamma}}$$

with $\gamma = E_b/N_0$ and K the Rice factor. Finally, the bit error probability is computed by averaging over the lognormal pdf: $P_b = \int_0^{+\infty} P_{b|S}p_S(S)dS$. The integral can be solved numerically by means of Gauss-Hermite quadrature forms (no more than 12 points are recommended). Results for $N = 1$ can be found in [197]. Figure 8.6 shows performance results for the Rice-lognormal channel with $K = 4$, $\sigma_S = 4$ and the lognormal shadowing perfectly correlated over all paths.

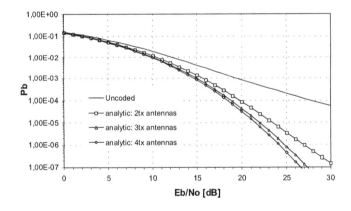

Figure 8.6. Average bit error probability, P_b, vs. E_b/N_0[dB] with transmit antennas number as a parameter. STBC scheme is considered in a Rice-lognomal channel with $K = 4$ and $\sigma_S = 4$. QPSK data modulation and ideal CSI are considered.

The analytical STBC system performance considering the Alamouti scheme ($\mathcal{O}_2$) for Binary Phase Shift Keying (BPSK) and QPSK modulation with one RX antenna can be provided in the presence of Rayleigh fading channel when the channel and oscillator phase rotation is not perfectly recovered [347].

Performace analysis of STTC systems. The well-known Transfer Function Bound (TFB) and improved bound extended to the space diversity domain allow to achieve an analytical expression for STTC performance in the presence of Rayleigh-lognormal channel with M-PSK modulation scheme.

In order to achieve an analytical expression of STTC performance, the interleaving depth has been assumed infinite to ensure uncorrelation of adjacent demodulated signals and no truncation in the Viterbi algorithm has been considered. Let $\mathbf{c}(k)$ and $\hat{\mathbf{c}}(k)$ be the transmit and the decided codeword, respectively, and being $\left| \sum_{i=1}^{N} \alpha_{ij}(k)[\hat{c}_i(k) - c_i(k)] \right|^2 = d^2(\mathbf{c}(k), \hat{\mathbf{c}}(k))\Psi_j^2(k)$, ($\Psi_j$ is a Rayleigh distributed random variable) it can be shown [349] that the Pairwise Error Probability (PEP) can be expressed in terms of d_k and Ψ_j:

$$P(\mathbf{C}_L \to \hat{\mathbf{C}}_L | \Psi_j(k), S_j(k)) = \frac{1}{2}\text{erfc}\left(\sqrt{\sum_{j=1}^{M}\sum_{k=1}^{L} \delta_k^2 \rho_{jk}^2}\right) \qquad (8.15)$$

where $\delta_k^2 = \frac{1}{N}\frac{E_s}{4N_0}d_k^2$ and $\rho_{jk}^2 = S_j^2(k)\Psi_j^2(k)$. Eq.(8.15) is useful to achieve an upper bound for the bit error probability, applying the Union Bound definition (see Section 2.4.1.3). Averaging eq.(8.15) over the rv's ρ_{jk}, we pursue a bounding approach obtaining:

$$P(\mathbf{C}_L \to \hat{\mathbf{C}}_L) \le \text{EB}(\mathbf{C}_L, \hat{\mathbf{C}}_L) = \frac{1}{2}\prod_{j=1}^{M}\prod_{k=1}^{L} \mathrm{E}_{\rho_{jk}}\{Z^{d_k^2\rho_{jk}^2}\} \qquad (8.16)$$

where $Z = \exp(-\frac{1}{N}\frac{E_s}{4N_0})$. Eq. (8.16) is useful in order to bound the bit error probability through the Transfer Function method. The augmented Transfer Function, $T(D, I)$, enumerates all $A(\beta)$ error events whose distance from the *zero-sequence* is β. Setting

$$D = \exp(-\frac{1}{N}\frac{E_s}{4N_0}) \qquad (8.17)$$

and

$$\beta = \sum_{j=1}^{M}\sum_{k=1}^{L} d_k^2 \rho_{jk}^2, \qquad (8.18)$$

averaging $T(D, I) = \sum_{\beta=\beta_{min}}^{\infty} I^\nu A(\beta)D^\beta$ over the path gain r.v.s and deriving respect to the symbolic variable I, P_b is provided by means of TFB

$$P_b \le \text{TFB} = \frac{1}{2b}\frac{\partial}{\partial I}\overline{T}(D, I)\Big|_{I=1, D=\exp\left(-\frac{1}{N}\frac{E_s}{4N_0}\right)} \qquad (8.19)$$

Following [350] and [351], we can also find an Improved Upper Bound (IUB) which supplies a more accurate approximation of simulation curves. Using

TFB and the set $\mathcal{D}$ of dominant errors, we achieve:

$$\text{IUB} = (\text{TFB} - \Delta_2) + \Delta_1 \qquad (8.20)$$

where $\Delta_2 = \sum_{\hat{C}_L \in \mathcal{D}} \nu(\mathbf{0}, \hat{\mathbf{C}}_L) \text{EB}(\mathbf{0}, \hat{\mathbf{C}}_L)$, which means that Δ_2 is the contribution given to TFB by $\mathcal{D}$. On the other hand, Δ_1 is the sum of PEPs of dominant error events calculated either by an exact procedure [351] or by a bound tighter than EB [350]. In essence, IUB is obtained replacing in TFB the contribution Δ_2 with the more accurate Δ_1. Figure 8.7 shows P_b vs. E_b/N_0 for Rayleigh channel in case of QPSK modulation. Increasing the Time Diversity (TD), i.e. the number of states, improves the performance. When E_b/N_0 is low, IUB and TFB performance is not sufficiently accurate since the approximation of erfc $(\cdot)$ in (8.16) is weak. For higher signal to noise ratios, when dominant error events are the only ones that occur, the distance between TFB and IUB reaches its maximum value of almost 2 dB and IUB curves are less than 1 dB away from Monte-Carlo simulations.

Sequential decoding algorithms, like the Fano and the stack algorithms [88], can be modified to allow for STTC decoding, using the metrics according to (8.11). Both algorithms lead essentially to similar performance to that of the vector Viterbi algorithm when critical parameters are appropriately chosen: the threshold modifier in the case of the Fano algorithm and stack size n of the stack algorithm. The stack algorithm stores paths in the trellis of different

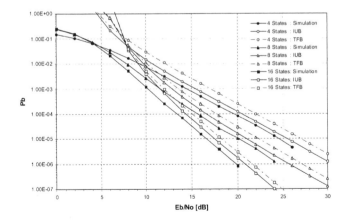

Figure 8.7. Comparison of TFB, IUB and simulation results for STTC scheme over a Rayleigh channel with 4, 8 and 16 trellis states. QPSK data modulation and ideal CSI are considered. $N = 2$.

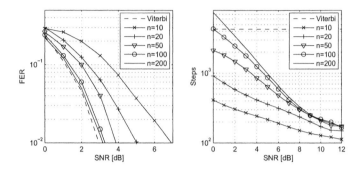

Figure 8.8. Performance and computational complexity of the stack algorithm for a 64-state STTC, depending on the stack size n

length with their metrics. Always the path with minimal metric is on the top of the stack. This can be realized by a storage structure called priority queue. (The original stack algorithm always ordered the paths in descending order by their metrics, but this requires unnecessary steps of ordering. Avoiding the stack ordering offers substantial savings.) In each step we extend this path on the top with each possible branch that can appear at the end of this path. We calculate the metrics of these new paths and put them back to the stack after we have removed the path on the top. In order to avoid the stack filling up, the paths with large metrics must be removed. If the stack is large enough the probability of removing a good path is small. The stack algorithm is known for its non-deterministic processing time. As demonstrated in Figure 8.8, the modified stack algorithm can outperform the Viterbi algorithm in terms of computational requirements, whilst maintaining virtually the same Frame Error Rate (FER) performance if the stack size is large enough. QPSK modulation and a frame length of 130 symbols was chosen in the simulations.

8.2.2.2   Adaptive Modulation over MIMO links .   Space-time coding or multiplexing of data cannot fully exploit the channel capacity , due to channel variation in time. The channel capacity can only be utilized efficiently by dynamic adjustment of the transmission parameters to the instantaneous channel characteristics. In a frequency flat fading wireless channel a straightforward method to combat channel attenuation is to increase the transmitter power. The second method, not only limited to the flat fading channels is to adjust the coding and modulation scheme to the instantaneous channel characteristic. In a system with multiple transmit and receive antennas, there exists a possibility to increase the system reliability by the same signal transmission in the parallel subchannels and combining those subchannels at the receiver.

If the gains of orthogonal subchannels are known at the transmitter, coding, modulation, and transmit power can be adjusted to the orthogonal subchannel attenuation. Due to the orthogonal subchannels the same technique can be applied on each subchannel. Although an optimum transmission scheme is selected for each subchannel, additional throughput can be gained by combing subchannels.

For that reason the subchannels are ordered by decreasing order of subchannels gains $\lambda_1 \geq \lambda_2 \geq ... \geq \lambda_m$. The subchannels are then classified into reliable and unreliable subchannels. In reliable subchannels the target Bit Error Rate (BER) value can be achieved, while in unreliable subchannels the propagation distortion is too severe to meet the target BER. Some heuristic strategies, which adjust the coding and modulation scheme and the data transmitted in each subchannel, are tested in [352]:

- With the *basic channel mapping strategy*, data are transmitted only in reliable subchannels, while no information is transmitted in unreliable subchannels.

- With the *advance channel mapping strategies*, the same data sequence is transmitted in a selected reliable and one or more unreliable subchannels. Assuming statistically independent channel distortions in each active channel, by combining all subchannels carrying the same information, the SNR of the combined signal is increased. Consequently a new, more power efficient coding and modulation mode may be chosen for transmission. When no reliable subchannel is found, the same data is transmitted in all unreliable subchannels. The block diagram of the advanced channel mapping strategy is shown in Figure 8.9.

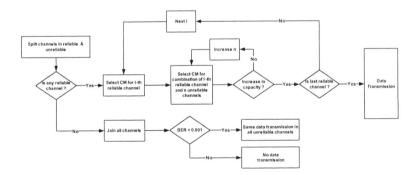

Figure 8.9. Advanced channel mapping strategy block diagram.

The proposed channel mapping strategies are analyzed for a MIMO system with eight transmit and eight receive antennas ($N = 8$, $M = 8$). Maximum likelihood sequence estimation is assumed at the receiver. The set of modulation schemes available for transmission consists of eight Quadrature Amplitude Modulation (QAM) modulation schemes, namely 2-QAM, 4-QAM, 8-QAM, ..., 256-QAM. The target value of BER is set to 10^{-3}. Simulation results show that the basic channel mapping strategy, where no data is transmitted in unreliable channels, provides an increase in system capacity similar to advanced channel mapping strategies at high SNR, while at moderate SNR, the combination of one reliable and one unreliable channel significantly increases the system capacity in comparison to basic channel mapping strategy. Even higher average system capacity is achieved when two or more unreliable subchannels are combined with one reliable subchannel [352]. At low SNR, when no reliable channels are available an increase of system capacity is achieved by combining all unreliable subchannels.

The availability of the system is shown in Figure 8.10 for moderate SNR and Equal-Gain Combining (EGC). A basic channel mapping strategy guarantees, on average, up to one bit lower link capacity than the advanced scenarios. In correspondence of a system capacity of 4 bits per symbol, the basic channel mapping strategy guarantees the 50% of the link availability, while combining one reliable and one unreliable channel increases the link availability to 80%. Additional 12% can be gained by using more unreliable subchannels. The EGC was applied for results in Figure 8.10. The results of system availability strongly depend on SNR. The described channel mapping strategies can be applied in satellite systems using satellite diversity techniques and with the terminal equipped with multiple antennas.

8.2.2.3   Applications of STC in satellite communications.   The advantages of the MIMO approach are proven for communication systems where a rich scattering medium exist at the transmitter and at the receiver. MIMO systems are successfully employed in terrestrial indoor and non-line-of-sight communications.

In satellite communication systems there are no scatterers at the satellite side. This fact could lead at first glance to the conclusion that the MIMO approach is not appropriate for satellite systems. However, transmitting from multiple satellites introduces this artificial scattering. Such "distributed" space-time schemes have to solve the problem of synchronization between the satellites, and to introduce some form of equalization in the receiver to cope with the different and possibly continuously varying delay difference among the signals.

Furthermore one can combine STC schemes with polarization diversity. In such systems the coded signal is transmitted from one satellite, using two an-

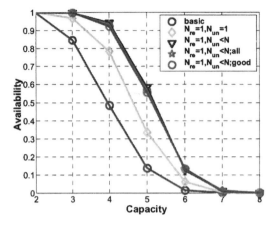

Figure 8.10. Availability of different channel mapping strategies for SNR = 5dB. EGC is considered.

tennas with orthogonal polarization. For instance, the Alamouti scheme is well suited to this purpose.

When the satellite signals are re-transmitted in urban areas through gap-fillers/IMRs (Intermediate Module Repeaters), application of STC schemes offer similar advantages for satellite communications as they proved in conventional terrestrial systems. An example of their applicability is provided by the Satellite Digital Multimedia Broadcasting (S-DMB) scenario.

8.2.3   Site Diversity

The site diversity rationale is that the link availability increases because the joint outage probability of two stations located at a certain distance is much lower than the single site case (equal to the product of the probabilities of each station in case of statistical independence). As a consequence the margin can be much lower. As main drawbacks, the station must be duplicated (at least a part of the equipment) and a dedicated terrestrial link must be set up.

This concept can be implemented with a more sophisticated and evolutionary architecture including a set of cooperating stations, working separately with the same satellite [353]. In this case if one of the stations needs assistance, its traffic flow can be re-routed to one of the others, selected on the basis of proper algorithms, through a new communication link for the needed time interval, as depicted in Figure 8.11. The stations can be interconnected through terrestrial public networks (e.g. ISDN) utilized only in case of necessity on a call basis.

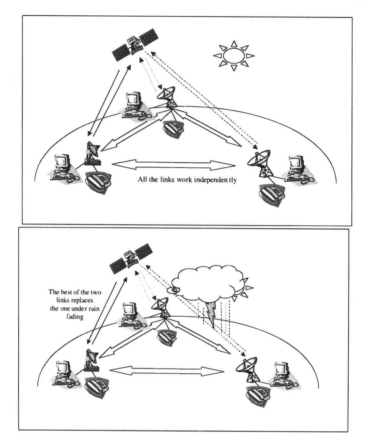

Figure 8.11.   Multiple site diversity set up.

With the multiple site diversity, when the architecture is considered, stations and equipments no longer need to be duplicated and dedicated terrestrial links no longer need to be set up. Concerning performance, the availability improvement is significant, and depending on the number of stations the outage probability reduces about to zero.

As a drawback, this system architecture is applicable only to a subset, yet meaningful, of scenarios, i.e.

- when the satellite connection is more convenient than the terrestrial links;

- when the set of stations belongs to a proprietary closed network for security reasons, aiming at utilizing the terrestrial networks only in case of necessity;

- if the provided services are more suitable to satellite transmission;

- if the stations are of transportable type and can occasionally utilize the diversity advantage in case they are close to an access point to the terrestrial network;

- when the satellite is preferred in normal working conditions because it provides more efficient interconnection with other systems.

8.2.3.1   Statistical models for site diversity .   To analyze and evaluate performance of multiple site diversity the joint probability of rain events of a set of stations must be carried out [354] taking into account the statistical characteristics of the rain events [355]. In the literature, results from measurement campaigns [353][356] and some statistical models using log-normal statistics for the single site distribution of rainfall intensity and rain attenuation [357] [358] are available. For low values of rain attenuation, this model is still widely accepted and the assumption of log normality in the case of many sites appears a quite natural extension of this model [359] [360] [361]. Low rain intensities are more representative of the intrinsic spatial nature of the phenomenon, because they can occur simultaneously over wide areas, thus allowing the description of the space correlation, while heavy rains occur within limited cells.

In [353], for the specific case of multiple station configurations, the evaluation of the outage probability or margin reduction as a function of the number of involved stations is proposed and carried out through the solution of multidimensional integral equations. Performance evaluation is carried out implementing the theoretical models through Montecarlo simulations. The main hypothesis of the model utilized in [353] is that the single site and the joint probability of rain attenuation for N-locations have log-normal behaviors [357] [358] [362].

The state of the N locations is described by means of an N-dimensional binary variable ("rainy" or "non rainy") with the associated multivariate probability function. It is in fact assigned by means of the probability density of a N-dimensional continuous variable and by a N-dimensional threshold chosen in such a way that the probabilities to be exceeded or not assume the wanted values.

According to the model presented in [353], the joint probability that a set of thresholds A_i ($i = 1$ to N) are simultaneously exceeded is then given by the product of two joint probabilities:

1 the *conditioning* one, which gives the "common rainy time", is obtained integrating the joint normal density of the continuous "rainy state variable" from the set of thresholds to infinity;

2 the *conditioned* one, similarly obtained by integrating the joint lognormal density of the normalized attenuation A_i from the set of the reduced attenuation thresholds to infinity.

The rain attenuation distribution for each site has been calculated using a prediction model (EXCELL [363]) where the rain intensity distribution is obtained via the ITU-R global map of rain intensity [364].

8.2.4   Frequency Diversity

Frequency diversity is based on a common resource shared among the Earth stations, utilized for those links which need assistance because of the local high value of rain fading. A frequency slot represents the common resource, of course allocated at a lower range than the slot utilized in normal conditions. In fact, with frequency diversity we use a higher band to carry most of the traffic and a lower band (typically much smaller than the former) to assist those links in the time intervals during which the rain attenuation overcomes a certain threshold. As a matter of fact the link at the higher frequency is dimensioned with a greater outage probability, allowing a smaller rain margin. When the rain attenuation overcomes the rain margin, the link will work at the lower frequency.

The main impairments due to raining are at implementation level. In fact, the partial duplication of RF equipment is needed and, as well as the other fade compensation techniques, the switching delay, which is greater than zero, can cause unexpected outage or false alarm states, if in the meanwhile eventual changes of weather occurred.

As a practical example, a portion of Ka band can represent the higher band while a portion of Ku band can represent the lower band. The traffic would be usually carried in Ka band, but when the margin on a particular link would become unable to overcome rain attenuation the link would be switched to Ku band, if available. The capacity in Ku band represents the shared resource [365].

8.2.4.1   Outage probability for frequency diversity systems.   In case shared resource methods are utilized, the outage probability for a given link is [365]:

$$P_{out} = P(A_a > M_a) + P(A_0 > M_0, E \text{ unavailable}) \qquad (8.21)$$

where A_a and M_a are the attenuation and the power margin in assisted condition (in the lower band) respectively, A_0 and M_0 are the attenuation and the power margin in unassisted conditions (in the higher band) respectively and E is the common resource. Equation 8.21 shows that the outage probability is the sum of two contributions: the probability that in the lower band link the attenuation overcomes the power margin and of the probability that the common resource is unavailable when the link requires assistance ($A_0 > M_0$), because already utilized by other links.

To evaluate outage probability the joint attenuation distribution among the stations of the system would be necessary. The analysis can be simplified assuming the following hypotheses:

1 the capacity and the outage probability, both in assisted conditions (P_a) and in unassisted conditions (P_0), are the same for each of the stations

2 all links are statistically independent [366].

In this way, the probability, P_{ra}, that one station would require assistance and E would be unavailable, can be associated to a random binomial variable, i.e. to the probability that the event would occur for a greater number of stations than N_a (maximum number of stations that is possible to assist simultaneously). This number is determined by the available bandwidth in the lower band and by the capacity required by each station [367].

In these conditions P_{ra} is

$$P_{ra} = \left(\begin{array}{c} N_s - 1 \\ k \end{array} \right) (P_0)^k (1 - P_0)^{N-1-k} \tag{8.22}$$

being N_s the total number of stations and k the number of assisted stations.

Then, being $k + 1$ the number of stations that require assistance, $k + 1 - N_a$ is the number of stations that cannot be assisted. Thus the probability that a station would not receive assistance is

$$P_{na} = \frac{k + 1 - N_a}{k + 1} \tag{8.23}$$

Regarding the statistical independence among the stations, it is likely if the distance between each station and the others is remarkable (> 1000 km); otherwise it is possible to take into account the statistical dependence by means of a coefficient $1/\delta_T$ multiplied to each probability factor in eq.(8.21) (as shown by the experimental results obtained with data collected at Spino d'Adda and Lario in [365], [367] [366]). This coefficient represents the fraction of time in which fading is active.

Finally, eq.(8.21) can be written as:

$$P_{out} = P_a + P_0 \sum_{k=N_a}^{N_s-1} \left(\begin{array}{c} N_s - 1 \\ k \end{array} \right) \frac{P_0}{\delta_T}^k \left(1 - \frac{P_0}{\delta_T} \right)^{N_s-1-k} \frac{k + 1 - N_a}{k + 1}$$

$$\tag{8.24}$$

In this way, the model obtained in case of statistical independence can be still used when statistical independence cannot be assumed. Notably, statistical dependence among stations impacts system performance.

Performance analysis using the introduced model shows that many system parameters are of relevant importance: working frequencies, number of stations, degree of statistical dependence among the stations, amount of shared resources. In a study case assuming 170 stations evenly distributed over Europe, using Ka band as main frequency and Ku band as shared bandwidth, with even a small shared bandwidth (10% of the main) the improvement in terms of rain margin to dimension the Earth stations can reach 15 dB, while with greater amount of shared bandwidth does not imply worth further advantage [368].

8.2.5   Power Control in CDMA-based systems

In a CDMA-based satellite mobile communications systems Power Control (PC) is defined as an adaptive process to correct the mobile transmit power in order to keep the equivalent Signal to Noise plus Interference Ratio (SNIR) at the receiver side as close as possible to a target (nominal) value.

PC operation is not ideal and it has been measured [369] [370] that the power controlled signal follows a log-normal distribution with a mean around the target power and a given standard deviation. It has also been reported that standard deviation, σ, ranges from 0 dB (ideal PC) to 3 or 4 dB. However, even with this limited range its effect is dramatic in performance [371] [372]. A number of analytical formulations have been developed for performance analysis of CDMA satellite systems that include the effect of PC. However, numerical results generally assume intuitive values for σ, e.g. larger for shadowed than for unshadowed links since the signal dynamic affects the power control imperfections.

In [373] realistic values of σ are derived by applying a PC algorithm upon received signal timeseries generated by a 3-state Markov channel model. Two steps are followed for the PC error estimation. First, generation of the received signal for a given environment, mobile speed and time sampling. Second, operation of the PC algorithm upon the received signal timeseries is computed through simulation. The PC algorithm is modelled as follows. First the link setup is assumed to be successfully accomplished in such a way that only an adaptive closed-loop PC need to be applied to correct the transmitted power to keep the SINR at the receiver as close as possible to the target value. It is also assumed that an open-loop PC compensates for path loss. A 7-level PC is used and the power correction step size defining the change introduced in the transmission power are 0 dB, ± 0.75 dB, ± 1.10 dB and ± 1.9 dB as in [78]. A simple up/down adjustment or several power adjustment levels are possible. PC adjustment is always relative to the previous power setting, since absolute power setting would not be realistic due to the expensive circuitry needed. Several PC updates rates were obtained, namely, 5, 10, 15 and 20 times the frame length that it is consider to be 10 ms. Note that PC command is foreseen to be only one per frame in order to compensate slow variations only. However also larger updating rates accounting for round-trip delay up to 200 ms (for LEO systems is on the order of 10 to 20 ms) were used. The time resolution was set to 10 ms which is coincident with the frame length. Considered mobile speed were of 1 m/s, 15 m/s and 30 m/s which means that channel coherence times are about 70 ms, 5 ms and 2.5 ms respectively. This way time variability for 15 m/s and 30 m/s is undersampled, however PC command cannot be activated at so high rate.

Figure 8.12 shows the PC performance with updating time of 10 ms for the urban environment at a mobile speed of 15 m/s. It is observed that PC is not able to counteract fast fading due to the low PC updating compared to the coherence time (time series are normalized for target value of 0 dB). It is also shown the fitting of power controlled timeseries to lognormal pdf.

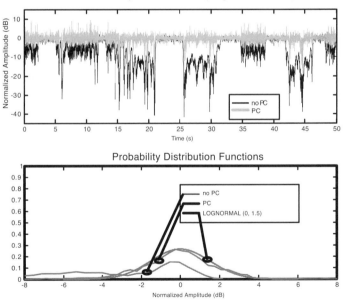

Figure 8.12. Example of power controlled timeseries (urban environment) and fitting to lognormal pdf [373] ©2002 IEEE

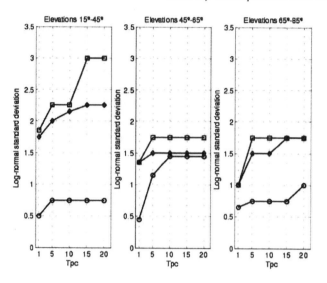

Figure 8.13.   PC error σ values for suburban environment [373] ©2002 IEEE

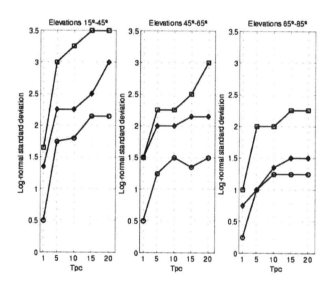

Figure 8.14.   PC error σ values for urban environment [373] ©2002 IEEE

Figure 8.13 and Figure 8.14 show the effect of the delay for suburban and urban environments respectively by plotting standard deviations for different PC

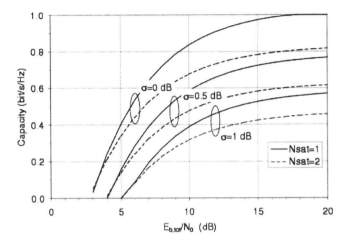

Figure 8.15. PC error effect on CDMA system capacity [328] ©2001 IEEE

updating time. It is apparent the worsening effect of the delay and user speed, which decreases with elevation. The effect on system capacity is derived in [374] and it is shown in Figure 8.15.

8.2.6   Polarization Diversity

With polarization diversity there are two orthogonally polarized (receive) antennas. In most cases, this is the simplest way to realize the diversity concept. In contrast to space diversity, the antennas need not to be multiplied. In the case of a reflector antenna the primary feed only must be more complex than its non-diversity counterpart. Dual-polarized radiators must be used in array or patch antennas. An economic advantage is that in either case antennas are virtually co-located, making the whole setup cheaper than with two distinct antennas. In contrast to frequency diversity, polarization diversity does not need a wider frequency band than a non-diversity system. On the other hand, the number of diversity routes is definitely two – in contrast to space or frequency diversity in which any number of routes are in principle available. Moreover, polarization diversity is slightly less effective than its counterparts.

In this subsection a method for characterizing polarization states, and the concept of Stokes space are introduced. Random polarization and quantitative characteristics of polarization diversity are given in the sequel. Finally the theory is verified by comparing theoretical and experimental/simulation results.

Note that the conventional diversity concept is applied here according to which one signal is transmitted (via a Single Input-Multiple Output (SIMO)) to two receiver antennas.

8.2.6.1   Characterization of polarization states.   Given a polarized
electromagnetic wave of arbitrary polarization propagating in the z direction
electric field-strength is

$$\mathbf{E} = E_x \boldsymbol{x_e} + E_y \boldsymbol{y_e} \tag{8.25}$$

with $\boldsymbol{x_e}$ and $\boldsymbol{y_e}$ be the unit vectors in the appropriate directions. Its polarization
state can be represented in the three-dimensional *Stokes space* (see e.g. [375])
where the coordinates (Q_1, Q_2, Q_3) and the square absolute value Q_0 are

$$Q_1 = E_x^2 - E_y^2; \quad Q_2 = 2E_x E_y \cos\delta \tag{8.26}$$
$$Q_3 = 2E_x E_y \sin\delta; \quad Q_0 = E_x^2 + E_y^2$$

where δ is the phase difference between the two components, and Q_0 gives the
intensity of the wave. Notably, it holds

$$Q_0^2 = Q_1^2 + Q_2^2 + Q_3^2 \tag{8.27}$$

The polarization state of a fully polarized wave is represented by a point in the
Stokes space located on a sphere of radius Q_0 (the so called Poincaré sphere).
Waves of orthogonal polarization lay in antipodal points. In the case of partially
polarized waves in the definitions (8.26) the operation of *averaging* must be
added. In this case the relationship between parameters is

$$Q_0^2 \geq Q_1^2 + Q_2^2 + Q_3^2 \tag{8.28}$$

If the polarization has *completely random* components

$$Q_1 = Q_2 = Q_3 = 0 \tag{8.29}$$

Polarization and power are *jointly* represented by a four-vector $\mathbf{Q}$. And, scatterers or any object or component sensitive to polarization are represented by
the 4×4 Stokes-scattering-matrix $\underline{\underline{\Sigma}}$:

$$\underline{\mathbf{Q}}_s = \underline{\underline{\Sigma}}\,\underline{\mathbf{Q}}_{\text{in}} \tag{8.30}$$

where subscript **in** means the incident and s the scattered Stokes-vectors; the
underlining is applied to make distinction between fictive and true vectors (the
latters being represented with not underlined bold letters. Matrices in the Stokes
space are represented by double-underlined bold letters).

8.2.6.2   Polarization mismatch loss.   While the transmitter antenna
output field is completely polarized (either linearly or circularly), the polarization state of the received wave may significantly differ from this. In the case of
multipath due to multiple scattering it is very likely that polarization becomes

random. On the other hand the receiver antenna receives polarized waves, and this can cause polarization mismatch loss. To elaborate on this let us designate the Stokes vector of the receive antenna polarization by $\underline{q}$ and let us assume that the antenna represents an *ideal polarization filter* i.e. its cross-polar isolation is infinite. As it is shown [376] the Stokes scattering matrix of an ideal lossless polarization filter is

$$\underline{\underline{\Sigma}} = \frac{1}{2}\underline{q} \cdot \underline{q}^T \qquad (8.31)$$

with $\underline{q}$ the Stokes vector of the polarization state of this filter, superscript T meaning the transpose of a matrix or vector and for being specific we take component q_0 of $\underline{q}$ as $q_0 = 1$.

Based on eq.(8.27), (8.30), and (8.31), the received intensity can be given as

$$Q_{s0} = \frac{1}{2}\underline{q}^T \underline{Q}_{in} \qquad (8.32)$$

and the loss due to polarization mismatch as

$$L_P = \frac{Q_{in0}}{Q_{s0}} \qquad (8.33)$$

With eq.(8.32)-(8.33), the average polarization loss can be computed if the polarization of the received wave is known. If it is randomly polarized, its Stokes vector transpose is $(1, 0, 0, 0)$. Thus $Q_{out0} = Q_{in0}/2$ and the average polarization loss becomes 3 dB, not depending on the polarization of the receive antenna.

8.2.6.3   Probability distribution of the polarization loss.   As an example we assume that any polarization state of the received field is equally likely. Then $\underline{Q}_{in}$ is uniformly distributed over the Poincaré sphere, i.e. its probability density is:

$$p(2\gamma, 2\psi) = \frac{1}{4\pi} \qquad |2\gamma| < \frac{\pi}{2}, 0 \le 2\psi \le 2\pi \qquad (8.34)$$

with 2γ and 2ψ the spherical polar co-ordinates in the Stokes space.

In this case, due to the spherical symmetry the polarization mismatch loss distribution vs. $\underline{q}$ will be the same, whatever the actual $\underline{q}$ of the receiver antenna is. So for sake of simplicity circular antenna-polarization will be assumed. In this case according to (8.26)

$$\underline{q}^T = (1, 0, 0, 1) \qquad (8.35)$$

and to (8.32)

$$\frac{1}{L_p} = \frac{1}{2}(1 + \sin 2\gamma) \qquad (8.36)$$

Assuming infinite cross-polar isolation I_p, and applying (8.34) and (8.36) it turns out that $1/L_p$ is uniformly distributed in $(0, 1)$. In the case of finite cross polar isolation, $1/L_p$ is uniformly distributed between $1/I_p$ and 1.

8.2.6.4   Error probability: polarization diversity gain.   Applying the above result, the effect of random polarization on the probability of error can be determined. Take into consideration that in *one realization* the received signal energy is attenuated by *fading loss* and also by *polarization mismatch loss*. The conditional error probability, conditioned on these losses, can be written as

$$P_E\left(L_f, L_p\right) = f\left(\frac{E_s}{N_0} \times \frac{1}{L_f} \times \frac{1}{L_p}, \dots\right) \qquad (8.37)$$

with E_s/N_0 the symbol energy over noise spectral density with no fading and no polarization mismatch; L_f the loss due to fading; and the dots represent possible further quantities having effect on the error probability, although not relevant in our discussion.

The total probability of error can then be determined in the usual way. The probability distribution of $1/L_f$ is assumed to be known.

As an example of the effect of random polarization on error probability let us consider uniform distribution of $\mathbf{Q_{in}}$, the case of Rayleigh fading, AWGN, and binary PSK modulation. Note that in the present case the term *Rayleigh fading* can only be applied as a designation. Namely the signal energy is a sum of the squares of *four* independent Gaussian random variables (i.e. the real and imaginary parts of the – say – x and y components of the field). Thus, it has a chi-squared distribution with four degrees of freedom, rather than having exponential distribution, that is the case in true Rayleigh fading. As it is known [17], in this case the expression of error probability under the condition of matched polarization is

$$P_E(Z) = \frac{1}{4}\left(2 - 3\sqrt{\frac{Z}{1+Z}} + \sqrt{\left(\frac{Z}{1+Z}\right)^3}\right) \qquad (8.38)$$

where the designation Z is introduced for the *average* bit energy over noise spectral density.

Thus, for random polarization the total probability of error is

$$P_E(Z) = \frac{1}{2} - \frac{3}{4}\sqrt{1 + \frac{1}{Z}} + \frac{1}{4}\frac{1}{\sqrt{1+1/Z}} + \frac{3}{4Z}\frac{1}{\sqrt{1+1/Z}}$$

that simplifies for high Z to

$$P_E(Z) \approx \frac{1}{2Z} \qquad (8.39)$$

As in the case of Rayleigh-fading and polarization matching $P_E \approx 1/4Z$, eq.(8.39) shows that due to random polarization a 3 dB increase of the energy is needed.

In switching-type polarization diversity, we choose the higher one among the orthogonally polarized antennas. Then $1/L_p$ is uniformly distributed in $(1/2, 1)$, leading to

$$P_E \approx \frac{3}{2Z^2} \tag{8.40}$$

Given the error probability and the loss distribution, the diversity gain and the improvement ratio (IR) can be determined. Of course, the former depends on maximal error probability, $P_{E,out}$, where the latter on E/N_0, Z_{out} yielding $P_{E,out}$. Comparing eq.(8.39) and (8.40) we get for diversity gain

$$G_D = -5 \log P_{E,out} - 4\text{dB} \tag{8.41}$$

and for high Z

$$IR = \frac{Z}{2Z_{out}} \tag{8.42}$$

8.2.6.5 Polarization diversity: Numerical results. *Random polarization* in heavily multipath conditions was shown in a measuring campaign (see [59] for a detailed description). A 1.6 GHz transmitter was placed on board a helicopter (simulating the satellite) and a computer-controlled receiver was placed in the center of a hall on the sixth floor of the seven-storey office building. While being close to the window, the difference in the received signals did more or less correspond to the gain and polarization of the antennas, at the hall center it did not depend on antenna characteristics.

Concerning polarization *diversity*, a few measurement and simulation results are available in the literature. A general experience is that polarization diversity performs approximately as well as space diversity. In the literature, very few researchers deal with comparing spatial and polarization diversity – being an essential point in the present discussion. The study in [377] investigates, via simulation, DS-CDMA uplink performance with and without interference cancellation. In both cases gain of space diversity with matched transmit-receive antenna polarizations was 1.5-3 dB higher than that of polarization diversity. [378] got similar results (experimentally, in various environments) with *matched* polarization antennas (horizontal or vertical). They investigated also the case with one of the antennas being *inclined*. In that case there was a constant additional loss both in the space diversity and in polarization diversity, the latter one being lower. This seems to prove the fact that in none of their environments was received field polarization completely random.

8.2.7 Adaptive Coding and Modulation

In conventional communication systems the communication link is designed according to the worst channel conditions to provide required reliability of the transmission. However, due to the channel variations, the transmitter and receiver are not always optimized for current channel conditions, and thus fail to exploit the full potentiality of the satellite channel. Several adaptive transmission techniques [379], [380], [381], [174], [382], [28], which adjust the coding and modulation scheme to the instantaneous channel condition, have been studied recently to increase the system reliability and throughput in the communication systems with time varying channel. The approach is known as the Adaptive Coding and Modulation (ACM). These Fade Mitigation Techniques (FMTs) could be split into Adaptive Coding (AC), Adaptive Modulation (AM) and Data Rate Reduction (DRR).

The introduction of redundant bits to the information bits when a link is experiencing fading, allows detection and correction of errors (i.e. Forward Error Correcting (FEC) techniques, as described in Chapter 4) caused by propagation impairments and leads to a reduction of the required energy per information bit. AC consists in implementing a variable coding rate matched to impairments originating from propagation conditions.

Higher system capacity for a given bandwidth can be achieved with spectral efficient modulation schemes depending on the available power link budget (clear sky conditions). As AC, the aim of AM is to decrease the required energy per information bit required corresponding to a given BER, which translates into a reduction of the spectral efficiency as SNR decreases. The reduction of the spectral efficiency is the result of the use of lower level modulation schemes.

Further, energy per information bit reduction can be obtained by a decrease of the information data rate at constant BER. This technique is called DRR. Here, user data rates should be matched to propagation conditions: nominal data rates are used under clear sky conditions, whereas reduction is introduced according to fade levels. A first method consists in using spreading sequences [383]. In that case, the used bandwidth remains constant. The main limitation of this approach is that the link throughput is strongly reduced. An alternate way to limit this drawback is to match the bandwidth to the adapted useful data rate, in order to either maintain or less reduce the throughput [384]. In the latter case, resource management is more complex.

Moreover, the ACM schemes can be classified into two groups:

- In *pre-estimated* ACM schemes, the modulation type, burst structure and data transmission rate are assigned at call setup, and the transmission parameters do not adapt to the variation of the channel characteristic during the connection.

- In *dynamic* ACM schemes, the modulation parameters are controlled slot-by-slot and can be changed adaptively during the connection.

The dynamic ACM scheme, the more suitable for time varying channel, is usually combined with MAC layer algorithm, for example with packet reservation multiple access (PRMA) [385]. Suppose that two signals with spectral efficiency of 1 and 2 bits per symbol are used in the adaptive communication system. In case of a good channel, two bits are transmitted in one symbol interval, while in case of a bad channel only one bit per symbol interval is transmitted. In order to obtain the constant data throughput required for some telecommunication services, only one time slot in the frame is reserved when the signal with two bits per symbol is transmitted. Conversely, two time slots are required in the frame when signals with only one bit per symbol is transmitted.

The availability of CSI at the transmitter is needed to implement the adaptive transmission methods. CSI can be estimated only from the received signal. In the case of Time Division Duplex (TDD)), downlink and uplink signals exhibit the same radio channel. Consequently the downlink transmission mode can be selected on the basis of the uplink received signal and viceversa. Conversely, in Frequency Division Duplex (FDD) systems, the downlink and uplink signals exhibit different radio channels, and, therefore, either the complete CSI or only transmission mode, i.e. the modulation and coding scheme and transmit power, have to be sent from the receiver to the transmitter via a return channel. Long propagation delays in satellite systems restrict the usage of TDD mode. The propagation delay is a serious limiting factor for ACM implementation in satellite systems with FDD, but for slowly time-varying channels and with efficient prediction techniques ACM can still be applied. Figure 8.16 shows an example of the FDD ACM mobile satellite system. At the satellite segment, the uplink channel coding and modulation scheme is selected by observing the uplink channel, while in the land segment the downlink channel coding and modulation scheme is determined from the received signal.

As already anticipated, in order to select the transmission mode for the next frame or burst, the channel state shall be estimated in ACM systems. Various criteria have been proposed for channel estimation, namely:

- Received Signal Strength Indicator (RSSI) [174], [381],

- eye closure [380],

- bit error rate calculated from the Bose-Chaudhuri-Hocquenghem (BCH) block codes [174],

- Euclidean distance,

- delay spread [386].

The RSSI approach is widely used because of its simplicity and accuracy for flat fading quasi static radio channels. Other more complicated methods can be applied to frequency selective fading. The studies with flat fading quasi static channels [174, 381] assume the constant amount of noise generated at the

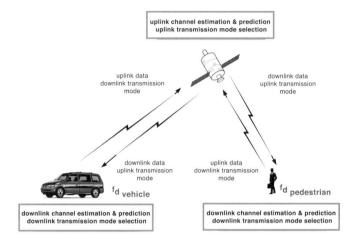

Figure 8.16.   Land Mobile Satellite system with Adaptive Coding and Modulation.

receiver. The coding and modulation scheme is accordingly selected from an available set considering the achievable performance of the modulation scheme in Gaussian channel. The results plotted in Figure 8.17 illustrate the procedure of determining the thresholds and ranges for transmission different coding and modulation schemes.

The thresholds for switching between two coding and modulation schemes complying the constant BER criterion is obtained by drawing horizontal line at the target BER value. The line intersects the BER curves. The projections of the intersection points on the abscissa define the thresholds for coding and modulation schemes interchanging. The most robust coding and modulation scheme ($r = 1/4$ convolutional encoded, QPSK modulation) is used when the SNR is between the projection of the first and the second intersection point. The information is carried by the second coded modulation scheme ($r = 1/2$ convolutional encoded QPSK) when SNR is between the second and the third projection of intersection points. The coding modulation scheme with the highest spectral efficiency (16-QAM) is used for SNR higher than the projection of the last intersection point on abscissa. When the SNR ratio is lower than the first intersection point, the target BER cannot be achieved and no information is transmitted. The BER curves have been generated by computer simulations.

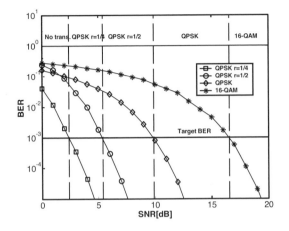

Figure 8.17. Properties of the transmitted coding and modulation schemes and switching thresholds.

8.2.7.1 An ACM scheme for the narrow band LMS system. This subsection presents numerical results of ACM techniques performance compared to conventional approaches for the narrowband LMS system.

In LMS systems, the efficient use of the available spectrum can be achieved by adapting coding and modulation scheme to the propagation conditions. A two state narrowband channel model [387], [45] is widely used to model the LMS narrowband channel (see Section 3.2.2). The channel is in good state when the signal does not exhibits shadowing and in bad channel state when the satellite is not visible from the mobile terminal.

The signal reflection from the surrounding obstacles is modelled as a complex Gaussian process. When the received signal exhibits no shadowing, the fading coefficients obey to the Rice distribution, where K is Rice factor. When no LOS exists, the multipath fading has a Rayleigh distribution. Slow channel variations are modelled by multiplying the Rayleigh distributed coefficients by coefficients which obey to the lognormal distribution with parameters μ and σ, where μ is mean power level and σ^2 is the variance of power level due to shadowing (see Section 3.2.2). The resulting distribution is Rayleigh/lognormal distribution. The time, when either Rice or Rayleigh/lognormal distribution is used, is determined by a two-state Markov model. The typical parameter values for city and rural environments can be found in [387, 45]. The propagation conditions are worse in city environment with high buildings, which causes more severe mobile terminal shadowing than in the highways where the mobile terminal is rarely shadowed, but the mobile terminal velocity is higher.

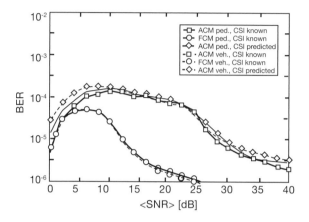

Figure 8.18.   Average BER of the satellite system with ACM.

ACM is applied to the system, which is similar to Shiron [388] adaptive rate VSAT modem in L-band. Depending on channel properties the following data rates are available: 0 kbits/s, 64 kbits/s 128 kbits/s 256 kbits/s and 1 Mbits/s. The duration of a time slot is 1000 symbols. Computer simulations were performed for city and rural environment at pedestrian speed of 1m/s and vehicle speeds of 30m/s. Two scenarios were simulated. In the first one the channel characteristic is perfectly known at the transmitter side, while in the second one the channel characteristic is predicted by linear predictor. In both cases the Fixed Coded Modulation (FCM) and ACM scheme has a BER below the target value. At low SNR the throughput of the ACM scheme is similar to FCM; while at high SNR, ACM outperforms FCM. Comparing the results for rural and urban (city) areas, in spite of difference in mobile speed, the system throughput in rural area is higher than in urban area. When the channel state information is predicted for next transmissions by the linear predictor, no significant difference in average BER and average system throughput is observed. The simulation results for rural area are plotted in Figure 8.18 and 8.19.

Concluding, the ACM seems to be a promising method for better exploitation of the available resources of the mobile radio channel even in a satellite system for the vehicle users in rural environment and the pedestrians at the urban environment. The efficient prediction technique is necessary for channel prediction even for slow varying mobile channels.

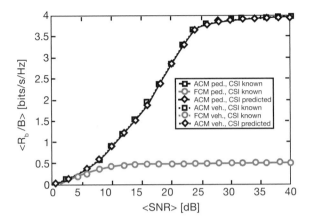

Figure 8.19. Average system bandwidth efficiency of the satellite system with ACM.

8.3 Blockage/Multipath mitigation

In satellite systems, the propagation path is a very important factor that mainly affects channel quality. In land mobile satcoms, perhaps the most serious propagation problem is the blockage effect, caused by buildings and surrounding objects, which cause signals from the satellite to shut down completely. The second problem is multipath fading, which causes amplitude and phase fluctuations, and time delay in the received signal, leading to the degradation of system performance. To face the undesired effects of blockage/multipath in satcom systems, several mitigation techniques are proposed:

- Joint fading mitigation techniques,

- Intermediate Module Repeater (IMR) multipath diversity and combining schemes.

8.3.1 Joint fading mitigation techniques

The use of FMTs aims at matching in real time the link budget to the propagation conditions through some specific parameters such as power, data rate, coding, etc. However, this real time adaptivity has an impact not only on carrier-to-noise ratio, but also on carrier-to-interference ratios. Both aspects have therefore to be carefully studied.

Several publications have been written up to now on the subject [389], [390], [391], [392], [393] and a review of FMTs has been realized in the framework of COST 255 [70]. As just previously described in Section 8.2, FMT at physical layer can be divided into Power Control (PC), transmitting power level fitted to propagation impairments, Adaptive waveform, compensating fade by a more efficient modulation and coding scheme, and Layer 2 techniques, coping with the temporal dynamics of the fade.

PC, adaptive waveform and layer 2 FMTs are taking advantage of unused in-excess resource of the system, whereas diversity FMT implies a re-routing strategy. Techniques sharing unused resource aim to compensate fading occurring on a given link in order to maintain or to improve the link performance (required C/N_0). Diversity techniques maintain link performance by changing the frequency band or the link geometry.

8.3.1.1   Power Control for fade mitigation.   As just stated in section 8.2.5, four types of carrier PC FMTs can be considered: Up-Link Power Control (ULPC), End-to-End Power Control (EEPC), Down-Link Power Control (DLPC) and On-Board Beam Shaping (OBBS).

With ULPC, the output power of the transmitting Earth station is matched to uplink impairments. Transmitter power is increased to counteract fade or decreased when favorable propagation conditions are available so as to limit interference in clear sky conditions and therefore to optimism satellite capacity. In the case of transparent payloads , ULPC can prevent from reductions of satellite EIRP caused by the decreased uplink power level that would occur in the absence of ULPC, if no on-board Automatic Level Control (ALC) is implemented.

EEPC can be used for transparent payload configuration. Thanks to the dependency of the downlink budget with the uplink budget, the adjustment of output power of the transmitting Earth station can be used to mitigate impairments occurring on either or both uplink and downlink. In the case of regenerative repeaters, up and down links budgets are independent, so the concept of EEPC can not exist anymore. EEPC aims at keeping a constant margin on the overall link budget. As for ULPC, transmitter power is increased to counteract fade or decreased when more favorable propagation conditions are recovered to limit interference and optimism satellite capacity.

With DLPC, the on-board channel output power is adjusted to the magnitude of downlink attenuation. DLPC aims to allocate a limited extra-power on-board in order to compensate a possible degradation in term of down-link C/N_0 due to propagation conditions on a particular region. In this case, all Earth stations in the same spot beam benefit from the improvement of EIRP.

OBBS technique is based on active antennas, which allows gain/power in specific beams of a multibeam antenna to be adapted to propagation conditions. Actually, the goal is to increase the selected beam EIRP, with the objective to compensate rain attenuation only on beam coverage where rain is occurring.

8.3.1.2   Layer 2.   FMT at layer 2 level are techniques which do not aim at mitigating a fade event but instead rely on the re-transmission of the message. Two different techniques can be envisaged at layer 2 : Automatic Repeat reQuest (ARQ) and Time Diversity (TD). With ARQ, the message is resent until the message reaches successfully the receiver. ARQ with a random or predefined time repetition protocol are alternate solutions.

Time Diversity can be considered as a FMT that aims to re-send the information when the state of the propagation channel allows to get through. This technique can be considered when there is no need to receive the data file in real time and it is acceptable for the user point of view to wait for the end of the propagation event (in general some tens of minutes) or during a decrease of traffic. This technique benefits from the use of propagation mid-term prediction model in order to estimate the most appropriate time to re-sent the message without repeating the request.

8.3.1.3   Joint FMTs.   Previous works performed in the frameworks of [394] and [70] pointed out two important aspects for new systems operating at Ka and Q/V-bands:

- On the one hand, it appears that each FMT is more or less adapted to a specific range of availability. Then these fade mitigation methods are quite complementary and can be implemented simultaneously (joint FMT) to extend the availability range of requirements.

- On the other hand, such methods implemented individually can only relatively compensate small propagation impairments. It will be possible to regularly improve the performance of the mitigation to a large extent, by carrying out a combination of different kinds of FMTs when there is a need to transmit high priority information.

8.3.2   IMR multi-path diversity

With reference to Chapter 4 and 10, satellite systems are promising candidates in the provisioning of Digital Multimedia Broadcast (DMB) services, as well as of a fast downlink channel for the download of Internet content. The main reason to support the use of satellite for these services is the better efficiency of satellites in the delivery of the same content over a vast service area. However, for the economical success of these services, it is essential that the most of potential users can be actually reached, i.e. that the satellite signal reception be possible even in large cities and densely populated areas, where the presence of buildings greatly increases the link obstruction probability, i.e. the absence of a LOS link between the satellite and the User Equipment (UE). This problem is exacerbated in the case of GEO satellites (usually preferred for economical reasons too). To this regard, it is worth reminding that at high

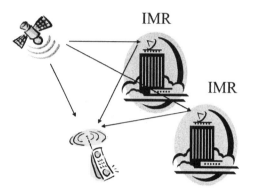

Figure 8.20.   Direct and indirect (through IMRs) satellite reception.

latitudes, such as in central Europe as well as in North America, where most of potential users are located, the elevation angle of GEO satellites is quite low, increasing the probability of satellite link obstruction in presence of buildings other obstacles.

To overcome this problem, the introduction of IMRs, or Gap Fillers, seems a viable and promising choice, as illustrated in Figure 8.20. By providing the user equipments (UEs) with boosted local replicas of the satellite signal, IMRs can improve the coverage of densely populated areas, as well as being indispensable to permit the satellite signal reception inside buildings, cars, ships and

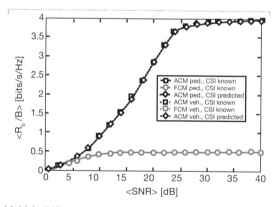

Figure 8.21.   Multiple IMRs reception: example of power delay profile. Hexagonal layout, Lat. 51 North (Central Europe) distance from the reference cell center (normalized to the cell radius) d=0.86, chip rate $R_c = 3.84$ Mchip/s).

other means of transport. However, if on one hand IMRs help the reception by strengthening the satellite signal, on the other hand, as a side effect they introduce a sort of "artificial" multipath, whose impact on the signal reception cannot be neglected, especially when dealing with the coverage of a densely populated area, where it is likely that the signal is received through multiple IMRs. In particular, as both IMRs and terrestrial multipath introduce many replicas of the satellite signal, the number of signal components and the amount of delay spread are largely increased. In order to exploit this additional multipath, the adoption of a Direct Sequence Spread Spectrum rake receiver is clearly instrumental. Provided that the multipath components are resolvable and there are enough fingers in the rake receiver to collect them, an increased level of diversity is actually obtained. In the opposite case, multipath components result in an additional interference, with detrimental effects. A preliminary analysis of the impact of IMRs on the power delay profile of a satellite downlink reception is reported in [395]. Results highlight the influence on the received signal of many factors, included the latitude of the serviced areas, the coverage layout, the IMR coverage radius and the UE location. An example of the received power delay profile is reported in Figure 8.21, distinguishing the contributions of the direct reception (i.e. from the satellite) and the indirect one (through seven different IMRs, a reference one plus those belonging to the surrounding six hexagonal cells). The complexity of the achieved power delay profile, as well as the possibility to take advantage of the increased diversity introduced by IMRs, is apparent.

8.3.3   Combining Techniques

Diversity combining is one of the most practical, effective and widely used technique in digital communications receivers for mitigating the effects of multipath fading and improving overall wireless systems performance. The most popular diversity techniques are EGC , Maximal-Ratio Combining (MRC) , Selection Combining (SC) and a combination of MRC and SC, called Generalized-Selection Combining (GSC) . The performance of EGC diversity receivers has been extensively studied in the open technical literature for several well-known fading statistical models, such as Rayleigh and Nakagami-m assuming independent or correlative fading (see [396], [397], [398], [399], [400], [401], [402], [403] and references therein). However, the performance of EGC receivers operating over Weibull, Nakagami-n (Rice) and Nakagami-q (Hoyt) fading channels, has not yet received as much attention as the Rayleigh and Nakagami-m fading channels, mainly due to the complex form of their pdfs, despite the fact that these models exhibit an excellent fit to experimental fading channel measurements for land mobile terrestrial and satellite telecommunications. More specifically, the Weibull model, exhibits an excellent fit to experimental fading channel measurements, for both indoor, as well as for outdoor environments [9], [10], [11], [12]. Nakagami-n (Rice) distribution [404] contains Rayleigh distribution as a special case and provides the optimum fits to collected data in indoor [405], outdoor [406] and mobile satellite applications [407]. Nakagami-q (Hoyt) [396] distribution is normally observed on satellite links subject to strong ionospheric scintillation and ranges from one-sided Gaussian fading to Rayleigh distribution [408], [409]. Next, we call Rice and Hoyt the two fading models under investigation.

8.3.3.1   Moments of the Output SNR for the EGC receiver.   We consider an L-branch EGC receiver with statistically independent but not necessarily identically distributed input branches, operating in a flat fading environment. Such a channel model covers the case of antenna diversity, where the input channels tend to be identically distributed as well as multipath diversity frequency-selective fading channels where the input Power Delay Profile (PDP) tends to be non-uniform. The output SNR, γ_{out}, of the receiver is given by

$$\gamma_{out} = \frac{E_s}{L\,N_0} \left(\sum_{i=1}^{L} a_i \right)^2 \tag{8.43}$$

where a_i is the envelope of the i-th input path, modelled as Rice, Hoyt or Weibull rv, E_s is the symbol energy and N_0 is the one-sided power spectral density of the Additive White Gaussian Noise (AWGN).

By definition, the nth moment of the output SNR is

$$
\begin{aligned}
\mathrm{E}\left[\gamma_{out}^n\right] &= \mathrm{E}\left[\left(\frac{E_s}{L\,N_0}(a_1 + \cdots + a_L)^2 \right)^n \right] \tag{8.44} \\
&= \left(\frac{E_s}{L\,N_0} \right)^n \mathrm{E}\left[(a_1 + \cdots + a_L)^{2n} \right].
\end{aligned}
$$

Expanding the term $(a_1 + \cdots + a_L)^{2n}$, using the multinomial identity [179, eq. (24.1.2)], (8.44) can be written as

$$\mathrm{E}\left[\gamma_{out}^n\right] = \left(\frac{E_s}{LN_0} \right)^n (2n)! \sum_{\substack{k_1,\ldots,k_L=0 \\ k_1+\cdots+k_L=2n}}^{2n} \left\{ \frac{\mathrm{E}\left[a_1^{k_1} \cdots a_L^{k_L} \right]}{\prod_{j=1}^{L} k_j!} \right\} \tag{8.45}$$

Assuming that the branches of the EGC are uncorrelated and in terms of the instantaneous SNR of each diversity path, $\gamma_i = a_i^2 E_s / N_0$, (8.45) can be rewritten as

$$\mathrm{E}\langle\gamma_{out}^n\rangle = \frac{(2n)!}{L^n} \sum_{\substack{k_1,\ldots,k_L=0 \\ k_1+\cdots+k_L=2n}}^{2n} \left[\prod_{j=1}^{L} \frac{\mathrm{E}\left\langle \gamma_j^{k_j/2} \right\rangle}{k_j!} \right]. \tag{8.46}$$

When the receiver operates over Rice fading channels, the SNR of each diversity path is distributed according to a noncentral chi-square distribution. Using the definition for the moments of a noncentral chi-square rv [396, eq. (2.18)] with (8.46), the moments of the EGC output SNR can be written in a simple and closed-form expression given by

$$E\left[\gamma_R^n\right] = \frac{(2\,n)!}{L^n} \tag{8.47}$$

$$\times \sum_{\substack{k_1,\dots,k_L=0 \\ k_1+\dots+k_L=2\,n}}^{2\,n} \left[\prod_{j=1}^{L} \frac{\overline{\gamma}_j^{k_j/2}\,\Gamma(1+k_j/2)}{k_j!\,(1+K_j)^{k_j/2}} \times {}_1F_1\left(-\frac{k_j}{2},1;-K_j\right) \right]$$

where $\overline{\gamma}_j = \Omega_j\,E_s/N_0$ is the average SNR per symbol of the i-th branch with $\Omega_j = a_j^2$, $\Gamma(\cdot)$ is the Gamma function [179, eq. (6.1.1)], ${}_1F_1(\cdot,\cdot;\cdot)$ is the confluent hypergeometric function of the first kind [179, Ch. (13)], and K_j is the Rice factor of the jth input path, defined as the ratio of the signal power in dominant component over the scattered power. For $K_j = -\infty$ (dB) the Rayleigh fading is described, while $K_j = \infty$ (dB) represents the no-fading situation. Values of Rice factor in land mobile terrestrial (outdoor and indoor) and satellite applications usually range from $0-12$ dB [406]. Into the following and without loss of generality we assume that the Rice factor takes the same value for all diversity paths, i.e. $K_1 = \dots = K_L = K$.

When the receiver operates over Hoyt fading channels, the moments of the output SNR can be obtained substituting the moments of the input paths SNRs [396, eq. (2.13)] in (8.46) resulting in

$$E\left[\gamma_H^n\right] = \frac{(2\,n)!}{L^n} \tag{8.48}$$

$$\times \sum_{\substack{k_1,\dots,k_L=0 \\ k_1+\dots+k_L=2\,n}}^{2\,n} \left[\prod_{j=1}^{L} \frac{\overline{\gamma}_j^{k_j/2}\,\Gamma(1+k_j/2)}{k_j!} \, {}_2F_1\left(-\frac{k_j-2}{4},-\frac{k_j}{4},1;\left(\frac{1-q_j^2}{1+q_j^2}\right)^2\right) \right]$$

where ${}_2F_1(\cdot,\cdot;\cdot;\cdot)$ is the Gauss hypergeometric function [179, eq. (15.1.1)] and q_i is the Nakagami-q fading parameter of the i-th branch, which ranges from 0 (one-sided Gaussian fading) to 1 (Rayleigh fading). Again, we will assume into the following, without loss of generality, that q_i takes the same value for all diversity paths, i.e. $q_1 = \dots = q_L = q$.

For the Weibull fading case, the moments for a_i are given by

$$E\left[a_i^n\right] = \omega_i^n\,\Gamma(d_n) \tag{8.49}$$

while $E\langle\gamma_i^n\rangle$ is obtained as

$$E\gamma_i^n = \frac{\Gamma(d_{2n})}{\Gamma^n(d_2)}\,\overline{\gamma}_i^n \tag{8.50}$$

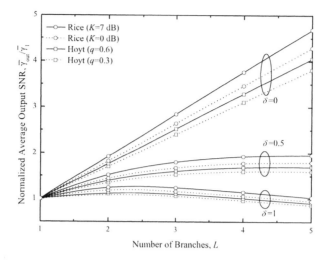

Figure 8.22. First branch normalized average output SNR of EGC versus L, for Rician and Hoyt fading with exponentially decaying PDP.

due to the fact that the instantaneous SNRs follow also a Weibull pdf [410], where β and ω_i are the fading and scaling parameters of the distribution, respectively and $\overline{a_i^2}$ is the average power of fading. The scaling parameter is connected with the average power of fading as $\omega_i = \sqrt{\overline{a_i^2}/\Gamma(d_2)}$, where $d_\tau = 1 + \tau/\beta$, τ is a real constant. Substituting (8.50) into (8.46) the moments for the output SNR for the Weibull case are given by

$$E\langle\gamma_{out}^n\rangle = \frac{(2n)!}{\Gamma^n(d_2)\,L^n} \sum_{\substack{k_1,\dots,k_L=0 \\ k_1+\cdots+k_L=2n}}^{2n} \prod_{j=1}^{L} \frac{1}{k_j!}\,\Gamma(d_{k_j})\,\overline{\gamma_j}^{\frac{k_j}{2}}. \qquad (8.51)$$

The moments of the output SNR can be used to study significant performance criteria such as the average output SNR and the Amount of Fading (AoF) at the output of the receiver. Also, using the moments, the outage and the error performance are studied, approximating the moments generating function (mgf) with the Padé approximants theory.

8.3.3.2   Average Output SNR and AoF for the EGC receiver.   Using (8.47), (8.48) and (8.51) for $n = 1$ the average output SNR is obtained in closed-form for the Rice, Hoyt and Weibull fading case.

Assuming that the receiver operates with an exponentially decaying PDP $\left(\overline{\gamma}_i = \overline{\gamma}_1\,e^{[-\delta(i-1)]}\right)$, Figure 8.22 plots the first branch normalized average output SNR of EGC, as a function of L, for Rice ($K = 0$ dB and $K = 7$ dB) and Hoyt fading ($q = 0.3$ and $q = 0.6$) and several values of the power decay factor δ.

The AoF is a unified measure of the severity of fading, which is typically independent of the average fading power and is defined as [396]

$$AoF \triangleq \frac{\text{var}\,(\gamma_{\text{out}})}{\overline{\gamma}_{out}^2} = \frac{\text{E}\left[\gamma_{out}^2\right]}{\overline{\gamma}_{out}^2} - 1. \qquad (8.52)$$

Using (8.47), (8.48) and (8.51) for $n = 1$ and $n = 2$ the AoF of the EGC receiver can be evaluated for Rician, Hoyt and Weibull fading, respectively.

8.3.3.3   Average Symbol Error Probability.

The mgf based approach presented in [396, Ch. 1] is a unified method to calculate error rates for several modulation schemes. However, for the EGC receiver of interest, a useful expression for the mgf of the output SNR is not available. For this reason, we propose in this paper the Padé approximants theory as an alternative and simple way to approximate the mgf. For the reader's convenience we explain briefly how the Padé approximants theory can be applied, in order to find an accurate rational approximation to the mgf of the output SNR, $\mathcal{M}_{\gamma_{out}}(s)$. By definition, the mgf is given by [1]

$$\mathcal{M}_{\gamma_{out}}(s) \triangleq \text{E}\left[\exp\{s\,\gamma_{out}\}\right] \qquad (8.53)$$

and can be represented as a formal power series (e.g. Taylor) as

$$\mathcal{M}_{\gamma_{out}}(s) = \sum_{n=0}^{\infty} \frac{1}{n!}\,\text{E}\left[\gamma_{out}^n\right]\,s^n. \qquad (8.54)$$

We cannot conclude that the power series in eq.(8.54) has a positive radius of convergence and where or whether it is convergent. To overcome this problem, the Padé approximants theory [411] is proposed, as a simple and alternative way to approximate the mgf. A Padé approximant, is that rational function approximation to $\mathcal{M}_{\gamma_{out}}(s)$ of a specified order B for the denominator and A for the nominator, whose power series expansion agrees with the $A + B$ order power expansion of $\mathcal{M}_{\gamma_{out}}(s)$, i.e.

$$R_{[A/B]}(s) \equiv \frac{\sum\limits_{i=0}^{A} a_i\,s^i}{1 + \sum\limits_{i=1}^{B} b_i\,s^i} = \sum_{n=0}^{A+B} \frac{\text{E}\left[\gamma_{out}^n\right]}{n!} s^n + O\left(s^{N+1}\right) \qquad (8.55)$$

with $O\left(s^{N+1}\right)$ being the remainder after the truncation. Hence, the first $(A + B)$ moments need to be evaluated in order to construct the approximant

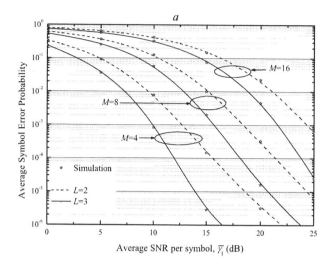

Figure 8.23. Error performance of M-DPSK in a Rician fading environment with $K = 7$ dB.

$R_{[A/B]}(s)$. Next, $\mathcal{M}_{\gamma_{out}}(s)$ is approximated using sub-diagonals Padé approximants $\left(R_{[A/A+1]}(s)\right)$, since it is only for such approximants that the convergence rate and the uniqueness can be assured [411]. With the aid of Padé approximants the error-rate expressions can be calculated directly for non-coherent and differential binary signaling (non-coherent BFSK, DBPSK), since for all other cases single integrals with finite limits and integrands composed of elementary functions have to be readily evaluated via numerical integration.

Into the following, some numerical results are presented to illustrate the proposed mathematical analysis. Figure 8.23 plots the Average Symbol Error Probability (ASEP) of 4-DPSK, 8-DPSK, and 16-DPSK, employing EGC over Rice fading with $K = 7$ dB for $L = 2, 3$. In the same figure, computer simulation results are also plotted in order to check the accuracy of the proposed Padé approximants approach. As it is clear, a very good match between computer simulation and analytical results is observed.

8.3.3.4   Outage Probability.   In addition to the average error rate, outage probability is another standard performance criterion of communication systems operating over fading channels. It is defined as the probability that the combined SNR, γ_{out}, falls below a specified threshold γ_{th}, i.e. [396, Ch. 1]

$$P_{out} = F_{\gamma_{out}}(\gamma_{th}) = \mathcal{L}^{-1}\left[\mathcal{M}_{\gamma_{out}}(s)/s\right]\Big|_{\gamma_{th}} \qquad (8.56)$$

where $F_{\gamma_{out}}(\cdot)$ and $\mathcal{L}^{-1}(\cdot)$ denote in our case the cumulative distribution function (cdf) of the EGC output SNR and the inverse Laplace transform, respectively. Due to the Padé rational form of $\mathcal{M}_{\gamma_{out}}(s)$,

$$\mathcal{M}_{\gamma_{out}}(s) \cong \frac{\sum_{i=0}^{A} a_i s^i}{1 + \sum_{i=1}^{B} b_i s^i} = \sum_{i=0}^{B} \frac{\lambda_i}{s + p_i} \tag{8.57}$$

and using the residue inversion formula, the outage probability can be easily evaluated from (8.56) as

$$P_{out} = \sum_{i=1}^{B} \frac{\lambda_i}{p_i} \exp\{-p_i \gamma_{th}\} \tag{8.58}$$

where p_i and λ_i are the poles and the residues of the approximant, respectively.

8.4   Atmospheric effects mitigation: Implementation issues

To implement Fade Mitigation Techniques (FMTs), it is necessary to monitor in real time the dynamic behavior of the propagation channel. This subsection discusses the way of detecting and quantifying a possible fade and the best location to make the decision (Earth Station (ES), satellite, or Network Control Center (NCC)). The way the signalling information is disseminated to the system components which play an active role in the mitigation (transmitting Earth station, satellite, control station, etc.) is also of great importance.

From the operational point-of-view, a key issue is to evaluate when and for how long time an outage of the system due to propagation conditions is going to happen. More precisely, it implies to be able to detect a propagation event and to quantify its depth in order to estimate if the current system mode (or margin) is going to be sufficient or not. It is therefore necessary to evaluate the depth of the event from a measurement performed on the particular link affected by this event. Whereas for transparent repeater the ground segment (either the transmitting or the receiving Earth station) is the most appropriate to carry out this function, for regenerative repeaters this function could be carried out onboard also.

However, in satellite communication systems, three kinds of phenomena impact the performance of the physical layer: interference contributions (internal or external to the considered system, atmospheric propagation impairments, or hardware issues such as satellite antenna gain roll-off and mispointing as well as RF chain degradations. When it comes to operation of satellite systems, it is assumed that hardware issues are dealt with separately. It is therefore necessary to separate propagation impairments and interference contributions, as well as uplink and downlink propagation impairments.

From this assumption, three kinds of detection schemes can be designed [412], [413], [414], [415]:

- open-loop: assessment of propagation impairments from direct measurements,

- closed-loop: Quality of Service (QoS) estimation from the monitoring of the physical layer performance at the receiving side,

- hybrid-loop: combination of open-loop and closed loop detection schemes.

In complement to these detection schemes, the decision function can be either distributed (e.g. for mesh networks of gateways) or centralized (e.g. for star networks of user terminals), whether the detection is performed at the ground segment level or on-board the satellite [416].

8.4.1   Distributed Detection and Decision schemes

In distributed Detection and Decision (D&D) schemes, the decision is directly performed by the Earth Stations. This configuration is therefore possible if the Earth Stations have been allocated some amount of autonomous operation, which is not the case for low-cost terminals. This scheme may concern mainly gateway ESs.

An illustration of this principle is given in Figure 8.24 for open-loop, closed-loop and hybrid-loop D&D schemes. In such distributed D&D schemes, the measurement is performed in a first step:

- either by both transmitter and receiver Earth Station (for instance from the measurement of the attenuation on a downlink beacon) in open-loop,

- or only by the receiver Earth station (from a measurement of the degradation of the received communication signal) in closed-loop,

- or both by the transmitter Earth Station (downlink beacon) and the receiver Earth station (communication signal) in hybrid-loop.

In a second step, the measurement carried out by the receiver Earth station is sent to the transmitter Earth station. In a third step, the decision is taken by the transmitter Earth station on a possible selection of a different FMT mode.

This configuration is therefore possible only if the user Earth station has their relevant level of autonomy, which is not the case for low-cost terminals and concerns mainly gateway Earth stations.

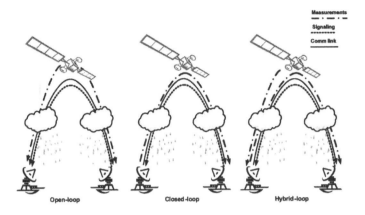

Figure 8.24.   Example of distributed D&D schemes.

8.4.2   Centralized Detection and Decision schemes

In centralized D&D schemes, the decision is performed by a central component of the system (for example the NCC), which is the only component allowed to take decisions on the Earth stations mode of operation. The D&D scheme is different according to the satellite payload configuration (regenerative or transparent).

Transparent payload. With transparent payload, the detection can be carried out either by each system component (User Earth Station (UES), Gateway Earth Station (GES) and NCC) or by the NCC only. As far as decision process is concerned, it can be either distributed between the gateways or concentrated in the NCC.

An illustration of this principle for the transparent payload configuration is given in Figure 8.25) for open-loop, closed-loop and hybrid-loop D&D schemes. In such distributed schemes, the detection is performed in a first step:

- either by UES and GES and by the NCC (for instance from the measurement of the attenuation on a downlink beacon) in open-loop,

- or only by the NCC (from a measurement of the degradation of the received communication signal) in closed-loop,

- different types of hybrid-loop schemes can be designed, depending whether a downlink beacon is monitored only by the NCC (as in Figure 8.25), or by the UES and the GES too, as for the distributed D&D hybrid-loop scheme.

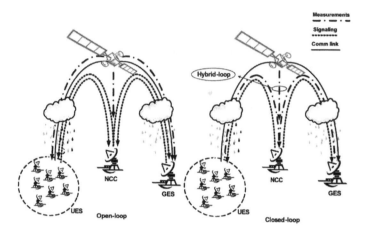

Figure 8.25.   Example of centralized D&D schemes for transparent payloads.

In all these configurations, the D&D functions can be implemented either in the NCC or in the GES. In the latter, the amount of signalling is lower because there would be no need to have exchanges between Gateways and NCC.

Then, the decision process is carried out in three steps:

- firstly, the measurements collected by UES and GES are sent to the v (needed for open-loop and hybrid-loop only),

- secondly, from the measurements at its disposal, the NCC takes a decision about a possible adaptation of the link budgets,

- thirdly, the NCC notifies its decision by sending a command to all the concerned Earth stations (UES and GES).

Regenerative payloads. With regenerative payloads, uplink detection can be performed at satellite level and downlink detection at user Earth station (downlink) level. As far as decision is concerned, it can be taken either by the satellite or by the NCC.

An illustration of this principle for the regenerative payload configuration is given in Figure 8.26 for open-loop and closed-loop D&D schemes. In such distributed schemes, the detection is performed in a first step:

- either by both transmitter and receiver Earth station (for instance from the measurement of the attenuation of a downlink beacon) in open-loop,

- or by the satellite (from measurements of the demodulated carriers) in closed-loop,

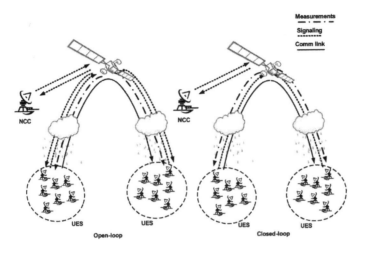

Figure 8.26. Example of centralized D&D schemes for regenerative payloads.

■ or both by transmitter and receiver Earth stations (downlink beacon) and by the satellite (on-board demodulated carriers) in hybrid-loop.

Then, the decision process is carried out in three steps:

■ firstly, the measurements collected by UES are sent to the NCC via the satellite (needed for open-loop and hybrid-loop only),

■ secondly, from the measurements at its disposal, the NCC takes a decision about a possible adaptation of the link budgets,

■ thirdly, the NCC notifies its decision by sending a command via the satellite to all the concerned Earth stations (UES and GES).

For regenerative configurations, the decision function can be implemented either in the NCC or in the satellite. In the latter, the amount of signalling is lower at the expense of a higher on-board complexity.

8.5   Conclusions

The objective of this chapter was to present a review on diversity and fade mitigation techniques, as well as on the blockage/multipath and atmospheric effect mitigation techniques for satellite communication systems. Several issues related to diversity domains, as MIMO and STBC systems, PC and ACM where also analyzed and discussed. Recent and novel results for joint fading mitigation, IMR multipath diversity and combining techniques have been presented.

Chapter 9

MULTIUSER SATELLITE COMMUNICATIONS

M. A. Vázquez Castro[1], J. Bitó[2], J. Ebert[3], B. Héder[2], N. V. Kokkalis[7],
O. Koudelka[3], P. T. Mathiopoulos[7], S. Morosi[4], C. Novák[2], A. Quddus[5],
G. Seco Granados[1], A. Vanelli-Coralli[6]

[1] *Independent University of Barcelona, Spain*

[2] *Budapest University of Technology and Economics, Hungary*

[3] *Graz University of Technology, Austria*

[4] *University of Florence, Italy*

[5] *University of Surrey, England*

[6] *University of Bologna, Italy*

[7] *Institute for Space Applications and Remote Sensing, National Observatory of Athens, Greece*

9.1   Introduction

Point-to-point communications occur between a single transmitter and a single receiver. In this chapter, we focus on multiuser communications, that occur between a network of fixed and/or mobile users and a common network center probably having access to a wireline network infrastructure. Multiuser communications in a satellite scenario usually refer to a star topology consisting of one or more *gateways* and a large number of small user terminals. Gateways communicate with the user terminals through the forward link and terminals may communicate with the Gateways through the return link, both links having a multiuser uplink channel and a multiuser downlink channel. Broadly speaking multiuser communications may refer either to a one-way communication as radio or TV broadcasting or to a two-way multiuser communication where each user is interested in a message specific to herself. In general, multiuser communications usually refer to the latter case.

In order to support multiple users, a multiuser channel needs to be allocated to the different users. Given that an infinite number of possible allocations exist, multiuser channel capacity cannot be expressed by a single number but by a set of rates. The set of rates that can be supported simultaneously by the channel with an arbitrarily small error probability is called capacity region and its information theoretic limits are the subject of study of networking information theory [2]. A multiuser system achieves a specific point in the capacity region depending on how the multiuser channel is shared by the users, which depends on the multiple access technique or Multiple-Access Channel (MAC) protocol used. MAC protocols are designed to coordinate multiuser transmissions as well as eventual retransmissions or resolution of collisions.

In this chapter, we introduce a number of basic multiple access techniques and MAC protocols that are specific to the satellite scenario. This scenario poses some major constraints on a MAC protocol performance such as long propagation delay, remote control of on-board processing capabilities or power limitation that preclude the use of some protocols developed for a terrestrial scenario. A classification according to the most efficiently supported types of traffic is presented. The hybrid solution based on demand assignment is also discussed.

We also present a brief introduction to some basic results of multiuser capacity discussing the effect of having channel side or state information. Finally, techniques able to mitigate or cancel the interference arising from spectrum sharing are also introduced and their applications to current satellite networks are presented.

9.2   Connection-Oriented Multiple Access

A connection-oriented multiple access establishes a dedicated channel (or circuit) for the duration of a transmission. This solution is appropriate for communications that require data to be transmitted in real-time, i.e. connection-oriented traffic. This type of communication is different to both contention-oriented communications, addressed later in this chapter, and packet switching, which divides messages into packets and sends each packet individually. Packet switching is more efficient for the so-called elastic traffic for which some amount of delay is acceptable. Internet Protocol (IP) traffic is connectionless and packet oriented, and in a wireline network it is transmitted through packet switching. However, a connection-oriented channel can be set up in a satellite sub-network (or wireless) even if the end-to-end communication is IP.

IP traffic may be sent over a satellite link after having established a connection in order to access the satellite system air-interface.

The connection or circuit can be established by static or dynamic allocation of the channel. A static allocation of the channel assumes a fixed assignment of the transmission resources. A dynamic allocation assumes that channel allocation is granted to users dynamically over time usually depending on traffic requirements. Channels to be allocated to users can be created by dividing radio resources orthogonally or non-orthogonally along available axis: time, frequency, and/or code. In general, channels obtained from time or frequency division are orthogonal while code division can be non-orthogonal depending on the code design.

Note that in general the uplink multiuser channel is usually called multiple access channel while the multiple access for the downlink channel is usually referred to as multiplexing. For satellite networks, the multiple access occurs in the uplink of the return link only.

9.2.1   Frequency Division Multiple Access (FDMA)

In FDMA the frequency axis is divided into non overlapping channels and each user is assigned a different frequency channel. FDMA was the first technique used in early multiple access for satellite communications. It has been also extensively used in telephony, commercial radio, television broadcasting and in cellular mobile systems.

This access method is efficient if the user has a steady flow of information to send (digitized voice, video, transfer of long files) and uses the system for a long period of time but it can be very inefficient if the user data are sporadic in nature, as is the case with bursty data or short message traffic. In such a scenario, it can be effectively applied only in hybrid forms, as described hereinafter. In FDMA, the total common bandwidth is B Hz and K users are trying to share it. Each of the K users (transmitting stations) can transmit all of the time, or at least for extended periods of time, but using only a portion B_i (sub-channel) of the total channel bandwidth B, so that $B_i = B/K$ Hz. If the users generate constantly unequal amounts of traffic, one can modify this scheme to assign bandwidth in proportion to the traffic generated by each one of them.

Adjacent users occupy different carriers of B_i bandwidth, with guard channel D Hz between them to avoid interference. Then, the actual bandwidth that is available to each station for information is $B_i = [B - (K + 1)D]/K$. The input of each source is modulated over a carrier and transmitted to the common channel. Consequently, the common medium serves several carriers simultaneously at different frequencies. At the receiving end, the user band pass filters select the designated channel out of the composite signal and a demodulator obtains the transmitted baseband signal.

Instead of transmitting one source signal on the carrier, a multiplexed signal can be sent on it, as it is the case with a satellite feeder link. Depending on the multiplexing and modulation techniques used, several transmission schemes can be considered.

9.2.1.1   Bandwidth and Power Efficiency of FDMA.   The required FDMA bandwidth assuming a single carrier is

$$B_F = N_c(B_c + B_g) \qquad (9.1)$$

where N_c is the number of FDMA channels supported by the carrier, B_c is the channel bandwidth and B_g is the guard band . The guard bands cause needless waste of the available bandwidth. Whatever filters are used to obtain a sharp frequency band for each carrier, part of the power of a carrier adjacent to the one considered will be captured by the receiver of the latter. To determine the proper spacing between FDMA carrier spectra, this adjacent channel interference (crosstalk power) must be carefully calculated. Spacings can then be selected for any acceptable crosstalk level desired. Common practice is to define the guard bands equal to around 10% of the carrier bandwidth (for carriers equal in amplitude and bandwidth). This will keep the noise levels below the ITU-T requirements.

When multiple FDMA carriers pass through non-linear systems, notably power amplifiers in satellites, two basic effects occur; a) the non-linear device output contains not only the original frequencies but also undesirable frequencies, i.e. unwanted Intermodulation (IM) products which fall within the FDMA bands as interference and b) the available output power decreases as a result of conversion of useful satellite power to intermodulation noise. Both of these effects depend on the type of non-linearity and the number of simultaneous FDMA carriers present, as well as their power levels and spectral distributions. This makes it necessary to reduce the input to the amplifier from its maximum drive level in order to control intermodulation distortion. In satellite repeaters, this procedure is referred to as *input backoff* and is an important factor in maximizing the power efficiency of the repeater. The throughput capability of an Frequency Division Multiplex (FDM)/FM/FDMA scheme has been studied as a function of the number of carriers taking into account the carrier-to-total noise (C_T/N_T) factor, where the carriers are considered to be modulated by multiplexed signals of equal capacity. As the number of carriers increases, the bandwidth allocated to each carrier must decrease and this leads to a reduction of the modulating multiplexed signal. As the total capacity is the product of the capacity of each carrier and the number of carriers, it would seem logical that the total capacity would remain constant, but it is not; the total capacity decreases as the number of carriers increases. This results from the fact that each carrier is subjected to a reduction in the value of C/N since the back-off is large when the number of carriers is large (extra carriers bring more IM products). Another reason is the increased need for guard bands.

9.2.1.2   Pros and Cons of FDMA.   FDMA it is a technique that offers itself to both analogue and digital communications. Since each user takes a part of the spectrum for extended periods of time, low symbol rates are employed, a fact that makes FDMA especially robust to narrowband fading and noise. In cases where such phenomena are present, only a few carriers are affected, while all other users remain unharmed. The low symbol rate also makes this technique relatively insensitive to multipath fading.

On the other hand, frequency division techniques are based on analogue components, a fact that is linked with increased cost and reliability problems. Also, FDMA is considered as a relatively inflexible technique regarding bandwidth allocation, since it is difficult and technically complex to acquire more than one subchannels for transmission. Furthermore, as was commented above, these techniques require linear components, since any non-linearities introduce intermodulation products, which have an adverse effect on system throughput. The need for linear amplifiers increases both the cost of the design and also reduces the transmission power.

9.2.2   Time Division Multiple Access (TDMA)

In TDMA the time axis is divided into non overlapping time slots and each user is cyclically assigned a different time slot. Time slots are organized in a higher level time structure unit, called a frame; the length of a frame in slots defines how frequently a station gains channel access, since, at least in the basic form of TDMA, each station transmits in one slot per frame. According to the TDMA technique, the total frame length is T_f and K users are sharing it, i.e. there are K slots per frame. Each of the K users (transmitting stations) can transmit for a fraction of the frame duration - a slot duration T_s -, but gain all the available system bandwidth during this interval. Consequently, a station that transmits in one slot per frame gets a fraction of the system capacity that is equal to $B_i = B * T_s/T_f$, where B is the total system capacity. If the users generate constantly unequal amounts of traffic, one can modify this scheme to assign bandwidth in proportion to the traffic generated by each one of them by allowing certain users to transmit in multiple slots per frame.

9.2.2.1   Efficiency of TDMA.   Even though the previous equation shows the gross capacity that is allocated to each user, this value does not coincide with the capacity that will actually be available for data transmissions. Indeed, in each slot a number of symbols are experienced as overhead, while also guard times are used to separate consecutive slots. TDMA demands that all users are synchronized, so that each one of them is able to transmit in the time interval that is exclusively allocated to him. Since time clock drifts are very common in distributed systems, these guard times provide some error resilience against synchronization errors in the transmitting stations, while they also allow the synchronization of the receivers between different slots and frames. Furthermore, since power amplifiers in communication systems are not perfect, the transmitters fall time is not zero. The guard times ensure that the transmitted power of the previous user has fallen to levels that do not disrupt transmissions in the current slot.

Besides the guard times, a number of bits within a slot are experienced as overhead and consequently are subtracted by the available net capacity for data transmissions. One of the major reasons behind the TDMA overhead is the fact that according to this technique each station gains the whole bandwidth for a finite amount of time, characterizing thus the system as wideband. For this reason, a training sequence is included in the preamble of the slot for configuring the equalizers of the receiver. Further, the slot preamble is used to synchronize the transmitting and receiving stations. Also, another source of overhead is the reference burst that is transmitted per slot by the central controller and helps in synchronizing the transmitting stations. An example of the structure of a TDMA frame and slot is provided in Figure 9.1. The number of bits that are experienced as overhead may be calculated by using the following expression:

$$b_{OH} = N_r b_r + N_t b_p + (N_t + N_r) b_g$$

where N_r number of reference burst per frame
        N_t number of traffic bursts (slots) per frame
        b_r number of overhead bits per reference burst
        b_p number of overhead bits per preamble per slot
        b_g number of equivalent bits in each guard time interval

The total number of bits per frame is: $b_T = T_f R_{rf}$

where T_f frame duration
           R_{rf} bitrate of the radio frequency channel

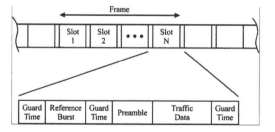

Figure 9.1.   Frame and slot structure

Then the frame efficiency is: $\eta_F = (1 - b_{OH}/b_T)$

Good TDMA schemes demonstrate frame efficiencies that may reach as high as 0.9.

9.2.2.2   Pros and Cons of TDMA.   One of the most positive character-istics of TDMA is the flexibility that it offers regarding bandwidth allocation. Simply by allowing a station to transmit in more than one slot per frame, the offered bandwidth is a multiple of the basic rate employed in the system. Also, since it employs no guard bands, TDMA may exploit the whole available spec-trum attaining much higher spectral efficiency figures, compared to FDMA. TDMA is particularly appropriated for cellular systems; the transmitter of each station is active for small amounts of time only (up to a few slots per frame) and so stations are able to listen to the common medium for other base stations in cases when a handover must be executed. Regarding robustness to noise, this scheme is particularly robust to impulse noise and interference. Intense, short spikes essentially affect only one user, leaving all others unharmed.

On the downside, each user gains the whole spectrum for finite amounts of time, thus the system may be characterized as wideband. Consequently, TDMA systems are sensitive to multipath and selective fading. Furthermore, this same reason demands that TDMA receivers are equipped with complex equalizers. Finally, time division techniques requires network wide synchronization, since each station transmits in globally unique time intervals.

9.2.3   Code Division Multiple Access (CDMA)

In CDMA both time and frequency axis are used simultaneously by all users. In order to allow distinction among users a third dimension is incorporated given by spreading codes assigned to each user. The codes can be orthogonal to each other, in which case there is an exact number of CDMA channels as in TDMA and FDMA. The codes can also be non-orthogonal, in which case there is no hard limit on the number of users that can access the channel simultaneously as long as the overall performance degradation remains acceptable. By con-sidering a performance level for the system the capacity can be determined. This effect is also referred to as CDMA soft capacity. The performance degra-dation arises from the non-zero cross correlation between the non-orthogonal spreading codes which cause mutual interference among users. In general, downlink transmission use orthogonal codes (such as Walsh-Hadamard) since in this case it is feasible to synchronize codes. Uplink transmissions generally use non-orthogonal codes allowing for unsynchronized user access.

9.2.3.1   Direct Sequence CDMA (DS-CDMA).   Direct Sequence CDMA is the most widely used CDMA technology. It is utilized by most of the Third Generation (3G) standards such as IS-95A/B, cdma2000, UMTS-UTRA, S-UMTS, W-CDMA, TD-SCDMA, etc. It consists of multiplying the baseband signal by a pseudo-random code that spreads de signal bandwidth. In the receiver, the received signal is multiplied again with the same code. A code sequence has N elements or chips. The ratio of transmission and information bandwidth is referred to as processing gain while the ratio of transmission and symbol bandwidth is referred to as spreading factor.

All users are assigned a distinct signature code sequence to achieve multiple accesses within the same frequency band, and allow signal separating at the receiver. Under multipath fading environment with near-far effect, CDMA systems employ RAKE receivers, to enhance the system performance taking advantage of frequency selective channels. Satellite channels hardly show frequency selectivity and capacity is enhanced through satellite diversity. Assuming L resolvable multipath; the spreading gain N, at the receiver, after conventional chip-matched filtering and sampling at the chip rate, collected samples during the i-th symbol interval in the form of an N-dimensional vector can be expressed as follows

$$\mathbf{r}(i) = \sum_{k=1}^{K} A_k b_k(i) \sum_{l=1}^{L} h_{k,l} \mathbf{s}_{k,l} + n(i) \tag{9.2}$$

where K is the total number of active users; K are the received amplitude, $b_k(i)$ the symbol stream, and $h_{k,l}$ the channel response of the lth path of the kth user, respectively. $\mathbf{s}_{k,l} = [0 \ldots 0 \; c_{k,1} \; c_{k,2} \ldots c_{k,N-l+1}]^T$ is the normalized spreading sequence of the the user.

When non-orthogonal codes are used in a DS-CDMA system, users located close to the base station (satellite) may exceed the received signal of those users located further away thus masking their signals. With a proper power control this effect, known as near-far effect, can be reduced thus increasing the capacity of the system. Additionally, scrambling codes separate users from different base stations (satellites) thus reducing interference and allowing uncoordinated code management within one satellite beam per base station cell. In conventional systems, the multiuser interference is treated as background noise and contributes negatively to the Signal to Interference plus Noise Ratio (SINR). Multiuser detection and interference cancellation techniques aim at reducing the interference effects by exploiting its structure at the receiver side. Section 9.6.1 will describe in detail these techniques and their application to satellite systems.

9.2.3.2   Frequency Hopping CDMA (FH-CDMA).   Another CDMA technology is based on allowing the carrier frequency to "hop" according to a known sequence. This method is called Frequency Hopping (FH-CDMA) . In this way the bandwidth is also increased. With Slow Frequency Hopping (SFH) one or more data bits are transmitted within one frequency hop. With Fast Frequency Hopping (FFH) one data bit is divided over more Frequency Hops. In both cases the near-far effect is less significant than in DS-CDMA.

A disadvantage of FH-CDMA compared to DS-CDMA is that it is more difficult to obtain a high processing gain. A frequency synthesizer is required that is capable of rapidly hopping over a set of carrier frequencies. The more frequencies required, the higher the processing gain and the more demanding the frequency synthesizer becomes.

9.2.3.3 Pros and Cons of CDMA. CDMA combines the transmission from an individual user with a faster signal. This process allows CDMA systems to operate with low transmit powers leading to smaller terminals and smaller batteries and longer life. As far as other users are concerned, such signals simply appear as low-level noise. So while there is the potential for increased co-channel interference, the spread spectrum technique actually makes the CDMA air interface more robust than other air interfaces.

CDMA is able to increase capacity more easily than TDMA by exploiting variable data rates and/or voice activity detection.

CDMA possesses many intrinsic advantages over the earlier access techniques such as time-division multiple access TDMA and frequency-division multiple access FDMA. However, it has fundamental difficulties in a nearfar situation when transmission power from one user overwhelms signals of the others. In practice, power control schemes are employed to maintain a balanced power distribution among the users. However, the full benefit of power control is exploitable only for stationary and slow-moving mobile stations.

In general it is difficult to determine cell capacity in CDMA networks. The term soft capacity refers to the fact that capacity is dependent on the average level of mutual interference. The more users access the system the higher the level of interference which in turn affects the performance of every user connected to the network. This is a feature of CDMA networks known as "cell breathing": the effective service area expands and contracts according to the number of users connected. Overlap-regions between cells (which are known as "handover areas") need careful planning and management. It has also been argued that CDMA is appropriate for low bit rate services while high bit rates services will require some hybrid access technology.

9.2.3.4 CDMA Sectoring. In this section, first we give an overview of the sectoring problem, and next we focus on the particular case of sectoring for CDMA satellite networks.

Sectoring. Sectoring aims at reducing interference and improving capacity by dividing the coverage area into separate zones, spots radiated by different beams of the satellite antenna systems, called sectors. Sectors are arranged so that the subscribers in the system generate reduced interference compared to the non-sectored case. Basically, sectoring is achieved by sectored antennas, i.e. multiple beams of the satellite antenna systems which results in *space sector-*

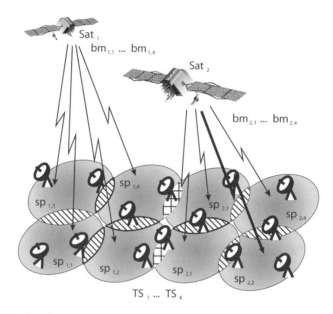

Figure 9.2. Interference scenarios in multibeam satellite coverage with two satellites (Sat1, Sat2) deploying 2x4 spots, assuming single-frequency operation

ing. In satellite systems, the term sectoring is interpreted by applying multiple beams. Sectors coincide with the separate spots of a multibeam network (see Sec. 9.2.4 for multibeam systems overview). Inter-sector interference does, however, occur due to the overlapping of different satellite spots. Figure 9.2 shows a multibeam coverage including two satellites, deploying 2×4 spots, operating at the same frequency. Interference situations may occur at sector (spot) borders, depicted by highlighted areas in the figure. Inter-sector interferences can originate from the same (hatched areas) or the adjacent (squared areas) serving satellite.

Inter-sector interference can be reduced by assigning separate resources to different sectors. The most common method is *frequency sectoring*, which is usually combined with frequency reuse according to the cellular principle. Other resources can be combined with the mentioned sectoring methods: In CDMA systems, different spreading codes can be assigned to the sectors, i.e. spots, resulting in *code sectoring*.

CDMA for sectoring networks. The model given in Figure 9.2 introduces multi-satellite, multibeam sectoring network. Multibeam spots of a satellite coincide with the sectors of a terrestrial base station. In terrestrial networks more accurate antenna sectoring can be implemented using sectored antennas.

Satellite spots, however, have overlapping when the whole area is needed to be covered (see Figure 9.2). Spot overlapping induces significant interference if single-frequency operation is assumed. In sectoring networks in addition to pure antenna space sectoring the most common solution is frequency sectoring. Considering downlink transmission, data streams of a certain source can be synchronized easily. Hence, TDMA is a straightforward multiple access method within the spots. In this case frequency sectoring provides inter-sector interference protection between the subscribers transmitting in the same time slot in different spots, whereas TDMA gives adequate intra-sector interference protection. Considering CDMA, either synchronous or asynchronous sources can be separated, provided enough spreading codes are available with appropriate cross-correlation properties. Frequency sectoring can be avoided, allowing a single-frequency approach for the coverage area that can make onboard satellite hardware simpler. In this case, of course, the number interferer sectors will increase. Investigations in [417] [418] of CDMA code sectoring methods applied on terrestrial Point-to-Multi-Point (PMP) networks showed that significant interference protection can be obtained by choosing the appropriate system parameters. To suppress the increased amount of interference, CDMA code utilization plays an important role. In addition to space sectoring, further sector separation can be realized by applying code sectoring. Hereunder different code sectoring schemes are considered with the following notations. Assuming an asynchronous CDMA system with K users transmitting continuously the received signal is given by eq. (9.3).

$$r(t) = \sum_{i=-\infty}^{\infty} \sum_{k=1}^{K} A_k B_k(i) s_k(t - iT - \tau_k) + n(t) \qquad (9.3)$$

where the kth user is identified by spreading waveform s_k, $b_{k(i)}$ denotes the sent i^{th} bit with the duration T, A_k is the received amplitude of the kth user and n is the white Gaussian noise. Code sectoring means the appropriate assignment of spreading sequences s_k to the K users in the service area. This is practically done by combined code systems, similarly as that of UMTS [419], in which the combination of orthogonal and quasi-orthogonal codes is applied. On the downlink, because of different delays of the received signal components from different sectors, Pseudo Noise (PN) codes should be used for sector separation, which provide acceptable cross correlation properties even in asynchronous case. Sector synchronization therefore can be avoided. Users of a certain sector are separated by orthogonal Walsh codes. Spreading waveform in this case reads:

$$s_k(t) = PN_c(t) \cdot w_u(t) \qquad (9.4)$$

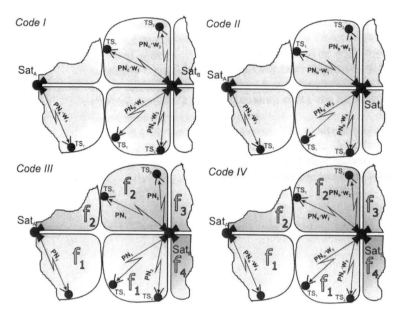

Figure 9.3. Code assignment of Code I to IV schemes

where PN_c is a pseudo-noise sequence assigned to sector c, w_u is a sequence
from the family of orthogonal Walsh-codes assigned to Terminal Station (TS)
u located in sector c. As in the downlink, the BS separates the users' signals
by orthogonal Walsh-codes, intra-sector interference at the TS is totally eliminated. Because the sector identifier PN codes are not orthogonal, inter-sector
and inter-cell interference must be taken into account.

Four code sectoring strategies are considered: *Code I* and *II* for single-, and
Code III and IV for four-frequency approaches. Figure 9.3 illustrates the discussed code assignment of schemes in consecutive quadrants of the figure. A
multibeam satellite network is considered similar to that of Figure 9.2, with
each satellite radiating four beams. The spot arrangement and the satellites are
illustrated from top view, similarly to a terrestrial network with base stations
and rectangular sectors.

Code I :

A unique identifying code (i.e. PN_A is assigned to each sector. Because
of different delays of the received signal components from different sectors, PN codes should be used, which provide acceptable cross correlation
properties. Sector synchronization therefore can be avoided. Users of
a certain sector are separated by orthogonal Walsh codes (W_i). A certain TS receives the desired signal spread by the PN code identifying

its serving sector (i.e. satellite beam) and also multiplied by its unique Walsh code, identifying the desired downlink connection within the radiated beam, similarly as in UMTS [419]. As in the downlink, the satellite separates the users' signals by orthogonal Walsh-codes, intra-sector interference at the TS is totally eliminated. However, the sector identifier PN codes are not orthogonal, inter-sector and inter-cell interference must be taken into account. In the first quadrant of Figure 9.3 the same tone of the sectors means that single-frequency operation is applied. Downlink connections to TSs are depicted by TS_1 and TS_2, sector identifier PN codes are also distinguished by PN_A, PN_B, and PN_C. Intra-sector separation Walsh-codes are depicted by W_1 and W_2.

Code II :

The synchronization of the four spots belonging to one BS is easy to implement. In this case orthogonal Walsh codes can separate all subscribers of the cell coverage area. One PN code is assigned to each satellite, therefore each beam of a certain satellite applies the same PN code. Four orthogonal Walsh code subsets are assigned to the four sectors of a cell, eliminating intra-satellite interference. The assignment of the four Walsh code sets to the sectors corresponds the reuse pattern of the frequency sectoring. Sectors of Sat_B are identified by PN_B, intra-satellite separation is done by $w_1, w_2, \ldots w_4$ Walsh sequences.

Code III :

Reduced inter-sector interference by the four-frequency sectoring allows simpler code utilization, exploiting the possibility of code sectoring is not necessary. In this approach the same PN code set is applied in each sector of the service area. (See Figure 9.3 for illustration: sectors illustrated with letters in the background letters represent frequencies $f_1, f_2, \ldots f_4$).

Code IV :

In this second four-frequency scheme unique PN codes are assigned to each satellite (PN_A and PN_B in Figure 9.3). TSs of each sector are separated by the same Walsh code set. Therefore code system IV is also the combination of PN+Walsh codes. Bit-asynchronism between the sectors is allowed in contradiction to the similar single frequency code system II.

The code sectoring schemes presented above do not consider additional sophisticated CDMA interference suppression techniques (i.e. multiuser detection). Studies of [417][420], explore possibilities of such techniques, resulting in further sub-sectoring effects of interference cancellation (See Sec. 9.6.2 for details).

Example. Inter-cell and intra-cell interference can be reduced by using code sectoring. Recent investigations [417] lead to the experience that Code sectoring is superior to frequency sectoring in terms of interference protection. The most feasible sectoring method in case of synchronized sectors of a satellite is the scheme Code II. 4-frequency solutions are inferior to single frequency approaches and cannot provide acceptable Bit Error Rate (BER) performance without interference cancellation (i.e. multiuser detection).

A simulation example presented hereunder compares the BER vs. Signal to Noise Ratio (SNR) analysis of a network applying the introduced code systems and different CDMA sectoring methods. Considering CDMA codes, PN codes of code system I and II are 64 chip long extended Gold codes. Users of a certain sector are separated by orthogonal 64 chip long Walsh codes, as described above. For comparison, in 4-frequency sectored approaches to preserve signal bandwidth, code length divided by four was applied in code systems III and IV.

The BER performance was investigated in the bottom left corner location of the examined sector. BER vs. SNR curves are depicted in Figure 9.4 showing the effect of different sectoring methods. As we can see applying single frequency rectangular sectoring with code system II and (1-fr. code II) the performance is close to the theoretical BPSK situation. The result is slightly worse in the case with four frequencies applying code system III (4-fr. code III) and results in 1 dB loss at BER $= 10^{-4}$. Poor behaviour of code system IV can be observed (4-fr. code IV). The loss is more than 2 dB at 10^{-4} BER compared to the theoretical BPSK curve. Because in code system IV the correlation properties of the combined PN+Walsh codes are worse than those of the PN codes applied in Code III, furthermore, adjacent frequency beam interference and lower processing gain results in increased interference compared to the single-frequency case. (1-fr. Code I) indicates poor BER conditions due to heavy interference, which is especially critical at the sector borders. In this approach non-orthogonal codes are assigned to the asynchronous sectors, therefore adjacent sector interference due to the spot overlapping is critical. In the best performing single frequency approach (1-fr. Code II) effective code sectoring is provided by orthogonal subsets of Walsh codes assigned to synchronized spots of a satellite. Four-frequency sectored approaches with matched filter detector are inferior to Code II scheme. Due to constant bandwidth, in this case the applied processing gain is the quarter of that applied in single frequency case. Not perfect frequency channel separation cannot be compensated with the reduced processing gain. This shows that the interference suppression capability of code sectoring by code system II is more effective than that of frequency sectoring.

Our downlink BER performance analysis of a Broadband Fixed Wireless Access (BFWA) sector revealed that code sectoring is superior to frequency sectoring in terms of interference protection. Code sectoring in CDMA-based

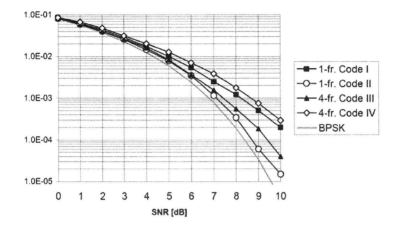

Figure 9.4. Downlink BER vs. SNR analysis for a fixed location of the simulated CDMA sectored network for different code sectoring schemes

multibeam networks allows the effective control of interference. According to our simulations, a single-frequency approach with a UMTS-like code system (Code II) has proved to be the most feasible sectoring in case of synchronized sectors of a cell. Since applying frequency separation does not utilize the whole available frequency band optimally, higher spectral efficiency can be achieved by applying code separation. The implementation of CDMA sectoring in multibeam satellite coverage does not require complex onboard signal processing capabilities. The transmission of the CDMA modulated data stream with layered code system can transparently be implemented with the present equipment. Reception of CDMA signals, however, requires increased number of matched filters and signal processing complexity. Furthermore, it is essential for each network entities to know the relevant spreading code information. This requires the implementation of the reliable broadcasting of code information in a dedicated channel.

9.2.4 Space Division Multiple Access (SDMA)

An additional domain that can be exploited to provide multiple access is space, which can be used in combination with time, frequency and/or code division. Space division is provided via beamforming , which consists of combining multiple antenna elements capable of changing the directionality of its radiation patterns according to a given signal scenario. The analog architecture of a beamformer is usually based on RF phase shifters which are expensive devices. Digital beamforming systems make use of a Digital Signal Processor (DSP) that computes and applies weight vectors to each antenna element. The digital approach has the advantages of simplicity, flexibility and lower power dissipation. Because of the limitations of analog beamforming, digital beam-forming is preferable. Adaptive beamforming networks are able to dynamically change the directionality of the radiation pattern and several methods can be applied to maximize Signal to Noise plus Interference Ratio (SNIR), such as the Minimum Mean Square Error (MMSE) by which the complex weights are adjusted to minimize the MSE between the beamformer output and the expected signal waveform, or the Minimum Variance (MV) by which the weights are adjusted to minimize the noise on the beamformer output. Adaptive beamforming systems for communications are sometimes referred to as smart antenna systems.

Space division enables to reduce the system interference thus allowing for an increase of the overall throughput. In satellite systems, the space dimension is particulary appealing due to the unique position in space of a satellite. On-board beamforming allows efficient frequency re-use and satellite systems employing beamforming are usually referred to as multibeam satellite systems. Space division combined with adaptive transmission techniques, (like Adaptive Coding and Modulation), have been shown to be capable of providing substantial capacity improvements for the time-varying satellite channel specially at Ka band [245].

9.2.5   Orthogonal Frequency Division Multiple Access (OFDMA)

An interesting variation of FDMA is the OFDMA scheme proposed for wideband communications. In OFDMA, multiple access is achieved by providing each user with a number of the available sub-carriers, which are now orthogonal to each other. The orthogonality of this scheme stems from the fact that the peak of any given subcarrier coincides with the nulls of all the other subcarriers. This property of Orthogonal Frequency Division Multiplex (OFDM) is depicted in Figure 9.5. The most important advantage of OFDM/OFDMA compared to FDM/FDMA is that according to the former there is no need for the relatively large guard bands that were necessary in the case of FDM/FDMA. Consequently, OFDM/OFDMA retains all the advantages of the traditional scheme, while enhancing it in terms of spectral efficiency and thus, throughput.

The main property of OFDM is the guard time and the cyclic extension that is applied whenever this technique takes place. A guard time is inserted in each OFDM symbol and its duration is chosen larger than the expected delay spread, such that multipath components from one symbol cannot interfere with the next symbol. It is required to cyclically extend the OFDM symbol in this period of time rather than to have no signal at all in it, which would cause intercarrier interference (ICI). This ensures that delayed replicas of the OFDM symbol

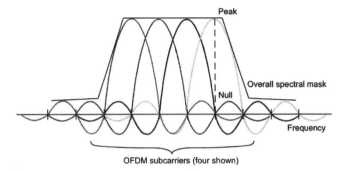

Figure 9.5. Orthogonality of OFDM subcarriers

always have an integer number of cycles within the Fast Fourier Transform (FFT) interval, as long as the delay is smaller than the guard time. As a result, the main advantage of this technique is that the multipath signals with delays smaller than the guard time cannot cause ICI.

Regarding OFDM and OFDMA, many of the advantages and disadvantages that were shown in the case of FDM and FDMA apply as well. The orthogonality of this scheme, however, is linked with certain characteristics that are unique to OFDM and OFDMA. With OFDM and OFDMA, high data rates may be achieved using low symbol rate modulations, because information is sent in parallel on many carriers that are closely spaced in the frequency domain. Also, as was the case with FDM and FDMA, the orthogonal variants are robust to narrowband fading and noise, since they affect only a few carriers, while leaving the others unharmed. Furthermore, information is spread across many carriers and redundancy and Forward Error Correcting (FEC) may easily be built in. Also, it is easy to implement a high order modulation scheme (e.g. multi level QAM) to adapt to the frequency selective channel SNR providing highly spectral efficient transmission.

On the downside, highly fluctuating (Gaussian noise like) envelope and sensitivity to non-linear distortion require highly linear analogue processing in both transmitter and receiver. Furthermore, in impulsive noise channel conditions, OFDM and OFDMA show inferior performance compared to coded single carrier schemes. Finally, OFDM and OFDMA are extremely sensitive to frequency error in the system clock, since in cases of frequency drifts the orthogonality between the carriers may be lost.

9.2.6   Hybrid Schemes and Advanced Topics

Any multiple access technique based on one type only of orthogonal or non-orthogonal signal dimension division shows advantages and disadvantages. Hybrid schemes try to combine the advantages of several techniques. The most commonly used hybrid technique in satellite systems combines FDMA with TDMA and is called Multi Frequency TDMA (MF-TDMA), which is widely described herinafter. Other hybrid techniques combining TDMA, FDMA and/or CDMA exist that are mostly used in terrestrial systems such as Multi-Carrier CDMA (MC-CDMA). A conceptually different scheme is based on Ultra Wide Band (UWB) transmission, which is also discussed in this section as satellite communications using UWB signals are possible.

9.2.6.1   Multi Frequency TDMA (MF-TDMA).   In a conventional single-frequency time-division multiple access Single Frequency TDMA (SF-TDMA) system, capacity is assigned in terms of time slots. For the duration of a time slot, a single user station has access to the whole bandwidth, where the maximum of SNR is available. If the network capacity has to be increased in an SF-TDMA system, the information rate of all stations must be increased. This requires a higher Effective Isotropic Radiated Power (EIRP) and G/T [1] for *all* terminals, even for those whose traffic rate may be low. This is a major drawback, which can be avoided by a Multi Frequency TDMA (MF-TDMA) approach .

In an MF-TDMA system, the capacity is assigned two-dimensionally in terms of time and frequency. A station can now transmit within an assigned time slot on one of several frequency channels. In order to provide the necessary interconnectivity among the stations, the transmitters and receivers must be frequency-agile. Ideally, they should be able to hop between channels from burst to burst. This requires modulators and demodulators which can change frequency rapidly. Typically, numerically-controlled oscillators (NCOs) are implemented for this purpose.

In contrast to SF-TDMA, the transmission rate per terminal can remain the same as soon as the network capacity is increased by adding further frequency channels. Thus, the terminal parameters (EIRP, G/T) need not be changed. Compared to a single-frequency system, MF-TDMA provides a *cost-efficient* and *scalable* solution since terminal costs in satellite communications depend largely on antenna size and available transmission power.

In *meshed satellite systems*, any terminal can communicate directly with any other terminal in a *single* hop via a geostationary satellite with *transparent*

---

[1] Given a satellite, G/T represents the ratio among the satellite gain and its thermal noise temperature. For this reason it depends on satellite components.

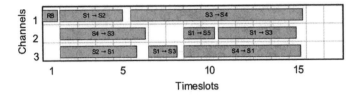

Figure 9.6. Allocation plan for meshed MF-TDMA systems

transponder. In many implementations, stations may not transmit or receive bursts on multiple channels at the same time. This restriction has to be taken into account by an appropriately designed burst scheduling algorithm. Figure 9.6 shows the two-dimensional time-frequency structure (*allocation plan*) as it is necessary for MF-TDMA in a meshed satellite system . Similar to SF-TDMA, the time axis is subdivided into time slots and frames. In addition, several frequency channels are available. The *reference burst* (RB), sent by the reference station (master) at the beginning of a frame on a dedicated channel, allows all stations to obtain as well as maintain timing synchronization. Furthermore, it contains the allocation plan telling all stations which time slot and frequency channel they have to use for transmission (TX) as well as reception (RX) of data bursts. The burst descriptor in the allocation plan contains source and destination information. In Figure 9.6, S1 $\rightarrow$ S2 defines a burst transmitted by station 1 and received by station 2. For adequate frequency hopping of their modems, both stations use this burst descriptor together with the information about timing synchronization.

For scalability reasons, MF-TDMA has been adopted for access in the return link of DVB-RCS systems [120], whereas the forward link uses DVB-S or DVB-S2 as transport mechanism for IP packets. MF-TDMA is also implemented in many VSAT systems, e. g., the SkyWan VSAT or the L*IP system [421] [422].

Figure 9.7 shows the error performance of an MF-TDMA modem, realized via an Field Programmable Gate Array (FPGA) platform. Assuming an Additive White Gaussian Noise (AWGN) channel, the dots indicate the test results if the carrier frequency/phase offset is estimated by appropriately selected algorithms; the timing is controlled through six samples per symbol, i. e., the sample next to the computed timing instant is employed for further processing. Compared to theory [17], a performance loss of less than 1 dB is observed in the SNR range of 5 - 10 dB as it is relevant in practice. A negligible degradation of not more than 0.2 dB, mainly due to phase noise effects, must be taken into account when operating via an L-band converter (diamonds).

As a figure of merit for data throughput in MF-TDMA systems, however, the BER at the output of the demodulator is not very helpful. Since data transmis-

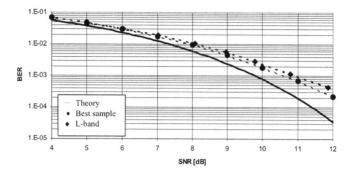

Figure 9.7.   BER performance of an MF-TDMA modem (QPSK/AWGN)

sion is basically burst-oriented, the *packet error rate* PER is the most important parameter in this context – no matter if the communications system uses FEC or not, although FEC is mandatory for an efficient satellite communications system (see Chapter 4).

Given a QPSK/AWGN channel, Figure 9.8 shows the simulation results for $1-$ PER (i.e. the correct packet reception probability) as a measure for data throughput. It is assumed that both symbol timing and carrier frequency/phase are recovered in the demodulator as described previously. For FEC, an eight-state Turbo codec with rate $R \in \{1/2, 2/3, 3/4, 4/5\}$ and a pseudo-random interleaver (2040 bits burst length) is used. Because of inevitable overheads (guard time between bursts, preamble for data-aided acquisition, signaling between master and slave stations, wasted slot fractions caused by variable burst lengths as sketched in Figure 9.6), an additional loss of approximately 10 % must be taken into account which has been verified by tests of a *real* system [422]. Finally, using the MF-TDMA modem presented before, about 0.5 dB of the SNR is lost due to acquisition failures.

9.2.6.2   Ultra Wide Band (UWB).   UWB technology is based on the transmission of baseband pulses with sub-nanosecond duration which results in a signal that typically spans over several GHz of bandwidth. It is thus a carrier-free communication system, contrary to conventional systems that employ a sinusoidal carrier. Although such a signal potentially overlaps several existing conventional systems, they can coexist without significant interference because its energy is distributed in a very large bandwidth, resulting in small Power Spectral Density (PSD) [423].

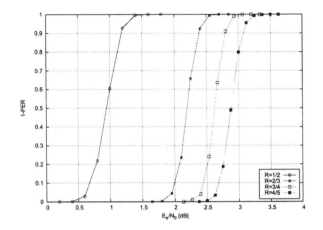

Figure 9.8. Throughput performance for Turbo-coded MF-TDMA systems (QPSK/AWGN, 2040 bits burst length)

At the time of this writing, the Federal Communications Commission (FCC) has authorized the unlicensed operation of UWB devices in the 3.6 - 10.1 GHz band [424] and other major regulatory organizations in Europe and Japan are considering similar regulations. The FCC ruling limits the PSD according to a spectral mask shown in Figure 9.1, which allows UWB devices to operate without causing excessive interference to existing systems. Critical applications such as the Global Positioning System (GPS) are protected by imposing stricter PSD limits below 2 GHz. The main consequence of the spectral mask is that UWB systems are allowed only very low transmit power which limits their applications to short ranges. However, their use in satellite applications should be considered as a possibility if these UWB regulations are appropriately expanded in the future. Moreover, some implementations considerations would be in order in the space context, since conventional satellite transponders are bandlimited and designed to operate close to saturation.

An important property of an UWB system is the pulse shape, which specifies the spectral characteristics of the signal. Of the several pulses that have been proposed, the most commonly employed for theoretical analysis has been the Gaussian monocycle presented in Figure 9.10, which is obtained by differentiation of the Gaussian function. However, such a waveform is not suitable for the FCC spectral mask since it has significant energy from near-DC up to the system bandwidth. The need to develop alternate pulse shapes has led to pulses which can provide more control over the transmitted spectrum [425].

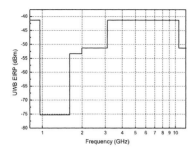

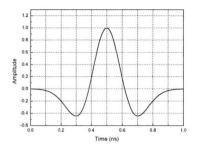

Figure 9.9.   FCC Spectral Mask for UWB

Figure 9.10.   Gaussian monocycle waveform with 1ns duration

The power constraint is the primary factor limiting UWB system performance. As a result, the employed modulation format must be optimized for power efficiency. This makes binary modulations the most obvious choice for this purpose, and mainly antipodal Pulse Amplitude Modulation (PAM) is being used in practice. Alternatively, M-ary orthogonal Pulse Position Modulation (PPM) can be a likely candidate. A combination of the two, namely Pulse Position and Amplitude Modulation (PPAM) has been proposed as well [426].

The impulsive nature of the signal facilitates the development of new somehow hybrid multiple access methods that take advantage of its unique characteristics. For example, the spreading of transmitted pulses in time allows multiplexing of users provided that the pulses at the receiver do not collide. The design of a multiple access scheme for UWB is primarily targeted towards an asynchronous network with dynamic topology. There are two likely candidates suitable for this purpose: time-hopping and direct sequence, which will be briefly examined in the following paragraphs.

Time hopping. Time Hopping (TH) was the first multiple access scheme to be proposed for modern UWB communications [427]. Its operation relies on the impulsive nature of the signal, and thus is not applicable on conventional systems. The modulation format that is typically used in conjunction with time hopping is either binary or M-ary PPM, combined into a scheme known as Time-Hopped PPM (TH-PPM).

In TH-PPM, each symbol is spread over several pulses, though not necessarily combined with a chip code as with CDMA systems. The pulses are transmitted successively, each one within a fixed time window (frame) which is much larger than the pulse duration. The frame is divided into slots, and the slot used to transmit each pulse is determined by a sequence unique for each user (TH sequence). This way, if the appropriate TH sequences are well chosen, the pulses of different users do not frequently collide and the receiver is able to correctly decode the received signal.

Although the pulse collisions can be minimized by selecting the appropriate TH sequence, it is not possible to perfectly isolate the signals of different users, mainly for two reasons. First, the dynamic topology of the network favors an asynchronous operation mode which prohibits the assignment of orthogonal sequences that could be implemented in a synchronous TDMA-like model. Second, the extremely short duration of the pulses results in multipath components that have delay several orders of magnitude larger than the pulse duration, making the perfect separation of multiuser pulses impossible.

In the analysis of TH-PPM systems, the TH sequence typically employed for analysis and performance evaluation purposes is the ideal uniformly distributed random sequence. In practice, however, a true random sequence is not realistic. A maximal-length pseudo-random sequence is a possible solution, although others have been suggested, including sequences used in frequency hopping systems. Aperiodic sequences have been proposed as well, such as pseudo-chaotic time hopping [428].

The performance evaluation of these systems is usually made by considering multi user interference (MUI) as Gaussian. This approach makes closed-form analysis possible and is known as Gaussian approximation (GA). Unfortunately, studies have shown that TH-PPM signals do not conform well to the GA. This has led to the use of computer simulation methods for performance evaluation. A few non-GA semi-analytical methods have been developed as well [429] [430].

Direct Sequence. The direct sequence (DS) approach is more akin to the traditional CDMA scheme [431]. The transmitted data is modulated by a higher-rate code sequence (chip code) and the resulting data is transmitted using binary PAM pulses. This scheme was able to outperform the TH-PPM scheme in terms of multiple access capabilities for high data rate systems [432]. However, when multiuser detection is employed, both systems perform similarly. The TH-PPM scheme could be preferable in this case, due to reduced near-far effect.

Multiband OFDM. The difficulty in constructing UWB pulses that are suitable for the spectral mask imposed by the regulation was the main motivation for developing multiband schemes [433] [434]. Instead of impulse-based radio that uses the entire band simultaneously, the multiband design divides the available bandwidth into several subbands of at least 500 MHz to meet the FCC definition. Each subband is used sequentially for transmission, in a manner similar to traditional frequency hopping. The main advantage of this system is that it is able to take full advantage of the whole available bandwidth, thus maximize the output power while meeting the FCC spectral mask. The data is transmitted using a direct sequence spread-spectrum OFDM scheme.

9.3   Contention-Oriented Multiple Access

In a contention-oriented or random access system, terminals access the common shared channel without any previous assignment. Random access techniques are usually employed in networks where most sources are bursty since they are easy to implement and adapt well to traffic fluctuations.

It is normally assumed that throughput is the figure of merit of a contention architecture. The throughput is defined as [435] the fraction of time during which the channel can be used to transmit data, i.e., it gives the mean number of successful attempts.

In the following the most representative random multiple access protocols for satellite communications are described including their throughput and examples are given.

9.3.1   Random Access

Random access techniques are usually employed in networks where most sources are bursty since they are easy to implement and adaptive to traffic fluctuations. However, collisions may result in a waste of capacity. In the following the most representative random multiple access protocols for satellite communications are described and examples are given.

9.3.1.1   Narrowband ALOHA.   The most well-known multiple access protocol for satellite communications is ALOHA which was initially conceived by Norman Abramson and his colleagues at the University of Hawaii in 1970. Although ALOHA was developed for ground-based radio broadcasting, its basic principles can be applied to any system in which a large number of users are contending for access to a single shared channel as it is depicted in Figure 9.11. There exist two versions of ALOHA, the pure and the slotted ALOHA. The difference between them is that in the latter one time is divided into slots, and therefore, global time synchronization is required.

If the bandwidth of a narrow band ALOHA channel is increased by a factor F, we have a High Bandwidth ALOHA multiple access scheme, where the packets will be shorter thus decreasing the number of collisions. There is however a practical problem that excludes ALOHA for high bit user data rates: to maintain E_b unchanged (digital quality does not depend on power but on bit energy), power during the packet burst must be increased by F (since the packet has shorter length due to the larger data rate). For values of F in the order of 100 and above the power to be transmitted can be extremely high. This fact limits pure ALOHA to narrowband operation (up to few tens of kbit/s). For broadband systems, different solutions can be therefore evaluated as spread spectrum technology on top of ALOHA to provide a high bandwidth interactive return channel as it is described in the following section.

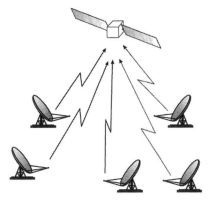

Figure 9.11. Random Multiple Access

Pure ALOHA. Pure ALOHA is the basic contention scheme proposed by Abramson in 1970. Pure ALOHA represents perhaps the simplest multiple access protocol since it is predicated upon the simple idea of letting users transmit whenever they have data to be sent; therefore, there is no need for synchronization. Although collisions are not avoided, a sender can always find out if the data has been transmitted by receiving an acknowledgement in the form of a short return packet. In the case of a GEO satellite the delay before the sender is informed whether the transmission was successful is around 250msec. When a collision occurs, the data need to be retransmitted and the sender waits an exponentially distributed random time interval before it retransmits its data in order to avoid another collision. A sketch of a collision is shown in Figure 9.12. *User A* attempts to occupy the channel at time t_1, however, *User B* has also data to be send at time t_2. Therefore, the collision is unavoidable and both users should retransmit their data later. A collision occurs even if the last bits of one frame overlap with the first bits of another frame.

Except for the advantage of simplicity, another advantage of pure ALOHA is that it is also possible to have unequal frame lengths. Notwithstanding, the throughput of an ALOHA system is maximized when all frames are of the same length, and hereafter we consider that all frames have the same lengths. If we assume an infinite number of users the arrival process of new frames is Poisson-distributed. Let us further assume that the probability P_{ta} of k transmission attempts per frame time (both original and retransmitted frames) is Poisson-distributed. Then, with mean value G per frame time.

$$P_{ta}[k] = \frac{G^k \, e^{-G}}{k!} \tag{9.5}$$

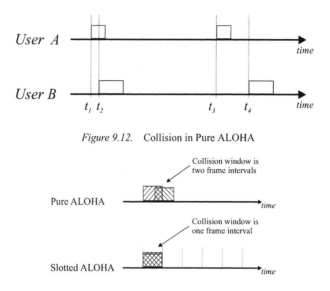

Figure 9.12. Collision in Pure ALOHA

Figure 9.13. Collision windows in Pure ALOHA and Slotted ALOHA

The probability of zero packets in the frame time is e^{-G}. In pure ALOHA, the vulnerable period is equal to the time interval of two frames as it can be seen in Figure 9.13 and the probability of zero packets in this interval is e^{-2G}. Thus, since throughput is given by $S = GP_o$ we get

$$S = G e^{-2G} \qquad (9.6)$$

Slotted ALOHA. This technique was initially proposed by Roberts in 1972 in order to increase the capacity of an ALOHA system. In slotted ALOHA time is divided in slots and transmissions from users are synchronized. Hence, partial collisions do not occur (Figure 9.13) and throughput is increased.

Since the collision window is equal to one frame time, throughput is given by $S = Ge^{-G}$. Slotted ALOHA peaks at $G = 1$, with a throughput about 0.368, while pure ALOHA peaks at $G = 0.5$ with a throughput around 0.184.

While in pure ALOHA each user transmits without any synchronization with other users, in slotted ALOHA terminals are only allowed to transmit within a certain time pattern, the throughput increases to

$$S_{\text{Slotted ALOHA}} = G e^{-G} \qquad (9.7)$$

In this case the maximum normalized throughput is $1/e$ occurring at $G^* = 1$.

Multiple ALOHA Multiple Access. If instead of one single narrow band channel of pure ALOHA, independent channels can be chosen at random, the

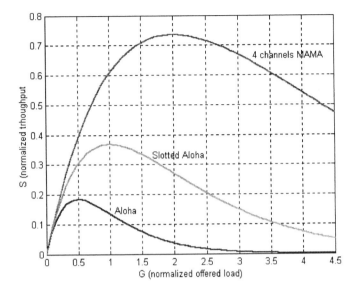

Figure 9.14. Throughput of ALOHA systems

random access is called Multiple ALOHA Multiple Access (MAMA) . The term was introduced by the inventor of ALOHA, Abramson. In this case the throughput is

$$S_{\text{MAMA}} = G\,e^{-2G/N}\,, \quad S_{\text{Slotted MAMA}} = G\,e^{-G/N} \qquad (9.8)$$

which means that N ALOHA channels have N times the throughput of one single channel when serving N times the offered load. Note that for an arbitrary number of channels N, the maximum normalized throughput is $N/(2\,e)$ occurring at $G^* = N/2$.

Figure 9.14 shows a comparison of pure ALOHA, slotted ALOHA and MAMA throughputs.

9.3.1.2   Spread ALOHA.   In this contention-oriented multiple access protocol, a unique code is used by all users. The receiver distinguishes between the received packets if there is a sufficient time offset between them. Note that Spread ALOHA can be viewed as a version of CDMA that uses a single, common code for all remote transmitters in the multiple access channel. The use of an unique code implies that the channel is still a contention one, similar in nature to ALOHA but different in two essential factors: the probability of having contention collisions is drastically reduced and there can be several users simultaneously on the channel as long as the multiuser interference keeps below

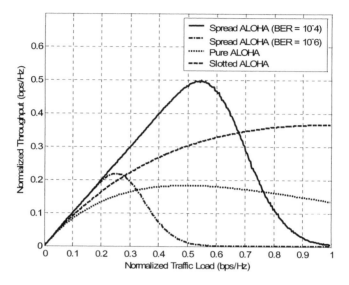

Figure 9.15. Throughput comparison for ALOHA and Spread ALOHA 1ns duration

a given level. Note that because of the elimination of multiple codes, many of the most complicated features required in a CDMA receiver can be removed. Channel coding and Automatic Repeat reQuest (ARQ) strategies can be used to mitigate error due to the interference of the multiple packets on-the-air.

There can still be two essentially different ways to design spread spectrum ALOHA with one single code [436]: Spread ALOHA Multiple Access (SAMA) and Code Reuse Multiple Access (CRMA). SAMA repeats the code or spread sequence every symbol packet. Conversely, CRMA uses one single code that is longer than the data symbol. However, the use of a single long code may not really bring advantage over repeating the code every symbol. The reason is that even though the collision probability decreases; the Spread ALOHA throughput is limited by interference rather than by collision as it is shown next.

Throughput performance of CDMA ALOHA has been well analyzed. [437] investigates the performance analysis of slotted ALOHA DS-CDMA when each user has been assigned a spreading code. In [438] the performance is analyzed with the same assumptions but including the effect of block FEC coding. Moreover not only the multiuser interference but also the effect of random access has been analyzed by assuming that a user randomly chooses one of the available spreading codes when it has a packet to transmit. In this case, the normalized throughput was found to be around 0.55, which doubles that for the bandwidth equivalent multi-channel slotted narrowband ALOHA.

The performance of the Spread ALOHA scheme can be estimated assuming Poisson packet generation. Considering unslotted transmission and no power control error, packet collision vulnerable zone is the first chip of every symbol. It can be shown [439] that packet collision probability is very low compared to packet discarding probability due to multiuser interference. Therefore, bit errors are caused by multiuser interference and noise. The bit error probability for this system considering both effects can be computed by the model in [440] which is a function of the system central parameters: E_b/N_o, Processing Gain , G_p, and number of simultaneous transmitting users, N_{INT}. If the required signal-to-noise plus interference ratio is computed as follows

$$\left[\frac{E_b}{N_o + I_o}\right] = \frac{\dfrac{E_b}{N_o}}{1 + \dfrac{N_{\text{INT}}\dfrac{E_b}{N_o}}{G_p}} \tag{9.9}$$

where E_b is the bit energy, N_o is the power noise density and N_{INT} is the number of interfering users. If throughput is normalized respect to the utilized bandwidth for a fair comparison, the resulting throughput is shown in Figure 9.15 compared to that of pure ALOHA for realistic system values. It can be observed that Spread ALOHA throughput maintains a linear relation up to the vicinity of its maximum throughput, for high processing gains can achieve as high as 0.70 throughput without FEC. It is important however to take into consideration the respective thresholds requirements. Specially, this argument is fundamental in systems where power is a scarce resource like in satellite systems. Since Spread ALOHA is limited by interference, receivers with multiuser detection capabilities will allow throughput to increase.

A number of refinements are possible to further increase the throughput like the use of packet combining and hybrid use of CDMA ALOHA and Spread ALOHA.

Packet combining consists of not discarding erroneous packets but retaining them at the receiver in order to combine them with successively received copies. Combination of corrupted packets at the codeword level allows to form a single more reliable packet, thus minimizing the average number of retransmissions as well as increasing throughput.

An example of hybrid use of CDMA ALOHA and Spread ALOHA is given in [441]. It consists of using Spread ALOHA in the packet header while CDMA ALOHA in the data. The receiver needs then to operate in two-stages. A first stage will detect the header detection using conventional spread spectrum receivers. A second stage will detect the data using a multiuser detector to allow for decoding of simultaneous users. It is shown that throughput can approach the same capability of joint detection that can be achieved by coordinated access (CDMA).

9.3.2   Contention-Based Reservation

Some multiple access protocols are based on channel reservation in order to avoid collision. Channel reservation is made via a subchannel such that only one station can access the channel at a time. Most reservation protocols adopt either a fixed assigned TDMA protocol or some variation of the Slotted ALOHA protocol. There is a trade-off between the channel stability and the channel control mechanism. While a TDMA protocol performs poorly for large number of users with bursty traffic, the Slotted ALOHA protocol is independent of the number of users. However Slotted ALOHA needs to be adaptively controlled for stable operation. Contention-based reservation protocols may increase significantly end-to-end delay. The minimum delay is more than twice the channel propagation time. This is an important consideration for satellite channels.

9.4   Demand Assignment Multiple Access (DAMA)

In a DAMA system , the network allocates system resources on a demand-assigned basis. With DAMA, satellite capacity is dynamically assigned to capacity requests coming from users and therefore capacity is made available when needed. DAMA multiple access is also referred to as Bandwidth on De-mand (BoD) . Note that the demanded resources can be frequency, time and/or codes. The demand assignment can be performed either by centralized or dis-tributed control. In a centralized architecture, the controller can be located at an Earth station or at a satellite with On-Board Processor (OBP). Note that DAMA techniques may introduce significant delay especially when the control is lo-cated on earth. In this case, the minimum delay is two round-trip times before the reservation request is granted. By using on-board controller, the minimum delay can be halved, but the satellite payload capacity is limited compared to a ground-based controller. It is possible to allow distributed control, in which capacity requests are broadcasted so as each station decides on whether demand capacity or not.

The two most commonly DAMA techniques used in satellite communica-tions systems are Packet Reservation Media Access (PRMA) and Combined Free - Demand Assignment Multiple Access (CF-DAMA).

9.4.1   Packet Reservation Media Access (PRMA)

The PRMA technique combines the ALOHA technique with TDMA on a reservation basis. It is primarily intended to be used in mobile satellite systems based on LEO satellite constellation. The multiple time slots of TDMA are assigned dynamically. The main problem of PRMA is the inherent Round Trip Delay (RTD) that typically varies from 5 to 30 ms, depending on the satellite constellation altitude and the minimum elevation angle acceptable by mobile terminals for reliable communications.

A PRMA carrier is divided in slots which are grouped together to form a frame. Each packet sent by the terminals has a header containing synchronization data as well as the identification of both the source terminal and the destination one.

The efficiency of the PRMA approach in managing the TDMA slot assignment is based on the use of a voice activity detector, which separates silent periods and active talking periods within a voice conversation. Only during the active periods, the terminals can transmit on a given slot of one frame. When there is a silent period, this slot can be assigned to another active source. As soon as a time slot is assigned to a given terminal, it becomes as unavailable. The system controller broadcasts the state of each slot to the terminals.

The access to an idle slot is based on ALOHA: as soon as a terminal needs to transmit its packets, it attempts to transmit a packet in the first available slot, according to some probability. If two or more terminals collide by trying to access to the same slot; no reservation is obtained and the involved terminals randomize their following transmission attempts. If the transmission attempt of a terminal on a slot has no collision, the terminal attains a reservation for the exclusive use of this slot in subsequent frames.

9.4.2   Combined Free-Demand Assignment Multiple Access (CF-DAMA)

A refinement of the DAMA techniques is CF-DAMA , which distributes unrequested bandwidth to terminals following a certain algorithm. This method is called free capacity assignment and improves the delay performance in low traffic conditions. Usually and in the absence of capacity requests from active users, the unused capacity will be freely assigned to terminals. This free assignment is granted to the terminal at the head of some free assignment table. In order to provide some fairness access to the free capacity, users that have not been served for a long time are tracked and given a better chance of obtaining a free capacity slot.

For low traffic load, the probability for a terminal to receive free capacity assignment is high and therefore the end-to-end delay can be improved. For heavy traffic load, this method performs like DAMA.

9.5   Multiuser Capacity

In wireless communications systems both overall and individual system throughput fully depend on the radio resource management strategies that are considered. The radio resources to be managed are essentially transmission power and bandwidth. One of the challenges that wireless systems introduce on system design is how to optimize such radio resource management given the time-varying nature of the impairments of the signal due to propagation, which can produce attenuations in the order of tens of dBs. These impairments are produced by the path loss, the multipath fading and the shadowing (see Chapter 3).

This time variability must be followed by the resource management strategies so as optimizing resource utilization (whenever economical considerations are left out). For example, dynamic bandwidth allocation techniques aim at adaptively optimizing the per-user and total bandwidth utilization while maximizing throughput.

From a theoretic information perspective, it is possible to find the limits of achievable set of rates of a multiuser system that any resource management strategy is able to obtain. In this section we briefly describe such limits, which are different for the uplink and the downlink multiuser channel . Both channels however can be seen as a dual of the other.

9.5.1   Downlink

The downlink channel can also be referred to as Broadcast Channel (BC). As said in the introduction, the multiuser capacity is not generally given by a single number but by the set of achievable rates or capacity region. The BC capacity is known for the degraded version of the channel, i.e. when the quality of the received power can be ordered.

9.5.1.1   Gaussian Channel Capacity.   Assuming AWGN and power spectral density of $N_o/2$ and degraded channel attenuation model for user k, a_k can be obtained by dividing the noise instead of multiplying the signal amplitude for which is possible to compute the capacity regions for system using Time Division Multiplexing (TDM), FDM and Code Division Multiplex (CDM). These regions are bounded by the set of rates achieved when all the resources are allocated to one user only, such bounding set is illustrated in Figure 9.16 and can be expressed by a vector of zeros except for the rate of that user, which achieves the single-user Shannon capacity

$$C_{\text{AWGN}}^{\text{BC}_{\text{lim}}} = B \log_2 \left(\frac{P}{n_k B} \right), k = 1, 2, ..., K \qquad (9.10)$$

An optimal use of resources will achieve any vector rate belonging to this bounding set. Assuming that power can vary with a total power constraint of P, the achievable rate region for TDM can be expressed as [21]

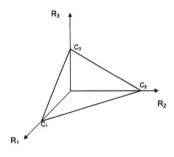

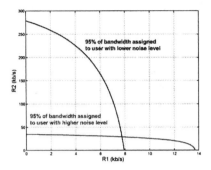

Figure 9.16. Set of bounding achievable rates (3 users) users

Figure 9.17. Capacity region for FDM (two users).

$$C_{\text{AWGN}}^{\text{BC,TDM}} = \bigcup_{\substack{\sum_k \tau_k = 1 \\ \sum_k \tau_k P_k = P}} \left\{ (R_1, R_2, ..., R_K) : \right.$$

$$\left. R_k = \tau_k B \log_2 \left(1 + \frac{P_k}{n_k B} \right) \right\} \qquad (9.11)$$

where P_k is the power allocated to user k, n_k is the kth user attenuated noise and τ_k is the percentage of time the channel is used by user k. For FDM, the capacity region can be expressed as

$$C_{\text{AWGN}}^{\text{BC,FDM}} = \bigcup_{\substack{\sum_k \alpha_k = 1 \\ \sum_k P_k = P}} \left\{ (R_1, R_2, ..., R_K) : \right.$$

$$\left. R_k = \alpha_k B \log_2 \left(1 + \frac{P_k}{n_k \alpha_k B} \right) \right\} \qquad (9.12)$$

where α_k is the percentage of bandwidth used by user k. Figure 9.17 shows an example of capacity region for FDM. In particular, it can be observed that assigning most of the power to the user with the worst channel the capacity region is considerably reduced. It can be shown that the CDM (Code Division Multiplexing) capacity region is the same as the FDM capacity region when the bandwidth is equally divided among all users.

By noting that in TDM power can be increased respect to FDM proportionally to the fraction of time the user accesses to the channel, 9.11 and 9.12 define exactly the same region. These capacity regions are actually suboptimal solutions since the optimal solution can be achieved with superposition coding and successive cancellation

$$
C_{\text{AWGN}}^{\text{BC}} = \bigcup_{\left\{ P_k : \sum_{k=1}^{K} P_k = P \right\}} \left\{ (R_1, R_2, ..., R_K) : \right.
$$

$$
\left. R_k = B \log_2 \left(1 + \frac{P_k}{n_k B + \sum_{j=1}^{K} P_j 1_{[n_k > n_j]}} \right) \right\} \qquad (9.13)
$$

where $1_{[x]}$ equals one if condition x is satisfied. An alternative expression can be found in Chapter 3.

9.5.1.2   Fading Channel Capacity.   When the channel is time variant the capacity can be defined either considering delay constraints or not. In either case Channel State Information (CSI) must be assumed and the capacity is defined in terms of the set of achievable rates averaged over all fading states for a given power policy.

If delay is not an issue, the capacity region (that may be called ergodic or throughput capacity) is given by

$$
C_{\text{fading}}^{\text{BC}}(P_m) = \bigcup_{P \in F_P} C^{\text{BC}}(P) \qquad (9.14)
$$

where F_P denotes the set of all policies that satisfy average power constraint P_m, and $C_{BC}(P)$ is the set of achievable rates averaged over all fading states for power policy P. If delay is an issue, this C cannot be achieved since it implies the use of long codes. A way of limiting the delay is to find the maximum rate that can be maintained over all fading states (also called outage capacity or delay-limited capacity).

Note that these capacities assume that power can be adapted to channel conditions. For satellite systems, it is common to assume constant power given the limitations imposed by the on-board amplifier. Adaptation however has been assumed for SDM (Space Division Multiplexing) satellite communications, also called multibeam systems. In a multibeam satellite system, adaptivity is enabled at transmission level by allowing a range of feasible spectral efficiencies (corresponding to different combinations of modulation formats and coding protection levels). In this case, the average spectral efficiency assuming a given frequency re-use factor can be expressed as

$$\eta\left(x,y\right) = \int_0^1 \eta\left(x,y,a\left(x,y\right)\right) p\left(a\left(x,y\right)\right) da\left(x,y\right) \qquad (9.15)$$

$$\bar{\eta} = \frac{1}{N_B} \sum_{b=1}^{N_B} \int \int_{(x,y)\in B} \eta\left(x,y\right) p\left(x,y\right) dx\,dy \qquad (9.16)$$

where $a(x,y)$ is the channel attenuation, N_B the total number of beams, and (x,y) is the user location. For the spectral efficiencies optimization both time and space probability of outage must be defined.

9.5.2   Uplink

The uplink channel can also be referred to as MAC. This channel consists of multiple transmitters sending data to one receiver.

9.5.2.1   Gaussian Channel Capacity.   Assuming AWGN and power spectral density of $N_0/2$ the uplink multiuser capacity region of a multiuser channel with two users is a pentagon (see Chapter 2). As it can be observed in Figure 3.4, surprisingly each of the two users is able to transmit at its single-user capacity up to a given point each. These two points enclose the optimal operation points of the multiple access. This capacity region can be achieved, as in the downlink, by a successive cancellation receiver [34] and the general expression for K users is given by

$$C_{\text{AWGN}}^{\text{MAC}} = \left\{ (R_1, R_2, ..., R_K) : \sum_{K\in S} R_k \leq B \log_2 \left(1 + \frac{\sum_{K\in S} h_k P}{N_0 B}\right) \right\}$$
$$(9.17)$$

where h_k is the channel attenuation. Note that an alternative expression can be found in Chapter 2. This region has exactly $K!$ corner points. As in the case of BC, the capacity region of orthogonal multiple access schemes are in general suboptimal from a theoretic information perspective. Moreover all orthogonal multiple access schemes have the same capacity region (as in the downlink).

9.5.2.2   Fading Channel Capacity.   As introduced in the downlink, the time variability of the uplink channel calls for different definitions of capacity. The same two definitions introduced for the downlink are also applicable for the uplink. The ergodic or throughput capacity assumes that rate vectors are averaged over all fading states and therefore this capacity usually refers to a fast-fading channel. Alternatively, the outage or delay-limited capacity is the maximum rate that can be maintained over all fading states above a certain threshold with a given probability (outage probability). The ergodic capacity region can be easily obtained by averaging the set of achievable rates over all fading states under a given power policy P [21]

$$C_{\text{fading}}^{\text{MAC}}\left(\mathbf{P}_m\right) = \bigcup_{P \in F_{\mathbf{P}_m}} C_{\text{MAC}}\left(P\right) \qquad (9.18)$$

where $\mathbf{P}_m$ is the vector of users average power constraint, and $F_{\mathbf{P}_m}$ is the set of power policies satisfying average per-user constraint.

The uplink outage capacity is more difficult to obtain. Moreover, while an outage is simple to define for a point-to-point transmission, a multiuser uplink outage consists of any realization of the fading state vector, where one or more sates are below the threshold defining an outage. In terms of power, this means that assuming CSI at both transmitter and receiver, when a user is in outage power can be saved by not transmitting at all. For the general case of an outage probability larger than zero, the multiuser capacity is obtained in [442].

9.6   Capacity Enhancement Techniques

This section provides in the first place an overview of the multi-user detection problem, placing some emphasis on interference cancelation techniques, both applicable in a general case and in code-sectored networks. Next, it dwells on single user detection techniques, and it finishes with a description of the turbo concept applied to multi-user detection.

9.6.1   Multi-User Detection (MUD)

A clear definition of the Multi-User Detection (MUD) problem is provided in [443]. It is defined as the process of detection of data from multiple users when observed in a non-orthogonal multiplex. Madhow [444] provides a generic definition of MUD that is slightly modified over here: *"Given multiple digitally modulated signals being heard simultaneously at a receiver, how does the receiver reliably demodulate a particular user or users of interest?"*

Typically, for the uplink of a wireless communication system, multiuser detection involves the detection of all the active users, may be in a joint manner resulting in very good performance but computationally complex detectors. For the downlink, the information about other users is usually not available in the mobile terminal, hence the detection is single user in nature but the detectors that take into account the multi-access interference outperform those that simply ignore it and are called as enhanced single-user detectors [445].

In the context of CDMA, it had long been thought that when sufficiently large numbers of users are sharing the channel and their received powers are almost the same then the applicability of the central limit theorem dictates that the multiple access interference can be modelled as additive white gaussian noise and hence the optimal receiver for CDMA in an AWGN channel is a matched filter detector or a Rake receiver in a multipath fading channel. This led to the reasoning that CDMA systems are inherently interference limited due to the poor performance of the matched filter detector in multi-access interference. This view was proved incorrect when in 1984, Verdu [446] showed that the received CDMA signal inherently consists of structured interference which could be exploited in an intelligent way such as in the manner of maximum a posteriori or maximum likelihood sequence detection to obtain much better performance than the matched filter detector. Thus theoretically speaking, CDMA systems are not interference limited if the optimal data detector [447] is employed. But the problem with optimal detector lies in the fact that its complexity grows exponentially with the number of active users in the system. This has led to a flurry of papers in the last 20 years on sub-optimal detectors having less complexity than the optimal detector yet obtaining appreciable performance gain over the traditional matched filter detector. A good deal of information on multiuser detectors can be found in [447] and particularly the ones suitable for mobile terminal implementation in [448]. A comprehensive review of multiuser detection schemes is presented in a review paper in [445].

9.6.1.1   Single Stage Interference Cancellation (IC).   The essence of interference cancellation techniques is the fact that if the multi-access as well as the multipath induced interference were known, it could be subtracted from the received signal and only the desired signal would remain. The fundamental blocks that comprise an IC unit are the interference estimation, regeneration and cancellation. The interference cancellation could be carried in a serial manner called as Serial Interference Cancellation (SIC) or in a parallel manner called as Parallel Interference Cancellation (PIC). If the process of interference cancellation is carried out only once, it constitutes a single stage IC.

Typically, multi-access interference cancellation is carried out only in the uplink where the parameters of all the active users are known. For the downlink, multipath interference cancellation for a single user could be carried out in a similar manner as the uplink multi-access interference cancellation.

9.6.1.2   Multistage IC.   The interference estimates that are available in the first stage of an IC scheme are not very reliable as the data estimate of each user is influenced by the multi-access and multi-path interference. So if a wrong estimate of interference is made, cancellation process actually increases the interference. To reduce this effect, interference cancellation is done in stages. In the first stage, full interference is not cancelled; rather only a part of it is cancelled. The amount of interference cancellation is determined by a factor called as partial interference cancellation coefficient that varies between 0 (no interference cancellation) and 1 (full interference cancellation). In the next stage, the interference estimates are relatively more reliable than the previous stage, as part of the interference is already cancelled in the earlier stage, so the value of interference cancellation coefficient may be increased. This process is repeated for a few stages where in the final stage, full interference may be cancelled. As the number of stages increase, the interference estimation process gets more and more correlated. Typical values for the number of stages are three or four.

9.6.1.3   Iterative IC.   The main problem in IC is the reliable interference estimation. The interference estimates that are provided by the front end, that usually consists of a bank of matched filters, are not very accurate. On the other hand, channel decoders can provide fairly good estimates of the transmitted data. So the interference cancellation and channel decoding can be combined in an iterative manner whereby each helps in the operation of the other. Interference estimates are generated after the channel decoding and interference cancellation helps in channel decoding. Such a scheme referred to as Iterative IC.

9.6.2   Interference Cancellation in Code-Sectored CDMA

In CDMA-based networks it is a straightforward solution to apply interference cancellation on the uplink. Central station receivers (which can be terrestrial base stations or satellite receivers) can incorporate the necessary signal processing elements. Furthermore, the central receiver knows the spreading codes of the served terminals, which is a need for interference cancellation. It is also possible to avoid power calibration of the terminals, because IC provides near-far resistant detection. Downlink IC requires the terminals to have information about the interference received from downlink data streams towards other terminals. Hence, knowledge of other spreading codes in the system is needed, which would lead to the increase of terminal receiver complexity. To overcome this problem, reduced complexity IC receivers can be applied, as considered in the studies [417] [449]. The main limiting factor of receiver complexity is the number of available matched filters (MFs). IC algorithms (as mentioned above) have filter banks matched to the interferers, then the output of these MFs are used for interference estimation and cancellation. If we limit the IC algorithm to the most common interferers, fewer MF banks are needed. Investigating a CDMA system given in Section 9.2.3.4, and considering a PIC receiver, the algorithm detects all users at the same time with a first bank of matched filters (MF). The output vector of the first MF bank $\hat{b}^{(0)}$ is used for respreading and subtracting from the delayed received signal, which then will be the input of a second MF bank to perform the final decision, which gives eq. (9.19) using the notations of Section 9.2.3.4:

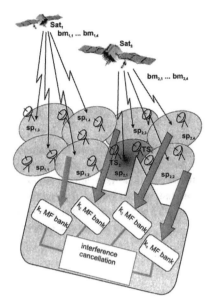

Figure 9.18. Allocation of MF banks in reduced complexity downlink PIC receiver

$$\hat{b}_k(i) = sgn\left[\int_{J_T}\left(r(t) - \sum_{j=1,j\neq k}^{K} A_j s_j(t-iT-\tau_j)\hat{b}_j^{(0)}(i)\right)s_k(t-\tau_k)dt\right]$$

(9.19)

Reduced Complexity PIC. Investigating the source of interference in a single-frequency PMP network, it can be revealed, that only a few sectors cause most of the interference in a given point of the service area. This way the investigated sector can be divided into sub-sectors comprising separate zones, which have common main interferers.

Receiver complexity can be reduced by assigning only a few branches of correlators (i.e. MF banks) to the most dominant interferers, without the significant decrease of interference suppression efficiency, as examined in [449]. In Figure 9.18 a reduced complexity PIC receiver is illustrated. The allocation of available MF banks is optimized in a way that only the most severe interferers are included in the interference cancellation algorithm. In the example of Figure 9.2 the receiver of TS_1 located in the spot $sp_{2,1}$ is detailed: the dominant interferers are spots $sp_{2,2}$, $sp_{2,3}$, and $sp_{2,4}$. In the case of TS_2 the neighboring spot $sp_{1,2}$ would produce interference and this spot would take part in the cancellation algorithm. This optimum allocation of signal processing capacity means a location dependent receiver algorithm. Interference zones of a sector will lead

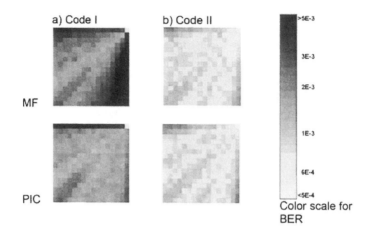

Figure 9.19. Downlink BER maps for the examined CDMA-based network applying Code system I and II, single user (MF) and PIC multiuser detectors, SNR=7 dB.
Original source: [417] (© IEEE 2003)

to subsectors in terms of multiuser detection algorithm, this way a further sectoring procedure is obtained with the aid of CDMA code sectoring combined with reduced complexity IC (for basic sectoring methods see Sec. 9.2.3.4).

Example. To demonstrate the effect of multiuser detection in CDMA code sectored multibeam networks, single frequency schemes are considered. Only the downlink direction is shown hereby. PIC is simulated with reduced complexity scenario (see above). In Figure 9.19 results obtained with matched filter (MF) and PIC receivers are depicted. MF results serve for comparison. Applying location dependent PIC receiver in a sub-sectoring scenario could significantly improve BER conditions of the border zone. However, the PIC receiver could not improve BER conditions significantly in the case of code system II, furthermore, worsening can be observed at the corners of maps b/PIC in Figure 9.19. In heavy interference situations PIC detector might even increase interference, as the algorithm may use the result of wrong initial decisions; this means a $\hat{b}^{(0)}$ vector with too many wrong decisions in eq. (9.19).

9.6.3   Enhanced Single User Detection for DS/CDMA Downlink

The enhanced single user detection schemes consist of chip level multipath equalizers, multipath interference cancellers and symbol level adaptive interference mitigating filters. The chip level equalizers operate on the assumption that if orthogonal walsh-hadamard codes are used in the downlink then the only source of interference comes from multipath propagation that destroy the orthogonality of the codes, hence if equalizers are used to used to restore the orthogonality of the codes, interference can be reduced. Numerous papers have appeared in the literature on this theme, such as [450], [451], [452], [453], [454], [455], [456]. A review on the adaptive methods for chip level equalization is presented in [457]. In [458] and [459], the authors present a SINR maximizing Rake receiver and its variants for downlink equalization. In [460], a generalized Rake receiver is presented for interference suppression.

The multipath interference cancellers (MPIC) operate on a similar principle to the parallel multiuser interference cancellers [461] but they cancel only the interference caused by multipath propagation. A detailed investigation of the multipath interference cancellation for High Speed Downlink Packet Access (HSDPA) mode of UMTS is carried out in [462] and it is shown that by using the MPIC the maximum peak throughput can be increased by a factor of about 2.1. In [463] and [464], MPIC and common channel interference cancellation in Rel. 99 UMTS and in HSDPA is investigated.

For low earth orbit satellite systems, successive interference cancellation has been studied in [465] and [466] and an adaptive parallel interference cancellation scheme in [467].

The symbol level interference mitigating detectors are typically adaptive filters that exploit the cyclostationary properties of the received signal at the symbol level (obtained by the use of short spreading codes) for suppressing the multi-access interference. Rapajic and Vucetic [468] proposed adaptive linear and decision feedback structures based on the use of known training sequences for multiple access interference elimination.

A breakthrough in adaptive interference mitigating detectors was the introduction of the blind adaptive detector [469], [470] and [444] that does not need the training sequence for its adaptation and needs only the signature sequence and timing of the desired user, i.e. the same knowledge that a conventional receiver requires. Further improvements of the original blind orthogonally anchored algorithm [469] were presented in [471], [472] by making the detector invariant to phase error on the desired carrier and resistant to large frequency shifts on the interfering carriers which may occur for signals coming from different LEO satellites. The blind detector was suggested to be quite suitable for satellite personal communication systems in [473] and in [474] where improvement in quality of service was shown in terms of outage probability. This analysis was performed in a simple AWGN channel. The performance of the blind detector in the satellite channel model [475] of the suburban and rural environments was investigated in [476]. Also an ASIC implementation of extended complex blind adaptive interference mitigating detector Extended Complex Blind Adaptive Interference Detector (EC-BAID) of [472] has been presented in [477] and is suggested for integration into the hand held user terminal for 3G mobile satellite communications. Further improvements in the blind Land Mobile Satellite (LMS) algorithm are reported in [478] and in [479] where filter-tap averaging is introduced and the step size of the LMS algorithm is optimised.

It was pointed out in the original paper [469] on the blind multiuser detector that it suffers from a mismatch problem in multipath fading channels or when the timing information about the signature sequence is not accurate, which could result in the suppression of the desired user. In [469], it was proposed to put a constraint on the Euclidean norm of the detector that acts additional background noise and prevents suppression of the desired user to some extent. This problem of mismatch has also been addressed in [480], [481], [482], [483], [484] and [485] where quite a few possible solutions are presented. Most of these rely in one form or another applying constraint(s) on the detector.

Particular attention has been directed in literature at the adaptive versions of the MMSE detector. The performance of adaptive detectors based on the MMSE criterion in flat-Rayleigh fading channels has been reported in [486], [487] and in frequency-selective fading channels in [488], [489]. Performance comparison of the linear minimum mean square error (LMMSE) detector before Rake combing and post Rake combining is presented in [490] where it is shown that the post Rake combining LMMSE has potentially larger capacity but it can not be used in fast fading channels whereas the precombining LMMSE has slightly worse capacity than the postcombining LMMSE but significantly larger than the conventional Rake receiver at the signal to noise ratios of interest. Among other available adaptive methods, a generalized side lobe canceller based partially adaptive receiver has been proposed for the downlink of mobile satellite systems in [491].

Another important development related to the blind detector was the signal sub-space concept investigated in [492] and it was shown that under this scheme, both the decorrelating and MMSE detectors could be obtained blindly and performed somewhat better than the Minimum Output Energy (MOE) based blind detector of [469] under the mismatch conditions, nevertheless the complexity of the subspace adaptive detectors is higher than the adaptive MOE detectors due to the estimation of signal subspace components. A comparison of the adaptive interference mitigating detectors including the subspace based ones has been presented in [493].

9.6.3.1 Downlink Synchronous CDMA Signal Model.

We consider a K-user, time-invariant, synchronous downlink CDMA system. The signal model, we use in the following, is similar to the one presented in [476]. The continuous-time received waveform in one symbol interval can be represented as

$$r(t) = \sum_{k=1}^{K} A_k b_k s_k(t) + \eta(t), t \in [0, T_s] \qquad (9.20)$$

where t represents continuous time, T_s is the symbol interval, $\eta(t)$ is the AWGN, A_k is the received amplitude of kth user that may be complex, b_k is the transmitted PSK symbol of the kth user, $s_k(t)$ is the real-valued spreading waveform associated with the kth user and is given by

$$s_k(t) = \sum_{j=0}^{N-1} c_k[j] \Phi(t - jT_c) \qquad (9.21)$$

where T_c is chip interval, $N = T_s/T_c$ is the spreading gain, $\Phi(t)$ is the chip pulse shape, which we assume a square pulse with support $[0, T_c)$ and $c_k[j] \in \pm 1/\sqrt{N}, j = 0, 1, ..., N-1$ is the normalized spreading sequence. We use short spreading sequences, i.e. the same spreading sequence is used to modulate each symbol. After passing the received signal through a chip matched filter and a chip rate sampler, an N vector r_i is obtained that consists of the samples of the output of the chip matched filter within the symbol T_s. Hence the discrete time representation of the received signal at time i (corresponding to i-th transmitted symbol) is

$$r_i = \sum_{k=1}^{K} A_k b_{i,k} s_k + n_i \qquad (9.22)$$

where $b_{i,k}$ is the i-th symbol transmitted by kth user, s_k is the normalised spreading vector of kth user and n_i is a complex baseband noise vector with NxN covariance matrix $\sigma^2 I$, where σ^2 is the variance of noise samples and I is identity matrix.

9.6.3.2   Linear Receivers for Synchronous CDMA.   The output of a linear receiver or a linear filter at time i can be written as $y_i = c_i^H r_i$, where r_i is the received vector and c_i is the filter's impulse response vector. For the matched filter or the conventional detector, $c_i = s_1$, where user 1 is the desired user. In the following, we describe the MMSE solution and its blind implementations.

MMSE Solution. The MMSE solution is obtained by minimising the mean squared error, defined as $e = E(|b_{i,1} - c_i^H r_i|^2)$ and is given by $c_i = R^{-1} p_i$ where

$$R = E[rr^H] = \sum_{k=1}^{K} A_k^2 s_k s_k^H + \sigma^2 I \qquad (9.23)$$

$$p_i = E[b_{i,1} r_i] = s_1 \qquad (9.24)$$

Direct Matrix Inversion based MMSE Detector. In direct matrix inversion method [494], the correlation matrix is approximated by time averaging and is then directly inverted for obtaining the MMSE solution for a block of received signals in a batch mode as shown below

$$R = \frac{1}{M} \sum_{j=0}^{M-1} r_j r_j^H \qquad (9.25)$$

$$c_i = \hat{R}^{-1} s_1 \qquad (9.26)$$

Blind Adaptive MOE Algorithms. In this approach, the detector's impulse response is decomposed into two vectors and is given as $c_i = s_1 + x_i$ where $c_i^H s_1 = 1$. It was shown in [469] that by keeping the s_1 (anchor) fixed and selecting x_i to minimise the MOE is proportional to minimising the Mean Square Error (MSE) and results in minimizing the interference plus noise at the detector output. The Blind Least Mean Squares (B-LMS) algorithm results in a simple adaptation rule for updating x_i and is given below

$$x_i = x_{i-1} - \mu[(s_1 + x_{i-1})^H r_i]^*[r_i - (s_1^H r_i)s_1] \qquad (9.27)$$

where μ is the step size of the algorithm. The complexity of the B-LMS algorithm is $O(N)$, where N is the spreading factor.

For the Blind-Recursive Least Square (B-RLS) algorithm [470], the following equations result

$$c_i = \frac{R_i^{-1} s_1}{s_1^H R_i^{-1} s_1} \qquad (9.28)$$

$$R_i = \sum_{j=1}^{i} w^{i-j} r_j r_j^H \qquad (9.29)$$

where w is the exponential weight factor, needed to discount the past data in a non-stationary data. It is such that $(0 < w < 1)$ and $(1 - w << 1)$. The Recursive Least Square (RLS) algorithm has higher complexity than the LMS algorithm and is $O(N^2)$ when the inverse of the matrix is calculated through the matrix inversion lemma [470].

9.6.3.3   Numerical Performance Comparison.   We simulate an interference limited synchronous CDMA system with binary antipodal modulation. The spreading factor is chosen to be 10 and there are 7 users in total. User 1 is the desired user and each of the 6 interfering users is 6 dB stronger than the desired user. The E_b/N_0 is 10 dB and the spreading sequences are chosen

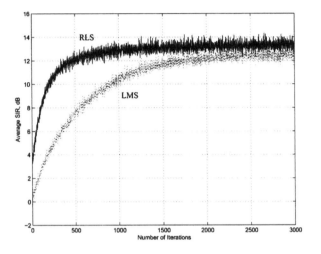

Figure 9.20. Comparison of blind LMS and RLS Algorithms

randomly. Figure 9.20 shows the averaged SIR (averaged over 400 simulation runs) of the B-LMS and B-RLS algorithms.

It is seen that the standard blind LMS detector reaches its steady-state SIR of about 12 dB in about 2000 iterations whereas the B-RLS does that in about 500 iterations. For these simulations, the step-size of the LMS algorithms was chosen to be 0.001 and the forgetting factor of the RLS algorithm was 1.0. Both were found to be optimal after some experimentation. These results illustrate the faster convergence of RLS algorithm over the LMS algorithm at the cost of higher complexity.

9.6.4 Turbo MUD

Turbo MUD gets inspiration from the "Turbo Principle" of the exchange of soft information between the multiuser detector and the channel decoders iteratively for joint multiuser detection and channel decoding [443]. The soft information is typically in the form of posterior symbol probabilities based on the given prior probabilities as well as the structure of the signal. Near-single user performance can be achieved by Turbo MUD though the complexity of these schemes is somewhat high. Typically Turbo MUD is used for the uplink whereas for the downlink, similar principle can be used for Turbo Equalization.

Since the Turbo MUD receivers have the drawback of the complexity, the attention has been addressed to IC schemes [495].

As it is known, as the number of decoding iterations increases, the coding gain offered by a turbo decoder becomes larger. However, the performance improvement obtained by turbo codes is remarkable in the first iterations, and more and more negligible in the successive ones. Hence, it is better to concentrate the significant part of interference-cancellation in the first iterations: for the same reason many IC based iterative receiver with a first linear stage have been proposed [496]. Nevertheless, a linear MUD has the drawback of an extremely high computational complexity.

A common feature of these algorithms is that single-user Soft-Input Soft-Output (SISO) decoders provide at each iteration an estimate of the *a posteriori* probabilities for the user coded symbols, which are used to form the soft estimates of interference to be subtracted from the received signal. As a result, the contribution of each user is effectively subtracted from the signal only if its symbol decisions are sufficiently reliable. For the sake of the error reduction in the estimation of the channel parameters, iterative interference cancellation schemes can be combined with iterative parameter estimation in order to improve the estimates with the iterations, as long as the signal is cleaned-up from interference.

In [497] the PIC detector is broken up so that it is possible to perform IC after each constituent convolutional decoder. Due to the tight relationship between the proposed receiver and the MAP decoders, it is defined as *MAP decoder aided PIC (MPIC)*: this solution aims to profit by IC introduction from the first iterations. Moreover, the variance of the noise-plus-(residual-)interference is determined by a new algorithm: particularly, only the most reliable symbols are used in variance determination, neglecting all the others. The performance of this Turbo-MUD in a synchronous AWGN channel is reported in Fig.9.21: the system has 10 equal-power users with PN short codes, processing gain $G=16$ and frame length 800. The quantized Log-MAP algorithm is used for the decoding. It is shown the performance improvement obtained using the enhanced variance estimator in comparison with the basic one. For the sake of the error reduction in the estimation of the channel parameters, iterative interference cancellation schemes can be combined with iterative parameter estimation in order to improve the estimates with the iterations, as long as the signal is cleaned-up from interference.

Turbo MUD in HAPs/Satellite Channel. Recently, a new iterative multiuser detector [498] has been proposed which is based on the utilization of a PIC and a bank of Turbo decoders coupled with a low-complexity iterative soft-PIC algorithm for the estimation of the channel parameters. In order to achieve a complexity which is polynomial in the number of users, the Expectation Maximization (EM) algorithm is applied locally [499], i.e. the true *a-posteriori* distribution of the missing data, given the observation and the current parameter estimate, is replaced by the product distribution induced by the a-posteriori

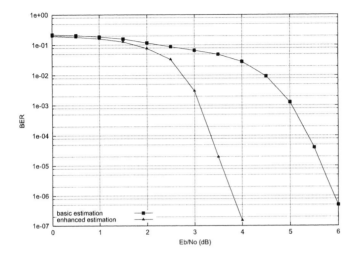

Figure 9.21. Performance comparison of the iterative PIC receiver with basic and enhanced variance estimator in a synchronous AWGN channel, with 10 equal-power users and processing gain equal to 16.

marginal probabilities which are the outputs of the the SISO decoders at each receiver iteration. The proposed approach is proved to be effective in the High Altitude Platforms (HAPs) and satellite communication channels [48], [67]. In these environments, the propagation conditions and the multipath components change quickly though only one signal component can be discriminated by the receiver. Hence, an accurate channel estimation becomes even more important.

The proposed system is is frame-oriented, i.e., encoding and decoding is performed frame-by-frame and users are synchronous also at frame level. The insertion of the training sequence in each frame takes to obtain frames whose length is equal to $L+T$, where L and T denote the number of symbols included in the block and in the training sequence, respectively. We assume also that the channel parameters remain constant over each frame.

In order to demonstrate the performance of the proposed receiver, we considered the following simulation settings, strictly inspired by the UMTS-TDD system:

- Spreading factor $SF = 16$;

- Rate $R_c = 1/2$ turbo code, composed by two 8-state RSC codes with generator polynomials $G_0 = (13)_8$, $G_1 = (15)_8$;

- Code block length $L = 1600$ coded symbols, corresponding to 800 information bits per frame;

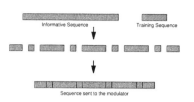

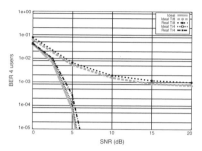

Figure 9.22. Satellite training sequence insertion inside the informative frame.

Figure 9.23. Performance of the iterative PIC with Ideal and Real knowledge of the channel parameters in satellite environment, 4 equalpower users, elevation angle equal to 60°.

- Training sequence length $T = 32$;

Moreover a fixed number of PIC iterations equal to 18 has been considered since this value permit to study the asymptotic behaviour.

Firstly the satellite channel model [48] has been considered with the following values: log-normal shadowing bandwidth $B_L = 0.8Hz$, fading bandwidth $B_R = 50Hz$, elevation angles of 60 and 20 degrees, best and worst cases respectively for 4 active users.

In order to follow the channel variations the training sequence has been divided on the overall frame length: particularly, since channel coefficients cannot be assumed constant over the entire frame in satellite channel, the 32 bit training sequences have been divided in 4 and 8 parts (see Fig. 9.22) in order to determine these parameters more effectively: for each of the training sequence segments, different value of channel parameters have been obtained. The relative results are shown in Fig. 9.23 and Fig. 9.24: as it is evident, better performance is achieved when the training sequence is divided in 4 parts: this behaviour indicate that too short training segments, i.e., to split the training sequences and the frame in too many parts, do not afford reliable estimates of the channel parameters.

Finally, we have considered the HAP channel model which is described in [67]: in this model the LOS component is assumed to be much stronger than the multipath ones and time-varying phenomena are nearly negligible. Therefore, the channel estimation results to be more reliable and interference cancellation can be performed efficiently. The results are reported in Fig. 9.25: better performance is achieved when the training sequence is divided in 4 parts.

As a conclusion, the iterative Turbo MUD approach with soft turbo decoder outputs for channel estimation can be beneficial even for challenging chan-

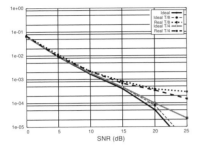 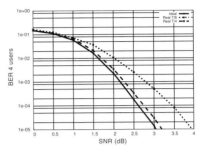

Figure 9.24. Performance of the iterative PIC with Ideal and Real knowledge of the channel parameters in satellite environment, 4 equal-power users, elevation angle equal to 20°.

Figure 9.25. Performance of the iterative PIC receiver with Ideal and Real knowledge of the channel parameters in HAPS environment for 4 equal-power users.

nels. Nevertheless fast varying channel conditions can reduce the Turbo Mud performance because of channel estimation limitation.

9.7 Conclusions

In this chapter an overview of digital communications under the perspective of a multiuser scenario has been presented. Conventional and recently proposed multiple access techniques for resource sharing have been described as well as some of their variations and new developments. Additional insight is provided by introducing a brief discussion on recent results on the limits of multiuser communications from an information theoretic perspective. Constraints and specific developments for satellite systems have been introduced for each of the techniques introduced.

In general, research on multiuser communications is a constantly evolving research area where cross-fertilization between signal processing, information theory and networking are yielding emerging communication technologies that nearly achieve the information theoretic limits of wireless multiple-access channels.

Chapter 10

SOFTWARE RADIO

L. S. Ronga[1], A. Cardilli[2], B. Eged[4], P. Horvath[4], W. Kogler[3], M. Wittig[5]

[1]*Consorzio Nazionale Interuniversitario per le Telecomunicazioni (CNIT), Italy*

[2]*University of Florence, Italy*

[3]*Graz University of Technology, Austria*

[4]*Budapest University of Technology and Economics, Hungary*

[5]*Guest author of "Phased array antenna systems" European Space agency (ESA)*

10.1   Introduction

An exact definition of Software Radio (SR) or Software-Defined Radio (SDR) does not exist yet. Also the vision, outlined in the first publications, changed from just defining a widely sensed architecture denoted as canonical software radio [500], over an arbitrarily reconfigurable hardware programmed with a high-level programming language [501] to a cognitive device reacting on environmental impacts like vibrations, velocity, position, speech and so on [502]. The common denominator of all definitions is a fairly *open hardware platform*, which can be easily reconfigured. Some applications, e.g., mobile radios, need to be reconfigured very frequently, in contrast to satellite terminals, which have to be adapted only a few times during their lifetime. For the latter, a kind of boot-cycle solution could be appropriate, but mobile radios might be adapted on the fly.

An interesting aspect is where the code for the software radio comes from. The air interface, for instance, needs a well defined configuration channel with an appropriate bandwidth to keep the configuration time low, which is not an issue for a simple I/O interface. However, the associated protocols and the reconfiguration process itself are out of scope with respect to this chapter; it is assumed that all Field Programmable Gate Array (FPGA) and/or Digital Signal Processor (DSP) components are loaded properly.

The chapter provides a comprehensive overview of the State-Of-The-Art and future trends in software radio concepts for space communications. Section 10.2

reports advances on analog signal processing and the conversion solutions that support the implementation of software radio. Section 10.3 present solutions for the implementation of complex computational structures on configurable devices. Payload architectures based on SR technologies are in Section 10.4 while earth terminals architectures are considered in 10.5. Section 10.6 illustrates the benefits of a network-driven configuration of terminals and payloads.

10.2 Antenna, Analog Processing and A/D Conversion

The general wireless transceiver architecture can be divided into two main parts: the digital back-end and the analog front-end. These parts are divided by the domain converters.

The currently available conversion technology usually doesn't permit the full digital implementation of a practical system. Therefore, the main functionality of the analog radio front-end is establishing the connection between the digitally processable and practically radiatable signals. This task is performed in both receive and transmit directions. Such transformations affect signal level and carrier frequency as well: the front-ends have to perform amplification and frequency extension tasks. The objectives for both transmit and receive direction are summarized in the following:

- Reject as many undesired signals as possible.

- Convert the desired signal to a frequency range which is compatible with that of the converters. The conversion should be done with minimal distortion.

- Amplify the desired signal to the level required by the converters with minimal distortion.

- Achieve a dynamic range that is compatible with the dynamic range of the converters.

In the transmit direction, the digitally produced and digital to analog converted signal should be translated to the output frequency and it should be amplified to the required level. In receive direction the low-level input signal should be translated to the frequency range of the analog to digital converter and the level should be amplified into the input level range of the converter. In both directions the analog front-end has to remove as much as possible of the unwanted signal components after and before the conversion in the transmitter and receiver, respectively. This means providing enough suppression of spurious components including noise. This process is easier in the transmitter, where all the spurious components are generated by the device itself and can therefore be controlled by the designer. The receiver path in the analog front-end is more complex in order to provide the required quality of analog signal processing. The main purpose of the receiver front-end is to isolate the desired signal from interference components (including out-of-band noise) prior to A/D conversion. This is the channelization behavior of the analog front-end. Generally we can find an analog front-end in a SDR with a bandwidth equal to the Nyquist bandwidth of the converters. This is the optimum as converters can neither generate nor process any wider bandwidth and the analog front-end doesn't limit the capabilities of the converters. If the system channels are narrower than this bandwidth, further channelisation (i.e. up/downconversion and sample rate conversion) is done in the digital domain. This is a typical Intermediate Frequency (IF) digitized SDR architecture .

Besides bandwidth, another important parameter is the actual frequency where the conversion is done. As we will see the value of the intermediate frequency is limited by the selectivity of the analog filters and the bandwidth of converters. Usually, the available converters allow for an input bandwidth as high as four to five times the Nyquist bandwidth (much less for very high-speed converters). This enables using sub-Nyquist sampling in the conversion stage. The frequency extension is realized by mixers and filters. In receiver stages the incoming wanted signal is mixed with a signal generated by a local oscillator to produce the intermediate frequency signal. The filtering is required to suppress the image signal which produces the same mixing product as the desired signal. If we want to place the input frequency and intermediate frequency far from

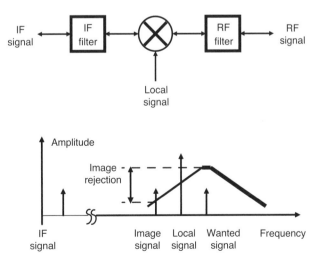

Figure 10.1. Frequency extension by mixing

each other, the requirements on the selectivity or shape factor becomes more and more stringent. To overcome this problem, practical systems could utilize many mixing stages to achieve enough image suppression between the IF and Radio Frequency (RF) bands.

In the transmit path the output RF filter is used to suppress the high level unwanted local spurious signals which would produce spurs in the final output signal. In this case the requirements on the output RF filter are more stringent as an unwanted local signal at the output of the mixer is stronger and closer to the wanted signal.

In receivers, the IF filters have to remove the high-level local spurious output of the mixers to avoid overload of the IF amplifier. In transmitters, the output images of the digital to analog converters in higher Nyquist bands have to be suppressed.

Generally one can say that employing a higher IF facilitates the implementation of the analog front-end, therefore the evolution of conversion technology will probably not eliminate the necessity for analog front-ends, but makes them simpler and cheaper.

Having defined the frequency plan of the analog front-end the next challenge is providing an adequate dynamic range. The dynamic range of present-day multi-bit converters can be as high as 80-90dB, mainly limited by the internally generated spurious signals of the converter itself instead of noise. If the input signal possess a wider dynamic range, the analog stages must limit the signal hitting the converter. In the design and implementation of the front-end the optimization of gain-distribution method can be used [503].

10.2.1   Phased Array Antenna Systems

The advantage of phased array antennas is that by simply changing the phase of the signal fed to the radiating element the overall radiation pattern can be changed. The attractiveness is that the beam steering is performed electronically, meaning fast and without all the caveats of mechanical beam steering mechanisms. Another advantage is that interfering signals can be reduced by creating nulls toward the direction of the interfering signal. Phased array antennas are implemented in large quantities for RADARs, both civilian and military. However, they are also very attractive for mobile satellite communications for both on-board an on-ground applications.

On 12 March 2005 the first satellite for mobile communications with digital beamforming networks on board, Inmarsat IV, was launched by an Atlas V from Cape Caneveral, a illustration is shown in figure 10.2. This satellite is dominated by its large antenna, a 9.6 m diameter reflector for the L-band signals. A feed array consisting of 120 helical feeds shown in figure 10.3 generate the mobile link beams. A C-band link provides the uplink from the gateways and after

Figure 10.3. Implementation of an 4 antenna element

downconversion to L-band a digital signal processor performs channel filtering, channel switching and finally digital beamforming. More than 200 user beams can be generated by the digital beamforming network. The return link from the user towards the gateway is the mirror image, first channel filtering, followed by a Digital BeamForming Network (DBFN) and channel switching. Direct user-to-user links are also possible, as well as direct feeder to feeder links in C-band. The DSP consists of 25 modules integrated onto a baseplate which has the main task to transport the heat of about 900 W to the satellite radiators

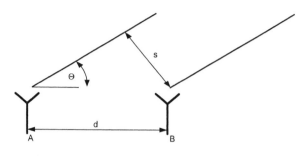

Figure 10.4. Two radiating elements

via heat pipes. Each module contains several multi-chip modules with ASICs dedicated to the signal processing functions.

The DBFN ASIC's receive the signals in a Time Division Multiplexing (TDM) frame from the digital channel switch. For each frame the complex multiplication with a weighting function is performed, followed by an accumulation. Channels that have the same frequency in different beams will have different weights applied to them but are required to be in the same TDM slot on the mobile side. Therefore after the weighting function, these channels samples need to be summed. The output of the accumulator is stored and then output at the appropriate time. This is achieved by using a traffic memory with independent write and read functions.

Another interesting development is a Ka-band phased array antenna for mobile applications. The vision for this development was to provide a modular electronically steerable antenna with digital beamforming and tracking for mobile Ka-band users, such as aircrafts. The antenna consists of an antenna array, the RF and IF components and than the transition into the digital domain. A digital downconverter is followed by the Direction Of Arrival (DOA) estimation and the digital beam-former. The next module is the conventional communication chain of demodulation and decoding. A four by four transmit and receive element antenna was realized and extensively tested. The digital beamforming is performed by using complex multipliers for the I and Q signals. All the digital signal processing functions are implemented in the DSP located at the main board of the antenna module. One interesting feature is that the signal from the satellite is used to calibrates the array using the DBFN. The realized four by four building block allows to assemble larger phased array antennas to achieve the required EIRP and G/T needed for broadband communication with Ka-band satellites.

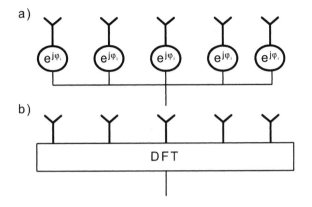

Figure 10.5.   a) Conventional beamforming network implementation and b) the digital representation

10.2.1.1   Basic Theory.   If the beam is inclined by an angle Θ, the path difference s is $s = d \cdot \sin(\Theta)$. The resulting phase shift of the wave emitted by A compared with the wave emitted by B is denoted in equation (10.1) and visualized in figure 10.4.

$$\varphi = 2\pi \sin(\Theta)\frac{d}{\lambda} \tag{10.1}$$

For a phase difference $\varphi = \pi$ destructive interference occurs, resulting in a cancelation of the emitted wave by the radiating elements A and B. The angle Θ_0 were this happens can be approximated for small values described in equation (10.2)

$$\Theta_0 = \arcsin\left(\frac{\lambda}{2d}\right) \approx \frac{\lambda}{2d} \tag{10.2}$$

The beam direction of the emitted wave can be changed, by introducing a phase shift φ between the signals emitted by A and B. This is the basic principle of a phased array antenna, which allows to steer the beam into any desired direction. For M radiating elements as shown in figure 10.5, the emitted wave is obtained by weighting the individual radiating pattern with the phase of the signal fed to the radiator. Equation (10.3) points out to become the Fourier transform:

$$E = \sum_i x_i e^{j2\pi ft} \cdot e^{j2\pi \sin(\Theta)\frac{d}{\lambda}} \tag{10.3}$$

This consideration leads to two beamforming network implementations, as shown in figure 10.5, where each radiating element is preceded by a phase shifter or a Fast Fourier Transform (FFT) is calculated for each beam. One analogue implementation of a beamforming network is using the Butler matrix. Such implementations are applied for transmit stages.

More efficient implementations of beamforming networks are possible by applying digital signal processing. The basic building blocks for a beam-former are multipliers and adders/accumulators. One 8 bit × 8 bit multiplier can be implemented by using 150 logical cells of an FPGA, a 16 bit × 16 bit multiplier requires about 600 logical cells. A 16 bit pipelined adder is implemented by 110 logical cells. For Application-Specific Integrated Circuit (ASIC) implementation the complexity can be estimated by assuming that one logical cell consists of about 20 gates.

A phase shifter can be realized with two multipliers, one look-up table and one adder. The input signal must be available as in-phase and quadrature-phase signal and the phase shifted output signal is a real signal. If a complex output signal is required, the phase shifter consists simply of four multipliers and two adders. For narrow band signals, phase shifters can be realized by delay lines. To compute the discrete Fourier transform, two multipliers two accumulators are necessary.

10.2.2   Sampling Principles

The conversion from the analog to digital domain is performed by sampling and quantization. In the previous chapter, all aspects of the conversion process are discussed. Basically sampling and mixing are related processes, therefore in some cases they are merged.

To gain advantage of this relation, the property of the mixer is pointed out. Its task is to shift the modulated signal down to baseband, and reconstruct the original complex baseband signal. This performs a complex mixer, which basically consists of two real mixers, where the carrier frequencies have a phase offset of $\pi/2$. This phase shift represents a cosine and a sine wave, which are orthogonal signals, and represents the real and imaginary axis of the resulting complex signal.

This separation can also be performed by a special non-uniform sampling. All possible methods of generating the complex baseband signal are illustrated in Figure 10.6 and discussed in the following chapters. Digital receivers must derive sufficient statistics of the received signal to derive the necessary synchronization parameters and to be able to perform detection tasks. Some bandlimiting must be done preceding the Analog-to-Digital Converter (ADC) to prevent aliasing in the discrete-time signal. It was derived in [504] that sufficient statistics can be obtained from sampling this analog prefilter as long as

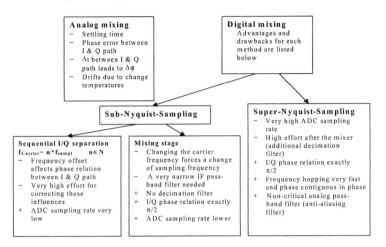

Figure 10.6. Diagram of the sampling methods

the following relationships for its transfer function is met:

$$H(f) = \begin{cases} \text{arbitrary} \neq 0, & |f| < B \\ \text{arbitrary}, & B \leq |f| < B_F \\ 0, & B_F < |f| < f_{\text{samp}} - B \end{cases} \quad (10.4)$$

where B is the useful signal bandwidth and B_F is the prefilter bandwidth. In this case, there always exists a realizable discrete time filter which compensates for the in-band distortion caused by $H(f)$ (which can be for instance incorporated into the digital matched filter). Another finding is that even if f_{samp} is incommensurate with the symbol rate, a discrete time interpolator can reconstruct the optimum sampling instants. This means that a fixed sampling clock can be used in the receiver without loss of performance. Practical receivers have a prefilter bandwidth slightly higher than necessary to allow for distortionless transmission even if the received signal has some frequency offset.

10.2.3 Sub-Nyquist Sampling

The sampling theorem is only satisfied for the bandwidth to be converted, and not for the highest frequency component contained. This implies that the sampling frequency must be at least twice the bandwidth to be observed. Sampling is mathematically performed by multiplying the analog signal with an infinitely long series of equally spaced Dirac pulses. Obviously, the distance between these pulses is determined by the sampling frequency. Multiplications in the time domain correspond to the convolution in the frequency domain.

Applying this to the sampling process, the result is a periodically repeated frequency spectrum folded around multiples of the sampling frequency. To be specific, a frequency located at $(n \cdot f_{sampl} \pm f_1)$ will also appear at f_1 in baseband. This shows obviously the need of a narrow IF-band filter, which might be tuneable to select the correct carrier. Figure 10.7 illustrates the IF-band around a multiple of the sampling frequency, which is folded into the baseband. But not only the IF-band, also the aliased frequency bands would be mixed down. This effect can be used for two different kinds of obtaining the complex baseband signal, described in detail in the next two subsections.

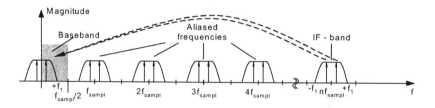

Figure 10.7.   Spectrum due to equally spaced sub-Nyquist sampling

10.2.3.1   Mixing Stage.   The intuitive solution is to use sampling as a frequency shift, such that the multiple of the sampling frequency behaves equal to the local oscillator of a conventional mixer. The upper or lower sideband has to be selected by an analog bandpass filter. Note that the passband of this filter must satisfy the sampling theory.

After sampling the modulated signal is still a passband signal, merely the center frequency is much smaller. For this new constellation, the modulated signal is super-Nyquist sampled, hence to obtain the complex baseband signal an additional I/Q mixer has to be used. The new digital IF frequency is determined by equation (10.5). Figure 10.8 illustrates all aliased frequencies and the transfer function of the analog passband filter. The drawback of this solution for a Multi Frequency TDMA (MF-TDMA) modem is obviously the tight analog filter, which must be tuneable in case of frequency hopping. Additionally, the demand of a fixed oversampling rate for the following matched filter enables only a limited values of possible $n \cdot f_{sampl}$ mixing frequencies., which furthermore, changes with the symbol rate of the modulated signal. All together, this is technically very interesting, but has tremendous drawbacks in flexibility.

$$f_{IFD} = f_{Carrier} - n \cdot f_{sampl} \quad \left| \begin{array}{l} f_{IFD} + B < f_{sampl}/2 \\ f_{IFD} - B < 0 \end{array} \right. \quad (10.5)$$

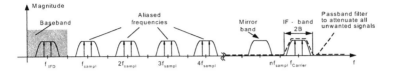

Figure 10.8. Sampling process as mixer stage

10.2.3.2   Sequential I/Q Separation.   Sub-Nyquist sampling maps all frequency located around $(n \cdot f_{sampl} \pm f_1)$ to f_1 in the baseband. Therefore, a modulated signal having a symmetrical spectrum, which is located around $(n \cdot f_{sampl})$, can be directly demodulated if the carrier frequency is equal to $(n \cdot f_{sampl})$. Figure 10.7 illustrates the spectrum of a BPSK modulated signal, which has symmetrical spectrum.

Complex symbols, which are used to form QPSK signals have an odd symmetrical spectrum. A closer look to the time signal gives an idea to obtain the complex baseband samples with sub-Nyquist sampling methods. The information of the symbols is carried in particular phases of the carrier frequency; in case of a rectangular pulse used for the baseband symbols, the carrier phase will be constant over the entire symbol as defined in equation (10.5). But in case of a Nyquist filtered pulse, the phase coincides with the ideal phase constellation only at the sampling instant. Equally spaced sampling obtains information of the sine wave only at one specific phase, which is not sufficient to reconstruct all points of the phase constellation. Therefore, a second sample with different phase, (different time instant) provides information about a second axis in the complex plain. These two samples can be considered as vectors in the complex plain. To obtain orthogonal vectors, the second sampling instant must be spaced $\pi/2$ radians apart. In this case, the obtained vectors represent the complex plain, which contains the modulated symbols. Figure 10.9 illustrates the sampling instants obtained by non-uniform sampling. If the two samples are spaced exactly $\pi/2$ radians apart, the complex plain is correctly defined, as indicated in plot b). The phase offset between recovered I/Q path and the ideal Im/Re axis of the modulated signal can be easily corrected. In contrast, any variation of the sampling instants produces intersymbol interference, and

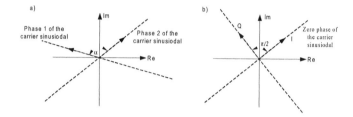

Figure 10.9.   Complex plain defined by non-uniform sampling

therefore a degradation of the bit error rate visualized in plot a).

$$s(t) \;=\; A \cdot \cos\left(2\pi f t + \varphi_i \cdot \prod \frac{(i \cdot T_s - t)}{T_s}\right)$$

$$\varphi_i \;\in\; \left[\frac{\pi}{4}, \frac{3\pi}{4}, \frac{5\pi}{4}, \frac{7\pi}{4}\right]$$

$$\prod \frac{(i \cdot T_s - t)}{T_s} \;=\; \begin{cases} 0 & t < i \cdot T_s \\ 1 & i \cdot T_s \le t \le i \cdot T_s + T_s \\ 0 & t > i \cdot T_s \end{cases} \qquad (10.6)$$

The time difference for the second non-uniform samples is described in equation (10.7). Note that the time difference is determined by the IF carrier, which leads to a very small value. Additionally, this value and also the sampling frequency itself should match exactly, which is nearly impossible due to variations of the carrier frequency. Moreover, the ideal sampling instant must be estimated out of the IF-band to avoid another intersymbol interference. The major drawback of this method is the dependency of the sampling frequency and the carrier frequency, which makes frequency hopping nearly impossible

$$\Delta t_{sampl} = \frac{\varphi}{2\pi f_{Carrier}} = \frac{\pi/2}{2\pi \cdot f_{Carrier}} = \frac{1}{4 \cdot f_{Carrier}} \qquad (10.7)$$

10.2.4   Super-Nyquist Sampling

The relevant frequency for determining the sampling rate is the highest frequency component of the modulated signal, which is approximately determined by the carrier frequency plus half the bandwidth of the signal. This forces the ADC to operate at very high sampling rates.

The result of the sampling process is the original IF signal in digitized form, hence a digital complex mixer must shift the signal down to baseband and recover the complex signals. The digital frequency for the mixer is produced by a numerical controlled oscillator, where the frequency can easily be changed via a tuning word. Furthermore, the sampling frequency of the mixer (also ADC) and the matched filter have to be matched. This decimation process operates at very high frequencies, which is only possible by using FPGA solutions. The main advantage of super-Nyquist sampling is the non-critical analog filter, which preselects the complete hopping range of the modem.

10.2.5   High-speed ADC Technologies

It has been proven that ADCs are more critical devices than DACs and their performance can be a real bottleneck in the system. Today's conversion technology used for wireless RF/IF applications uses conventional purely electronic devices. State-of-art ADC sampling rates reach to hundreds Ms/s and even up to a few Gs/s providing $10 \dots 14$ bit resolution. These converters are prevalently flash, pipelined or folding & interpolating designs [505].

- *Flash* converters have highly parallel architectures. They simultaneously compare the voltage to $2^n - 1$ reference levels, separated by the voltage corresponding to 1 LSB. This can be realized principally using $2^n - 1$ comparators and some reference array like a resistor ladder. The output of the comparators are potentially error-corrected and converted to the appropriate representation. These converters can be very fast but their resolution is limited by the exponential growth of the comparator number and by the associated problems. Flash ADCs are often embedded in other types of converters.

- *Pipeline* technology employs a "divide and conquer" approach by using subranging. A pipeline ADC consist of a series of low-resolution ADCs and DACs. In each stage, the ADC quantizes its input signal and the Digital-to-Analog Conversion (DAC) converts this coarse estimate into an analog waveform. The output signal, called residue, the difference between the input voltage and the quantized voltage enters the next stage after appropriate amplification to preserve dynamic range. ADC outputs are combined by a calibration and correction logic. Each sample progresses one stage per clock cycle, hence the name of this architecture. Pipelining means introducing intermediate registers into the signal path which can reduce the timing requirements at the expense of increased latency. Pipeline converters provide reasonable resolution and conversion rate while maintaining small power consumption and chip area.

- *Folding and interpolating* ADCs are basically two-stage pipeline converters but they solve the subranging without the need for DACs and subtractors. The input voltage is *folded* by using a nonlinear circuit into folds. Two coarse low-resolution flash ADCs are embedded in the device. The MSBs are obtained by quantizing the original voltage, and the LSBs are resolved using the folded voltage. Folding multiplies the frequency of the input signal, considerably increasing the operating frequency requirement. The number of LSBs can by increased by an interpolation operation without introducing more folding stages.

There are commercial modules which utilize multirate signal processing techniques in order to deliver a very high speed converter by postprocessing the signal of multiple lower speed ADCs.

Electronic ADCs have inherent limitations [506] that render considerable increase of their sampling rate and resolution unfeasible. Thermal noise, aperture jitter of the sampling clock and comparator ambiguity pose practical limits on resolution and conversion rate. Considerable research is done on optoelectronic ADCs, dating back to the seventies. Although they face many practical problems to be solved, they might allow feasible direct sampling of millimeter-wave RF signals in the future.

It was demonstrated that optical pulses from a mode-locked laser source exhibit far less jitter when compared to electrical clock sources. There are efforts to build hybrid (optic-electronic) devices and fully optical ADCs.

Hybrid devices use the optical pulses for sampling the input voltage and subsequently quantize it by electronic means. The samples are converted into the electrical domain and distributed among many slow (in terms of the sampling frequency) ADCs in some interleaved manner. The interleaving can be done in the time domain or in the frequency domain by using standard wavelength-division techniques.

Optic devices do the quantization in the optical domain as well, they provide digitally encoded outputs. The first devices were folding ADCs where the folding was achieved by parallel Mach-Zehnder interferometers with different (originally always doubling) electrode length. These interferometers exhibit periodic raised-cosine-like nonlinearity between their electric ports with a period inversely proportional to the electrode length. Thus, the bank of interferometers with subsequent photodetectors and comparators yields directly a Gray-encoded ADC output. Recent developments made such structures more feasible and a wide variety of other working principles appeared as well [507].

10.3   Optimized Computational Structures

The main enabling technologies for implementation of software radio architectures are FPGA and DSP. Both have peculiarities that can be efficiently exploited to obtain high performance devices. The following sections provide optimized implementation solutions for the two technologies and design indication for a feasible distribution of the required computational power.

10.3.1 Implementation for FPGA

In this section, design structures for FPGA solutions are investigated, in particular with respect to Time Division Multiple Access (TDMA) modems. Of course, some communication units might be suitable for realization in DSPs as well, which depends mainly on the application itself. In contrast to FPGA solutions, DSPs are more flexible in terms of reconfiguration; on the other hand, they are much slower concerning the number of operations per second, which would have an impact on the achievable symbol rate. With DSPs, the implementation follows a serial pattern, whereas massive parallel computing is possible with FPGAs.

10.3.1.1 Matched Filtering and Baseband Processing. For AWGN channels, matched filters provide the maximum Signal to Noise Ratio (SNR) in the optimum sampling instant [508]. Wherever possible, Finite Impulse Response (FIR) structures [509] are used in digital communications, in particular if raised-cosines are used for baseband shaping as it is most times the case with satellite systems. With respect to FPGAs, the delay line is implemented with logic elements, whereas the multipliers (taps) for weighting each sample are dedicated DSP blocks; the latter are already available with the newest FPGA products, e. g., the Stratix series from Altera.

If multipliers are to be avoided at any cost, a kind of noise shaping could be performed [510]. The basic idea is to use the $\Sigma\Delta$ principle to reduce the resolution of the sampled signal to, say, 4 bits, such that the SNR remains very high in the band of interest and shrinks dramatically out of this band. The overall SNR coincides with the theoretically achievable SNR of 4 bit but, due to noise shaping, it does not show a white distribution. The following matched filter attenuates the unwanted band. Because of the reduced resolution, the multipliers can be replaced by look-up tables containing the pre-computed filter taps. The same idea can be applied to matched filters in the transmitter path. The data source for this filter is the symbol mapper. The resolution of the data flow depends on the modulation scheme. For example, QPSK symbols have 1 bit per dimension (real and imaginary part), 16-QAM already 2 bit and so on. Another simplification may be exploited if the impulse response exhibits an even symmetry like raised-cosines, which halves the number of LUT's, which hold the pre-computed filter weights. Figure 10.10 depicts the block diagram of the reduced FIR filter. The main benefit of this structure is the extremely efficient implementation, which allows the design of FIR structures up to 64 taps with a maximum clock speed of 100MHz with state-of-the-art FPGAs.

The discrepancy between theory and implementation of filter functions depends on the length of the FIR device in terms of symbols. Hence, a 64 tap FIR can store about 21 symbols in case three times oversampling is assumed.

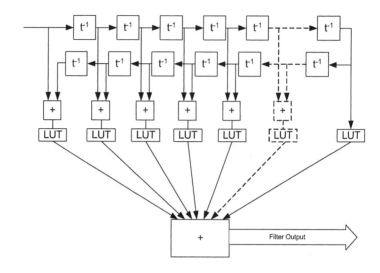

Figure 10.10. *FIR* structure with multipliers replaced by look-up tables

Furthermore, this yields a stopband attenuation of more than 60 dB. The latter is an important figure of merit in the design of digital modems, in particular with respect to the TX path, because the out-of-band power is limited and well defined in nearly all the standards for digital communications [511].

In FPGAs, any mathematical functions may be realized with look-up tables. But some functions, like the arcus-tangent or the square root would definitely exceed the internal memory of current FPGAs. In the following, two algorithms are discussed, which are not so much part of up- and down-converters; instead, they are frequently used for acquisition and tracking purposes. For the implementation of an arcus-tangent function, as it is employed to evaluate the argument of a complex number (required by many algorithms for carrier synchronization), an iterative solution known as the so-called CORDIC [512] is widely applied . This algorithm determines in each cycle a bit of the argument of a complex number.

Also for square-root functions, as they are necessary for the computation of correlation thresholds or the magnitude of a complex number during baseband processing, many iterative algorithms do exist. But sometimes the result must be finished within one clock cycle, where accuracy is not so important. The basic idea for this sort of approximation is that the input number $x > 0$ is

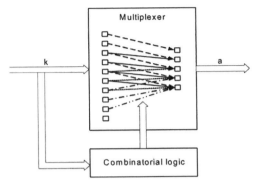

Figure 10.11. Block diagram of the square-root approximation

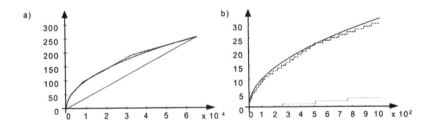

Figure 10.12. Square-root approximation performances

stepwise approximated by

$$y \quad = \quad \frac{x}{2^{2 \cdot \lceil ld \frac{x}{4} \rceil}} + 2^{2 \cdot \lceil \left(ld \frac{x}{4} \right) - 1 \rceil}$$

y ... approximated square root of x

x ... input number   (10.8)

Figure 10.11 visualizes the implementation basically consisting of a combinatorial logic and a multiplexer. The former determines the size of the input number in order to initialize the latter, which is more or less a simple divider. Figure 10.12 shows the result of the algorithm for two different ranges of x. The solid line indicates the ideal square root function approximated by the dotted one. In order to visualize the simple truncation of the least significant bit, corresponding to a static division, the straight line is drawn.

10.3.2   Implementation for DSP

DSPs are usually high-speed Harvard architecture processors capable of doing various fixed-point and/or floating-point arithmetic operations very efficiently, integrated with on-chip I/O-controller, direct memory access (DMA) interface and memory. The computing core provide instructions and addressing features specific to signal processing algorithms. These include multiply-accumulate for filtering, add-compare-select for Viterbi decoding, bit reversal, circular buffers, specific arithmetic modes like saturation etc.

DSPs offer some additional flexibility in the sense that it can dynamically balance the processing power between different tasks. It can, for instance, afford a better channel equalizer if working under difficult conditions with reduced functionality in other stages.

Performance comparison with FPGAs and even among different DSP families is not obvious. To contrast the computing power (expressed as the number of instructions or floating-point operations per second) is unfair as many processors utilize single instruction, multiple data (SIMD) architecture and highly parallelized computational units. Signal processing specific benchmarks like the number of multiply-accumulate operations (MACs) per second, number of FFTs with given complexity, number of matrix multiplications etc. can be compared. The best insight can be gained by comparing the performance in specific applications (equalization, demodulation, Viterbi decoding etc.)

As DSPs are often bandwidth-limited, they are often augmented with down or upconverter ASICs ([513][514][515]) or functionally equivalent FPGAs. These devices interface directly to high-speed IF sampling ADCs and DACs, respectively, and perform common channelization tasks. Using this approach means that the complex baseband signal is generated and processed in the DSP. The ASIC performs usually sample rate conversion employing a chain of fixed and/or programmable digital filters and digital mixing using a Numerically Controlled Oscillator (NCO) based digital frequency synthesizer. A downconverter is shown schematically in Fig. 10.13

The filter chain commonly consists one or more of the following filter types:

- Stages running at high sample rates use cascaded integrator-comb (CIC) filters [516] which are multiplierless structures with the drawback of monotonically decreasing transfer function.

- Half-band filters which are fixed-coefficient FIR filters with near half of their coefficients being zero. They bandlimit the input signal to the half of the Nyquist frequency, hence their name.

- Programmable-coefficient FIR filters in low sampling rate stages.

The DSP runs in typical applications at few times the symbol/chip rate. Sample rate conversion in the ASIC is mostly programmable but restricted to rational

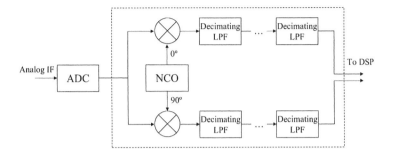

Figure 10.13. Block scheme of a digital downconverter

factors; resampling can be done in the DSP at a low sample rate if necessary. The partitioning approach offers a less general computing scheme but this is adequate for many common standards. Recent ASICs contain a few uniform parallel branches with internal routing capabilities allowing for multirate processing and load balancing.

10.3.3   Distribution of Computational Tasks to DSP and FPGA

The highest degree of flexibility would be obtained if both the ADC and the DAC could be placed as near to the antenna as possible. In this case, only analog components, such as amplifiers and anti-aliasing filters, are needed. It is to be noticed, however, that satellite communications systems use carrier frequencies in the GHz range, which requires very fast AD/DA conversion. Super-Nyquist sampling is then out of scope for state-of-the-art converters, not at least, because also part of the data processing has to run at this speed.

For direct AD conversion, sub-Nyquist sampling, where the sampling process acts as a mixing stage, is sometimes applied to cover the L-band (1 - 1.5 GHz). Tuneable synthesizers and analog anti-alias filters guarantee the proper operation of the AD converter with respect to the band of interest.

Clearly, the same idea can be used for DA conversion because the sampling process generates a periodical frequency response, where the sampling frequency defines the period. Unfortunately, the frequency response is weighted by the sinc-pulse. The latter stems from the rectangular shape of the sample-and-hold device and must be equalized adequately. Anyway, tuneable synthesizers and analog filters are needed in this case as well.

The sub-Nyquist concept is mainly restricted by the tuneable analog components mentioned previously. Therefore, it is normally suggested to shift the

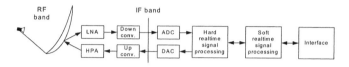

Figure 10.14.   Software radio communications system

band of interest down to an appropriate IF of, say, 70 MHz, where super-Nyquist sampling can be implemented.

According to Figure 10.14, the units on the right hand of the AD/DA conversion are denoted as *hard and soft realtime signal processing*. The definition of hard realtime signal processing is such that the results are generated every rising or falling edge of the sampling clock; data are read from or written to a FIFO buffer to be processed by subsequent soft realtime processing. In this context, soft realtime processing means that the calculations are finished on the average within a clock cycle without occurrence of buffer underflows or overflows. The phase noise performance of the AD/DA conversion depends on the quality of the sampling clock; this is particularly true for the TX path, where it should be as low as possible. Therefore, the realtime processing unit must be operated with "high-quality" clocks.

From these considerations, it is obvious that FPGA solutions can serve both processing types since all necessary components are available like dual-port memories, cells for sequential logics and dedicated DSP blocks; in addition, all FPGA units can be performed in parallel. The interface to the soft realtime processing are FIFO's, which can buffer clock variations. The best solution for soft realtime processing are DSPs, usually running at higher clock speeds and providing powerful mathematical functions in high-level programming language. An exact time response, however, as it is needed for hard realtime processing, cannot be guaranteed. In the following, implementation of downconversion filters realized in FPGA/DSP technology are discussed.

In the open literature, a cascade of integrated comb filters is suggested for rate conversion problems [517]. They work without multipliers, but they might get corrupted by frequency hopping as needed in MF-TDMA modems. A clever alternative is the use of *cascaded moving averages*, where cascading is nothing else than the convolution of the impulse responses of each filter involved. As an example, the overall impulse response of a four-stage filter with 4, 5, 6 and 7 taps is given in (10.9). With this type of filter, a decimation factor of four can be obtained as verified in Figure 10.15, which depicts the related frequency response. It is to be noticed that a stopband attenuation of more than 60 dB is achieved, whereas a simple moving average provides just 13.2 dB (magnitude of the first side lobe). The distortion of the amplitude due to the transition

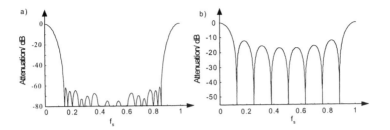

Figure 10.15. a) Overall transfer function of 4 cascaded moving average filters with 4, 5, 6 and 7 taps and b) a single moving average filter with 7 taps.

between passband and stopband is usually equalized by the matched filter.

$$
\begin{aligned}
h(t) &= h1(t) * h2(t) * h3(t) * h4(t) \\
&= [1,1,1,1] * [1,1,1,1,1] * [1,1,1,1,1,1] * [1,1,1,1,1,1,1] \\
&= [1,4,10,20,34,51,69,85,96,100,96,85,69,51,34,20,10,4,1]
\end{aligned}
$$
(10.9)

10.4   Software Radio Onboard

Reconfigurability onboard must deal with several technological challenges, the most important being the space qualification of signal processor to be used on spacecrafts. The section shows to which extend the current signal processing technology can be used for payload design.

10.4.1   Payload Reconfigurability

With the evolution of communication system towards multimedia services, the required data-rates, bandwidth and Quality of Service (QoS) get higher. In this scenario, satellites can play a major role as they offer the unique advantages of a wide area coverage, also in areas where terrestrial infrastructures are not economically effective, and a rapid deployment. Payloads with on-board processing are particularly effective for multimedia services as they improve the link budget and decrease the transport delay when on-board packet switching is achieved. The lifetime of a communication satellite however, along with high deployment costs, make the transparent payload solution still attractive, since all smart functions are located on the earth terminals. As the satellite lifetime increases (15 years typical for a geostationary satellite) and communication standards and services evolve faster, the technological requirements for new multimedia services cannot be predicted accurately at the payload design time. Conventional regenerative payloads provide a poor level of flexibility because

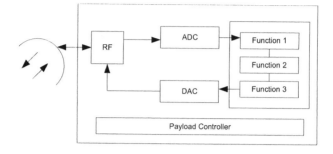

Figure 10.16.   Improved payload with digital signal processing functions

of a large use of ASIC technology for the digital hardware equipment. The potential solution is to provide the payload with a reconfigurable architecture that starting from a conventional fully transparent behavior, can host more sophisticated functions during the satellite lifetime. A scheme of such a improved semi-transparent solutions is shown in figure 10.16.

When flexibility is of primary importance, Software Defined Radio is the best evolutive solution to increase satellite capacity and efficiency with a cost effective investments. The key advantages obtained are an improvement of the payload functionalities, possibility of upgrades to existing communication standards, adaptive modifications of the mission of the payload and the development of new access schemes for satellite services (i.e. adaptive modulation and coding).

Communication satellite payloads can be seen as several functional blocks interconnected to each other in order to process communication streams and signaling information. Typical digital implementing functional blocks are multiplexers and demultiplexers, coders and decoders, beam formers, packet or circuit switches. In theory, all of them could be SDR implemented, but a relevant number of technological challenges are to be addressed, the most relevant being the partition between analogue and digital baseband sections, the analogue-to-digital converter performances, available technologies for the space segment (radiation hardened or tolerant), the definition and management of the configuration process (protocols, signalling, security) and finally the prototyping of a a software radio application. In order to fulfil all the requirements of efficiency for a specific mission scenario the following points must be analyzed in the SDR design phase:

- assess what functions might be software implemented and how these can be implemented,

- the state of the art of space qualified technologies,

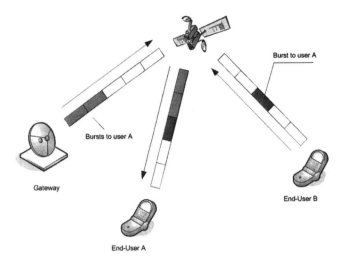

Figure 10.17. Regenerative mobile satellite system

- preliminary definition of operational implementation including payload architecture description, reconfiguration process and required computational power,

- benefit and tradeoffs of a SDR implementation.

As an example, a potential SDR implementation of a mobile satellite system is considered. In a mobile satellite network the main links are the gateway (GW) to end-user (EU), EU to GW and EU to EU, as shown in figure 10.17.

The processing functions to be implemented on-board depend on the particular link.

- GW to EU: a regenerative processor is required with burst switching capabilities,

- EU to GW: the link is considered transparent with a circuit switch functionality,

- EU to EU: a specific regenerative processor is required for instant routes between the two cited processors.

The burst switching concept stands for the possibility of routing bursts sent by a user B to user A on the carrier from a gateway to the user A using the vacant bursts. A SDR approach can be used to implement all the payload processors, as illustrated in figure 10.18, with different technologies. Digital semi-transparent

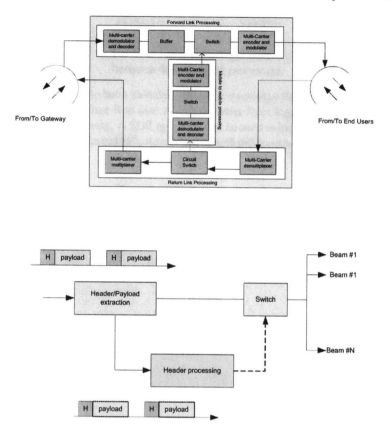

Figure 10.19.   Packet Header Processing Payload

functions of the EU-to-GW and EU-to-EU processors can be easily provided with fast FPGAs while the GW-to-EU regenerative functions can be hosted on higher level programmable devices, such as DSPs.

Another example is represented by a SDR implementation of a broadband satellite system. Satellites are known to be an attractive solution for broadcasting, multicasting or point-to-point communications. A broadband meshed satellite network requires a full connectivity among a large number of beams, thus on-board switching at packet level is envisaged to provide an high level of efficiency. One of the most promising architectural solutions is represented by a regenerative payloads with switching capability driven by the packet header processing. An example of header processor is shown in figure 10.19. In this case SDR technology allows several possible signal processing solutions depending on elements such as network dimension, user quality requirements, user terminal capabilities, propagation conditions and other factors which may have a significant impact on the service delivery. Each customer can receive a different configuration by loading the appropriate profile on the payload.

10.4.2   Space-Qualified Signal Processing

The space-qualified processing technologies that enable the realization of SDR based equipment in space segment are mainly three: FPGA, DSP and array processors.

- *FPGA* is made of numerous programmable configuration logic blocks (CLB) interconnected through programmable devices. Each programmable logic block contains small group of logic gates and one or more memory elements. Loading configuration data into the internal memory cells customizes the FPGA. FPGAs are programmed in VHDL and automatic place and route software delivers good efficiency use of the device so that the development process flow is similar to ASICs. FPGA offers higher level of flexibility than ASIC and good efficiency in term of speed and capacity (Gb/s interface speed and million gates density). Atmel and Xilinx have already developed specific devices for space applications (AT40KEL and XQVR1000 respectively).

- *Digital Signal Processor (DSP)* offers a higher degree of flexibility but they generally suffer from a reduced speed capacity. TSC21020 and ERC32 are two processors developed by Atmel for space applications. The LEON2 processor has been developed for future space mission and is a synthesisable VHDL model of 32-bit processor compliant with the SPARC V(architecture. It can be implemented on any kind of FPGA or ASIC technology.

- *Array Processors (APs)* consist in a large number of "appropriately sized" devices on a single die, interconnected by a very fast on-chip fabric. This structure avoids performing parallel-to-serial and reverse conversion for executing a function. This approach is recent and different components are provided: PulseDSP (by Systolix), CS2112 (by Chameleon), ACM (Adaptive Computing Machine by QuickSilver), Parallel processor provided by PicoChip, XPP (eXtreme Processing Platform by PACT).

Numerous criteria have to be taken into account in order to select the best-suited devices for a SR technology realization. The design of a SDR architecture for payloads must consider the proper balance between flexibility and functional complexity. The goal is to have a device with sufficient processing power to implement complex algorithm, but within real technology constraints. Another issue is the tolerance to radiations: latch-up immunity and SEU tolerance is to be considered in the design phase. Mass and power consumption is to be tuned; SR implementation must remain compatible with power and mass dissipation granted by satellite platform to payload. To be economically attractive, the speed and processed bandwidth of an SR solution must remain as close as possible to traditional implementations. Also the design of system interfaces is of great importance. Only a part of the payload is expected to be implemented with SR technology, so it is mandatory the interoperability with the other parts.

10.5   Multistandard Multiservice Reconfigurable Terminals

The principal rationale for SR design of satellite terminals is investigated here, exploring the potential benefits of reconfigurability in the present panorama of satellite services.

10.5.1   Service and Technologies in Next Generation Terminals

Future satellite applications will aim to the distribution of many types of traffic with different quality, security requirements and transmission techniques.

Starting form the upcoming DVB-S2 standard, new concepts of signal modulation adaptivity has been inserted in the space paradigm. Modulation and Coding adaptivity can improve dramatically the performance of a forward link satellite system, by choosing the most appropriate modulation scheme on various signal-to-noise ratios. The increase of efficiency can allow more throughput or better fade margins. Both effects can be obtained with an accurate design of the adaptivity procedure.

Space communications suffer from vulnerability issues due to large uplink areas. Security at physical level is sometimes required for certain applications. Diversity and spread spectrum technologies can be integrated to obtain secure link-level satellite communications. By increasing the reliability of the satellite link, the overall transport quality of service is improved. When a differentiated level of QoS is required, an interaction between available resources (spectrum allocation), modulation and coding configurations, information protection and redundancy is activated, following the so-called "cross-layer approach".

The new technologies require the optimization in terms of bandwidth and power allocation with a greater flexibility with respect to traditional transmission systems. This point of view increases the possibility of integration for heterogeneous systems (e.g., UMTS, for third generation mobile radio system, will use terrestrial and satellite networks together). The target of the software radio technology is the implementation of a multi-mode and multi-bandwidth telecommunication system with characteristics defined by software over all protocol layers. This means a multi-mode radio with software dynamic characteristics defined in all protocol stack layers, included the physical one. It allows a large flexibility in terms of the particular coding, modulation, multiple access techniques adopted. Moreover, it grants a long system life through remote updating and re-configuration. This issue is of key importance when the advanced transmission and reception features are located on earth terminals (transparent and translucid satellite payloads).

In these last years, the coming of software radios has revolutionased the design of digital communication systems. The use of software radio technologies allows one to design flexible and reconfigurable digital transmission architectures. New and different communications standards and protocols can be simply implemented through reprogramming and software downloading operations, thus giving rise to the possibility of having multimode terminals. In such way, software radios can substantially reduce the costs of manufacturing and testing, while providing a suitable way of upgrading the communication system to take advantage of innovative signal processing techniques and new applications. Therefore, software radio technologies can be regarded as basic tools for chasing the continual evolution of the communication standards.

With the increasing processing power, helped by the growing availability of high speed sampling devices, the RF sampling becomes possible also for broadband wireless communications. Depending on the market deployment of such devices, the DSP/FPGA "Software Radio" application could became even more effective than the competing ASIC solution, this is mainly due to the high level of standardization that a "Software Radio" hardware platform can reach.

10.5.2 Software-Radio Architectures for MF-TDMA Modems

Starting with the modulator, the matched filter (TX filter) shown in Figure 10.20 is the most resource-consuming part of the baseband processing. The stopband attenuation depends on the length of the impulse response in terms of symbols, where the number of taps might get amazingly large for larger sampling rates. Through oversampling the baseband is separated from the aliased frequency band. The higher the sampling rate, the easier is the task to suppress the alias effects in case of upsampling or analog anti-alias filtering. Hence, the trade-off is the implementation effort between matched and anti-alias filter. A good approach is to choose three times oversampling, which leads to a 10 symbols impulse response for state-of-the-art FPGA solutions and a stopband attenuation of 40 dB.

In the sequel, upsampling is performed such that the sampling rate of the matched filter is aligned with that of DA conversion. Especially when using variable symbol rates, the upconversion filter has to cope with the variable sample rate of the matched filter and the fixed conversion rate of the ADC. The latter is normally fixed at very high values (200 MHz) to provide a standardized intermediate frequency of 70 MHz, which is supported by any satellite earth station. All in all, the digital implementation of the upsampling filter combined with the mixing stage enables both burst-to-burst frequency and symbol-rate hopping. The frequency agility is achieved by a numerical controlled oscillator, which switches phase immediately and continuously between carriers. The settling time of analog components is equivalent to the calculation delay and the flash-in / flash-out of the shift registers used for digital filtering. The sum of these delays in both transmitter and receiver gives the minimum guard time between adjacent bursts; additional guard time is needed for tracking the time variation due to Doppler shifts.

The demodulator has to shift the passband signal such that the baseband symbols can be recovered appropriately. After AD conversion and frequency shifting, the sampling rate must be reduced to meet that of the matched filter. The latter implies the same criteria as in the transmitter, but the impulse response can be much shorter due to the relaxed requirements for stopband attenuation. In the transmitter, the matched filter must satisfy the ETSI standard, where out-of-band-power and spurious are defined. In contrast, the matched receiver filter optimizes the SNR, which is normally smaller than 16 dB; hence, a stopband attenuation of 30 dB is sufficient, which leads to a impulse response of only 5 symbols.

In the sequel, carrier frequency/phase and symbol timing must be estimated during initial acquisition. For symbol timing, in particular, a suitable oversampling of the baseband signal is needed. If it is lower than a factor of six, the correct sampling instant must be interpolated; otherwise, the sample closest to the optimal sampling instant is taken.

Carrier frequency and phase offset are inherently coupled, i.e., a frequency offset causes a permanent rotation of any symbol constellation. Therefore, the frequency offset must be accurately estimated so that the residual frequency offset is small enough to be tracked in the subsequent recovery loop. This phase tracker is normally implemented as a second-order device with appropriately selected loop bandwidth: for a value too small compared to the frequency offset, lock is lost quite easily, whereas in the opposite case the cycle-slip rate increases because of too much loop noise. Both phenomena have a disastrous impact on the error performance.

In MF-TDMA modems, parameter estimation has to be done for each burst separately since every station in a network has different transmitter as well as receiver conditions, in general. Non-data-aided algorithms are less suitable (accuracy, ambiguity). The better way is to use data-aided solutions working on a preamble, which ought to have good auto-correlation but low cross-correlation properties with respect to noise and data. Optimal sequences known from open literature are recommended in this case. The baseband processing part shown

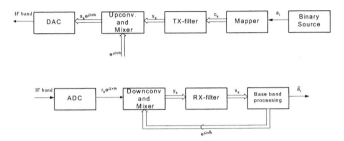

Figure 10.20. Schematic block diagram of the MF-TDMA modulator (the upper part indicates the modulator, the lower one the demodulator)

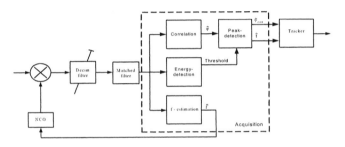

Figure 10.21. Detailed block diagram of the acquisition process used in the MF-TDMA modem

in Figure 10.20 is provided in more detail in Figure 10.21. The acquisition process is split into two parts: (i) frequency estimation and (ii) correlation unit providing both the symbol timing and the phase rotation for the tracking device; the peak of the correlation indicates the perfect match of the preamble and, therefore, the beginning of the burst.

10.6 Remote Configuration of Physical Layer

Remote configuration capability is a relevant element enabled by SR technology. The configuration process is obtained by delivering to the remote SDR device a set of configuration commands able to properly define the physical layer of the remote device[1]. The level of details used by the configuration commands represents a complex trade-off between two elements: the required detail in the description of the radio components and the level of abstraction required for

---

[1] Upper layers are already software defined, so a remote firmware upgrade is sufficient to modify their behaviour.

SOFTWARE RADIO TERMINAL   BASE STATION

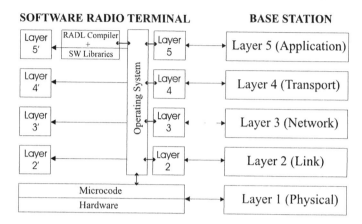

Figure 10.22. Software architecture for remote configuration of SDR terminal and payloads.

hardware-independent remote configurations. A specific on-board processor controller is required to manage the remote configuration process. This processor is implemented with a general purpose Central Processing Unit (CPU) (with good performance and low power consumption). The FPGA/DSP blocks represent the reconfigurable strata of the hardware. The Read Only Memory (ROM) banks are used to store the firmware of the correspondent DSP/FPGA block. The general purpose CPU runs an embedded-Operative System (OS) (stored in Flash memories) to perform all the required control functions. The OS is also used to generate/modify the firmware which will be loaded into the DSP/FPGA blocks. A possible software architecture is shown in fig. 10.22. It derives from the well known TCP/IP suite. The main difference is in physical layer of the SR device which is split into two sub-layers: the re-configurable hardware and the microcode (the SW part) which implements the target communication protocol. The SR terminal/satellite has two different states: an operational mode and a setup mode. The OS assists the device in its normal function during the operational mode and performs the hardware reconfiguration during the setup mode.

Different approaches can be followed in the design of the configuration language. *CORBA middleware* can provide a complex structured layering of software enabling the target device to request signal processing modules not initially loaded in the configuring device. *JAVA based meta-languages* provide a platform independent support to remote configuration, particularly useful for upgrading the physical layer of multi-vendor earth terminals. Custom specific meta-languages, such as *Radio Access Definition Language* (RADL), provide high level hierarchical configuration commands, indicated for complex configuration functions where standard access schemes are frequently subject to low-level design an modifications.

To ease in-service radio reconfiguration over the air interface, the partial reconfiguration features of some modern FPGAs can be used. This allows partially changing the functionality of the device while other parts are capable of working uninterrupted whereas full reconfiguration is highly time-consuming, would break service continuity and can be impractical due to other reasons as well. There are, however, some limitations of this technique, such as fixed device boundaries, communication issues between reconfigured and static blocks and I/O and placement constraints.

10.7 Conclusions

Software Radio paradigm is rapidly entering in the design of modern advanced communication devices, supported by the progresses of analog-to-digital conversion technologies, signal processing, software architectures and available computational power at reduced costs.

Also in space communications, software radio approaches are favorably taken into consideration for earth terminals and payload design. In earth terminals SR technologies allow a dramatic reduction of mass production costs, provide easy upgrade procedures and give the opportunity of network-driven user terminal reconfiguration.

On payloads the main benefit of SR architecture is the ability to follow all possible evolutions of access technologies that may occur during the satellite lifetime.

Chapter 11

SYSTEMS AND SERVICES

A. Duverdier[1], C. Bazile[1], M. Bousquet[2], B. G. Evans[7], I. Frigyes[3], M. Luglio[4], M. Mohorcic[5], R. Pedone[6], A. Svigelj[5], M. Villanti[6], A. Widiawan[7]

[1] *Centre National d'Etudes Spatiales (CNES), France*

[2] *Supaéro, France*

[3] *Budapest University of Technology and Economics, Hungary*

[4] *University of Roma 'Tor Vergata', Italy*

[5] *Jožef Stefan Institute, Slovenia*

[6] *University of Bologna, Italy*

[7] *University of Surrey, U.K.*

11.1   Introduction

After several science fiction papers and books, satellite telecommunications appeared with the paper titled "Extraterrestrial Relays" authored by Arthur Clarke and published on Wireless World in October 1945. In that paper, the orbital configuration of a satellite constellation utilizing a Geostationary Earth Orbit (GEO), an orbit located at 35800 km over the Earth surface coplanar with the equatorial plane, was proposed. In that configuration, three manned satellites, fixed with respect to a point on the Earth surface, were equally spaced in angle of 120° each and covered one third of the world. The second milestone was represented by the paper published in 1954, titled "Telecommunication Satellites", authored by John Pierce where also Low Earth Orbit (LEO) were proposed using unmanned satellites. Latter, Highly Elliptical Orbit (HEO) and Medium Earth Orbit (MEO) have been considered.

The first space mission was the Sputnik satellite launched in 1957 by the former URSS which was not a telecommunication mission but was extremely useful for mechanical validation. After that a real run to the space started. In the USA, the "Policy Statement on Communications Satellites" was issued in 1961 and the law "Communications Satellite Act" was issued in 1962 (birth of

Comsat). Then, the inter-governmental agreement "Interim Arrangements for a Global Commercial Communications Satellite System" was issued in 1964 (birth of ICSC, Interim Communications Satellite Committee) and Intelsat (International Telecommunication Satellite), an intergovernmental organization, was founded moving forward the satellite telecommunication age.

The first missions were launched on an experimental basis using LEO, HEO or GEO constellation (Score in December 1958, Courier in October 1960, ECHO I in August 1960, TELSTAR I in July 1962, RELAY I in December 1962, TELSTAR II in May 1963, SYNCOM II in July 1963, ECHO II in January 1964, RELAY II in January 1964, SYNCOM III in August 1964, MOLNIYA I in April 1965).

The first commercial satellite, Early Bird, was launched in April 1965 in geostationary orbit by Intelsat to provide intercontinental fixed services. Marisat launched in 1976 was the first GEO system providing mobile services with a constellation of three satellites. In 1982, Inmarsat (International Maritime Satellite), an intergovernmental organization with the target to provide mobile satellite services, launched the first fully operational system targeted to maritime users (In more recent years, Inmarsat has also increased its offer with services on land masses and to aeronautical users.). In 1989, TDF1 was the first television direct broadcasting satellite. Omnitracs was launched in 1988 as the first system for land mobile services. Italsat, launched in 1991, represented the first satellite with a regenerative payload and multibeam antennas operating at higher frequency. ACTS (Advanced Communication Technology Satellite) was launched in 1993 as an experimental mission to further validate very high frequency band technology, on board processing, multibeam antennas, hopping beams and on board switching. In 1996, the first digital television bouquets were broadcasted in the USA with DirectTV and in Europe with Astra1E. Two systems, Iridium (1998) and Globalstar (2000) both deployed in LEO orbits, were the first satellite systems designed for personal communications. In 2004, the GEO MBSAT system introduced in Japan and Korea high quality digital broadcasting services to mobile users and allowed the access to the services by Third Generation (3G) mobile terminals. Multimedia satellite systems are also being developed to provide broadband services.

The objective of this chapter is to present systems and services by satellite in the present and for the future. First, we show how satellite telecommunications are built introducing a mission, a system architecture and designing a satellite for a given air interface. A general presentation of present and future fixed and mobile services follows, introducing the convergence of services in high layers. At the end, future solutions are presented with hybrid constellations, high altitude platforms, and terrestrial extensions.

11.2   Satellite telecommunication systems

The conception of a satellite telecommunication system is based on three main elements. A mission justifies the role of the satellite and its place in the system. An architecture is built to configure, manage, and use the system. A satellite is designed to be adapted to a given air interface and to make the system operational.

11.2.1   Mission

The mission for a satellite system must be based on clear opportunity, in a specified context, and in a significant market. These three aspects are dealt with in turn in the following.

11.2.1.1   Opportunity.   To carefully identify which kind of requirements can be satisfied and where the market can be addressed by satellite telecommunication systems, the main characteristics of such systems are listed and briefly discussed in the following.

- Cost almost uniform within coverage

    The cost of a satellite link is independent of the distance, by contrast with terrestrial networks. In fact, due to the broadcast nature of the signal transmitted by the satellite, it can be received in all the coverage area. For this reason, trade-offs may be searched to find the distance threshold above which it would be cheaper to use the satellite instead of the terrestrial system.

- Efficient and cost effective for collecting, broadcasting and multicasting

    Satellites are very suitable to efficiently distribute and collect signals for a very large population. In fact, due to the capability to cover very large areas, one repeater in space is able with just one carrier to reach theoretically infinite users. In case of terrestrial fixed networks as many connections as the number of users are required.

- Unique in areas with scarce or no infrastructures

    In scenarios such as developing countries either with unfriendly environments or with low traffic requirements, when the installation of terrestrial infrastructures is not economically feasible, the use of satellites, especially if already deployed, is the only possibility.

- Unique in case of disaster

    In case of natural disaster (earthquake, storm) or war when terrestrial infrastructures can be seriously damaged or even destroyed, the satellite facilities usually keep working. Even if an earth station may be damaged, other stations continue to provide service.

- Allowing very crowded terrestrial network bypass

  It is estimated that a typical Internet connection needs to pass through seventeen up to twenty routers in average. With satellites very far away servers can be reached with only one link.

- Suitable for long range mobility

  Due to the capability to cover very large areas (even one third of the world), the satellite is particularly suitable to ensure long-range mobile services, such as in the aeronautical or maritime case.

- Relatively shortly deployable

  If referred to quite large coverage areas, a satellite system can reach its target in very limited time if compared to a terrestrial infrastructure. For standard commercial satellites even about eighteen months can be sufficient to design, realize and launch the system. For more innovative solutions the time needed may be more.

- Flexible

  The satellite on board architecture allows an extremely flexible use of resources. In fact, in case of multibeam coverage, capacity can be allocated unevenly in time to the different beams according to actual requirements.

- Combining with the same terrestrial infrastructure both fixed and mobile services

  As any radio system the same infrastructure is able to provide both fixed and mobile services.

- Extremely suitable for localization services worldwide

  Last, because not telecommunication in strict sense, but not least, in terms of commercial importance, the satellite solution is largely the most efficient for localization and navigation services. In fact, both the present Global Positioning System (GPS) and the forthcoming Galileo, adopt this kind of architecture based on a limited number of satellites (twenty four and thirty respectively).

11.2.1.2   Context.   It is quite obvious that the technology advancement in telecommunications over the last two decades shaped the focus of satellite market and the new satellite system development. Therefore it is important to review the changes and the related impact on the satellite industries to determine the future position in the competitive telecommunication market. Some of the important events in the telecommunication market and the related issues in the satellite industry are listed below.

1 Introduction of TAT-8 digital submarine optical cable in mid 1980's. Hence satellite lost the dominant status in the over the ocean communication in terms of capacity and quality and became more television dominated.

2 Success of GSM in earlier 1990's and unprecedented interested in non-GEO based personal satellite systems like Iridium, Globalstar, etc. [518] and the high profile failure of these systems in late 1990's. Due to the above development, nowadays non-GEO solutions are treated with extra caution.

3 Internet based services boom in mid 1990's and more and more dependency on Internet for day-today activities. Based on the above fact, it has been argued that there is a high probability that the current business approach in the telecommunications industry will not survive the new Internet economy and hence there is urgent need for dramatic reform in the business approach. It further cited that there should be new World business models that rely on one network infrastructure to deliver multiple services over the Internet. Therefore flexible and scalable networks are to be designed and installed to support the emerging new World broadband fixed packet-based services. This approach has already been implemented in some sector of the telecommunication industry where a growing number of cable television companies are now bundling broadband Internet with television service. Several USA industry analysts have been recently quoted [519] as saying that satellite companies hoping to offer Internet services are most likely to succeed if they join forces with digital broadcast satellite companies on the basis that the latter already have developed relationships with consumers and have a consumer infrastructure in place that broadband companies can use to their advantage.

4 In general, all wireless technologies have broadcast/multicast potential, making them well suited to television distribution and multicast data distribution. The increasing storage capacity of PCs may lead to an increasing number of home servers, favoring the use of multicast and push technology. Satellite systems, which have low individual capacity per user within the area of coverage, have the advantage of total coverage independent of local population density. They are thus well suited to broadcast, updating of servers in more local networks, and broadband to rural areas and in areas with poor infrastructure.

5 80% of population lives in urban or sub-urban areas, that is 20% of the World surface area. Therefore satellite can target 20% of the population

if the satellite industry goes for competitive approach competing with terrestrial systems. This may be even less if we consider Europe since GSM reached 88% of the territory and address 97% of the total population in France, which is the widest and less dense country in Europe.

11.2.1.3   Market.   In general satellite systems provide three main types of services: Mobile Satellite Services (MSS), Fixed Satellite Services (FSS), and Broadcast Satellite Services (BSS). These services target today three different types of market: the maritime, aeronautical, and land sectors. Satellite can still maintain the monopoly status in the maritime and aeronautical market due to its unique coverage feature. However the story is different in the land case where satellite have to compete with the terrestrial mobile systems like Global System Mobile (GSM), UMTS, etc. and broadband access technologies like broadband wireless access and Digital Subscriber Line (xDSL).

For the land mobile case, there is general feeling that there is still a gap to fill in terms of coverage (complementary approach) where terrestrial mobile system would not have any coverage (during disaster relief, peace keeping and remote areas) as well as effectiveness of delivery mechanism (corporative approach) where terrestrial system is not effective (for example broadcasting/ multicasting).

For the fixed broadband case, satellite proved that cost-effective broadband interactive service could be provided for the following market sector:

- Broadband connectivity for enterprises and Small Office / Home Office (SOHO)

- Extending and complementing terrestrial networks for residential consumers

- Wide-area teleservices such as telemedicine, telecommuting, and tele-education

The figure 11.1 shows how the traditional satellite services are mapped on to different market sectors.

Nowadays, satellites offer also conventional services such as backbone or backup links for terrestrial networks, setting-up or interconnection of Local Area Networks (LANs), broadcasting, news gathering, etc. Moreover, they are viewed as a complimentary technology to other forms of access and used where its features give a distinct advantage over other technologies. In particular, satellite is the unavoidable solution for some public services (digital divide, emergency management, disaster relief, etc.).

The trend for the future is to offer multimedia services, to comply with Internet Protocol (IP) standard interface in order to easily achieve integration with terrestrial networks (UMTS, Wi-Fi, Tetra, etc.). Especially, the development

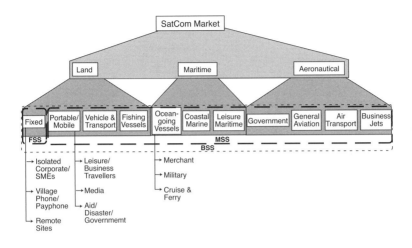

Figure 11.1. Satellite communication market sectors

of multimode terminals is studied. In such a scenario, satellite systems can provide an useful complement to reach the full mobility.

11.2.2   System Architecture

The satellite system architecture comprises a space segment, a ground segment, and a control segment, which are described in the following.

11.2.2.1   Generalities.   A generic satellite system architecture is shown in figure 11.2. It is composed of a space segment, a control segment and a ground segment:

- The space segment contains one or several active and spare satellites organized in a constellation.

- The control segment consists of all ground facilities for the control and monitoring of the satellites, also named Tracking, Telemetry and Command (TTC), and of the traffic, that is the Network Operation Centre (NOC).

- The ground segment consists of all the traffic earth stations. Depending on the considered type of service, these stations can be of different size, from a few centimeters to tens of meters.

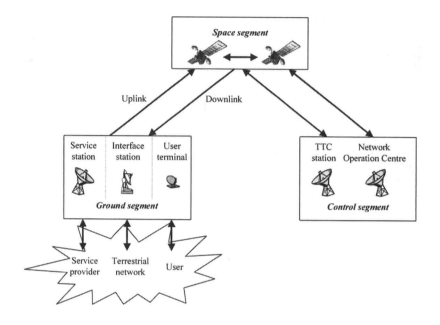

Figure 11.2.   Satellite system architecture

Communications between users are set up through user terminals which consist of equipments such as telephone sets or computers that are connected to the user stations or are part of the user station.

The path from a source user terminal to a destination user terminal is named a simplex connection. There are two basic schemes: Single Connection Per Carrier (SCPC), where the modulated carrier supports one connection only, and Multiple Connections Per Carrier (MCPC), where the modulated carrier supports several time or frequency multiplexed connections. Interactivity between two users requires a duplex connection between their respective terminals, i.e. two simplex connections, each along one direction. Each user terminal should then be capable of sending and receiving information. A connection between a service provider and a user goes through a hub (for collecting services) or a feeder station (for broadcasting services). A connection from a gateway, hub or feeder station, to a user terminal is called a forward connection. The reverse connection is the return connection. Both forward and return connections entail an uplink and a downlink, and possibly one or more intersatellite links.

Uplinks and downlinks consist of radio frequency modulated carriers, while intersatellite links can be either radio frequency or optical.

11.2.2.2 Space segment. The spacecraft contains a power section that consists of batteries and solar panels, a stabilization and propulsion section including propellant, a TTC section and a communication payload section.

The payload architecture can be transparent if acts as a bent pipe (just change frequency and amplify signals) or regenerative if implements some On Board Processing (OBP) features, that means at least demodulate and modulate again signals. In transparent configuration, it is in charge of providing radio relay for links between earth stations, to capture the carriers transmitted by the earth stations, to amplify the received carriers, to change carrier frequency, to radiate carriers back. In regenerative configuration it can also regenerate signals, change polarization, interconnect and route packets, switch messages. The on board switching can be implemented either at microwave or base band frequencies.

11.2.2.3 Control and ground segments. The control segment exchanges information with the TTC section on board and includes the NOC that manages the traffic and the associated resources onboard the satellite.

The ground segment includes a number of stations coming in three classes:

- Interface stations, known as gateways, which interconnect the space segment to a terrestrial network.

- Service stations such as hub/feeder stations which collect/distribute information from/to users via the space segment.

- User terminals, such as handsets, portables, mobile stations, Very Small Aperture Terminals (VSAT) or Ultra Small Aperture Terminals (USAT), which allow the customer a direct access to the space segment.

User terminals can be interconnected in full mesh topology or in star topology. In the former case the use of the bandwidth is minimized and the delay is the minimum (one hop) while in the latter case a hub is required to interconnect the terminals, thus needing double hop, maximizing delay and doubling the use of the bandwidth. Terminals suitable to the mesh topology are more complex and expensive than those utilized for star networks, if the space segment does not include on-board processing.

11.2.3 Satellite System Design

Here we concentrate on three specific aspects: orbital configuration, coverage, and air interface.

11.2.3.1   Orbit.   Orbits can be classified according to eccentricity (circular or elliptical with the Earth in one of the two focuses), to inclination with respect to the equatorial plane (equatorial, polar or inclined), to altitude. For telecom satellites, orbit is in general circular, the altitude allowing to classified LEO between 500 and 1700 km, MEO between 5000 km and 10000 km and over 20000 km, and GEO at 35800 km [72] [387]. Nevertheless, an elliptical orbit can be used with an apogee between 39000 km and 54000 km.

The main geometrical parameters are the minimum elevation angle and the maximum user-satellite distance. The former is important for the propagation channel, for both effects of the troposphere and mobility, the latter impacts performance in terms of both free space losses and propagation delay, which are the only really unavoidable impairments of satellite communications. The delay implies subjective disturb in case of real time interactive services (e.g. videoconference) and objective troubles to delay sensitive protocols (e.g. TCP).

The use of GEO orbit means to experience high delay (120-135 ms Earth-satellite) and high free space losses. The elevation angle decreases as the latitude increases. On the other hand the Earth can be covered with the simplest architecture composed of just three satellites, excluding the polar regions (above 70 latitude N and below -70 latitude S). The geostationary orbit theoretically allows to avoid tracking systems at the ground station. Actually, due to residual movement the satellite may require to be tracked.

The use of LEO orbits implies better performance than GEO in terms of propagation delay (20-40 ms) and free space losses but a large number of satellites is necessary to provide full coverage and real time interactive services [520]. Doppler effect and frequent handovers between spots and between satellites are the main drawbacks. Moreover, an efficient tracking system is necessary on the ground station. The elevation angle can be very high at any latitude and in general is time dependent.

MEO orbits show performance in between the two previously presented. They are actually not utilized for telecommunication systems but for localization and navigation systems such as GPS and the future Galileo.

An analysis of LEO and MEO constellations is reported for example in [81]. Notably, the choice of the altitude A has a strong impact on the constellation size, N_s. Considering the mere geometrical problem, by dividing the coverage area in sub-regions illuminated by a single satellite, it is possible to determine the number of satellites required to guarantee a minimum elevation angle. The result of this analysis is shown in figure 11.3, where a lower bound on N_s is reported as a function of A, with the minimum elevation angle, α, as a parameter. Note that the considered global coverage does not include polar regions. It is worthwhile noting that worldwide coverage with high elevation angles ($\alpha > 70$) is practically unfeasible.

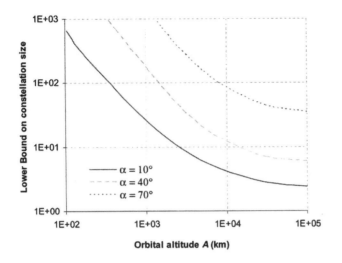

Figure 11.3. Constellation size for global coverage (polar regions excluded) vs. orbital altitude

Satellites in HEO go very slow around the apogee appearing almost fixed for users located at very high latitudes. Satellite handover, Doppler effect and zoom effect represent the main drawbacks but are not so critical. Delay and free space losses are comparable to GEO.

The different orbital configurations can be combined to realize a hybrid constellation that can be composed of one or more satellites in GEO orbit and one or more satellites in LEO orbits. The two components can either be interconnected to reciprocally extend coverage or be alternative point of access to improve availability [521] [522].

11.2.3.2 Coverage. Most telecommunication satellites are single beam satellites where each transmit and receive antenna generates one reconfigurable beam only. Nevertheless, one can also consider multiple beam antennas [523]. The payload has then as many inputs/outputs as upbeams/downbeams. Routing of carriers from one upbeam to a given downbeam implies either routing through different satellite channels with channel hopping or onboard switching at radio frequency [524].

For constellations, the instantaneous system coverage consists of the aggregation at a given time of the coverage areas of each individual satellite. The long-term coverage is the area on the Earth scanned over time by the antennas of all the satellites.

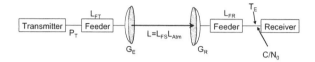

Figure 11.4.   Carrier-to-noise ratio for a satellite link

11.2.3.3   Air interface.   The satellite is generally a simple repeater
that amplifies and re-directs the signal, even if it can eventually also regenerate
the latter. Two links have to be established on uplink and downlink, the link
budget allowing to validate their availability. The carrier-to-noise ratio C/N_0
is expressed by:

$$\frac{C}{N_0} = P_T G_T \frac{1}{L} \frac{G_R}{T_E} \frac{1}{k} \qquad (11.1)$$

where P_T is the transmitter power, G_T is the transmitter gain, L the losses
due to feeder, free space and atmosphere, T_E is the system equivalent noise
temperature, G_R is the receiver gain, and k is the Boltzmann constant. These
elements are depicted in figure 11.4. The choice of the air interface is based
on the availability of the link budget. The unavailability factor indicates the
percentage of time per month during which the minimum values in forward and
return is not be met taking into account the average weather statistics (fading
margin). Service providers may offer different kind of terminals for different kind of required availability or performance according to a Service Level
Agreement (SLA) signed with the customer. For example, premium customer
terminal may guarantee an availability of 99.7% while other a basic terminal
may guarantee only 99.0% availability [525].

Following chapters show how to improve the air interface after presenting
theoretical background and satellite channel. Figure 11.5 summarizes where
coding/decoding, modulation, synchronization, and distortion/equalization are
needed in the transmission chain. In a higher point of view, the air interface can
be optimized introducing fade mitigation techniques at the higher frequency
bands and enhanced access techniques. For flexibilty, software radio enables
terminals automatical reconfiguration.

11.3   Fixed and mobile satellite services

Most existing satellite telecommunication systems try to propose integrated
services between broadcast, multicast and unicast. Nevertheless, such integration is an important challenge that involves high layers to optimise the layers
in a common IP environment. In fixed networks, the drive is broadcasting

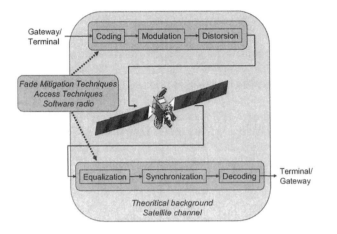

Figure 11.5. Transmission chain for a satellite link

with eventually an interactive channel. In mobile networks, point-to-point and terrestrial-satellite integrated solutions are encountered.

In the following, we describe issues related to higher layer protocols and standards, fixed networks and mobile networks.

11.3.1 Higher layers

Multiple-Access Channel (MAC) is the link between physical layer and high layers. It has to incorporate a Radio Resource Management (RRM) in a packet based environment to adapt user traffic.

The protocols of interest to support multimedia communications are:

- Moving Pictures Expert Group (MPEG),

- Asynchronous Transfer Mode (ATM),

- Internet Protocol (IP),

- Transfer Control Protocol (TCP) / User Datagram Protocol (UDP).

For any application, the user perceivable quality is given in packet or bit error rate, latency for real or non-real time and delay jitter. For interactive applications, an other important parameter defining the Quality of Service (QoS) is the nominal and minimum user data rate throughput. In the case of these applications, data rate variations above the minimum value and variable delays can be tolerated.

The MPEG-2 standard [262] was initially designed to provide digital video and audio services but is now considered as an attractive protocol to provide multimedia interactive applications due to its flexibility. Indeed, in the basic MPEG-2 standard, digital data streams of different bit rates corresponding to video, voice or data information are formatted into 188-byte fixed length Transport Stream (TS) packets, and then multiplexed. Moreover, TS packets can be used to carry any type of digital information using data piping, data streaming, multiprotocol encapsulation, data carousels and object carousels. The quality of the flow is acceptable when the transmission is Quasi Error Free (QEF) or when less than one TS packet in error is observed per hour.

ATM is a connection-oriented transmission protocol using fixed-size 53-byte cells as basic data units. Its inherent capacity is to handle various QoS data streams:

- The Constant Bit Rate (CBR), acting as a circuit emulation, is a class of service defined for traffic requiring a constant amount of bandwidth specified by the peak cell rate traffic descriptor, to be permanently available.

- The Variable Bit Rate (VBR) class is intended to bursty data streams with variable bit rate. The real-time VBR (rt-VBR) service corresponds to delay-sensitive traffics such as video and audio and makes use of the statistical multiplexing. The network attempts to deliver cells respecting a specified cell delay and cell delay variation. The non real-time VBR (nrt-VBR) class supports those traffics which are less sensitive to delay and delay variations.

- The Available Bit Rate (ABR) connections do not have significant time demands. Only an average cell rate is guaranteed by the network, but the bandwidth allocated to the connection may vary during the lifetime of the connection.

- The Unspecified Bit Rate (UBR) is intended for best effort applications and does not support any service guarantees. It is well suited for web browsing, emails, ftp transfers...

IP is a network layer protocol whose purpose is to enable data traffic to flow seamlessly between different types of transport mechanisms (ethernet, ATM, frame relay...). The protocol is implemented at the level of terminal devices and routers functioning as switches for routing datagrams towards their destinations, using an address field contained in the datagram. IP provides a global addressing scheme. It does the best effort to deliver packets but does not guarantee their safe delivery, or their arrival in the proper sequence.

TCP constitutes the end-to-end transport protocol for reliable sequenced data delivery (web, email, ftp transfers, videoconferencing, broadcast, remote login, etc.). The role of TCP is to ensure a proper delivery of a complete message. It resides at the end services. This delivery is performed by assigning each byte a unique sequence number. The receiver keeps track of this sequence numbers and sends acknowledgments to the sender whenever an entire message has been received. UDP is the same kind of protocol but is simpler as TCP and does not control the proper delivery of the message.

11.3.1.1   Convergence of different access systems on IP core network.
Naturally, people would like to have the same facilities and comfort as they have at home, wherever they go. Nowadays, people depend more and more on advance communication technologies like Internet, computer and mobile phone. Therefore they would like to have these technologies anywhere and any time. The requirements born from this specific demand can be listed as below:

- Access to different type of telecommunication services anywhere and any time

- Single device which can communicate with different networks

- Single connection number (say IP address)

- One single bill for all the services with reduced cost

- Reliable wireless access even with the failure of one or more networks.

These requirements push vertical network approach with single service towards horizontal network approach with multi-services as shown in figure 11.6.

In order to implement this horizontal approach, engineers and researchers in telecommunication industry are looking for a generic radio access network connected to the unified IP core network. As there are already several access technologies available developed independently for special purposes, inter-working between different access systems is becoming a key issue for the future systems.

11.3.1.2   Layered system structure.   Even though the above mentioned access systems are developed separately, they do complement each other under a layered concept as shown in figure 11.7 in terms of coverage and capacity.

Broadcast layer will have the capability to cover large area and handle the full mobility, however capacity is lower than other layers. The technologies used in this layer can vary from DVB/DAB-2 to 3GPP UMTS (system like Satellite Digital Multimedia Broadcasting (S-DMB) currently under study). Cellular layer includes 2G and 3G systems and supports up to 2Mbit/s depending on the mobility condition. This layer supports full mobility using handover techniques. Hot spot layer supports very high data rate, but the mobility is restricted. Personal network is meant for office and home to communicate with different appliances (refrigerators, toasters, washing machines, smart sensors

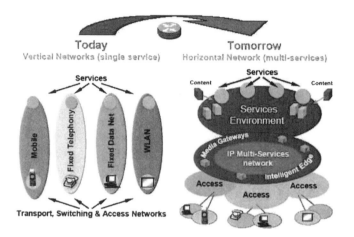

Figure 11.6.   Move from single service to IP based multi-services [526] ©2005 IEEE

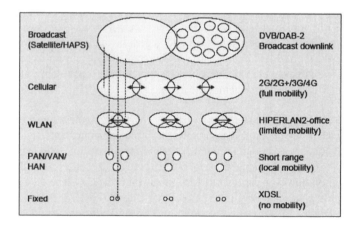

Figure 11.7.   Layered concept

etc.). Fixed layer contains optical fiber (e.g. FTTx), twisted pair systems (e.g., xDSL) and coaxial systems (e.g. CATV). Fixed wireless access systems also come under this layer.

In this vision, satellites are at the top layer. This scheme provides for global roaming across all layers and points the way to full integration of terrestrial and satellite systems at terminal and network levels with handover across all the networks. The terminal accesses any network and service to maximize the cost and efficiency. Such a scenario can only be possible if operators embrace the totality of the delivery mechanism and network providers (both satellite and terrestrial) embrace common standards. A key enabler to this scenario is also software radio with automatically reconfigurable terminals.

11.3.2 Fixed networks

11.3.2.1 Overview.
The FSS make use of the following bands:

- Around 6 GHz for the uplink and around 4 GHz for the downlink, there are systems described as 6/4 GHz or C band. These bands are occupied by the oldest systems (Intelsat, American domestic systems...) and tend to be saturated.

- Around 8 GHz for the uplink and around 7 GHz for the downlink there are systems described as 8/7 GHz or X band. These bands are reserved, by agreement between administrations, for government use.

- Around 14 GHz for the uplink and around 12 GHz for the downlink there are systems described as 14/12 GHz or Ku band. This corresponds to current operational developments (Eutelsat...).

- Around 30 GHz for the uplink and around 20 GHz for the downlink there are systems described as 30/20 GHz or Ka band. These bands are raising interest due to large available bandwidth and little interference due to present little use.

- Above 30 GHz will be used eventually in accordance with developing requirements and technology.

We show in figure 11.3.2.1 some of the key landmarks and the major FSS. Fixed satellite systems play an important role in the core network where on a point-to-point basis, they can still compete with terrestrial links in some areas in which their coverage and reduced ground infrastructure are advantages. Major international satellite operators such as Intelsat, SES Global, Eutelsat, etc. remain viable businesses. It is interesting to note that their business models have evolved, they have moved from international governamental organisations to private companies. They have moved from selling bandwidth to selling service

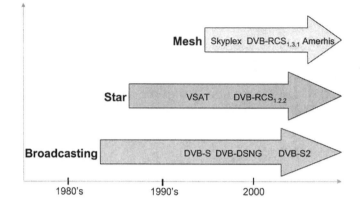

Figure 11.8. Fixed satellite development

connections (MHz to Mbit/s), and now have ground infrastructure as well as satellites amongst their assets. The industry has evolved very conservatively and the vast majority of satellites are still the transparent transponder type operating in C, Ku and Ka bands but with increasingly complex multibeams. The digital pipe remains the major success using Frequency Division Multiple Access (FDMA), with Time Division Multiple Access (TDMA), self-stabilizing TDMA (SS-TDMA) being introduced but not really catching on. Satellites have remained of the transparent type with the exception of some digital television broadcast satellites that have embraced limited on-board-switching. Full on-board-processing has been considered too risky due to lack of flexibility of the channel allocations and bit rates. On the other hand, traffic has changed, with IP now a major percentage of the whole via Internet Service Providers (ISPs). Satellites have remained low to medium power which has meant low efficiencies of usage of the radio spectrum compared to terrestrial systems.

11.3.2.2   Broadcasting systems.   An area in which satellites have been very successful is television broadcast, initially on an analogue basis to cable-heads but progressively digital direct to home. Around 80% of digital television is distributed via satellite in Europe.

Here, the Digital Video Broadcast (DVB) is considered in detail as it played a major role and allowed mass production of cheaper equipments. It has proposed different coding and modulation schemes to broadcast MPEG-2 TS packets depending on the specific features of the communications channel. For satellite transmissions, three international standards are available:

- DVB-S (Digital Video Broadcast - Satellite) that proposes a physical layer for satellite broadcasting [99]

- DVB-DSNG (Digital Video Broadcasting - Digital Satellite News Gathering) that completes previous standard for (trans)portable stations [527]

- DVB-S2 (Digital Video Broadcasting - Satellite 2nd generation) that is the second generation of previous standards allowing also point-to-point connections [101].

The new DVB-S2 standard incorporates Adaptive Coding and Modulation (ACM) schemes which allow to optimise each individual connection dependently on path conditions as described in figure 11.9. The new standard allows also a range of data inputs including IP. DVB-S2 reduces further the space segment costs and allows performance increasing.

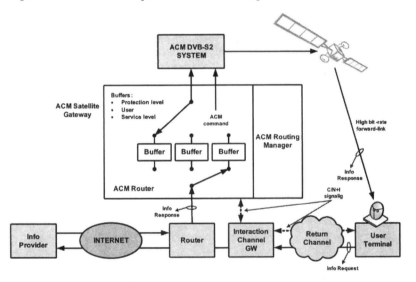

Figure 11.9. Example of IP services using a DVB-S2 ACM link [528] ©2004 John Wiley and Sons Limited. Reproduced with permission.

11.3.2.3 Interactive systems. The first interactive systems where based on Very Small Aperture Terminal (VSAT) networks. Nevertheless, in Europe, they have not really taken off as expected and they have not followed the size or volume of USA counterparts.

The success of DVB-S/S2 standards has then led to two-way systems incorporating VSATs at either Ku or Ka band return channel, based on Digital

Video Broadcast - Return Channel via Satellite (DVB-RCS) standard. The latter offers an interactive channel by satellite for an user receiving DVB-S/S2. In particular, DVB-S2 ACM schemes operated in connection with DVB-RCS allows transmission of parameters to provide a packet by packet optimisation to meet the adverse changing channel conditions. This alternative way of delivering IP services has however a limited interest in regards of terrestrial solutions as Asynchronous Digital Subscriber Line (ADSL). Nowadays, usage of fixed satellites in broadband access is limited to rural and sub-urban areas.

For the future, another usage for broadband access will grow. The satellite mesh network, combined with broadcasting when the satellite has an almost partly regenerative payload. Such a regenerative equipment has still been validated on an unique spot beam (the multiplexer Skyplex) and on four spot beams (the switch Amerhis). In the last case, a connection control protocol has entirely been designed to allow an efficient point to point connection.

The messages for future fixed satellite systems are:

- Greater integration with terrestrial systems

- Increase of system efficiency

- On-board processing that is flexible with large numbers of inter-connectable beams

- Scalable systems (maybe using clusters of small satellites).

11.3.3   Mobile networks

11.3.3.1   Overview.   The MSS make use of the following bands:

- Very High Frequency (VHF) with 137-138 MHz downlink and 148-150 MHz uplink and Ultra High Frequency (UHF) with 400-401 MHz downlink and 454-460 MHz uplink are for non-geostationary systems only.

- About 1.6 GHz for uplinks and 1.5 GHz for downlinks are mostly used by geostationary systems such as Inmarsat.

- About 2.2 GHz for downlinks and 2 GHz for uplinks are allowed for the satellite component of IMT2000.

- About 2.6 GHz for uplinks and 2.5 GHz for downlinks are available.

- Frequency bands have also been allocated at higher frequencies such as Ka-band.

MSS can be classified in three main classes: point-to-point systems, broadcasting systems, and integrated mobile satellite systems.

We show in figure 11.10 some of the key landmarks and the major MSS.

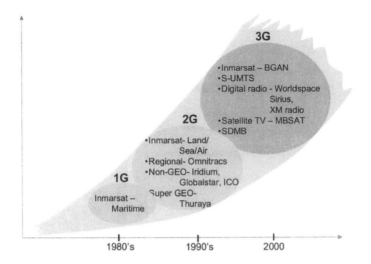

Figure 11.10. Mobile satellite development [526] ©2005 IEEE

11.3.3.2   Point-to-point systems.   It is interesting to note that the first satellite operator for mobile services, Inmarsat, came into existence at around the same time as the first cellular operators providing 1G analogue services. In its initial period, Inmarsat provided speech and low data rate services mainly to the maritime market of larger ships in L band using global beam coverage satellites. In 1990, Inmarsat added aeronautical services to passenger aircraft and to some land vehicles with the introduction of spotbeam higher power satellites. This was followed in 1997 with world wide spotbeam operation and the introduction of paging, navigation, higher rate digital to desktop sized terminals. Inmarsat have concentrated on the use of GEO satellites and in the mid 1990's several regional GEO systems emerged in competition e.g. OMNITRACS, EU-TELTRACS, AMSC and OPTUS concentrating on land vehicles and using both L and Ku bands. These were only moderately successfully whilst Inmarsat built its customer base to around 200,000. The major research in the late 1980's and early 1990's was in non-GEO constellations and this saw proposals for LEO and MEO satellite systems. Of these, Iridium and Globalstar came into services, but too late to compete with the spread of terrestrial GSM, and on business, rather than technological grounds went into Chapter 11 bankruptcy by the early 2000's. The lesson to learn was that constellations were too expensive, at up to $10B, to deploy unless markets had large initial growths to pay back. Both systems are in existence today but with a fewer customers than initially predicted. (Orbcomm, a little LEO provider mainly to fixed terminals suffered a similar fate.) ICO, the proposed MEO system got as far as launching one satellite before realising also that the business case was not there.

In the mid 1990's, larger so-called super GEO satellites were proposed, these being around 5kW with 100-200 spots rather than the earlier generation of GEO's 3-4kW with 5-10 spots. Several such systems were proposed, but the ones that has reached market in the early 2000's are AceS and Thuraya (which is based on ETSI GMR-1 standard [97]) providing GSM and GPRS like services covering Asia and much of Europe. This super GEO is successful, finding a niche with travellers, trucks and in areas where terrestrial mobile is expensive to deploy. Meanwhile Inmarsat is providing its own super GEO's- Inmarsat IV to take existing digital services from 64kbit/s up to 432kbit/s - from the Global Area Network (GAN) to the broadband GAN (BGAN). Despite the move by terrestrial mobile operators to CDMA, Inmarsat has continued to develop its proprietary TDMA system but deliver 3G equivalent packet services.

When we consider mobile broadband for satellite, the major market is for passenger vehicles (aircrafts [529], ships and trains/coaches) except the Inmarsat BGAN system, where customers are perhaps more likely to use broadband services. Connections by Boeing (CBB) [530] began operating broadband links to airplanes in 2002 and is now pursuing the maritime operators market. The technology here has been more akin to the VSAT model with local in vehicle distribution. CBB has already installed terminals with a number of airlines. VSAT systems started in the offshore oil business but have rapidly expanded to cruise liners and deep-sea ferry operators using Ku band and provide commercial, engineering and navigation services to passengers and crew. A number of satellite operators carry such services.

The above schemes still suffer from the poor efficiency of use of the satellite capacity which makes them expensive. A solution to this may be around the corner with DVB-RCS networks extended to mobility used with the new DVB-S2 standard on the forward link, ACM schemes and multi-spot Ka band satellites.

11.3.3.3   Broadcasting and integrated mobile satellite systems.   The Satellite Digital Audio Broadcasting (S-DAB) systems (DARS in the USA) have to be firstly mentioned in this context. The idea has been around since 1990 when digital radio first filed in the USA. Several systems have been proposed since the S-DAB standards were produced with WORLDSPACE [531] in the mid 1990's being perhaps the leading contender with the satellites to cover Asia, Caribbean and the Americas. In the USA in early 2000's two commercial systems have also become operational: Xm Radio using GEOs and SIRIUS radio using HEOs [531]. Both systems use terrestrial gap-fillers. The use of HEO satellites are interesting in that they achieve improved coverage in the urban area and reduce the number of gap-fillers required. Currently Xm Radio and SIRIUS together have around 6 million users in the USA.

The rapid advance in memory devices and their reduction in cost mean that personal terminals can store and play back large content volumes. Multimedia Broadcast Multicast Services (MBMSs) in nature, which well suits them to satellite delivery, are now integrated to services catered for cellular systems. It transpires that such services are not so efficiently delivered in a cellular environment where they severely reduce the capacity for basic services. This all points to an integrated system between satellite and cellular in which the services are divided to the delivery mechanism that best suits them.

The MBSAT system in Japan and Korea operational since 2005 is a S-DMB system similar to S-DAB systems in that it is content driven, being able to provide multimedia content in particular television and radio. S-DMB is nevertheless the only system that is specifically aimed at integration with cellular. Figure 11.11 shows MBSAT system.

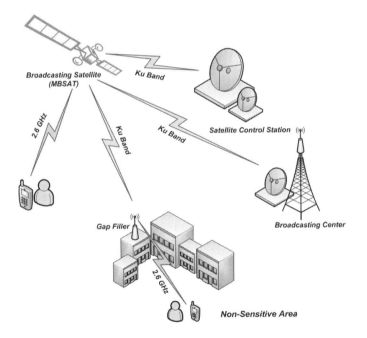

Figure 11.11. Japanese/Korean mobile broadcast satellite systems (MBSAT)

The architecture of another integrated satellite/terrestrial system based on UMTS is shown in figure 11.12 from an EU FP5 project (SATIN) [532] [533]

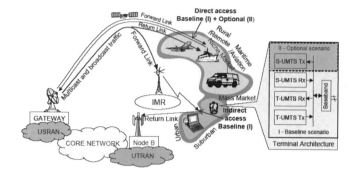

Figure 11.12. SATIN reference architecture

which demonstrated the feasibility of the system. A follow-on project MODIS [534] [535] has demonstrated in 2004 the integrated system using the Monaco 3G network. An FP6 project MAESTRO [536] is now in place which will take these proof of concepts to a fully operational satellite digital multimedia broadcast system. The architecture is characterised by gap-fillers or intermediate module repeaters located at some 3G base stations, which broadcast the MBMS signals terrestrially in the adjacent MSS band to allow in-building and urban presentation.

This integration concept is due to the three following lessons received in the past:

- Not to compete but collaborate with cellular networks,

- Use the wide coverage broadcast attribute of satellites,

- Select the service appropriate to the delivery mechanism.

It completes the picture and opens up a whole new future for satellites. Each of these systems has nevertheless regulatory issues to overcome in various regions of the world that is one of the most important determinant as to which succeeds in the future. They have competition from MBMS in 3G and from DVB for Handhelds (DVB-H) but offer a real new market for satellite and most importantly a first integrated system delivery.

Figure 11.13 shows the future seen for 3G International Mobile Telephony 2000 (IMT2000) and beyond 3G systems.

11.4   Complementary solutions

For satellite telecommunications, there are some trends that can already be seen which will apply not only to fixed but to mobile services as well. GEOs,

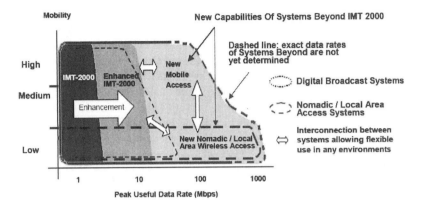

Figure 11.13. Data rate with mobility in the IMT-2000 3G and beyond 3G systems

LEOs, and HEOs can be mixed in the same network. There is much to be gained from use of hybrid orbits with interorbit links in terms of connectivity, coverage and service routing. Although not strictly satellite, High Altitude Platforms (HAPs) can also play a role in both mobile and fixed satellite service delivery. For the emerging network and security systems, they could standalone be more efficient than satellite and could also play a role (post 2010) in a hybrid network of satellites and HAPs. Satellites are not able to establish indoor links. To allow access of terminals to satellite networks a terrestrial extension is needed.

11.4.1   GEO and non-GEO hybrid constellation

Commercial satellite systems based on a GEO satellite provide large bandwidth while those based on LEO constellations can provide global coverage. To improve performance in terms of coverage over offered capacity ratio, hybrid constellation concepts have been proposed. It consists in combining at least two types of satellites (GEO, LEO, MEO, HEO) so that coverage or kind of service is extended with respect to single segment. In such cases, the different segments can be interconnected in the space or can represent an alternative point of access to the space segment.

The mostly often envisaged scenario proposes an hybrid GEO and LEO constellation, where the LEO segment provides an access point to the network for those users located in areas where GEOs do not cover or suffer from their typical limitations (high latitudes, mountain regions, shaded zones), may not be economically convenient (deserts, oceans), or might not satisfy temporary extra traffic (when occasional political, sport or disaster events occur).

In the forward link, the data gathered by the LEO component from one or more terminals at very high data rate (in the order of 100 Mbit/s) will reach the final destination through the Inter Satellite Links (ISL) with the GEO segment and the serving gateway. The gateway ensures the interconnection with terrestrial networks. In the return link, the data uploaded on the GEO component from the gateway will be downloaded to the final user through the LEO component. In both cases, when a direct link is available, if on board memory is empty, direct transferring will be allowed. The payload architectures (both on LEO and GEO) must be regenerative, to allow storing data on board.

The total coverage and capacity increases and flexibility is ensured by adding just one, or a few, small and low cost LEO satellites to the classical GEO architecture. In particular, it allows modularity in setting up and deploying the constellation.

In figure 11.4.1, a configuration is shown with several LEO satellites and several GEO satellites, according to service and traffic requirements. ISL among GEO or LEO satellites can be foreseen too. Several papers [521][537][538] present the performance for such an architecture composed of one LEO satellite and one GEO satellite.

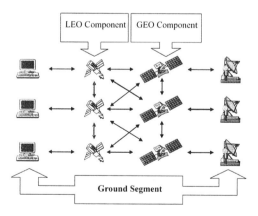

Figure 11.14. Hybrid LEO/GEO architecture

As concerns the double constellation conceived as alternative point of access efficient algorithms to perform handover must be designed and implemented [522]. This kind of hybrid constellation in line of principle could be implemented also with existing satellite systems, equipping user with proper dual mode terminals.

11.4.2   High Altitude Platforms

Recently HAPs are gaining much attention due to the growing communication needs of the modern world, particular in the provision of Internet access and broadband multimedia services to passengers travelling on-board public means of transport [539] or mobile users [540]. In 1995, SkyStation Inc. was created to start the development of an airship-based HAPs telecommunication system [541]. Other projects were subsequently proposed worldwide, including HALO, Helinet, SkyNet and SkyTower [542] [543] [524] [544] [545]. Two distinctive types of airborne platforms have been proposed for the provision of communication services from the stratosphere, unmanned airships and manned/unmanned aircrafts [546, 547]. Flying in the stratosphere just above the troposphere, at an altitude of about 17 km or higher, HAPs could fill the gap between satellite based communication systems on one hand, which are expensive and put high technical demands on the mobile units due to the large distance to the satellites, and terrestrial transmitters, on the other hand, which suffer from limited coverage.

HAPs combine advantages of both terrestrial and satellite communication systems, what make them attractive as an alternative telecommunications infrastructure. The main advantage of HAPs in comparison to terrestrial systems are large coverage area, simplified cell planning, small ground-based infrastructure, and the convenience to provide point-to-point, broadcast and multicast services to fixed or mobile users. In addition, HAPs can offer features of geostationary satellites (stationary service area, highly directive antennas, small Doppler shifts, etc.) with characteristics of low orbit satellites or terrestrial networks (low propagation delay, small size antennas, broadband capability, etc.). Furthermore, the position of HAPs can be shifted in accordance with changing communication demands, thus providing network flexibility and reconfigurability. Therefore HAPs are well suited for the instantaneous or gradual deployment of the network, as well as for temporarily provision of basic or additional capacity required (i) for the short time events with a large number of participants, (ii) in case of natural disasters, or (iii) in areas where the fixed infrastructure has suffered a major failure. Considering all this, HAPs represents a convenient infrastructure for telecommunication services of the future world.

HAPs communication system can be deployed as a stand-alone network providing access for the users in the coverage area of central ground station, or it can be connected to external communication networks via gateways providing internetworking functionality. In addition, HAPs communication system can consist of one or more platforms, which can be interconnected. However, the choice of network architecture has an impact on the wide range of services and consequently on the system penetration.

From the system architecture perspective HAPs can be used either as stand-alone platforms, providing broadband wireless access for single-user (mobile or fixed) or collective terminals (mobile or fixed) in the coverage area to the central ground station, or they can be interconnected via Terrestrial Network Links (TNLs) or Inter-Platform Links (IPLs) to form a network of platforms. Furthermore, such a telecommunication system can be deployed as a stand-alone network or it can be connected to external networks via gateways providing suitable internetworking functionality. Ground terminals communicate with platforms via user links, while ground stations, hosting gateways to external networks and different servers, are connected to platforms via backhaul links, together forming an Up/Down Link (UDL) segment.

The most important aspects of the architecture design phase are system coverage and types of services to be offered. To increase the system capacity the coverage area of each platform can be further divided into cells using multi-beam antennas. The size of the service area has a direct impact on the number of required HAPs and on their configuration. Similarly, the choice of network topology has an impact on the wide range of services that can be offered and consequently on the system penetration. Taking into account the location of switching equipment we can distinguish between the platforms without switching (transparent platform) and those with onboard switching (switching platform) capabilities. Onboard switching provides some gain in terms of QoSs parameters (mainly delay) to communicating parties within the same platform coverage area. In the case of switching in Ground Stations (GSs), however, the platform payload complexity, weight and power consumption can be significantly reduced, while backhaul requirements are more demanding and traffic from/to different cells needs to be efficiently aggregated/split on the platform. With respect to different possible interconnections of platforms, shown in figure 11.15, four network architectures can be defined [548][549]:

- In a **standalone platform scenario** the system coverage is limited to the cellular coverage of a single platform, enabling only communication between terminals within the coverage area, or with terminals in other networks using a gateway located in the ground station. In this scenario transparent or switching platform can be used.

- In a **scenario with the network of platforms connected via ground stations** the system coverage is no longer limited to single platform coverage but it heavily depends on ground facilities. Similarly, as in a standalone platform scenario, switching can be performed in the ground station or onboard.

- A **scenario with platform interconnection via IPLs** provides extended system coverage with significantly reduced terrestrial infrastructure. To support communication between adjacent platforms without any ground network elements each HAP payload includes a switching device and one or more IPL terminals. Depending on the link budget analysis we can choose between optical and radio frequency IPL terminals. In this scenario, ground stations are used mainly as gateways to other public and/or private networks, while providing also a backup interconnection between platforms in the case of IPL failure. Implementation of IPLs significantly reduces requirements for TNL and UDL segments, provides high flexibility of system coverage, and supports system operation independent of terrestrial network. On the other hand, IPL terminals represent additional weight and power consumption on the platform and require steerable IPL antenna to maintain permanent connection.

- A **scenario with Platform to Satellite Links (PSL)** can be used to integrate the HAP system into other non-local terrestrial or satellite networks, rather than as a prime mode of platform connection. It is mainly targeted for the use in areas with deficient (rural and remote areas) or non-existing terrestrial infrastructure. In addition to providing connection of the platform to other public or private networks via satellite, PSLs could also be used as a backup solution in the case when the connection with the rest of the network via IPL or GS is disabled due to a failure or extreme rain fading on UDL segment. The main drawback of PSLs is the need to use heavier terminals with higher power consumption in comparison to IPLs, due to the longer communication paths which result in higher attenuation. An important problem might be also interference with other satellite communication systems operating in adjacent frequency bands.

In addition to broadband wireless access at 47/48 GHz and 28/31 GHz, ITU has endorsed the use of HAPs in the IMT-2000/UMTS spectrum for the provision of 3G mobile services. HAPs UMTS systems will use the same Round Trip Times (RTTs) and provide the same functionality and meet the same service and operational requirements as traditional terrestrial tower-based UMTS systems. The HAP systems can be designed to serve as the sole station in a stand-alone infrastructure (essentially, replacing the tower base station network with a "base station network in the sky") or can be integrated into a system that employs traditional terrestrial base station towers, satellites and HAPs. Today, most of the major HAPs projects such as Sky Station, Sky Tower and SkyNet projects mentioned earlier include the delivery of IMT-2000 systems via HAPs.

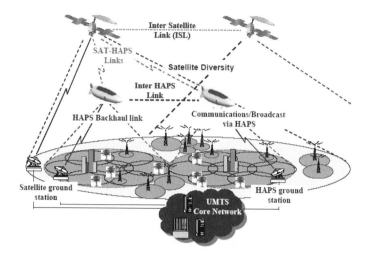

Figure 11.15. Possible B3G generic architecture based on HAPs [526] ©2005 IEEE

Recently, Space Data Corp. [550] has successfully deployed their Disposable HAPs covering the rural area of the USA. As one knows, the rural area of the USA is around 90% of the total USA territory. Deploying a large number of tower-based telecommunication infrastructures into such a huge area with less density will not be cost-effective. Therefore, Space Data has implemented the use of simple weather balloon to bring INTERACTIVE-TEXT-MESSAGING initially to ensure near-100% of USA territory are connected with telecommunication services. To guarantee most of the people anywhere in the USA can communicate easily to any destinations at anytime. Based on their simple architecture, now, Space Data has already gained the patent for such architecture and already operated extended services supporting location-based services, sensor networks, etc.

11.4.3 Terrestrial extension of satellite/HAP links for indoor penetration

Satellite-to-indoor links cause special problems: loss of a link joining the satellite (or a HAP) to a mobile terminal within a building can be very high. This excludes direct access of such terminals to the satellite network. As ubiquitous access is highly desirable, terrestrial extension can be needed. Relevant wave propagation characteristics are briefly listed below:

1 Indoor penetration of waves depends very much on the building design or the wall material, a rather self evident property; less self evident is

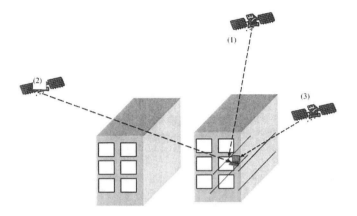

Figure 11.16. Satellite diversity

maybe that inner construction can be even more important than outer. This draws the attention to a theoretical consequence: electromagnetic investigation of a nontransparent box with transparent rectangular grid on the wall, applied sometimes for modeling the problem, has very limited or no relevance.

2 Windows of large office buildings are often covered by metallization, impeding wave penetration even further.

3 Dependence on the elevation angle of the satellite/HAP is rather special. In outdoor and also in suburban environment, higher altitude results in lower attenuation due to lower blockage probability. In multistory buildings higher altitude angle can result in higher or in lower loss as blockage is caused both by external objects and by upper floors.

4 Indoor electromagnetic field suffers from sever multipath if radiated along a slant path from above; as a consequence this field is diffuse. If receiving diffuse field, apparent gain of an antenna can be much less than its nominal value and equal to 0 dB in the worst case, whatever the nominal gain is.

All these have the consequence that large office buildings may need special architectures of the communication system if integrated cellular-HAP-satellite wireless networks are applied. A few possibilities are listed below which are applicable in different situations. For sake of completeness links without terrestrial extension are mentioned first:

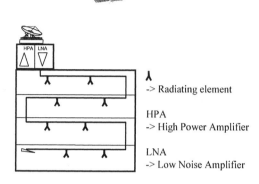

-> Radiating element

HPA
-> High Power Amplifier

LNA
-> Low Noise Amplifier

Figure 11.17. Satellite-to-indoor link based on distributed antenna

- In the simplest and most desirable case direct access via the usual ground terminal, without any additional infrastructure is possible, and, to enhance this, terminals contain a directive antenna with significant gain. However, due to points 1, 2 and 4 above, this may be effective at special locations of the building only; ubiquitous access may thus not be insured.

- Satellite diversity can increase the in-building access area. However, its efficiency is doubtful and to quantify or even to model diversity gain is rather difficult. In figure 11.16 an example is shown where the high altitude link (1) is blocked by the upper floors, the low altitude link (2) is blocked by neighboring objects (building) and the low altitude link (3) is unblocked.

- The building can form a (pico)cell of the terrestrial cellular network. Technically, this can be simply realized by placing an appropriate transponder (e.g. on the rooftop) and joining it to one or some base-stations. However, this is no direct access to the satellite or HAP.

- For direct access the satellite or HAP link must be lengthened into the building interior. At lower microwave frequencies a linear repeater with a distributed antenna can be applied (e.g. on the rooftop) as in figure 11.17.

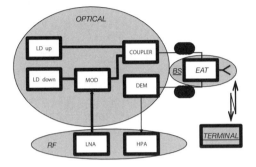

Figure 11.18. Satellite-to-indoor link based on fiber radio concept

- While a distributed antenna system cannot be applied at higher frequency bands, appropriate waveguides having very large attenuation (in the order of 1-3 dB/m), application of the fiber radio concept seems to be a viable solution. In fiber radio the modulated RF signal modulates light intensity of a laser diode. This signal is transmitted over an optical fiber and then reconverted to the radio frequency band and radiated by the antenna. An extremely simple version applies an electro-absorptive transceiver (EAT), that is a semiconductor able to intensity modulate certain optical frequencies and demodulate other ones. Figure 11.18 shows this concept applied to the linear repeater with one antenna where LD_{up} is a laser diode operating in the modulating band of the EAT, serving for uplink transmission, LD_{down}: is a laser diode operating in the demodulating band of the EAT serving for downlink transmission and MOD and DEM are optical co-located with RF parts.

11.5   Conclusion

It is very likely that satellites will continue to follow a conservative evolution over the next five years or so, the market being driven by GEO MSS satellites. Key technologies will be:

- Lighter and reduced volume components with on board wireless connections,

- Larger deployable reflectors up to tens of meters,

- Higher power multi-beam antenna with adaptive beam shaping via DSP,

- Scalable digital process enabling improved connectivity,

- On board regeneration and switching.

There will be a push to produce much cheaper earth-terminals and satellites that must approach the 50Gbit/s capacity using very large deployable antennas and more efficient and steerable beams. On board payloads with flexible filtering and bit rates and packet switching will emerge during this timeframe. There is also a need to provide much more efficient traffic routing, possibly extending the label switch path approaches and policy based routing algorithms. High speed flexible mesh connections of access points will emerge to augment the terrestrial networks.

Looking at the long term future, a prediction of future development can be done starting from the past. In the last 20 years, satellites have become thousand times more powerful and a similar times more cost effective. There does not seem anything in the laws of physics to prevent this trend continuing. The emergence of smaller and cheaper satellites will allow introduction of satellite clusters and swarms, with fragmented functions and connected together via an in-space ad-hoc network, or the resurgence of constellations, LEOs or MEOs being used to solve the latency problem.

Actual fixed and mobile services based on broadcast and point-to-point will be completely integrated with a layered network which will include GEO (for wide area broadcast and non latency -dependant services), non-GEO and HAPs (for local area concentrated services) with the terrestrial based cellular, Wi-Fi and Personal Area Networks (PANs). The drive will be broadband for non-broadcasting applications with particular emphasis on passenger vehicles—planes, ships and road vehicles. IPv6 services will continue to dominate requiring support of various QoSs, traffic profiles and the addition of security features that really work over satellite. Multicast service demand will increase and there will be a need to develop efficient and reliable IP native multicast structures. Client-server and peer-to-peer applications will continue to exist. There will be a need to look at cross-layer schemes in order to improve the efficiency of delivery within the IP network.

In the long-term future, it will make in particular possible an all IP network constellation with on-board routers switching IP (maybe improved versions and not IPv6).

We will see some moves towards integration as satellite and terrestrial operators merge and start to consider a more integrated provision. In this respect, interworking issues and standards are likely to be very important. The previous chapters of this book have presented several solutions for air interface techniques and reconfiguration that will remove the current limitations of the satellite transmission.

References

[1] A. Papoulis, *Probability, Random Variables, and Stochastic Procceses*, McGraw-Hill, New York, 3 edition, 2001.

[2] T.M. Cover and J.A. Thomas, *Elements of Information Theory*, John Wiley & Sons Inc., New York, 1991.

[3] H. L. Van Trees, *Detection, Estimation, and Modulation Theory. Part I: Detection, Estimation, and Linear Modulation Theory*, John Wiley and Sons, Inc., New York, 1968.

[4] I. S. Gradshteyn and I. M. Ryzhik, *Table of Integrals, Series, and Products*, Academic Press, Inc., New York, USA, 1973.

[5] W. Weibull, "A statistical representation of fatigue failures in solids," *Trans. Royal Inst. Techn.*, , no. 27, 1949.

[6] R. Ramakumar, *Reliability Engineering: Fundamentals and Applications*, Prentice Hall, 1993.

[7] R.B. Abernethy, *The New Weibull Handbook*, Barringer & Associates, New York, 4 edition, 2000.

[8] M. Sekine and Y. H. Mao, *Weibull Radar Clutter*, IEE, 1990.

[9] H. Hashemi, "The indoor radio propagation channel," *IEEE Proc.*, vol. 81, pp. 943–968, July 1993.

[10] F. Babich and G. Lombardi, "Statistical analysis and characterization of the indoor propagation channel," *IEEE Trans. Comm.*, vol. 48, no. 3, pp. 455–464, March 2000.

[11] N. S. Adawi, H.L. Bertoni, J.R. Child, W.A. Danial, J.E. Detra, R.P. Eckert, E.H. Flath, R.T. Forrest, W.C.Y. Lee, S.R. McConoughery, P.J. Murray, H. Sachs, G.L. Schrenk, N.H. Shepherd, and F.D. Shipley, "Coverage prediction for mobile radio systems operating in the 800/900 MHz

frequency range," *IEEE Trans. Vehicular Technology*, vol. 37, pp. 3–72, February 1988.

[12] N. H. Shepherd, "Radio wave loss deviation and shadow loss at 900 MHz," *IEEE Trans. Vehicular Technology*, vol. 26, pp. 309–313, 1977.

[13] M. D. Jacoub, "The α-μ distribution: A general fading distribution," in *Proc. of the International Symposium on Personal, Indoor, and Mobile Radio Communications*, Lisbon, Portugal, September 2002.

[14] N. C. Sagias, G. K. Karagiannidis, D. A. Zogas, P. T. Mathiopoulos, S. A. Kotsopoulos, and G. S. Tombras, "Dual Selection Diversity over Correlated Weibull Fading Channels," in *Proc. of the IEEE International Conference on Telecommunications*, Paris, France, June 2004, vol. 6, pp. 3384–3388.

[15] N. C. Sagias, G. K. Karagiannidis, D. A. Zogas, P. T. Mathiopoulos, and G. S. Tombras, "Performance analysis of dual selection diversity in correlated Weibull fading channels," *IEEE Trans. Comm.*, vol. 52, no. 7, pp. 1063–1067, July 2004.

[16] E. W. Weisstein, "The Wolfram Functions Site," From MathWorld Web Resource, 2004, http://mathworld.wolfram.com/.

[17] J. G. Proakis, *Digital Communications*, McGraw-Hill, 4 edition, 2001.

[18] G.E. Corazza and G. Ferrari, "New Bounds for the Marcum Q-Function," *IEEE Trans. Inform. Theory*, vol. 48, no. 11, pp. 3003–3008, November 2002.

[19] G. Taricco and M. Elia, "Capacity of fading channel with no side information," *Electronic Letters*, , no. 16, pp. 1368–1370, July 1997.

[20] I.B. Abou-Faycal, Trott M.D., and S. Shamai, "The Capacity of Discrete-Time Memoryless Rayleigh-Fading Channels," *IEEE Trans. Inform. Theory*, vol. 47, pp. 1290–1301, 2001.

[21] A. Goldsmith, *Wireless Communications*, Cambridge University Press, 2005.

[22] M.S. Alouini and A. Goldsmith, "Capacity of Rayleigh fading channels under different adaptive transmission and diversity-combining techniques," *IEEE Trans. Vehicular Technology*, vol. 48, no. 4, pp. 1165–1181, July 1999.

[23] W.C.Y. Lee, "Estimate of channel capacity in Rayleigh fading environment," *IEEE Trans. Vehicular Technology*, vol. 39, no. 3, pp. 187–189, August 1990.

[24] N. C. Sagias, G. S. Tombras, and G. K. Karagiannidis, "New results for the Shannon channel capacity in generalized fading channels," *IEEE Comm. Lett.*, vol. 9, no. 2, pp. 97–99, February 2005.

[25] N. C. Sagias, G. K. Karagiannidis, D. A. Zogas, P. T. Mathiopoulos, G. S. Tombras, and F.N. Pavlidou, "Second order statistics and channel spectral efficiency for selection diversity receivers in Weibull fading," in *Proc. of the International Symposium on Personal, Indoor, and Mobile Radio Communications*, Barcelona, Spain, September 2004, vol. 3, pp. 2140–2145.

[26] A. J. Goldsmith and P. P. Varaiya, "Capacity of Fading Channels with Channel Side Information," *IEEE Trans. Inform. Theory*, vol. 43, pp. 1986–1992, 1997.

[27] G. Caire and S. Shamai, "On the Capacity of Some Channels with Channel State Information," *IEEE Trans. Inform. Theory*, vol. 45, pp. 2007–2019, 1999.

[28] R. Rinaldo and R. De Gaudenzi, "Capacity analysis and system optimization for the forward link of multi-beam satellite broadband systems exploiting adaptive coding and modulation," *International Journal of Satellite Communications and Networking*, vol. 22, no. 3, pp. 401–423, June 2004.

[29] P.P. Bergmans and T.M. Cover, "Cooperative Broadcasting," *IEEE Trans. Inform. Theory*, vol. 20, pp. 317–324, 1974.

[30] W. Yu and J. Cioffi, "Trellis Precoding for the Broadcast Channel," in *Proc. of the IEEE GLOBECOM*, October 2001, pp. 1344–1348.

[31] L. Li and A.J. Goldsmith, "Capacity and Optimal Resource Allocation for Fading Broadcast Channels - Part I: Ergodic Capacity," *IEEE Trans. Inform. Theory*, vol. 47, pp. 1083–1102, 2001.

[32] L. Li and A.J. Goldsmith, "Capacity and Optimal Resource Allocation for Fading Broadcast Channels - Part II: Outage Capacity," *IEEE Trans. Inform. Theory*, vol. 47, pp. 1103–1127, 2001.

[33] N. Jindal, Vishwanath S., and A.J. Goldsmith, "On the Duality of Gaussian Multiple-Access and Broadcast Channels," *IEEE Trans. Inform. Theory*, vol. 50, pp. 768–783, May 2004.

[34] D. Tse and P. Viswanath, *Fundamentals of Wireless Communication*, Cambridge University Press, 2005.

[35] D. Tse and S. Hanly, "Multiaccess fading channels - Part I: Polymatroid structure, optimal resource allocation and throughput capacities," *IEEE Trans. Inform. Theory*, vol. 44, pp. 2796–2815, 1998.

[36] E. G. Larsson and P. Stoica, *Space-Time Block Coding for Wireless Communications*, Cambridge University Press, 2003.

[37] M.H.M. Costa, "Writing on Dirty Paper," *IEEE Trans. Inform. Theory*, vol. 29, pp. 439–441, 1983.

[38] H. Weingarten, Y. Steinberg, and S. Shamai, "The Capacity Region of the Gaussian MIMO Broadcast Channel," in *Proc. of Conference on Information Sciences and Systems*, March 2004.

[39] E. W. Weisstein, "Bonferroni's inequalities," MathWorld Web Resource, 2000.

[40] S. Yousefi, *Bounds on the performance of Maximum-Likelihood Decoded Binary Block Codes in AWGN Interference*, Ph.D. thesis, Waterloo, Ontario, Canada, 2002.

[41] R. Gallager, *Low-Density Parity-Check Codes*, MIT Press, Cambridge, 1963.

[42] R. Gallager, "A simple derivation of the coding theorem and some applications," *IEEE Trans. Inform. Theory*, vol. 11, no. 1, pp. 3–18, 1965.

[43] P. Bello, "Characterization of Randomly Time-Variant Linear Channels," *IEEE Trans. Comm.*, vol. 11, no. 4, pp. 360–393, December 1963.

[44] S. O. Rice, "Mathematical Analysis of Random Noise," *Bell System Technical Journal*, vol. 24, pp. 46–156, January 1945.

[45] E. Lutz, D. Cygan, M. Dippold, F. Dolainsky, and W. Papke, "The Land Mobile Satellite Communication Channel - Recording, Statistics, and Channel Model," *IEEE Trans. Vehicular Technology*, vol. 40, no. 2, pp. 375–386, May 1991.

[46] H. Suzuki, "A Statistical Model for Urban Radio Propagation," *IEEE Trans. Comm.*, vol. 25, no. 7, pp. 673–680, July 1977.

[47] C. Loo, "A Statistical Model for a Land Mobile Satellite Link," *IEEE Trans. Vehicular Technology*, vol. 34, no. 3, pp. 122–127, August 1985.

[48] G.E. Corazza and F. Vatalaro, "A Statistical Model for Land Mobile Satellite Channels and Its Application to Nongeostationary Orbit Systems," *IEEE Trans. Vehicular Technology*, vol. 43, no. 3, pp. 738–742, August 1994.

[49] S. R. Saunders and B. G. Evans, "Physical Model of Shadowing Probability for Land Mobile Satellite Propagation," *IEE Electron. Lett.*, vol. 32, no. 17, pp. 1548–1549, August 1996.

[50] E. Lutz, "A Markov Model for Correlated Land Mobile Satellite Channels," *Int. Journal of Satellite Communications*, vol. 14, pp. 333–339, 1996.

[51] Y. Karasawa, K. Kimura, and K. Minamisono, "Analysis of Availability Improvement in LMSS by Means of Satellite Diversity Based on Three-State Propagation Channel Model," *IEEE Trans. Vehicular Technology*, vol. 46, no. 4, pp. 1047–1056, November 1997.

[52] R.H. Clarke, "A Statistical Theory of Mobile-Radio Reception," *Bell System Technical Journal*, vol. 47, pp. 957–1000, 1968.

[53] W.C. Jakes, *Microwave mobile communications*, IEEE Press, New York, USA, 1994.

[54] F. Vatalaro and A. Forcella, "Doppler Spectrum in Mobile-to-Mobile Communications in the Presence of Three-Dimensional Multipath Scattering," *IEEE Trans. Vehicular Technology*, vol. 46, no. 1, pp. 213–219, February 1997.

[55] A. Jahn, "Propagation Considerations and Fading Countermeasures for Mobile Multimedia Services," *International Journal of Satellite Communications*, vol. 19, no. 3, May 2001.

[56] M.A.N. Parks, S.R. Saunders, and B.G. Evans, "A Wideband Channel Model Applicable to Mobile Satellite Systems at L-Band and S-Band," *IEE Colloqium on Propagation Aspects of Future Mobile Systems*, pp. 12/1–12/6, October 1996.

[57] J. Bito, "Adaptive Digitale Kanalmodelle für Mobilfunkkanäle," *Frequenz*, vol. 50, no. 11-12, pp. 261–267, Nov./Dec. 1996, In German.

[58] E.O. Elliott, "Estimates of error rates for codes on burst-noise channels," *Bell System Technological Journal*, vol. 42, pp. 1977–1997, 1962.

[59] Z. Bodnár, Z. Herczku, J. Bérces, I. Papp, F. Som, B. G. Molnár, and I. Frigyes, "A Detailed Experimental Study of the LEO Satellite to Indoor Channel Characteristics," *Intl. J. of Wireless Information Networks*, vol. 6, no. 2, pp. 79–91, Sept./Oct. 1999.

[60] M. A. Vázquez Castro, *Statistical and Deterministic Land Mobile Satellite Channel Modeling for non-urban environments*, Ph.D. thesis, Universidad de Vigo, Spain, 2002.

[61] F. Pérez-Fontán, M. Vázquez-Castro, C.E. Cabado, J.P. Garcia, and E. Kubista, "Statistical Modeling of the LMS Channel," *IEEE Trans. Vehicular Technology*, vol. 50, no. 6, pp. 1549–1567, November 2001.

[62] S. Scalise, J. Kunisch, H. Ernst, J. Siemons, G. Harles, and J. Horle, "Measurement Campaign for the Land Mobile Satellite Channel in Ku-Band," in *Proc. of the 5th EMPS Workshop*, September 2002.

[63] S. Scalise, M.A. Vázquez Castro, A. Jahn, and H. Ernst, "A Comparison of the Statistical Properties of the Land Mobile Satellite Channel at Ku, Ka and EHF Bands," in *Proc. of the 61st IEEE Vehicular Technology Conference*, May 2005, pp. 2687–2691.

[64] S. Scalise, R. Mura, and V. Mignone, "Air Interfaces for Satellite Based Digital TV Broadcasting in the Railway Environment," *IEEE Trans. on Broadcasting*, to appear in 2006 2006.

[65] A. Benarroch and L. Mercader, "Signal Statistics Obtained from a LMSS Experiment with the MARECS Satellite," *IEEE Trans. Comm.*, vol. 42, no. 2-3-4, pp. 1264–1269, 1994.

[66] Y. Karasawa and T. Shiokawa, "Characteristics of L-Band Multipath Fading Due to Sea Surface Reflection," *IEEE Trans. Antennas and Propagation*, vol. 32, no. 6, pp. 618–623, June 1984.

[67] F. Dovis, R. Fanini, M. Mondin, and P. Savi, "Small-Scale Fading for High-Altitude Platform (HAP) Propagation channel," *IEEE Journal on Selected Areas in Communications*, vol. 20, no. 3, pp. 641–647, April 2002.

[68] J.E. Allnutt, *Satellite-to-ground radiowave propagation - Theory, practice and system impact at frequencies above 1 GHz*, Peter Peregrinus Ltd, 1989.

[69] R. K. Crane, *Electromagnetic Wave Propagation Through Rain*, John Wiley & Sons, 1996.

[70] COST Project 255, "Radiowave propagation modelling for new satcom services at Ku-band and above," Final Report SP-1252, ESA Publications Division, March 2002.

[71] D. Dintelmann, "OPEX: Reference Book on Depolarization," Tech. Rep. Doc. ESA-ESTEC-WPP-083, ESA/ESTEC, Noordwijk, The Netherlands, November 1994.

[72] G. Maral and M. Bousquet, *Satellite Communications Systems: Systems Techniques and Technology*, John Wiley & Sons, 4 edition, May 2002.

[73] COST Project 280, "Propagation impairment mitigation for millimetre-wave radio systems," Final report, ESA Publications Division, 2005.

[74] COST Project 205, "Influence of the atmosphere on radiopropagation on satellite Earth paths at frequencies above 10 GHz," Tech. Rep. Report EUR 9923 EN, ISBN 92-825-5412-0, ESA Publications Division, 1985, Sections 5.3 and 5.6.

[75] J.P.V. Poiares Baptista and P. G. Davies, "OPEX: Reference Book on Attenuation Measurement and Prediction," Tech. Rep. Doc. ESA-ESTEC-WPP-083, ESA/ESTEC, Noordwijk, The Netherlands, November 1994.

[76] P. Loreti and M. Luglio, "Interference Evaluations and Simulations for Multisatellite Multispot Systems," *International Journal on Satellite Communications*, vol. 20, no. 4, pp. 261–281, Jul./Aug. 2002.

[77] G.E. Corazza and F. Vatalaro, "An Approach to Transmission Performance Evaluation for Satellite Systems Adopting Non-Geostationary Constellation," in *Proc. of the AIAA/ESA Workshop on International Cooperation in Satellite Communications*, March 1995, pp. 171–178, Noordwijk, The Netherlands.

[78] R. De Gaudenzi and F. Giannetti, "DS-CDMA Satellite Diversity Reception for Personal Satellite Communication: Satellite-to-Mobile Link Performance Analysis," *IEEE Trans. Vehicular Technology*, vol. 47, no. 2, pp. 658–672, May 1998.

[79] M. Moher, R.G. Lyons, and L. Erup, "Power and Bandwidth Tradeoffs for a Third Generation Satellite System," in *Proc. of the Inter. Mass Spectrometry Conference*, Pasadena, California, USA, June 1997, pp. 547–551.

[80] A. Jamalipour, M. Katayama, T. Yamazato, and A. Ogawa, "Performance of an Integrated Voice/Data System in Nonuniform Traffic Low Earth-Orbit Satellite Communication System," *IEEE Journal on Selected Areas in Communications*, vol. 13, no. 2, pp. 465–473, February 1995.

[81] F. Vatalaro, G.E. Corazza, C. Caini, and C. Ferrarelli, "Analysis of LEO, MEO, and GEO Global Mobile Satellite Systems in the Presence of Interference and Fading," *IEEE Journal on Selected Areas in Communications*, vol. 13, no. 2, pp. 291–300, February 1995.

[82] S. Blondeau, G. Maral, T. Roussel, and J. P. Taisant, "Self Interference in non Geostationary Satellite Systems," in *Proc. of the 10th Int. Conf. on Digital Satellite Comm.*, Brighton, UK, May 1995, pp. 290–297.

[83] F. Ananasso and F. Delli Priscoli, "The Role of Satellites in Personal Communication Services," *IEEE Journal on Selected Areas in Communications*, vol. 13, no. 2, pp. 180–196, February 1995.

[84] N. J. A. Sloane and A. D. Wyner, *Claude Elwood Shannon: Collected Papers*, IEEE Press, New York, 1993.

[85] W. Gappmair, "Claude E. Shannon: The 50th Anniversary of Information Theory," *IEEE Comm. Mag.*, vol. 37, pp. 102–105, April 1999.

[86] J. L. Massey, *Advanced Methods for Satellite and Deep-space Communications*, Springer, New York, 1992, pages : 1-17.

[87] G. C. Clark and J. B. Cain, *Error-Correction Coding for Digital Communications*, Plenum Press, New York, USA, 1981.

[88] S. Lin and D. J. Costello, *Error Control Coding: Fundamentals and Applications*, Prentice Hall, Englewood Cliffs, New Jersey, USA, 1983.

[89] S. B. Wicker, *Error Control Systems for Digital Communication and Storage*, Prentice Hall, Englewood Cliffs, New Jersey, USA, 1995.

[90] E. R. Berlekamp, *Algebraic Coding Theory*, McGraw-Hill, New York, 1968.

[91] G. D. Forney, "Convolutional Codes I: Algebraic Structures," *IEEE Trans. Inform. Theory*, vol. 16, pp. 720–738, November 1970.

[92] B. Vucetic and J. Yuan, *Turbo Codes: Principles and Applications*, Kluwer Academic Publishers, Boston, Massacchusetts, USA, 2000.

[93] J. B. Cain, G. C. Clark, and J. M. Geist, "Punctured Convolutional Codes of Rate $(n-1)/n$ and Simplified Maximum Likelihood Decoding," *IEEE Trans. Inform. Theory*, vol. 25, pp. 97–100, 1979.

[94] J. Hagenauer, "Rate-Compatible Punctured Convolutional Codes (RCPC Code) and their Applications," *IEEE Trans. Comm.*, vol. 36, pp. 389–400, April 1988.

[95] A. Vardy and Y. Be'ery, "More efficient soft decoding of the Golay Code," *IEEE Trans. Inform. Theory*, vol. 37, pp. 667–672, May 1991.

[96] L. Ping and K. L . Yeung, "Symbol by Symbol APP Decoding of the Golay Code and Iterative Decoding of Concatenated Golay Codes," *IEEE Trans. Inform. Theory*, vol. 45, pp. 2558–2562, November 1999.

[97] ETSI TS 101 376-5-3, *GEO - Mobile Radio Interface Specifications; General Packet Radio Service; Part 5: Radio interface physical layer*

specifications; Sub-part 3: Channel Coding, ETSI, 2.2.1 edition, March 2005.

[98] A. Ashikhmin and S. Litsyn, "Simple MAP Decoding of First-Order Reed-Muller and Hamming Codes," *IEEE Trans. on Information Theory*, vol. 50, pp. 1812–1818, August 2004.

[99] ETSI EN 300 421, *Digital Video Broadcasting (DVB); Framing structure, channel coding and modulation for 11/12 GHz satellite services*, ETSI, 1.1.3 edition, August 1997.

[100] CCSDS 101.0-B-6, *Telemetry Channel Coding Blue Book*, CCSDS, October 2002.

[101] ETSI EN 302 307, *Digital Video Broadcasting (DVB); Second generation framing structure, channel coding and modulation systems for Broadcasting, Interactive Services, News Gathering and other broadband satellite applications*, ETSI, June 2004.

[102] V. Guruswami and M. Sudan, "Improved Decoding of Reed-Solomon Codes and Algebraic-Geometriy Codes," *IEEE Trans. Inform. Theory*, vol. 45, no. 6, pp. 1757–1767, September 1999.

[103] R. Koetter and A. Vardy, "Algebraic Soft-Decoding of Reed-Solomon Codes," *IEEE Trans. on Information Theory*, vol. 49, no. 11, pp. 2809–2825, November 2003.

[104] H. J. Nussbaumer, *Fast Fourier Transform and Convolution Algorithms*, Springer-Verlag, Berlin, 1981.

[105] J. M. Pollard, "The Fast Fourier Transform in a Finite Field," *Mathematics of Computation*, vol. 25, pp. 365–374, April 1971.

[106] R. J. McEliece, *Finite Fields for Computer Scientists and Engineers*, Kluwer Academic Publishers, Boston, 1987.

[107] J. Hong and M. Vetterli, "Simple Algorithms for BCH Decoding," *IEEE Trans. Comm.*, vol. 43, pp. 2324–2333, August 1995.

[108] R. E. Blahut, *Algebraic Methods for Signal Processing and Communications Coding*, Springer-Verlag, New York, 1992.

[109] W. Gappmair, "An Efficient Prime-Length DFT Algorithm over Finite Fields $GF(2^m)$," *European Trans. Telecomm.*, vol. 14, pp. 171–176, March 2003.

[110] J. Hagenauer and P. Hoeher, "A Viterbi Algorithm with Soft-Decision Outputs and its Applications," in *IEEE Global Telecommunications Conference*, Dallas, Texas, USA, November 1989, pp. 1680–1686.

[111] L. Lin and R. Cheng, "Improvements in SOVA-Based Decoding for Turbo Codes," in *Proc. of the IEEE Int. Conference on Communications*, June 1997, pp. 1473–1478.

[112] L. Bahl, J. Cocke, F. Jelinek, and J. Raviv, "Optimal Decoding of Linear Codes for Minimizing Symbol Error Rate," *IEEE Trans. Inform. Theory*, vol. 20, pp. 284–287, March 1974.

[113] P. Robertson, E. Villebrun, and P. Hoeher, "A Comparison of Optimal and Sub-Optimal MAP Decoding Algorithms Operating in the Log Domain," in *IEEE Int. Conf. on Communications*, Seattle, USA, June 1995, pp. 1009–1013.

[114] A. J. Viterbi, "An Intuitive Justification and a Simplified Implementation of the MAP Decoder for Convolutional Codes," *IEEE Journal on Selected Areas in Communications*, vol. 16, pp. 260–264, February 1998.

[115] M. P. C. Fossorier, F. Burkert, S. Lin, and J. Hagenauer, "On the Equivalence Between SOVA and Max-Log-MAP Decodings," *IEEE Comm. Lett.*, vol. 2, pp. 137–139, May 1998.

[116] G. D. Forney, "The Viterbi Algorithm," *IEEE Proc.*, vol. 61, pp. 268–277, March 1973.

[117] C. Berrou, A. Glavieux, and P. Thitimajshima, "Near Shannon limit error-correcting coding and decoding: Turbo-codes," in *Proc. of the International Conference on Telecommunications*, Geneva, Switzerland, May 1993, pp. 1064–1070.

[118] C. Berrou and M. Jezequel, "Non-binary Convolutional Codes for Turbo Coding," *IEE Electron. Lett.*, vol. 35, pp. 39–40, January 1999.

[119] P. Elias, "Error-free coding," *IRE Trans. Inform.Theory*, vol. PGIT-4, pp. 29–37, September 1954.

[120] ETSI EN 301 790, *Digital Video Broadcasting (DVB) Interaction channel for satellite distribution systems*, ETSI, September 2005.

[121] C. Berrou, "The Ten-Year-Old Turbo Codes Are Entering Into Service," *IEEE Comm. Mag.*, vol. 41, pp. 110–116, August 2003.

[122] S. Benedetto, G. Montorsi, and D Divsalar, "Concatenated Convolutional Codes with Interleavers," *IEEE Comm. Mag.*, vol. 41, pp. 102–109, August 2003.

[123] Y. Wu, B. D. Woerner, and W. J. Ebel, "A Simple Stopping Criterion for Turbo Decoding," *IEEE Comm. Lett.*, vol. 4, pp. 258–260, August 2000.

[124] S. Ten Brink, "Convergence Behavior of Iteratively Decoded Parallel Concatenated Codes," *IEEE Trans. Comm.*, vol. 49, pp. 1727–1737, October 2001.

[125] L. Papke and P. Robertson, "Improved Decoding with SOVA in Parallel Concatenated (Turbo-Code) Scheme," in *Proc. of the IEEE Int. Conference on Communications*, July 1996, pp. 102–106.

[126] S. Papaharalabos, P. Sweeney, and B. G. Evans, "A New Method of Improving SOVA Turbo Decoding for AWGN, Rayleigh and Rician Fading Channels," in *IEEE Vehicular Technology Conference*, Milan, Italy, May 2004, vol. 5, pp. 2862–2866.

[127] F. Burkert and J. Hagenauer, "A Serial Concatenated Coding Scheme with Iterative Turbo and Feedback Decoding," in *Proc. of Int. Symp. on Turbo Codes*, Brest, France, 1997.

[128] H. Ebert, W. Gappmair, and O. Koudelka, "Decreasing the Error Floor of Low-Complexity Turbo Codes," in *Proc. Sixth Baiona Workshop on Signal Processing in Communications*, Baiona, Spain, 2003, pp. 337–342.

[129] S. Benedetto and G. Montorsi, "Unveiling turbo codes: some results for parallel concatenated coding schemes," *IEEE Trans. Inform. Theory*, vol. 42, pp. 409–429, March 1996.

[130] S. Benedetto, D. Divsalar, G. Montorsi, and D. Pollara, "Serial Concatenation of Interleaved Codes: Performance Analysis, Design and Iterative Decoding," *IEEE Trans. Inform. Theory*, vol. 44, pp. 909–926, March 1998.

[131] M. Breiling, "A Logarithmic Upper Bound on the Minimum Distance of Turbo Codes," *IEEE Trans. Inform. Theory*, vol. 50, pp. 1692–1710, August 2004.

[132] S. Benedetto, D. Divsalar, G. Montorsi, and D. Pollara, "Improved upper bounds on the ML decoding error probability of parallel and serial concatenated codes via their ensemble distance spectrum," *IEEE Trans. Inform. Theory*, vol. 46, pp. 231–244, January 2000.

[133] M. Chiani and A. Ventura, "Design and performance evaluation of some high-rate irregular low-density parity-check codes," in *Proc. of the IEEE GLOBECOM*, 2001, vol. 2, pp. 990–994.

[134] T. Richardson, "Error floors of LDPC codes," in *Proc. of the 41st Allerton Conference on Communication, Control, and Computing*, Allerton, Illinois, USA, October 2003, p. 1426–1435.

[135] J. Vogt and A. Finger, "Improving the Max-Log-MAP Turbo Decoder," *IEE Electr. Lett.*, vol. 36, pp. 1937–1939, November 2000.

[136] F. Chiaraluce, R. Garello, and M. Chiani, "Highly Efficient Channel codes for high data rate missions," Tech. Rep. 1504801DHK(SC), ESA/ESOC, December 2001.

[137] AHA, *United States Patent, Number 5,930,272: Block decoding with Soft Output Information*, July 1999.

[138] D.J.C. MacKay, "Good error-correcting codes based on very sparse matrices," *IEEE Trans. Inform. Theory*, vol. 45, pp. 399–431, 1999.

[139] R. Tanner, "A recursive approach to low complexity codes," *IEEE Trans. Inform. Theory*, vol. 27, pp. 533–547, 1981.

[140] T. J. Richardson and R. L. Urbanke, "Efficient encoding of low-density parity-check codes," *IEEE Trans. Inform. Theory*, vol. 47, pp. 638–656, 2001.

[141] Y. Mao and A.H. Banihashemi, "Decoding low-density parity-check codes with probabilistic scheduling," *IEEE Comm. Lett.*, vol. 10, pp. 414–416, 2001.

[142] Y. Mao and A.H. Banihashemi, "Decoding low-density parity-check codes with probabilistic schedule," in *Proc. of the IEEE Pacific Rim Conference on Communications, Computers and signal Processing*, 2001, vol. 1, pp. 119–123.

[143] J. Campello, D.S. Modha, and S. Rajagopalan, "Designing LDPC codes using bit-filling," in *Proc. of the IEEE International Conference on Telecommunications*, 2001, vol. 1, pp. 55–59.

[144] J. Campello and D.S. Modha, "Extended bit-filling and LDPC code design," in *Proc. of the IEEE GLOBECOM*, 2001, vol. 2, pp. 985–989.

[145] M. P. C. Fossorier, "Quasi-Cyclic Low-Density Parity-Check Codes From Circulant Permutation Matrices," *IEEE Trans. Inform. Theory*, vol. 50, pp. 1788–1793, August 2004.

[146] D. Divsalar, H. Jin, and R. J. McEliece, "Coding theorems for 'turbo-like' codes," in *Proc. of the 36th Allerton Conf. on Communication,*

Control, and Computing, Allerton, Illinois, USA, September 1998, pp. 201–210.

[147] K. Andrews, S. Dolinar, and J. Thorpe, "Encoders for Block-Circulant LDPC Codes," *Submitted to IEEE Int. Symp. on Information Theory*, 2005.

[148] S. Lin, J. Xu, I. Djurdjevic, and H. Tang, "Hybrid construction of LDPC codes," in *Proc. of the 40th Allerton Conference on Communication, Control and Computing*, Monticello, Illinois, USA, October 2002, pp. 1149–1158.

[149] W. W. Peterson and E. J. Weldon, *Error-Correcting Codes*, M.I.T. Press, Cambridge, Massachusetts, USA, 1972.

[150] H. Jin, A. Khandekar, and R. McEliece, "Irregular repeat-accumulate codes," in *Proc. of the International Symposium on Turbo codes and Related Topics*, September 2000, pp. 1–8.

[151] M. Yang, Y. Li, and W.E. Ryan, "Design of efficiently encodable moderate-length high-rate irregular LDPC codes," *IEEE Trans. Comm.*, vol. 52, pp. 564–571, April 2004.

[152] A. Abbasfar, K. Yao, and D. Disvalar, "Accumulate Repeat Accumulate Codes," in *Proc. of IEEE Globecomm*, Dallas, Texas, USA, November 2004.

[153] G. Liva, E. Paolini, and M. Chiani, "Simple Reconfigurable Low-Density Parity-Check Codes," *IEEE Comm. Letters*, vol. 9, pp. 258–260, March 2005.

[154] Y. Kou, S. Lin, and M. P. C. Fossorier, "Low-density parity-check codes based on finite geometries: a rediscovery and new results," *IEEE Trans. Inform. Theory*, vol. 47, no. 7, pp. 2711–2736, November 2001.

[155] B. Vasic and O. Milenkovic, "Combinatorial constructions of low-density parity-check codes for iterative decoding," *IEEE Trans. Inform. Theory*, vol. 50, no. 6, pp. 1156–1176, June 2004.

[156] I. Anderson, *Combinatorial Designs*, Ellis Horwood Limited, Chichester, 1990.

[157] E. F. Jr Assmus and J.D. Key, *Designs and their codes*, Cambridge University Press, Cambridge, 1992.

[158] S. Johnson, *Low-density Parity-check Codes from Combinatorial Designs*, Ph.D. thesis, The University Of Newcastle, Callaghan, N.S.W., Australia, August 2003.

[159] W.E. Ryan, *Coding and Signal Processing for Magnetic Recording Systems*, chapter X, p. Sec. VI: Iterative Decoding, CRC Press, New York, USA, 2004.

[160] J. Pearl, *Probabilistic Reasoning in Intelligent Systems: Networks of Plausible Inference*, Morgan Kaufmann, San Francisco, California, USA, 1988.

[161] F. R. Kschischang, B. J. Frey, and Loeliger H. A., "Factor Graphs and the Sum-Product Algorithm," *IEEE Trans. Inform. Theory*, vol. 47, pp. 498–519, 2001.

[162] H. Y. Hu, E. Eleftheriou, D. M Amold, and A. Dholakia, "Efficient implementation of the sum-product algorithm for decoding LDPC code," in *Proc. of the IEEE GLOBECOM*, 2001, vol. 2, pp. 1036–1036E.

[163] S. Y Chung, T. J. Richardson, and R. L. Urbanke, "Analysis of sum-product decoding of low-density parity-check codes using a Gaussian approximation," *IEEE Trans. Inform. Theory*, vol. 47, pp. 657–670, 2001.

[164] S. Litsyn and V. Shevelev, "On ensembles of low-density parity-check codes: asymptotic distance distributions," *IEEE Trans. Inform. Theory*, vol. 48, pp. 887–908, April 2002.

[165] G. Miller and D. Burshtein, "Bounds on the maximum-likelihood decoding error probability of low-density parity-check codes," *IEEE Trans. Inform. Theory*, vol. 47, pp. 2696–2710, November 2001.

[166] R. Gallager, *Information Theory and Reliable Communication*, John Wiley & Sons, New York, 1968.

[167] H. Ernst, S. Shabdanov, A. Donner, and S. Scalise, "Reliable Multicast for Land Mobile Satellite Channels," in *Proc. of the 21st International Communication Satellite Systems Conference & Exhibit*, 2003.

[168] H. Ernst, L. Sartorello, and S. Scalise, "Parity-Based Loss Recovery for Reliable Multicast Transmission," in *TVT*, Milano, Italy, 1998, vol. 6, pp. 349–361.

[169] ETSI EN 302 304, *Transmission System for Handheld Terminals DV B − H*, ETSI, 2004.

[170] M. G. Luby, M. Mitzenmacher, M. A. Shokrollahi, and D. A. Spielman, "Efficient erasure correcting codes," *IEEE Trans. Inform. Theory*, vol. 47, no. 2, pp. 569–584, 2001.

[171] D.J.C. MacKay, *Information Theory, Inference and Learning Algorithms*, Cambridge University Press, 2003.

[172] 3GPP TS 26.346, *Multimedia Broadcast/Multicast Service MBMS; Protocols and codecs*, 3GPP, 2005.

[173] F. Xiong, *Digital Modulation Techniques*, Artech House, Inc., Boston, 2000.

[174] W. T. Webb and L. Hanzo, *Modern Quadrature Amplitude Modulation*, Pentech Press Ltd., London, 1994.

[175] K. Feher, *Advanced Digital Communications*, Prentice Hall, Inc., Englewood Cliffs, 1987.

[176] C.M. Thomas, M.Y. Weidner, and S.H. Durrani, "Digital Amplitude-Phase Keying with M-ary Alphabets," *IEEE Trans. Comm.*, vol. COM-22, no. 2, pp. 168–180, February 1974.

[177] P. Salmi, M. Neri, and G.E. Corazza, "Design and Performance of Predistortion Techniques in Ka-band Satellite Networks," in *Proc. of the 22th International Communication Satellite Systems Conference*. AIAA, May 2004.

[178] J. B. Anderson, T. Aulin, and C-E. Sundberg, *Digital Phase Modulation*, Plenum Press Company, New York, 1986.

[179] M. Abramovitz and I. A. Stegun, *Handbook of Mathematical Functions with Formulas, Graphs and Mathematical Tables*, Dover, New York, USA, 9 edition, 1972.

[180] G. Ungerboeck and J. Csajka, "On improving data link performance by increasing the channel alphabet and introducing sequence coding," in *Proc. of the Int. Symp. Inform. Theory*, Ronneby, Sweden, June 1976.

[181] G. Ungerboeck, "Trellis Coded Modulation with redundant signal sets. Part I: Introduction, Part II: State of the Art," *IEEE Comm. Mag.*, vol. 25, no. 2, February 1987.

[182] G. Ungerboeck, "Channel Coding with Multilevel/Phase Signals," *IEEE Trans. Inform. Theory*, January 1982.

[183] G. Caire, G. Tarrico, and E. Biglieri, "Bit Interleaved Coded Modulation," *IEEE Trans. Inform. Theory*, vol. 44, pp. 927–946, May 1998.

[184] S. Le Goff, "Performance of Bit-Interleaved Turbo-Coded Modulations on Rayleigh Fading Channels," *IEE Electron. Lett.*, vol. 36, pp. 731–733, April 2000.

[185] S. Le Goff, A. Glavieux, and C. Berrou, "Turbo Codes and High Spectral Efficiency Modulation," in *Proc. of the IEEE Int. Conf. on Communications*, New Orleans, Louisiana, USA, May 1994, pp. 645–649.

[186] S. A. Barbulescu, W. Farrell, P. Gray, and M. Rice, "Bandwidth Efficient Turbo Coding for High Speed Mobile Satellite Communications," in *International Symposium on Turbo Codes and Related Topics*, Brest, France, September 1997, pp. 119–126.

[187] S. Papaharalabos, "Turbo Coding for High Data Rate Downlink in S-UMTS Air Interface," M.S. thesis, University of Surrey, Guildford, UK, 2002.

[188] D. Makrakis and P. Mathiopoulos, "Trellis coded noncoherent QAM: A new bandwidth and power efficient scheme," in *Proc. of the Vehicular Technology Conference*, San Francisco, California, USA, May 1989, pp. 95–100.

[189] F. Pingyi and X. Xiang-Gen, "A noncoherent coded modulation for 16-QAM," in *Proc. of Intern. Conf. Commun. Techn.*, October 2000, vol. 2, pp. 1347–1350.

[190] G. Colavolpe, R. Raheli, and G. Picchi, "Detection of linear modulatins in the presence of strong phase and frequency instabilities," in *Proc. of the IEEE International Conference on Telecommunications*, June 2000, vol. 2, pp. 633–637.

[191] H. Herzberg and G. Poltyrev, "The Error Probability of M-ary PSK Block Coded Modulation Schemes," *IEEE Trans. Comm.*, vol. 44, pp. 427–433, April 1996.

[192] P.F. Swaszek, "A Lower Bound on the error probability for signals in White Gaussian Noise," *IEEE Trans. Inform. Theory*, vol. 41, pp. 837–841, May 1995.

[193] D. Makrakis and P. Mathiopoulos, "Optimal decoding in fading channels: A combined envelope, multiple differential and coherent detection approach," in *Proc. of the IEEE GLOBECOM*, Houston, Texas, USA, November 1989, pp. 1551–1557.

[194] D. Makrakis, P. Mathiopoulos, and D. P. Bouras, "Optimal decoding of coded PSK and QAM signals in correlated fast fading channels: A combined envelope, multiple differential and coherent detection approach," *IEEE Trans. Comm.*, vol. 42, pp. 63–75, January 1994.

[195] R. Raheli, R. Schober, and H. Leib, "Guest editorial differential and noncoherent wireless communications," *IEEE Journal on Selected Areas in Communications*, vol. 23, no. 9, pp. 1693–1696, September 2005.

[196] P. S. Bithas, G. K. Karagiannidis, N. C. Sagias, P. T. Mathiopoulos, S. A. Kotsopoulos, and G. E. Corazza, "Performance Analysis of a Class of GSC Receivers Over Non-Identical Weibull Fading Channels," to appear in 2006.

[197] F. Vatalaro and G.E. Corazza, "Probability of Error and Outage in a Rice-Lognormal Channel for Terrestrial and Satellite Personal Communications," *IEEE Trans. Comm.*, vol. 44, pp. 921–924, August 1996.

[198] G. Ferrari and G. E. Corazza, "Tight Bounds and accurate approximations for DQPSK transmission bit error rate," *IEE Electron. Lett.*, vol. 40, pp. 1284–1285, September 2004.

[199] G.E. Corazza, "Analysis of Multidimensional Trellis-Coded MPSK in Rice-Lognormal Fading Channels," *IEEE Trans. Comm.*, vol. 45, pp. 5–8, January 1997.

[200] H. Meyr, M. Moeneclaey, and S. A. Fechtel, *Digital Communication Receivers: Synchronization, Channel Estimation and Signal Processing*, Wiley, New York, 1998.

[201] J. A. López-Salcedo and G. Vázquez, "Stochastic Approach to Non-Data-Aided Synchronization and Received Pulse Shape Estimation," in *Proc. of the 8th International Workshop on Signal Processing for Space Communications*, 2003, vol. 1, pp. 203–210.

[202] E. Casini, R. De Gaudenzi, and A. Ginesi, "DVB-S2 Modem Algorithms Design and Performance over Typical Satellite Channels," *Int. Journal on Satellite Communications and Networking*, vol. 22, pp. 281–318, May 2004.

[203] T. Pollet, M. Van Bladel, and M. Moeneclaey, "BER sensitivity of OFDM systems to carrier frequency offset and Wiener phase noise," *IEEE Trans. Comm.*, vol. 234, pp. 191–193, 1995.

[204] K. Bucket and M. Moeneclaey, "Effect of random carrier phase and timing errors on the detection of narrowband M-PSK and bandlimited DS/SS MPSK signals," *IEEE Trans. Comm.*, vol. 234, pp. 1260–1263, 1995.

[205] M. Morelli and U. Mengali, "Feedforward Frequency Estimation for PSK: a Tutorial Review," *European Trans. Telecomm.*, vol. 9, pp. 103–116, March 1998.

[206] D. C. Rife and R. R. Boorstyn, "Single-Tone Parameter Estimation from Discrete-time Observations," *IEEE Trans. Inform. Theory*, vol. 20, pp. 591–598, September 1974.

[207] M. Luise and R. Reggiannini, "Carrier Frequency Recovery in All-Digital Modems for Burst-Mode Transmissions," *IEEE Trans. Comm.*, vol. 43, pp. 1169–1178, February 1995.

[208] M. P. Fitz, "Further Results in the Fast Estimation of a Single Frequency," *IEEE Trans. Comm.*, vol. 42, pp. 862–864, March 1994.

[209] U. Mengali and M. Morelli, "Data-aided Frequency Estimation for Burst Digital Transmission," *IEEE Trans. Comm.*, vol. 45, pp. 23–25, January 1997.

[210] A. J. Viterbi and A. M. Viterbi, "Nonlinear Estimation of PSK-Modulated Carrier Phase with Application to Burst Digital Transmission," *IEEE Trans. Inform. Theory*, vol. 29, pp. 543–551, July 1983.

[211] W. Gappmair, "Analysis of Non-Data-Aided Carrier Frequency Recovery with Luise-Reggiannini Estimators Applied to M-PSK Schemes," *IEE Proc. Commun.*, vol. 152, pp. 415–419, August 2005.

[212] D. Efstathiou and A. H. Aghvami, "Feedforward Synchronization Techniques for 16-QAM TDMA Demodulators," in *Proc. of the International Conference on Telecommunications*, London, UK, 1996, pp. 1432–1436.

[213] U. Mengali and A. N. D'Andrea, *Synchronization Techniques for Digital Receivers*, Plenum Press, New York, USA, 1997.

[214] A. N. D'Andrea and U. Mengali, "Design of Quadricorrelators for Automatic Frequency Control Systems," *IEEE Trans. Comm.*, vol. 41, pp. 988–997, June 1993.

[215] M. P. Fitz and W. C. Lindsey, "Decision-Directed Burst-Mode Carrier Synchronization Techniques," *IEEE Trans. Comm.*, vol. 40, pp. 1644–1653, October 1992.

[216] W. Gappmair, "Open-Loop Characteristic of Decision-Directed Maximum-Likelihood Phase Estimators for MPSK Modulated Signals," *IEE Electron. Lett.*, vol. 39, pp. 337–339, February 2003.

[217] W. Gappmair, "Extended Analysis of Viterbi-Viterbi Synchronizers," *European Trans. Telecomm.*, vol. 16, pp. 151–155, Mar./Apr. 2005.

[218] M. Moeneclaey and G. De Jonghe, "ML-Oriented NDA Carrier Synchronization for General Rotationally Symmetric Signal Constellations," *IEEE Trans. Comm.*, vol. 42, pp. 2531–2533, August 1994.

[219] M. Oerder and H. Meyr, "Digital Filter and Square Timing Recovery," *IEEE Trans. Comm.*, vol. 36, pp. 605–612, May 1988.

[220] S J. Lee, "A New Non-Data-Aided Feedforward Symbol Timing Estimator Using Two Samples per Symbol," *IEEE Comm. Lett.*, vol. 6, pp. 205–207, May 2002.

[221] Y. Wang, E. Serpedin, and P. Ciblat, "An Alternative Blind Feedforward Symbol Timing Estimator Using Two Samples per Symbol," *IEEE Trans. Comm.*, vol. 51, pp. 1451–1455, September 2003.

[222] W. Gappmair, S. Cioni, G. E. Corazza, and O. Koudelka, "Symbol-Timing Recovery with Modified Gardner Detectors," in *Proc. of the Int. Symp. Wireless Commun. Systems*, Siena, Italy, 2005.

[223] R. De Gaudenzi and M. Luise, "Analysis and Design of an All-Digital Demodulator for Trellis Coded 16-QAM Transmission over a Nonlinear Satellite Channel," *IEEE Trans. Comm.*, vol. 43, pp. 659–668, 1995.

[224] R. Raheli, A. Polydoros, and C. K. Tzou, "Per-Survivor Processing: A General Approach to MLSE in Uncertain Enviroments," *IEEE Trans. Comm.*, vol. 43, February 1995.

[225] K.M. Chugg and A. Polydoros, "MLSE for Unknown Channel, Part II: Tracking Performance," *IEEE Trans. Comm.*, vol. 44, pp. 949–958, August 1996.

[226] G. Ungerboeck, "Adaptive Maximum Likelihood Receiver for Carrier Modulated Data Transmission Systems," *IEEE Trans. Comm.*, vol. 22, pp. 624–636, May 1974.

[227] A. Vanelli-Coralli, P. Salmi, S. Cioni, G. E. Corazza, and A. Polydoros, "A Performance Review of PSP for joint Phase/Frequency and Data Estimation in Future Broadband Satellite Networks," *IEEE Journal on Selected Areas in Communications*, vol. 19, no. 12, pp. 2298–2309, December 2001.

[228] W. Oh and K. Cheun, "Joint Decoding and Carrier Phase Recovery Algorithm for Turbo Codes," *IEEE Comm. Lett.*, vol. 5, pp. 375–377, September 2001.

[229] L. Zhang and A. Burr, "Phase Estimation with the Aid of Soft Output from Turbo Decoding," in *Proc. of the IEEE Vehicular Technology Conference*, Atlantic City, New Jersey, USA, October 2001, pp. 154–158.

[230] C. Morlet, I. Buret, and M. L. Boucheret, "A Carrier Phase Estimator for Multi-Media Satellite Payloads suited to RSC Coding schemes," in

IEEE Int. Conf. on Communications, New Orleans, USA, June 2000, pp. 455–459.

[231] G. Colavolpe, G. Ferrari, and R. Raheli, "Non-coherent Iterative Turbo Detection," *IEEE Trans. Comm.*, vol. 48, pp. 1488–1498, September 2000.

[232] A. Anastasopoulos and K. M. Chugg, "Adaptive Iterative Detection for Phase Tracking in Turbo-Coded System," *IEEE Trans. Comm.*, vol. 49, pp. 2135–2144, December 2001.

[233] S. Cioni, G. E. Corazza, and A. Vanelli-Coralli, "Turbo Embedded Estimation with imperfect Phase/Frequency Recovery," in *Proc. of the IEEE Int. Conf. on Communications*, Anchorage, Alaska, USA, May 2003, pp. 2385–2389.

[234] S. Cioni, G. E. Corazza, and A. Vanelli-Coralli, "Turbo Embedded Estimation for High Order Modulation," in *Proc. of the 3rd International Symposium on Turbo Codes and Related Topics*, Brest, France, September 2003, pp. 447–450.

[235] J.K. Cavers, "An Analysis of Pilot Symbol Assisted Modulation for Rayleigh Fading Channels," *IEEE Transactions on Vehicular Technology*, vol. 40, no. 4, pp. 686–693, November 1991.

[236] S. V. Vaseghi, *Advanced Digital Signal Processing and Noise Reduction*, Willey, Baffins Lane, Chichester, 2000.

[237] E. Biglieri, J. Proakis, and S. Shamai, "Fading Channels: Information-Theoretic and Communications Aspects," *IEEE Trans. Inform. Theory*, vol. 44, no. 6, pp. 2619–2692, October 1998.

[238] J. K. Tugnait and W. Luo, "Blind Identification of Time-Varying Channels Using MultiStep Linear Predictors," *IEEE Trans. Signal Process.*, vol. 52, no. 6, pp. 1739–1749, June 2004.

[239] G. B. Giannakis and C. Tepedelenlioglu, "Basis Expansion Models and Diversity Techniques for Blind Identification and Equalization of Time-Varying Channels," *IEEE Proc.*, vol. 86, no. 10, pp. 1969–1986, October 1998.

[240] B. Baykal, "Blind Channel estimation via Combining Autocorrelation and Blind Phase Estimation," *IEEE Transactions on Circuits and Systems*, vol. 51, no. 6, pp. 1125–1131, June 2004.

[241] E. Moulines, P. Duhamel, J. Cardoso, and S. Mayrargue, "Subspace Methods for the Blind Identification of Multichannel FIR Filters," *IEEE Trans. Signal Process.*, vol. 43, no. 2, pp. 516–525, February 1995.

[242] L. Tong, G. Xu, and T. Kailath, "Blind Identification and Equalization Based on Second-Order Statistics: A Time Domain Approach," *IEEE Trans. Inform. Theory*, vol. 40, pp. 340–349, March 1994.

[243] Z. Xu, P. Liu, and L.D. Zoltowski, "Diversity Assisted Channel Estimation and Multi User Detection for Downlink CDMA with Long Spreading Codes," *IEEE Trans. Signal Process.*, vol. 52, no. 1, pp. 190–201, January 2004.

[244] D. R. Pauluzzi and N. C. Beaulieu, "A Comparison of SNR Estimation Techniques for the AWGN Channel," *IEEE Trans. Comm.*, vol. 48, pp. 1681–1691, October 2000.

[245] S. Cioni, R. De Gaudenzi, and R. Rinaldo, "Adaptive coding and modulation for the forward link of broadband satellite networks," in *Proc. of IEEE GLOBECOM*, December 2003, pp. 3311–3315.

[246] S. Cioni, G. E. Corazza, and M. Bousquet, "An Analytical Characterization of Maximum Likelihood Signal-to-Noise Ratio Estimation," in *Proc. of the 2nd Int. Workshop on Satellite and Space Communications*, 2005.

[247] M.K. Simon, J.K. Omura, R.A. Scholtz, and B.K. Levitt, *Spread Spectrum Communications Handbook*, McGraw-Hill, New York, USA, revised edition edition, 1994.

[248] G.E. Corazza, "On the MAX/TC Criterion for Code Acquisition and Its Application to DS-SSMA Systems," *IEEE Trans. Comm.*, vol. 44, no. 9, pp. 1173–1182, September 1996.

[249] A. J. Viterbi, *CDMA, Principles of Spread Spectrum Communications*, Addison-Wesley Wireless Communications Series. Addison-Wesley Publishing Company, April 1995.

[250] R. De Gaudenzi, F. Giannetti, and M. Luise, "Signal Recognition and Signature Code Acquisition in CDMA Mobile Packet Communications," *IEEE Trans. Vehicular Technology*, vol. 47, no. 1, pp. 196–208, February 1998.

[251] G.E. Corazza, P. Salmi, A. Vanelli-Coralli, and M. Villanti, "Differential Post Detection Integration Technique in the Return Link of Satellite CDMA Systems," in *Proc. of the IEEE 7th International Symposium*

on Spread Spectrum Techniques and Applications Conference, Czech Republic, 2-5 September 2002, pp. 233–237.

[252] G.E. Corazza and R. Pedone, "Maximum Likelihood Post Detection Integration Methods for Spread Spectrum Systems," in *Proc. of the IEEE Wireless Communications and Networking Conference*, New Orleans, Louisiana, USA, 16-20 March 2003, pp. 227–232.

[253] G.E. Corazza and R. Pedone, "Generalized and Average Post Detection Integration Methods for Code Acquisition," in *Proc. of the IEEE International Symposium on Spread Spectrum Techniques and Applications*, 30 Aug./2 Sept. 2004.

[254] A. Polydoros and C.L. Weber, "A Unified Approach to Serial Search Spread Spectrum Code Acquisition - Part I: General Theory," *IEEE Trans. Comm.*, vol. 32, no. 5, pp. 542–549, May 1984.

[255] G.E. Corazza, C. Caini, A. Vanelli-Coralli, and A. Polydoros, "DS-CDMA Code Acquisition in the Presence of Correlated Fading - Part I: Theoretical Aspects," *IEEE Trans. Comm.*, vol. 52, no. 7, pp. 1160–1168, July 2004.

[256] J. L. Massey, "Optimum Frame Synchronization," *IEEE Trans. Comm.*, vol. 20, no. 2, pp. 115–119, April 1972.

[257] J. A. Gansman, M. P. Fitz, and J. V. Krogmeier, "Optimum and Suboptimum Frame Synchronization for Pilot-Symbol-Assisted Modulation," *IEEE Trans. Comm.*, vol. 45, no. 10, pp. 1327–1337, October 1997.

[258] Z.Y. Choi and Y.H. Lee, "Frame Synchronization in the Presence of Frequency Offset," *IEEE Trans. Comm.*, vol. 50, no. 7, pp. 1062–1065, July 2002.

[259] G.E. Corazza, R. Pedone, and M. Villanti, "Frame Acquisition for Continuous and Discontinuous Transmission in the Forward Link of Ka-band Satellite Systems," in *Proc. of the 6th European Workshop on Mobile/Personal Satcoms and 2nd Advanced Satellite Mobile Systems Conference*, ESA-ESTEC, Noordwijk, The Netherlands, 2004, pp. 211–218.

[260] U. Reimers, *Digital Video Broadcasting - The international Standard for Digital Television*, Springer-Verlag, 2001.

[261] A. A. D'Amico, A. N. D'Andrea, and R. Reggiannini, "Efficient Non-Data-Aided Carrier and Clock Recovery for Satellite DVB at Very Low

Signal-to-Noise Ratios," *IEEE Journal on Selected Areas in Communications*, vol. 19, no. 12, pp. 2320–2330, December 2001.

[262] ISO/IEC, "Information Technology - Generic coding of moving pictures and associated audio information - Part 1: Systems," Tech. Rep. 13818-1, International Organization for Standardization, 1996.

[263] M. Ibnkahla, Q.M. Rahman, A.I. Sulyman, H.A. Al-Asady, J. Yuan, and A. Safwat, "High-speed satellite mobile communications: Technologies and Challenges," *IEEE Proc.*, vol. 92, no. 2, pp. 312–339, February 2004.

[264] F. Xiong, "Modem Techniques in Satellite Communications," *IEEE Comm. Mag.*, vol. 32, no. 8, pp. 84–98, August 1994.

[265] T. Javornik and G. Kandus, "Solid state power amplifer impact on the satellite systems performance," in *Proc. of the 5th European Workshop on Mobile/Personal Satcoms*, Baveno, Italy, September 2002, pp. 231–238.

[266] P. Jia, *Broadband High Power Amplifiers Using Spatial Power Combing Technique*, Ph.D. thesis, University of California, Santa Barbara, 2002.

[267] K. J. Russell, "Microwave power combining techniques," *IEEE Transactions on Microwave Theory and Techniques*, vol. 27, no. 5, pp. 472–478, 1979.

[268] C.A.A. Wass, "A table of intermodulation products," *Journal of IEEE*, vol. 95, pp. 107–, 1948.

[269] A. M. Saleh, "Frequency independent and frequency dependent models of TWT amplifiers," *IEEE Trans. Comm.*, vol. 29, no. 11, pp. 1715–1720, November 1981.

[270] G. P. White, A. G. Burr, and T. Javornik, "Modelling of nonlinear distortion in broadband fixed wireless access systems," *Electron. Lett.*, vol. 39, no. 8, pp. 686–687, April 2003.

[271] N. Le Gallou, *Modelisation par series de Volterra dynamiques des phenomenes non lineaires pour la simulation susteme d'amplificateurs de puissance*, Ph.D. thesis, Universite de Limoges, 2001.

[272] C. Dominique, *Modelisation dynamique des modules actifs a balayage electronique par series de Volterra et integration de ces modeles pour une simulation de type systeme*, Ph.D. thesis, Universite de Paris VI, 2001.

[273] T. Reveyrand et al., "A Calibrated Time Domain Envelope Measurement System for the Behavioral Modeling of Power Amplifiers," in *Gallium Arsenide applications symbosium*, September 2002.

[274] F. Langlet, H. Abdulkader, D. Roviras, A. Mallet, and F. Castanie, "Comparison of neural network adaptive predistortion techniques for satellite down links," in *Proc. of the IEEE International Joint Conference on Neural Networks*, Washington DC, USA, July 2001.

[275] Y. Nagata, "Linear amplification technique for digital mobile communications," in *Proc. of the IEEE Vehicular Technology Conference*, 1989, vol. 1, pp. 159–164.

[276] J.K. Cavers, "Amplifier Linearization using a digital predistorter with fast adaptation and low memory requirements," *IEEE Trans. Vehicular Technology*, vol. 39, pp. 374–382, 1990.

[277] J. K. Cavers, "Optimum indexing in predistorting amplifier linearizers," in *Proc. of the IEEE Vehicular Technology Conference*, 1997, vol. 2, pp. 676–680.

[278] J. Y. Hassani and M. Kamarei, "A flexible method of LUT indexing in digital predistortion linearization of RF power amplifiers," in *Proc. of IEEE Intenrational Symposium on Circuits and Systems*, 2001, vol. 1, pp. 53–56.

[279] G. Karam and H. Sari, "A Data Predistortion technique with memory for QAM radio systems," *IEEE Trans. Comm.*, vol. 39, pp. 336–344, 1991.

[280] M. Schetzen, *The Volterra and Wiener Theories of Nonlinear Systems*, Wiley, New York, USA, 1980.

[281] E. Biglieri, S. Barberis, and M. Catena, "Analysis and compensation of nonlinearities in digital transmission systems," *IEEE Journal on Selected Areas in Communications*, vol. 6, pp. 42–51, January 1988.

[282] C.J. Clark, G. Chrisikos, M.S. Muha, A.A. Moulthrop, and C.P. Silva, "Time-domain envelope measurement technique with application to wideband power amplifier modeling," *IEEE Transactions on Microwave Theory and Techniques*, vol. 46, pp. 2531–2540, December 1998.

[283] S. Chang and E.J. Powers, "A simplified predistorter for compensation of nonlinear distortion in OFDM Systems," in *Proc. of the IEEE Global Communications Conference*, San Antonio, Texas, USA, November 2001, pp. 3080–3084.

[284] J. Kim and K. Konstantinou, "Digital predistortion of wideband signals based on power amplifier model with memory," *IEE Electron. Lett.*, vol. 37, no. 23, pp. 1417–1418, November 2001.

[285] G. Ding, G. T. Zhou, D. R. Morgan, Z. Ma, J.S. Kenney, J. Kim, and C. R. Giardina, "A robust digital baseband predistorter constructed using memory polynomial," *IEEE Trans. Comm.*, vol. 52, no. 1, pp. 159–165, January 2004.

[286] C. Eun and E. J. Powers, "A New Volterra Predistorter based on the Indirect Learning architecture," *IEEE Trans. Signal Process.*, vol. 45, no. 1, pp. 223–227, January 1997.

[287] F. Langlet, D. Roviras, A. Mallet, and F. Castanie, "Mixed analog/digital implementation of MLP NN for predistortion," in *Proc. of the International Joint Conference on Neural Networks*, Hawaii, USA, May 2002.

[288] F. Langlet, H. Abdulkader, and D. Roviras, "Predistortion of non-linear satellite channels using neural networks: Architecture, algorithm and implementation," in *Proc. of the European Signal Processing Conference*, Toulouse, France, September 2002.

[289] S. Haykin, *Neural Networks: A Comprehensive Foundation*, Prentice Hall, Upper Saddle River, New Jersey, USA, 2 edition, 1999.

[290] H. Abdulkader et al., "Natural gradient algorithm for neural networks applied to non-linear high-power amplifiers," *International Journal of Adaptive Control and Signal Processing*, vol. 16, no. 8, pp. 557–576, October 2002.

[291] S. Pupolin and L. J. Greenstein, "Performance Analysis of Digital Radio Links with Nonlinear Transmit Amplifiers," *IEEE Journal on Selected Areas in Communications*, vol. 5, pp. 534–546, 1987.

[292] D. W. Marquardt, "An algorithm for least-squares estimation of nonlinear parameters," *Journal of the Society for Industrial and Applied Mathematics*, vol. 11, pp. 431–441, 1963.

[293] S. Chen, F.N. Cowan, and P. M. Grant, "Orthogonal least squares learning algorithm for radial basis function networks," *IEEE Trans. on Neural Networks*, vol. 2, no. 2, pp. 302–309, February 1991.

[294] S. Haykin, *Adaptive Filter Theory*, Prentice-Hall, Upper Saddle River, NJ. 07458, 3rd edition, 1996.

[295] L. Hanzo, C. H. Wong, and M. S. Yee, *Adaptive Wireless Tranceivers*, John Wiley & Sons, Ltd, Chichester, 1st edition, 2002.

[296] E. Biglieri, M. Elia, and L. Lopresti, "The optimum linear receiving filter for digital transmission over nonlinear channels," *IEEE Trans. Inform. Theory*, vol. 35, no. 3, May 1989.

[297] S. A. Fredricson, "Optimum receiver filters in digital quadrature phase-shift-keying systems with a nonlinear repeater," *IEEE Trans. Comm.*, vol. 23, pp. 1389–1400, December 1975.

[298] M. F. Mesiya, P. J. McLane, and L. L. Campbell, "Optimal receiver filters for BPSK transmission over a bandlimited nonlinear channel," *IEEE Trans. Comm.*, vol. 26, no. 1, pp. 12–22, January 1978.

[299] S. Benedetto, E. Biglieri, and V. Castellani, *Digital Transmission Theory*, Prentice Hall, Englewood Cliffs, New Jersey, USA, 1987.

[300] M. F. Mesiya, P. J. McLane, and L. L. Campbell, "Maximum likelihood sequence estimation of binary sequences transmitted over bandlimited nonlinear channels," *IEEE Trans. Comm.*, vol. 25, no. 7, pp. 633–643, July 1977.

[301] S. Benedetto, E. Biglieri, and R. Daffara, "Modeling and performance evaluation of nonlinear satellite links - A Volterra series approach," *IEEE Transactions on Aerospace and Electronics Systems*, vol. 15, no. 4, pp. 494–507, July 1979.

[302] S. Benedetto and E. Biglieri, "Nonlinear equalization of nonlinear satellite channels," *IEEE Journal on Selected Areas in Communications*, vol. 1, pp. 57–62, January 1983.

[303] V. J. Mathews, "Adaptive polynomial filters," *IEEE Signal Process. Mag.*, vol. 8, no. 3, pp. 10–26, July 1991.

[304] E. Biglieri, A. Gersho, R. D. Gitlin, and T. L. Lim, "Adaptive cancellation of nonlinear intersymbol interference for voiceband data transmission," *IEEE Journal on Selected Areas in Communications*, vol. 2, no. 5, pp. 765–777, September 1984.

[305] R. Nowak and B. Van Veen, "Volterra filter equalization: a fixed-point approach," *IEEE Trans. Signal Process.*, vol. 45, pp. 377–388, February 1997.

[306] A. J. Redfren and G. T. Zhou, "A root method for Volterra system equalization," *IEEE Signal Processing Letters*, vol. 5, pp. 285–288, November 1998.

[307] G. Giannakis and E. Serpedin, "Linear multichannel blind equalizers for nonlinear FIR Volterra channels," *IEEE Trans. Signal Process.*, vol. 45, pp. 67–81, January 1997.

[308] G.M. Raz and D. Van Veen, "Blind equalization and identification of nonlinear and IIR Systems-A Least Squares Approach," *IEEE Trans. Signal Process.*, vol. 48, pp. 192–200, January 2000.

[309] R. Lopez-Valcarce and S. Dasgupta, "Blind equalization of nonlinear channels from second-order statistics," *IEEE Trans. Signal Process.*, vol. 49, pp. 3084–3097, December 2001.

[310] P. Chang and B. Wang, "Adaptive decision feedback equalization for digital satellite channels using multilayer neural networks," *IEEE Journal on Selected Areas in Communications*, vol. 13, no. 2, pp. 316–324, February 1995.

[311] S. Chen, G. J. Gibson, C. F. N. Cowan, and P. M. Grant, "Adaptive equalization of finite non-linear channels using multilayer perceptrons," *Signal Processing*, vol. 20, no. 2, pp. 107–119, June 1990.

[312] I. Cha and S. A. Kassam, "Channel equalization using adaptive complex radial basis function networks," *IEEE Journal on Selected Areas in Communications*, vol. 13, no. 1, pp. 122–131, January 1995.

[313] S. Chen, S. McLaughlin, and B. Mulgrew, "Complex-valued radial basis function network, Part I: Network architecture and learning algorithms," *Signal Processing*, vol. 35, no. 1, pp. 19–31, January 1994.

[314] S. Chen, S. McLaughlin, and B. Mulgrew, "Complex-valued radial basis function network, Part II: Application to digital communications channel equalization," *Signal Processing*, vol. 36, no. 2, pp. 175–188, March 1994.

[315] S. Bouchired, M. Ibnkahla, and W. Paquier, "A Combined LMS-SOM Algorithm for Time Varying non-linear Channel Equalization," in *Proc. of IX European Signal Processing Conference*, Rhodes, Greece, September 1998.

[316] S. Bouchired, M. Ibnkahla, D. Roviras, and F. Castanié, "Equalization of satellite mobile communication channels using combined self-organizing maps and RBF networks," in *Proc. of the International Conference on Acoustic, Speech and Signal Processing*, Seattle, Washington, USA, 12-15 May 1998, vol. 6, pp. 3377–3379.

[317] T. Kohonen, *Self-Organizing Maps*, Springer, Berlin, Germany, 1995.

[318] S. Bouchired, M. Ibnkahla, D. Roviras, and F. Castanié, "Equalization of satellite UMTS channels using neural network devices," in *Proc. of the*

International Conference on Acoustic, Speech and Signal Processing, Phoenix, USA, March 1999.

[319] S. Bouchired, D. Roviras, and F. Castanié, "Equalization of satellite mobile channels with neural network techniques," *Space Communications*, vol. 15, no. 4, pp. 209–220, 1999.

[320] C. Douillard, M. Jézéquel, C. Berrou, A. Picart, P. Didier, and A. Glavieux, "Iterative Correction of Intersymbol Interference: Turbo-Equalization," *European Trans. Telecomm.*, vol. 6, no. 5, pp. 507–511, Sept./Oct. 1995.

[321] M. Tüchler, R. Koetter, and A.C. Singer, "Turbo Equalization: Principles and New Results," *IEEE Trans. Comm.*, vol. 50, no. 5, pp. 754–767, May 2002.

[322] M. Tomlinson, "New Automatic Equaliser Employing Modulo Arithmetic," *IEE Electron. Lett.*, vol. 7, pp. 138–139, 1971.

[323] H. Harashima and H. Miyakawa, "A Method of Code Conversion for Digital Communications Channels with Intersymbol Interference," *Transactions of the Institute of Electronics and Communications Engineers of Japan*, vol. 52-A, pp. 272–273, 1969, (In Japanese.).

[324] H. Harashima and H. Miyakawa, "Matched-Transmission Technique for Channels with Intersymbol Interference," *IEEE Trans. Comm.*, vol. 20, pp. 774–780, 1972.

[325] J. E. Mazo and J. Salz, "On the Transmitted Power in Generalized Partial Response," *IEEE Trans. Comm.*, vol. 24, pp. 348–351, 1976.

[326] R. F. H. Fischer, *Precoding and Signal Shaping for Digital Transmission*, John Wiley & Sons, New York, USA, 2002.

[327] M. Alvarez-Diáz, M. Neri, C. Mosquera, and G. E. Corazza, "Joint precoding and predistortion techniques for satellite telecommunication systems," in *Proc. of the International Workshop on Satellite and Space Communications*, Siena, Italy, September 2005.

[328] C. Caini and G.E. Corazza, "Satellite Diversity in Mobile Satellite CDMA Systems," *IEEE Journal on Selected Areas in Communications*, vol. 19, pp. 1324–1333, 2001.

[329] P.P. Robet, B.G. Evans, and A. Ekman, "Land Mobile Satellite Communications Channel Model for simultaneous transmission from a Land mobile terminal via two separate satellites," *IEEE Journal on Selected Areas in Communications*, vol. 10, pp. 139–154, 1992.

[330] A. Jahn, H. Bischl, and G. Heib, "Channel Characterisation for Spread Spectrum Satellite Communications," *Proc. of the IEEE 4th Symposium on Spread Spectrum Techniques and Appl.*, pp. 139–154, 1992.

[331] R. Akturan and W. Vogel, "Path Diversity for LEO-Satellite-PCS in the Urban Environment," *IEEE Trans. on Antennas and Propagation*, 1997.

[332] W. J. Vogel, "Satellite Diversity for Personal Satellite Communications - Modelling and Measurements," in *Proc. of the 10th Int. Conference on Antennas and Propagation*, 1997, pp. 1269–1272.

[333] R. Akturan and W. Vogel, "Photogrammetric Mobile Satellite Service Prediction," *IEE Electron. Lett.*, vol. 31, pp. 165–166, 1996.

[334] C. Meenan, M. Parks, R. Tafazolli, and B.G. Evans, "Availability of 1st Generation Satellite Personal Communications network Service in Urban Environments," *IEEE Vehicular Technology Conference*, pp. 1471–1475, 1998.

[335] M. A. Vázquez-Castro, F. Pérez-Fontán, and S. R. Saunders, "Shadowing correlation assessment and modeling for satellite diversity in urban environments," *International Journal of Satellite Communications*, pp. 151–166, 2002.

[336] D. Gesbert, L. Haumonte, H. Bölcskei, R. Krishnamoorthy, and A. J. Paulraj, "Technologies and Performance for Non Line-of-Sight Broadband Wireless Access Networks," *IEEE Comm. Mag.*, vol. 40, no. 4, pp. 86–95, April 2002.

[337] G. J Foschini, "Layered space-time architecture for wireless communication in a fading environment when using multiple antennas," *Bell Labs Tech. J.*, vol. 2, no. 1, pp. 41–59, 1996.

[338] L. Zheng and D.N.C. Tse, "Diversity and multiplexing: a fundamental tradeoff in multiple-antenna channels," *IEEE Trans. Inform. Theory*, vol. 49, no. 5, pp. 1073–1096, May 2003.

[339] G. J. Foschini, G. D. Golden, A. Valenzuela, and P. W. Wolniansky, "Simplified processing for high spectral efficiency wireless communications employing multi-element arrays," *IEEE Journal on Selected Areas in Communications*, vol. 17, no. 11, pp. 1841–1852, November 1999.

[340] V. Tarokh, N. Seshari, and R. Calderbank, "Space-Time Codes for High Data Rate Wireless Communication: Performance Criterion and Code Construction," *IEEE Trans. Inform. Theory*, vol. 44, no. 2, pp. 744–765, March 1998.

[341] I. E. Telatar, "Capacity of Multiantenna Gaussian Channels," Tech. Rep., AT&T Bell Laboratories, June 1995.

[342] G. J. Foschini and M. J. Gans, "On Limits of Wireless Communications in a Fading Environment When Using Multiple Antennas," *Wireless Personal Communications*, vol. 6, pp. 311–335, March 1998.

[343] A.F. Naguib, N. Seshader, and A.R. Calderbank, "Space-Time Coding and Signal Processing for High Data Rate Wireless Communications," *IEEE Comm. Mag.*, pp. 77–92, 2000.

[344] V. Tarokh, H. Jafarkhami, and A.R. Calderbank, "Space-Time Block Codes from Orthogonal Designs," *IEEE Trans. Inform. Theory*, vol. 45, pp. 1456–1467, 1999.

[345] S.M. Alamouti, "A Simple Transmit Diversity Technique for Wireless Communications," *IEEE Journal on Selected Areas in Communications*, vol. 16, pp. 1452–1458, 1998.

[346] V. Tarokh, A.F. Naguib, N. Seshadri, and A.R. Calderbank, "Space-Time Codes for High Data Rate wireless Communication: Performance Criteria in the Presence of Channel Estimation Error, Mobility, and Multiple Paths," *IEEE Trans. Comm.*, vol. 47, pp. 199–207, 1999.

[347] G.E. Corazza, D. Bellini, P. Salmi, and A. Vanelli-Coralli, "Performance Analysis of Space-Time Block Coding Techniques in the presence of Imperfect Channel estimation," in *Proc. of the SPIE International Symposium*, August 2001, pp. 94–103.

[348] D. Bellini, G.E. Corazza, P. Salmi, and A. Vanelli-Coralli, "Space-Time Coding Techniques for Fourth Generation Systems," in *Proc. of the International Symposium on Signals, Systems, and Electronics*, July 2001, pp. 224–232.

[349] D. Bellini, G.E. Corazza, P. Salmi, and A. Vanelli-Coralli, "Improved Bounds for Space-Time Trellis Codes Performance in Mobile Fading Channels," in *Proc. of the 3rd Generation Infrastructure and Services*, July 2001, pp. 90–94.

[350] G.E. Corazza, "Analysis of Multidimensional Trellis-Coded MPSK in Rice-Lognormal Fading Channels," *IEEE Trans. Comm.*, vol. 45, pp. 5–8, January 1997.

[351] G. McKay, P.J. McLane, and E. Biglieri, "Error Bounds for Trellis-Coded MPSK on a Fading Mobile Satellite Channel," *IEEE Trans. Comm.*, vol. 39, no. 12, pp. 1750–1761, 1991.

[352] T. Javornik, G. Kandus, and S. Plevel, "Dynamic channel mapping strategies in adaptive MIMO systems," in *Proc. of the International Conference on Software, Telecommunications and Computer Networks*, Split-Dubrovnik, Croatia and Venice-Ancona, Italy, October 2003, pp. 69–73.

[353] M. Luglio, R. Mancini, C. Riva, A. Paraboni, and F. Barbaliscia, "Large scale site diversity for satellite communication networks," *Int. Journ. Satellite Commun.*, vol. 20, no. 4, pp. 251–260, Jul./Aug. 2002.

[354] F. Barbaliscia, G. Ravaioli, and A. Paraboni, "Characteristics of the Spatial Statistical Dependence of Rainfall Rate over Large Areas," *IEEE Trans. Antennas and Propagation*, vol. 40, no. 1, pp. 8–12, January 1992.

[355] R. K. Crane, "Space-time structure of precipitation," in *Proc. of the International Conference on Symbolic and Algebraic Computation*, Boston, Massachusetts, USA, 1987, American Metereological Society, pp. 265–268.

[356] J. Goldhirsh, B. H. Musiani, A. W. Dissanayake, and K. Lin, "Three-Site Space Diversity Experiment at 20 GHz Using ACTS in the Eastern United States," *IEEE Proc.*, vol. 86, pp. 970–980, June 1997.

[357] S.H. Lin, H.J. Bergmann, and M.V. Pursley, "Rain Attenuation on Earth Space Paths - Summary of 10-Year Experiment and Studies," *Bell Syst. Tech. J.*, vol. 59, pp. 183–228, 1979.

[358] COST Project 79, "Project on Radiopropagation above 10 GHz," Tech. Rep. 25/4, ESA Publications Division, 1979.

[359] K. Morita and I. Higuti, "Statistical Studies on Rain Attenuation Site Diversity Effect on Earth to Satellite Links in Microwave and Millimeter Wavebands," *Trans. IECE*, vol. E61, pp. 425–432, 1978.

[360] E. Matricciani, "Orbital Diversity in Resource-Shared Satellite Communication Systems above 10 GHz," *IEEE Journal on Selected Areas in Communications*, vol. SAC-5, pp. 714–723, 1987.

[361] F. Barbaliscia, A. Paraboni, and G. Ravaioli, "Analysis and Modelling of Rain Correlation Over Wide Areas," in *Proc. of the URSI F Open Symp.*, La Londe les Maures, France, 1989, pp. 6.1.1–6.1.5.

[362] F. Barbaliscia and A. Paraboni, "Modelling the multiple joint statistics of rain attenuation for resource allocation techniques," *IEEE Trans. Antennas Propagation*, 1997.

[363] C. Capsoni, F. Fedi, and A. Paraboni, "A Comprehensive meteorologically oriented Methodology for the prediction of Wave Propagation parameters in telecommunication Applications beyond 10 GHz," *Radio Science*, vol. 22, no. 3, pp. 387–393, 1987.

[364] ITU, "Characteristics of precipitation for propagation modelling, Propagation in Non-Ionized Media," Tech. Rep. P.837.2, International Telecommunication Union - Radiocommunication Sector, 1998.

[365] F. Carassa, G. Tartara, and E. Matricciani, "Frequency diversity and its applications," *Int. Journ. Satellite Commun.*, vol. 6, pp. 313–322, June 1988.

[366] F. Carassa, "Adaptive Methods to Counteract Rain Attenuation Effects in the 20/30 GHz Band," *Space Commun. and Broadcasting*, vol. 2, pp. 253–269, 1984.

[367] F. Carassa, "Methods to improve satellite systems performances in presence of rain," *Alta Frequenza*, vol. LVI, pp. 173–184, Jan./Apr. 1987.

[368] M. Luglio, "Application of Frequency Diversity for Dimensioning a Ka-band Satellite System," *Int. Journ. Satellite Commun.*, vol. 14, no. 1, pp. 53–62, Jan./Feb. 1996.

[369] S. Hamalainen, P. Slanina, M. Hartman, and A. Lappetelainen, "A Novel Interface between Link and System Level Simulations," in *Proc of the Advanced Communications Technology Satellite Conference*, 1997, pp. 599–604.

[370] T. Öjanperä and R. Prasad, *Wideband CDMA for Third Generation Mobile Communications*, Artech-House, Boston-London, 1998.

[371] F. Vatalaro, G.E. Corazza, F. Ceccerelli, and G DeMaio, "CDMA cellular systems performance with imperfect power control and shadowing," *International Journal of Satellite Communications*, vol. IEEE Vehicular Technology Conference VTC'96, pp. 874–878, May 1996.

[372] B. R. Voijcic, R. L. Pickloltz, and L. B. Milstein, "Performance of DS-CDMA with imperfect power control operating over low earth orbiting satellite links," *IEEE Journal on Selected Areas in Communications*, vol. 12, pp. 874–878, May 1994.

[373] M. A. Vázquez-Castro and F. Pérez-Fontán, "LMS Markov model and its use for power control error impact analysis on CDMA system capacity," *IEEE Journal on Selected Areas in Communications*, vol. 20, 2002.

[374] G.E. Corazza and C. Caini, "Satellite diversity exploitation in mobile CDMA systems," *IEEE Wireless Communications and Networking Conference*, pp. 1203–1207, 1999.

[375] M. Born and E. Wolf, *Principles of Optics*, MacMillan, New York, USA, 1964.

[376] R. A. Harris, "Radiowave Propagation Modelling for SatCom Services at Ku-Band and Above," Tech. Rep., COST Action 255, 2002.

[377] N. S. Correal and B. D. Woerner, "Enhanced DS-CDMA Uplink Performance through Base Station Polarization Diversity and Multistage Interference Cancellation," in *Proc. of the IEEE GLOBECOM*, 1998, vol. 4, pp. 1905–1910.

[378] A. M. D. Turkmani, A. A. Arowojolu, P. A. Kefford, and C. J. Kellett, "An Experimental Evaluation of the Performance of Two-Branch Space and Polarization Diversity Schemes at 1800 MHz," *IEEE Journal on Selected Areas in Communications*, vol. 44, no. 2, pp. 318–326, 1995.

[379] K. J. Hole, H. Holm, and G. E. Oiem, "Adaptive multidimensional coded modulation over flat fading channels," *IEEE Journal on Selected Areas in Communications*, vol. 18, no. 7, pp. 1153–1158, July 2000.

[380] T. Javornik and G. Kandus, "An Adaptive Rate Communication System Based on the N-MSK Modulation Technique," *IEICE Transactions on Communications*, vol. E84-B, no. 11, pp. 2946–2955, November 2001.

[381] T. Javornik, G. Kandus, and S. Plevel, "Adaptive modulation scheme for the land mobile satellite channel," in *Proc. of First International Conference on Advanced Satellite Mobile Systems*, Frascati, Italy, July 2003.

[382] A. J. Goldsmith and S. G. Chua, "Adaptive coded modulation for fading channels," *IEEE Trans. Comm.*, vol. 46, no. 5, pp. 595–602, May 1998.

[383] K. Kerschat, O. Koudelka, W. Riedler, M. Tomlinson, C. D. Hughes, and J. Hörle, "A variable spread-spectrum fade countermeasure system for the DICE video conference system," in *Proc. OF Olympus Util. Conference*, Seville, Spain, April 1993.

[384] K. Tanabe, K. Kobayashi, K. Ohata, and M. Ueba, "Multicarrier/multirate modem using time-division multiple processing," in *Proc. of the International Communications Satellite Systems Conference*, Montreal, Canada, May 2002, AIAA.

[385] J. Williams, L. Hanzo, and R. Steele, "Channel-adaptive modulation," in *Proc. of the 16-th IEE International Conference on Radio Receivers and Associated Systems*, Bath, UK, June 1999, pp. 144–147.

[386] U. Toyoki, S. Sampei, N. Morinaga, and K. Hamaguchi, "Symbol Rate and Modulation Level-Controlled Adaptive Modulation TDMA/TDD System for High-Bit-Rate Wireless Data Transmission," *IEEE Trans. Vehicular Technology*, vol. 47, no. 4, pp. 1134–1147, November 1998.

[387] E. Lutz, M. Werner, and A. Jahn, *Satellite Systems for Personal and Broadband Communications*, Springer, Berlin, Germany, 2000.

[388] SHIRON: Satellite Communications, ," From Shiron Web Resource, http://www.shiron.com/.

[389] M. J. Willis and B. G. Evans, "Fade countermeasures at Ka-band for OLYMPUS," *Int. Journ. Satellite Commun.*, vol. 6, pp. 301–311, June 1988.

[390] G. Tartara, "Fade countermeasures in millimetre-wave satellite communications: a survey of methods and problems," in *Proc. of the Olympus Utilization Conference*, Vienna, Austria, April 1989.

[391] A. P. Gallois, "Fade countermeasure techniques for satellite communication links," in *Proc. of the Int. Symp. on Commummications Theory and Applications*, July 1993.

[392] R. J. Acosta, "Rain fade compensation alternatives for Ka-band communication satellites," in *Proc. of the 3rd Ka-band Utilization Conference*, Sorrento, Italy, September 1997.

[393] L. Castanet, J. Lemorton, and M. Bousquet, "Fade Mitigation techniques for New SatCom services at Ku-band and above: a Review," in *Proc. of the 4th Ka-band Utilization Conference*, Venice, Italy, November 1998.

[394] COST Project 235, "Radiowave propagation effects on next generation fixed-services terrestrial telecommunications systems," Final Report EUR 16992 EN, ESA Publications Division, 1996.

[395] C. Caini and G.E. Corazza, "Satellite Downlink Reception through Intermediate Module Repeaters: Power Delay Profile Analysis," in *Proc. of the 6th European Workshop on Mobile/Personal Satcoms and 2nd Advanced Satellite Mobile Systems Conference*, ESA-ESTEC, Noordwijk, The Netherlands, September 2004, pp. 267–274.

[396] M. K. Simon and M.S. Alouini, *Digital Communication over Fading Channels*, Wiley, New York, 2 edition, 2004.

[397] V. A. Aalo, "Performance of maximal-ratio diversity systems in a correlated Nakagami-fading environment," *IEEE Trans. Comm.*, vol. 43, no.
8, pp. 2360–2369, August 1995.

[398] P. Lombardo, G. Fedele, and Rao M. M., "MRC performance for binary signals in Nakagami fading with general branch correlation," *IEEE
Trans. Comm.*, vol. 47, no. 1, pp. 44–52, January 1999.

[399] J. Luo, R. J. Zeidler, and S. McLaughlin, "Performance analysis of compact antenna arays with MRC in correlated Nakagami fading channels,"
IEEE Trans. Vehicular Technology, vol. 50, pp. 267–277, January 2001.

[400] N. C. Beaulieu and A. Abu-Dayya, "Analysis of equal gain diversity
on Nakagami fading channels," *IEEE Trans. Comm.*, vol. 39, no. 2, pp.
225–234, February 1991.

[401] A. Annamalai, C. Tellambura, and V. K. Bhargava, "Equal-gain diversity
receiver performance in wireless channels," *IEEE Trans. Comm.*, vol.
48, no. 10, pp. 1732–1745, October 2000.

[402] G. K. Karagiannidis, D. A. Zogas, and S. A. Kotsopoulos, "BER performance of dual predetection EGC in correlative Nakagami-m fading,"
IEEE Trans. Comm., vol. 52, pp. 50–53, January 2004.

[403] G. K. Karagiannidis, D. A. Zogas, and S. A. Kotsopoulos, "Statistical
properties of the EGC output SNR over correlated Nakagami-m fading
channels," *IEEE Trans. Wirel. Comm.*, vol. 3, pp. 1764–176953, September 2004.

[404] S. O. Rice, "Mathematical Analysis of Random Noise," *Bell System
Technical Journal*, vol. 23, pp. 282–332, July 1944.

[405] F. Wijk, F. Kegel, and R. Prasad, "Assesment of a pico-cellular system
using propagation measurments at 1.9 GHz for indoor wireless communications," *IEEE Trans. Vehicular Technology*, vol. 44, pp. 155–162,
February 1995.

[406] D. J. Parsons, *The Mobile Radio Propagation Channel*, Wiley, New
York, USA, 1 edition, 1992.

[407] W. W. Wu, "Satellite communications," *IEEE Proc.*, vol. 85, pp. 998–
1010, June 1997.

[408] B. Chytil, "The distribution of amplitude scintillation and the conversion
of scintillation indices," *J. Atmos. Terr. Phys.*, vol. 29, pp. 1175–1177,
September 1967.

[409] K. Bischoff and B. Chytil, "A note on scintillation indices," *Planet Space Sci.*, vol. 17, pp. 1059–1066, 1969.

[410] K. Bury, *Statistical Distributions in Engineering*, Cambridge Univercity Press, Cambridge, 1 edition, 1999.

[411] G. A. Baker and P. Graves-Morris, *Padé Approximants*, Cambridge University Press, Cambridge, 1 edition, 1996.

[412] S. Egami, "Individual closed-loop satellite access power control system using overall satellite link quality level," *IEEE Trans. Comm.*, vol. 30, July 1982.

[413] J. Horle, "Up-link power control of satellite Earth Stations as a fade countermeasure of 20/30 GHz communications systems," *Int. Journ. Satellite Commun.*, vol. 6, pp. 323–330, June 1988.

[414] S. Cacopardi, F. Martinino, and G. Reali, "Power control techniques performances evaluation in the SCPC/DAMA access scheme," in *European Conference on Satellite Communications*, Manchester, UK, November 1993.

[415] A. W. Dissanayake, K. T. Lin, C. Zaks, and J. E. Allnutt, "Results of an experiment to demonstrate the effectiveness of open-loop up-link power control for Ku-band satellite links," in *Proc. of the IEE International Conference on Antennas and Propagation*, Edinburgh, UK, Mar./Apr. 1993.

[416] L. Castanet, J. Lemorton, M. Bousquet, and L. Claverotte, "A Joint Fade Mitigation Technique applied to the regenerative packet switch payload of the GEOCAST system," in *8th Ka-band Utilization Conference*, Baveno, Italy, September 2002.

[417] C. S. Novák, T. Pálfalvi, and J. Bitó, "Code Sectoring Methods in CDMA-based Broadband Point-to-Multipoint Networks," in *Proc. of the IEEE Microwave and Wireless Components Letters*, August 2003, pp. 320–322.

[418] C. S. Novák, T. Pálfalvi, and J. Bitó, "Improved Antenna Sectoring Methods in Point-to-Multipoint Networks Applying CDMA with Multiuser Detection," in *Proc. of the Int. Conference on Antennas and Propagation*, Exeter, UK, Mar./Apr. 2003, pp. 824–827.

[419] 3GPP TS 25.213, *Spreading and Modulation (FDD)*, 3GPP, 2002.

[420] C.S. Novák and J. Bitó, "Sub-Sectoring by Multiuser Detection in Broadband CDMA-based Point-to-Multipoint Networks," in *Proc. of the 11th*

Microcoll. Conference, Budapest, Hungary, September 2003, pp. 285–288.

[421] M. Schmidt, J. Ebert, H. Schlemmer, S. Kastner-Puschl, W. Gappmair, W. Kogler, and O. Koudelka, "The L*IP Satellite Gateway," Tech. Rep. 16335/02/NL/US, European Space Agency, Netherlands, February 2002.

[422] O. Koudelka, M. Schmidt, J. Ebert, H. Schlemmer, S. Kastner-Puschl, and W. Riedler, "A Native IP Satellite Communications System," *Acta Astronautica*, vol. 55, pp. 255–259, October 2004.

[423] L. Zhao and A. M. Haimovic, "Performance of ultra-wideband communications in the presence of interference," *IEEE Journal on Selected Areas in Communications*, vol. 20, pp. 1684–1691, December 2002.

[424] FCC, *First Report and Order*, vol. 02-48, Federal Communications Commission, 2002.

[425] S. Roy, J. R. Foerster, V. S. Somayazulu, and D. G. Leeper, "Ultrawideband radio design: The promise of high-speed, short-range wireless connectivity," *IEEE Proc.*, vol. 92, pp. 295–311, February 2004.

[426] H. Zhang and T. A. Gulliver, "Pulse position amplitude modulation for time-hopping multiple access UWB communications," in *Proc. of the IEEE Wireless Comm. Networking Conference*, March 2004, pp. 895–900.

[427] R. A. Scholtz, "Multiple access with time-hopping impulse modulation," in *Proc. of the IEEE Military Comm. Conference*, October 1993, pp. 447–450.

[428] G.M. Maggio, N. Rulkov, and L. Reggiani, "Pseudo-chaotic time hopping for UWB impulse radio," *IEEE Trans. Circuits Syst.*, vol. 48, pp. 1424–1435, December 2001.

[429] G. Durisi and S. Benedetto, "Performance evaluation of TH-PPM UWB systems in the presence of multiuser interference," *IEEE Comm. Lett.*, vol. 7, pp. 224–226, May 2003.

[430] B. Hu and N. C. Beaulieu, "Exact bit error rate analysis of TH-PPM UWB systems in the presence of multiple-access interference," *IEEE Comm. Lett.*, vol. 7, pp. 572–574, December 2003.

[431] N. Boubaker and K.B. Letaief, "Ultra wideband DSSS for multiple access communications using antipodal signaling," in *Proc. IEEE Int. Conf. Comm.*, May 2003, pp. 2197–2201.

[432] V. S. Somayazulu, "Multiple Access performance in UWB systems using time hopping vs. direct sequence spreading," in *Proc. of the IEEE Wireless Comm. Networking Conference*, March 2002, pp. 446–449.

[433] A.H. Tewfik and E. Saberinia, "High bit rate ultra-wideband OFDM," in *Proc. of the IEEE Global Telecomm. Conference*, November 2002, pp. 2272–2276.

[434] J.R. Foerster, V. Somayazulu, and S. Roy, "A multi-banded system architecture for ultra-wideband communications," in *Proc. of the IEEE Military Comm. Conf.*, October 2003, pp. 903–908.

[435] N. Abramson, "Multiple access in wireless digital networks," *IEEE Proc.*, vol. 82, pp. 1360–1370, September 1994.

[436] C. Pateros, "Novel direct dequence direct spread spectrum multiple access technique," in *Proc. of the Military Communications Conference*, October 2000, vol. 1, pp. 1238–1239.

[437] D. Raychandhuri, "Performance Analysis of Random Access Packet-Switched Code Division Multiple Access Systems," *IEEE Trans. Comm.*, vol. 29, pp. 895–901, June 1981.

[438] R. K. Morrow and D. Raychandhuri, "Throughput of unslotted direct-sequence spread spectrum multiple-access channels with FEC coding," *IEEE Trans. Comm.*, vol. 41, pp. 1373 – 1378, 1992.

[439] Z. D. Belay and M. A. V. Castro, "ALOHA versus Single Code SPREAD ALOHA for Satellite Systems," in *Proc. of the IEEE Vehicular Technology Conference*, 2005.

[440] J. M. Holtzman, "A simple, Accurate Method to Calculate Spread-Spectrum Multiple Access Error Probabilities," *IEEE Trans. Comm.*, vol. 40, pp. 461–464, March 1992.

[441] P. Kota and C. Schlegel, "A Wireless Packet Multiple Access Method Exploiting Joint Detection," in *Proc. of the International Conference on Telecommunications*, 2003.

[442] L. Lifang, N. Jindal, and A. Goldsmith, "Outage Capacities and Optimal Power Allocation for Fading Multiple-Access Channels," *IEEE Trans. Comm.*, vol. 51, pp. 1326–1347, April 2005.

[443] H.V. Poor, "Turbo Multiuser Detection: An Overview," in *Proc. of the IEEE International Symposium on Spread Spectrum Techniques and Applications*, New Jersey, USA, September 2000, pp. 583–587.

[444] U. Madhow, "Blind Adaptive Interference Suppression for direct-sequence CDMA," *IEEE Proc.*, vol. 86, pp. 2049–2069, October 1998.

[445] D. Koulakiotis and A. H. Aghvami, "Data Detection Techniques for DS/CDMA Mobile Systems: A Review," *IEEE Pers. Comm.*, vol. 7, pp. 24–34, June 2000.

[446] S. Verdú, "Minimum probability of error for asynchronous Gaussian multiple-access channels," *IEEE Trans. Inform. Theory*, vol. 32, pp. 85–96, January 1986.

[447] S. Verdú, *Multiuser Detection*, Cambridge University Press, New York, USA, 1998.

[448] P. Castoldi, *Multiuser Detection in CDMA Mobile Terminals*, Artech House Books, Norwood, MA, 2002.

[449] C. S. Novák, T. Pálfalvi, A. Tikk, and J. Bitó, "Downlink interference analysis of CDMA-based LMDS networks applying multiuser detection," in *Proc. of the EURESCOM Summit*, October 2002, pp. 231–240.

[450] K. Li and H. Lie, "A new blind receiver for downlink DS-CDMA Communications," *IEEE Comm. Lett.*, vol. 3, pp. 193–195, July 1999.

[451] F. Petré, G. Leus, L. Deneire, M. Engels, M. Moonen, and H. De Man, "Space-Time Chip Equalizer Receivers for WCDMA Forward Link with Time-Multiplexed Pilot," in *Proc. of the IEEE Vehicular Technology Conference*, Atlantic City, New Jersey, USA, October 2001, pp. 1058–1062.

[452] F. Petré, G. Leus, L. Deneire, M. Engels, and M. Moonen, "Adaptive Space-Time Chip-Level Equalization for WCDMA Downlink with Code-Multiplexed Pilot and Soft Handover," in *Proc. of the International Conference on Telecommunications*, New York, USA, April 2002, pp. 1635–1639.

[453] T. P. Krauss, W. J. Hillery, and M. D. Zoltowski, "MMSE Equalization for forward link in 3G CDMA: Symbol-level versus chip-level," in *Proc. of the IEEE Workshop on Statistical Signal and Array Processing*, Pocono Manor, Pennsylvania, USA, August 2000, pp. 18–22.

[454] M. Heikkil´a, "A Novel Blind Adaptive Algorithm for Channel Equalization in WCDMA Downlink," in *Proc. of the IEEE Personal Indoor and Mobile Radio Communications Conf.*, San Diego, California, USA, October 2001, pp. A4 –A45.

[455] P. Komulainen, M. Heikkil'a, and J. Lilleberg, "Adaptive channel equalization and interference suppression for CDMA downlink," in *Proc. of the IEEE International Symposium on Spread Spectrum Techniques and Applications*, Parsippany, New Jersey, USA, September 2000, pp. 363–367.

[456] S. Chowdhury and M. D. Zoltowski, "Application of Conjugate Gradient Methods in MMSE Equalization for the Forward Link of DS-CDMA," in *Proc. of the IEEE Vehicular Technology Conference*, Atlantic City, New Jersey, USA, October 2001, pp. 2434–2438.

[457] K. Hooli, M. Juntti, M. Heikkil'a, P. Komulainen, M. Latva-Aho, and J. Lilleberg, "Chip-Level Channel Equalization in WCDMA Downlink," *EURASIP Journal on Applied Signal Processing*, pp. 757–770, August 2002.

[458] M. Lenardi, A. Medles, and D. T. M. Slock, "A SINR maximizing Rake Receiver for DS-CDMA Downlinks," in *Proc. of the Asilomar Conf. Signals, Systems and Computers*, Pacific Grove, California, USA, November 2000, pp. 1283 – 1287.

[459] I. Ghauri and D. T. M. Slock, "Linear Receivers for the DS-CDMA Downlink Exploiting the Orthogonality of Spreading Codes," in *Proc. of the Asilomar Conf. Signals, Systems and Computers*, Pacific Grove, California, USA, November 1998, pp. 650–654.

[460] G. E. Bottomley, T. Ottosson, and Y. P. E. Wang, "A Generalized RAKE Receiver for Interference Suppression," *IEEE Journal on Selected Areas in Communications*, vol. 18, pp. 1536–1545, August 2000.

[461] D. Divsalar, M. K. Simon, and D. Rapheli, "Improved Parallel Interference Cancellation for CDMA," *IEEE Trans. Comm.*, vol. 46, pp. 258–268, February 1998.

[462] K. Higuchi, A. Fujiwara, and M. Sawahashi, "Multipath Interference Canceller for High-Speed Packet Transmission With Adaptive Modulation and Coding Scheme in W-CDMA Forward Link," *IEEE Journal on Selected Areas in Communications*, vol. 20, pp. 418 – 432, February 2002.

[463] S. Raman and L. Yue, "Simulation Results and Interference Cancellation in UMTS Downlink Receiver," in *Proc. of the First 3G Mobile Communication Technologies Conference*, London, UK, March 2000, pp. 266–270.

[464] J. H. Horng, G. Vannucci, and J. Zhang, "Capacity Enhancement for HS-DPA in W-CDMA System," in *Proc. of the IEEE Vehicular Technology Conference*, Vancouver, Canada, September 2002, pp. 661–665.

[465] J. G. Andrews and T. H. Y. Meng, "Successive interference cancellation in a low-earth orbit satellite system," *Int. Journal Satell. Commun. Network*, vol. 21, pp. 65–78, Jan./Feb. 2003.

[466] J. Ma and H. Ge, "MMSE-based groupwise successive interference cancellation scheme for multirate CDMA systems," *Int. Journal Satell. Commun. Network*, vol. 21, pp. 127–142, January-February 2003.

[467] L. Zhang and J. Zhu, "Adaptive parallel interference cancellation receiver in CDMA satellite systems," *Int. Journal Satell. Commun. Network*, vol. 21, pp. 79–92, Jan./Feb. 2003.

[468] P. B. Rapajic and B. S. Vucetic, "Adaptive Receiver Structures for Asynchronous CDMA Systems," *IEEE Journal on Selected Areas in Communications*, vol. 12, pp. 685–697, May 1994.

[469] M. L. Honig, U. Madhow, and S. Verdú, "Blind Adaptive Multiuser Detection," *IEEE Trans. Inform. Theory*, vol. 41, pp. 994 – 960, July 1995.

[470] H.V. Poor and X. Wang, "Code - Aided Interference suppression for DS/CDMA Communications - Part II: Parallel Blind Adaptive Implementations," *IEEE Trans. Comm.*, vol. 45, pp. 1112–1122, September 1997.

[471] R. De Gaudenzi, F. Giannetti, and M. Luise, "Design of a Low Complexity Adaptive Interference-Mitigating Detector for DS/SS Receiver in CDMA Radio Networks," *IEEE Trans. Comm.*, vol. 46, pp. 125–134, January 1998.

[472] J.R. Garcia, R. De Gaudenzi, F. Giannetti, and M. Luise, "A Frequency Error Resistant Blind CDMA Detector," *IEEE Trans. Comm.*, vol. 48, pp. 1070–1076, July 2000.

[473] R. De Gaudenzi, F. Giannetti, and M. Luise, "Advances in satellite CDMA transmission for mobile and personal communications," *IEEE Proc.*, vol. 84, pp. 16–39, January 1996.

[474] R. De Gaudenzi, F. F. Giannetti, and M. Luise, "Capacity of a Multibeam, Multisatellite CDMA Mobile Radio Network with Interference-Mitigating Receivers," *IEEE Journal on Selected Areas in Communications*, vol. 17, pp. 204 – 213, February 1999.

[475] M. P'atzold, U. Killat, and F. Laue, "An Extended Suzuki Model for Land Mobile Satellite Channels and Its Statistical Properties," *IEEE Trans. Vehicular Technology*, vol. 47, pp. 617–629, February 1998.

[476] A. U. Quddus, S. Abedi, B. G. Evans, and R. Tafazolli, "Blind Multiple Access Interference Suppression in a Land Mobile Satellite Fading Channel," in *European Personal Mobile Communications Conference (EPMCC)*, Glasgow, Scotland, April 2003, pp. 402 – 405.

[477] L. Fanucci, E. Letta, R. De Gaudenzi, F. Giannetti, and M. Luise, "A Single Chip CDMA Blind Adaptive Interference Mitigating Detector for Space Applications," in *Proc. of the AIAA Int. Communication Satellite Systems Conference*, Toulouse, France, April 2001.

[478] V. Krishnamurthy, "Averaged stochastic gradient algorithms for adaptive blind multiuser detection in DS/CDMA systems," *IEEE Trans. Comm.*, vol. 48, pp. 125–134, January 2000.

[479] V. Krishnamurthy, G. Yin, and S. Singh, "Adaptive Step Size Algorithms for Blind Interference Suppression in DS/CDMA Systems," *IEEE Trans. Signal Process.*, vol. 49, pp. 190–201, January 2001.

[480] M.K. Tsatsanis and Z. Xu, "Performance Analysis of Minimum Variance CDMA Receivers," *IEEE Trans. Signal Process.*, vol. 46, pp. 3014–3022, November 1998.

[481] Z. Xu and M. K. Tsatsanis, "Blind Adaptive Algorithms for Minimum Variance CDMA Receivers," *IEEE Trans. Comm.*, vol. 49, pp. 180–194, January 2001.

[482] H. C. Hwang and C. H. Wei, "Blind adaptive algorithm for demodulation of DS/CDMA signals with mismatch," *IEEE Proc.*, vol. 146, pp. 29–34, February 1999.

[483] Z. Tang, Z. Yang, and Y. Yao, "Robust blind adaptive multiuser detector," *IEE Electron. Lett.*, vol. 35, pp. 384–385, March 1999.

[484] N. Neda and R. Tafazolli, "Simplified near-far resistant technique for joint time acquisition and demodulation of CDMA signals," *IEE Electron. Lett.*, vol. 37, pp. 65–66, January 2001.

[485] L. Mucchi, S. Morosi, E. Re, and R. Fantacci, "A New Algorithm for Blind Adaptive Multiuser Detection in Frequency Selective Multipath Fading Channel," *IEEE Trans. on Wireless Communications*, vol. 3, pp. 235–247, January 2004.

[486] M. L. Honig, M. J. Shensa, S. L. Miller, and L. B. Milstein, "Performance of Adaptive Linear Interference Suppression for DS-CDMA in the presence of Flat Rayleigh Fading," in *Proc. of the IEEE Vehicular Technology Conference*, Phoenix, Arizona, USA, May 1997, pp. 2191–2195.

[487] A. N. Barbosa and S. L. Miller, "Adaptive Detection of DS/CDMA Signals in Fading Channels," *IEEE Trans. Comm.*, vol. 6, pp. 115–124, January 1998.

[488] S. L. Miller, M. L. Honig, and L. B. Milstein, "Performance Analysis of MMSE Receivers for DS-CDMA in Frequency-Selective Fading Channels," *IEEE Trans. Comm.*, vol. 48, pp. 1919–1929, November 2000.

[489] M. L. Honig, S. L. Miller, M. J. Shensa, and L. B. Milstein, "Performance of Adaptive Interference Suppression for DS-CDMA in the presence of Dynamic Fading," *IEEE Trans. Comm.*, vol. 49, pp. 635–645, April 2001.

[490] M. Latva-Aho and M. J. Juntti, "LMMSE Detection for DS-CDMA Systems in Fading Channels," *IEEE Trans. Comm.*, vol. 48, pp. 194–199, February 2000.

[491] G. J. Lin and T.S. Lee, "A low complexity partially adaptive CDMA receiver for downlink mobile satellite communications," *Int. Journal Satell. Commun. Network*, vol. 21, pp. 23–38, Jan./Feb. 2003.

[492] X. Wang and V. Poor, "Blind Multiuser Detection: A Subspace Approach," *IEEE Trans. Inform. Theory*, vol. 44, pp. 677–690, March 1998.

[493] M. L. Honig, "Adaptive Linear Interference Suppression Techniques for Packet DS-CDMA," *European Trans. Telecomm.*, vol. 9, pp. 173–181, April 1998.

[494] X. Wang and V. Poor, *Wireless Communication Systems: Advanced Techniques for Signal Reception*, Prentice Hall Publishers, Upper Saddle River, New Jersey, USA, 2003.

[495] K. M. Wu and C. L. Wang, "An Iterative Multiuser receiver Using Partial Parallel Interference Cancellation for Turbo-Coded DS-CDMA Systems," in *Proc. of the IEEE GLOBECOM*, San Antonio, Texas, USA, November 2001, pp. 244–248.

[496] X. Wang and V. Poor, "Iterative (Turbo) Soft Interference Cancellation and Decoding for coded CDMA," *IEEE Trans. Comm.*, vol. 47, no. 7, pp. 1046–1061, July 1999.

[497] S. Morosi, E. Del Re, R. Fantacci, and A. Bernacchioni, "Improved iterative parallel interference cancellation receiver for DS-CDMA 3G Systems," in *Proc. of the IEEE Wireless Communications and Networking Conference*, New Orleans, Louisiana, USA, March 2003, pp. 877–882.

[498] S. Morosi, R. Fantacci, and P. Rufolo, "Iterative PIC detection and channel estimation for DS-CDMA 3G communications," in *Proc. of the IEEE Wireless Communications and Networking Conference*, Atlanta, Georgia, USA, March 2004, pp. 36–41.

[499] M. Kobayashi, J. Boutros, and G. Caire, "Successive interference cancellation with SISO decoding and EM channel estimation," *IEEE Journal on Selected Areas in Communications*, vol. 19, no. 8, pp. 1450–1460, August 2001.

[500] J. Mitola, "The Software Radio Architecture," *IEEE Comm. Mag.*, vol. 33, no. 5, pp. 26–38, May 1995.

[501] E. Buracchini, "The Software Radio Concept," *IEEE Comm. Mag.*, vol. 38, no. 9, pp. 138–143, September 2000.

[502] J. Mitola, "Signal Processing Technology Challenges of Cognitive Radio," in *Proc. of the Sixth Baiona Workshop*, September 2003, vol. 6, pp. 193–198.

[503] K. McClanning and T. Vito, *Radio Receiver Design*, Noble Publishing, 2000.

[504] H. Meyr, M. Oerder, and A. Polydoros, "On Sampling Rate, Analog Prefiltering, and Sufficient Statistics for Digital Receivers," *IEEE Trans. Comm.*, vol. 42, no. 42, pp. 3208–3214, December 1994.

[505] P. E. Pace, *Advanced Techniques for Digital Receivers*, Artech House, Norwood, 2000.

[506] R. H. Walden, "Analog-to-Digital Converter Survey and Analysis," *IEEE Journal on Selected Areas in Communications*, vol. 17, no. 4, pp. 539–550, April 1999.

[507] A. S. Daryoush, X. Hou, and W. Rosen, "All-optical ADC and its applications in future communication satellites," in *Proc. of the IEEE International Topical Meeting on Microwave Photonics*, October 2004, pp. 182–185.

[508] B. Sklar, *Digital Communications, fundamentals and applications*, Prentice Hall, Inc., Englewood Cliffs, New Jersey, USA, 2 edition, 2001.

[509] E. P. Cunningham, *Digital Filtering*, Prentice - Hall, 1996.

[510] C. Dick, "Configurable Logic or Digital Communications: Some Signal Processing Perspectives," *IEEE Comm. Mag.*, vol. 37, no. 8, pp. 107–117, 1999.

[511] ETSI EN 301 428, *Satellite Earth Stations and Systems (SES) - Harmonized EN for Very Small Aperture Terminal (VSAT)*, ETSI, 1.2.7 edition, February 2001.

[512] R. Andraka, "A Survey of Cordic Algorithms for FPGA based Computers," *Proc. of the Sixth ACM/SIGDA International Symposium on Field-Programmable*, 1998.

[513] Inc. Analog Devices, "AD6636 - 150 MSPS Wideband (Digital) Receive Signal Processor (RSP)," From Analog Devices Web Resources, http://www.analog.com/.

[514] Inc. Analog Devices, "AD6633 - Multi-channel (Digital) Transmit Signal Processor (TSP)," From Analog Devices Web Resources, http://www.analog.com/.

[515] Texas Instruments, "GC4016 - Quad Multi-Standard Digital Downconverter," From Texas Instruments Web Resources.

[516] A. S. Hogenauer, "An economical class of digital filters for decimation and interpolation," *IEEE Trans. Acoust., Speech, Signal Processing*, vol. 29, no. 2, pp. 155–162, April 1981.

[517] M. Moeneclaey and H. Wymeersch, "Low complexity multi-rate IF sampling receivers using CIC filters and polynomial interpolation," in *Proc. of the Sixth Baiona Workshop*, 2003, vol. 6, pp. 187–191.

[518] K. G. Johannsen, "Mobile P-Service Satellite System Comparison," *Int. Journal of Satellite Communications*, vol. 13, pp. 453–471, 1995.

[519] J. Yates, "Personal Broadband Satellite Systems and Services," in *Proc. of the IEE Seminar on Personal Broadband Satellites*, 2002, 059.

[520] G. Maral, J. De Ridder, B. G. Evans, and M. Richharia, "Low Earth Orbit Satellite Systems for Communications," *International Journal of Satellite Communications*, vol. 9, pp. 209–225, 1991.

[521] M. Luglio and W. Pietroni, "Optimisation of Double Link Transmission in case of Hybrid Orbit Satellite Constellations," *AIAA Journal on Spacecrafts and Rockets*, vol. 39, no. 5, 2002.

[522] M. Leo and M. Luglio, "Identification and Performance Evaluation of Intersegment Handover Procedures for Hybrid Constellation Satellite Systems," *Wireless Communications and Mobile Computing*, vol. 3, no. 1, pp. 87–97, February 2003.

[523] B. Demolder, G. Bekaert, G. Caille, and Y. Cailloce, "High-gain multi-beam antenna demonstrator for Ka-band satellites," *Alcatel Telecommunications Review*, December 2001.

[524] B. Trancart, J. Maurel, L. Pelenc, and P. Lepeltier, "Space antenna design for complex telecommunication payloads," *Alcatel Telecommunications Review*, December 2001.

[525] D. Breynaert, "IP over Satellite: DVB-RCS Digital Video Broadcast - Return Channel by Satellite," *ASBU/ITU sysmposium on satellite broadcasting and convergence of new multimedia services*, 1-3 November 2000.

[526] B. G. Evans, M. Werner, E. Lutz, M. Bousquet, G. E. Corazza, G. Maral, and R. Rumeau, "Integration of satellite and terrestrial systems in future multimedia communications," *IEEE Wireless Communications*, vol. 12, no. 5, pp. 72–80, October 2005.

[527] ETSI EN 301 210, *Digital Video Broadcasting (DVB); Modulation and channel coding system for Digital Satellite News Gathering (DSNG) and other contribution applications by satellite*, ETSI, 1.1.1 edition, 1999.

[528] A. Morello and U. Reimers, "DVB-S2, the second generation standard for satellite broadcasting and unicasting," *International Journal of Satellite Communications and Networking*, vol. 22, no. 3, pp. 249–268, 2004.

[529] A. Jahn, M. Holzbock, J. Muller, R. Kebel, M. de Sanctis, A. Rogoyski, E. Trachtman, O. Franzrahe, M. Werner, and F. Hu, "Evolution of aeronautical communications for personal and multimedia services," *IEEE Communications Magazine*, vol. 41, no. 7, pp. 36–43, July 2003.

[530] W. H. Jones and M. De La Chapelle, "Connexion by Boeing SM-broadband satellite communication system for mobile platforms," *IEEE Military Communications Conference*, vol. 2, pp. 755–758, October 2001.

[531] A. Hale and D. Ballinger, "Military applications for Digital Audio Radio Service (DARS)," in *Proc. of the IEEE Aerospace Conference*, March 2002, vol. 3, pp. 3–1039 – 3–1050.

[532] SATIN Project Consortium, "Satellite-UMTS IP-based Network," From SATIN Web Resorces, April 2004.

[533] K. Narenthiran, M. Karaliopoulos, B.G. Evans, and et al, "S-UMTS Access Network for Broadcast and Multicast Service Delivery: the SATIN Approach," *International Journal of Satellite Communications*, vol. 22, pp. 87–111, 2004.

[534] MODIS Project Consortium, "Mobile Digital Broadcast Satellite," From MODIS Web Resources, April 2004, http://www.ist-modis.org.

[535] K. Narenthiran, M. Karaliopoulos, R. Tafazolli, B. G. Evans, and et al, "S-DMB System Architecture and the MODIS Demo," in *Proc. IST Mobile and Wireless Summit*, Aveiro, Portugal, June 2003.

[536] MAESTRO Project Consortium, "Mobile Applications & sErvices based on Satellite & Terrestrial inteRwOrking," From MAESTRO Web Resource, April 2004, http://ist-maestro.dyndns.org/MAESTRO/index.htm.

[537] M. Luglio and W. Pietroni, "Design Methodology of Hybrid LEO-GEO Constellations for High Capacity Communications," *International Conference on Telecommunications*, pp. 1177–1180, 22-25 May 2000.

[538] M. Luglio and W. Pietroni, "The Use of Hybrid Orbit Satellite Constellations for High Capacity Communications," *Proc. of the 19th AIAA International Communications Satellite Systems Conference*, April 17-20 2001, Toulouse, France.

[539] CAPANINA Project Consortium, "CAPANINA: wireless and optical broadband technologies for use on High Altitude Platforms (HAPs)," http://capanina.org, 2003.

[540] HeliNet Project Consortium, "HeliNet: A Network of Stratospheric Platforms for Traffic Monitoring, Environmental Surveillance and Broadband Services, http://helinet.polito.it," http://helinet.polito.it, 2003.

[541] Y.C. Lee and H. Ye, "Sky station stratospheric telecommunications system, a high speed low latency switched wireless network," in *Proc. of the 17th AIAA International Communications Satellite Systems Conference*, 1998, pp. 25–32.

[542] J. Thornton, D. Grace, C. Spillard, T. Konefal, and T.C. Tozer, "Broadband communications from a high altitude platforms - the European Helinet programme," *IEE Elec. Comm. Eng. Journal*, pp. 138–144, June 2001.

[543] N.J. Colella, J.N. Martin, and I.F. Akyildiz, "The HALO network," *IEEE Commun. Magazine*, June 2000.

[544] Y. Hase, R. Miura, and S. Ohmori, "A novel broadband all-wireless access network using stratospheric platforms," in *Proc. of Vehicular Technology Conference*, 1998, pp. 1191–1194.

[545] SkyTower Telecomm., "SkyTower Stratospheric Telecommunications Network System," From SkyTower Web Resource, http://www.skytowerglobal.com/network.html.

[546] G. M. Djuknic, J. Freidenfelds, and Okunev Y., "Establishing Wireless Communications Services via High-Altitude Aeronautical Platforms: A Concept Whose Time Has Come?," *IEEE Comm. Mag.*, vol. 35, no. 9, pp. 128–135, September 1997.

[547] J. N. Pelton, "Telecommunications for the 21st Century," *Scientific American*, , no. 4, pp. 68–73, April 1998.

[548] M. Mohorčič, G. Kandus, and D. Grace, "Provision of fixed broadband wireless access over stratospheric platforms," *Electrotechnical Review*, vol. 71, no. 3, pp. 89–95, 2004.

[549] U. Drčić, G. Kandus, M. Mohorčič, and T. Javornik, "Interplatform Link Requirements in the Network of High Altitude Platforms," in *Proc. of the 19th AIAA International Communications Satellite Systems Conference and Exhibit*, Toulouse, France, April 2001, vol. 3, pp. 17–20.

[550] SpaceData Corporation, "Communications Where You Need It," From SpaceData Web Resource, http://www.spacedata.net.

Index

multiple access channel, 47
satellite, 315
 CDMA, 317
with outage, 41
carrier frequency
 error, 223
CDMA, 247, 373
 Blind Adaptive receiver, 410
 Direct Matrix Inversion receiver, 410
 Direct Sequence, 373
 Frequency Hopping, 374
 MMSE Receiver, 409
 sectoring, 375
 synchronous, 408
channel
 blind estimation, 244
 estimation, 243
 frequency selective, 243
 response, 221
channel coding theorem, 38
Chi-square distribution, 27
clustering effect, 267
code
 synchronization, 247
coded modulation, 190
 bit interleaved, 194
 bit interleaved turbo, 196
 Trellis, 191
coding theory, 117
coherence
 bandwidth, 68
 time, 68
Contention-Oriented Multiple Access, 390
continuous phase modulated (CPM), 178, 181,
 186
control segment, 457
convolutional codes, 121
Costas loop, 258
Cramer Rao Bound, 219
 Modified Cramer Rao Bound, 220
Cramer Rao bound, 62
cycle slips, 220, 231
cyclic codes, 120

delay
 spread, 68
delay and multiply method, 228
Demand Assigned Multiple Access, 396
Demand Assignment Multiple Access, 397
depolarization, 96
digital beamforming, 420
dirty paper coding, 51
diversity, 245, 314
 combining techniques, 355
 EGC, 355
 GSC, 355
 MRC, 355

SC, 355
 frequency, 314, 336
 outage probability, 336
 links correlation, 320
 polarization, 315, 341
 satellite, 314, 315
 CDMA, 316
 site, 314, 333
 statistical models, 335
Doppler
 shift, 68
 spectrum, 68
 spread, 68
DSP, 431
DVB-RCS, 260
DVB-S, 258
DVB-S2, 259
dynamic range, 420

efficiency, 176
 bandwidth, 176
 power, 176
equalization, 243
 blind, 295
 decision feedback, 292
 linear, 291
 nonlinear, 293
 turbo, 300
erasure decoding, 172
error propagation, 235
estimator, 219
 biased, 219
 carrier frequency, 224
 carrier phase, 229
 code aware, 224
 code unaware, 224
 feedback, 220
 feedforward, 220
 symbol timing, 232
 unbiased, 219
EXIT charts, 147
expanding window, 247
exponential distribution, 23

Factor graphs, 157
fading
 average duration, 71
 bandwidth, 69
 long term, 71
 short term, 71
 slope, 101
fading channels, 209
 mitigation techniques, 351
 Nakagami-m, 211
 Nakagami-q (Hoyt), 356, 357
 Rayleigh, 210
 Rayleigh-Lognormal, 215
 Rice, 211, 319, 357

Printed in the United States of America